De Gruyter Studies in Mathematical Physics 20

Valery A. Slaev
Anna G. Chunovkina
Leonid A. Mironovsky

Metrology and Theory of Measurement

De Gruyter

Physics and Astronomy Classification Scheme 2010: 06.20.-f, 06.20.F-, 06.20.fb, 03.65.Ta, 06.30.-k, 07.05.Rm, 07.05.Kf, 85.70.Kh, 85.70.Li

ISBN 978-3-11-028473-7
e-ISBN 978-3-11-028482-9
ISSN 2194-3532

Library of Congress Cataloging-in-Publication Data
A CIP catalog record for this book has been applied for at the Library of Congress.

Bibliographic information published by the Deutsche Nationalbibliothek
The Deutsche Nationalbibliothek lists this publication in the Deutsche Nationalbibliografie; detailed bibliographic data are available in the Internet at http://dnb.dnb.de.

Typesetting: PTP-Berlin Protago-TEX-Production GmbH, www.ptp-berlin.de
Printing and binding: Hubert & Co. GmbH & Co. KG, Göttingen
♾ Printed on acid-free paper

Printed in Germany

www.degruyter.com

"To measure is to know."
Lord Kelvin

"We should measure what is measurable
and make measurable what is not as such."
Galileo Galilei

"Science begins when they begin to measure …"
"Exact science is inconceivable without a measure."
D.I. Mendeleyev

"… for some subjects adds value only an exact match to a certain pattern. These include weights and measures, and if the country has several well-tested standards of weights and measures it indicates the presence of laws regulating the business relationship in accordance with national standards."
J.K. Maxwell

Preface

During the 125th anniversary of the Metre Convention in 2000, Steve Chu, now President Obama's Energy Secretary and a one-time metrologist, spoke and included the now famous quotation: "Accurate measurement is at the heart of physics, and in my experience new physics begins at the next decimal place".

Metrology, as the science of measures or measurements traceable to measurement standards [323], operates with one of the most productive concepts, i.e., with the concept of measurement accuracy, which is used without exception in all natural and technical sciences, as well as in some fields of social sciences and liberal arts.

Metrology, by its structure, can be considered a "vertically" designed knowledge system, since at the top level of research it directly adjoins the philosophy of natural sciences, at the average level it acts as an independent section of the natural (exact) sciences, and at the bottom level it provides the use of natural science achievements for finding solutions to particular measurement tasks, i.e. it performs functions of the technical sciences. In such a combination, it covers a range from the level of a knowl edge validity criterion to the criterion of correctness in the process of the exchange of material assets.

For metrology the key problem is to obtain knowledge of physical reality, which is considered through a prism of an assemblage of quantity properties describing the objectively real world. In this connection, one of the fundamental tasks of metrology is the development of theoretical and methodological aspects of the procedure of getting accurate knowledge relating to objects and processes of the surrounding world which are connected with an increase in the measurement accuracy as a whole. Metrology, as the most universal and concentrated form of an organizing purposeful experience, allows us to make reliability checks of the most general and abstract models of the real world (owing to the fact that a measurement is the sole procedure for realizing the principle of observability).

Metrology solves a number of problems, in common in a definite sense with those of the natural sciences, when they are connected with a procedure of measurement:

- the problem of language, i.e. the formalization and interpretation of measurement results at a level of uniformity;
- the problem of structuring, which defines the kind of data which should be used depending on the type of measurement problem being solved, and relates to a system approach to the process of measurements;

- the problem of standardization, i.e. determination of the conditions under which the accuracy and correctness of a measurement result will be assured;
- the problem of evaluating the accuracy and reliability of measurement information in various situations.

At present, due to the development of information technologies and intelligent measurement systems and instruments, as well as the growing use of mathematical methods in social and biological sciences and in the liberal arts, there were a number of attempts to expand the interpretation of metrology [451] not only as the science about measurements of physical quantities, but also as a constituent of "gnoseo-techniques", information science, "informology" and so on, the main task of which is "to construct and transmit generally recognized scales for quantities of any nature", including those which are not physical. Therefore, it remains still important to determine the place of metrology in the system of sciences and its application domain more promptly, i.e., its main directions and divisions [9, 228, 429, 451, 454, 503, 506, and others].

It is possible to analyze the interconnection of metrology and other sciences from the point of view of their interaction and their mutual usefulness and complementariness, using as a basis only its theoretical fundamentals and taking into account a generally accepted classification of sciences in the form of a "triangle" with vertexes corresponding to the philosophical, natural, and social sciences [257], but paying no attention to its legislative, applied, and organizational branches.

Among such sciences one can mark out philosophy, mathematics, physics, and technical sciences, as well as those divisions of the above sciences, the results of which are actively used in theoretical metrology, and the latter, in its turn, provides them with materials to be interpreted and given a meaning to.

It is known that in an application domain of the theoretical metrology two main subdivisions can be singled out: *the general theory of measurements* and *the theory of measurement assurance and traceability.*

The general theory of measurements includes the following directions:

- original regulations, concepts, principles, postulates, axiomatic, methodology, terms, and their definitions;
- simulation and investigation of objects, conditions, means, and methods of performing measurements;
- theory of scales, measures, metrics, references, and norms;
- theory of measurement transformation and transducers;
- theory of recognition, identification, estimation of observations, and data processing;
- theory of measurement result uncertainties and errors;
- theory of dynamic measurements and signal restoration;

- theory of enhancement of the measurement accuracy, sensitivity, and *ultimate capabilities taking into account quantum and other limitations;*
- automation and intellectualization of measurement information technologies, interpretation and use of measurement information in the process of preparing to make decisions;
- theory of the optimal planning for a measurement experiment;
- *theory of metrology systems*;
- theory of measurement quality estimation, as well as of technical and social-economic efficiency of metrology and measurement activities.

The theory of measurement assurance and traceability consists of the following directions:

- theory of physical quantities units, systems of units, and dimensionality analysis;
- theory of measurement standards;
- *theory of reproducing, maintaining, and transferring a size of quantity units;*
- theory of estimating normalized metrological characteristics of measuring instruments;
- methodology of performing metrology procedures;
- theory of metrological reliability and estimation of intercalibration (interverification) intervals;
- theory of estimating the quality of metrology systems, and the methodology of optimizing and forecasting their development.

Let the particular features of measurement procedures be considered in the following order: interaction of an object with a measuring instrument → recognition and selection of a measurement information signal → transformation → comparison with a measure → representation of measurement results.

Interaction of an object under investigation with a measuring instrument assumes searching, detecting, and receiving (reception) a quantity under measurement, as well as, if necessary, some preparatory procedures of the probe-selection or probe preparation type, or exposure of the object to some outside influence for getting a response (stimulation), determining an orientation and localization in space and time.

Discrimination or selection assume marking out just that property of an object to which the quantity under measurement corresponds, including marking out a useful signal against a noise background and applying methods and means for noise control.

Transformation includes changes of the physical nature of an information carrier or of its form (amplification, attenuation, modulation, manipulation, discretization and quantization, analog-digital and digital-analog conversion, coding and decoding, etc.), as well as the transmission of measurement information signals over communication channels and, if necessary, recording, storing, and reproducing them in memory devices.

Comparison with a measure can be realized both directly and indirectly with the help of a comparator or via some physical or technical mechanism. A generalization of this procedure is the information comparison with an image.

Representation of measurement results assumes data processing according to a chosen algorithm, evaluation of uncertainties (errors) of a measurement result, representation of measurement results with a digital display, pointer indicator device, analog recorder, hard copy (printing), or graphical data representation, their use in automatic control systems, semantic interpretation (evaluation) of the results obtained, identification, structuring, and transmission of the results into data and knowledge bases of artificial intelligence systems.

In performing a measurement procedure it is very important to use a priori and a posteriori information. Under a priori information one understands the models selected for an object, conditions, methods and measuring instruments, type of measurement scale, expected ranges of amplitudes and frequencies of a quantity to be measured, etc. Examples of using a posteriori information include improvement of the models being used, recognition of patterns (images) or their identification, determination of the equivalence classes to which they refer, as well as structurization for data base augmentation and preparation for making decisions.

The key procedures to which the measuring instruments are subjected are checks, control operations, tests, graduation, calibration, verification (metrological validation), certification, diagnostics, and correction. From the above, in the first turn, the calibration, verification, certification, and validation can be considered to be metrological procedures.

It should be noted that many repeated attempts to create a general theory of measurement have been made, using various approaches. Among them there are approaches of the following types [54, 84, 143, 164, 209, 216, 218, 256, 268, 280, 297, 320, 362, 368, 382, 384, 404, 412, 437, 438, 481, 487, 489, 493, 496, 510, 551, and others]: energetical, informational, thermodynamical, algorithmical, theoretical-set, theoretical–bulk, statistical, representational, analytical, quantum ones, as well as approaches using algebraic and geometric invariants, and others. Moreover, there are some particular theories of measurement which have been quite well developed and which presuppose an experimental determination of particular physical quantities in various fields of measurements.

Taking into account the technology of performing measurements, which includes the above indicated procedures of interaction between an object and measuring instrument, detection and selection of a signal, transformation of measurement information, comparison with a measure and indication of measurement results, it is possible to consider various scales used in performing measurements (and a corresponding axiomatic), i.e. from the nominal to absolute ones, and to discuss a relationship between a set of scales and a sequence of measurement procedure stages.

On the basis of an analysis of links of the general theory of measurements, as one of the main components of the theoretical metrology, with philosophy, mathematics, physics, and technical sciences, it is possible to determine divisions of mathematics [242, 450], the results of which are used in the theoretical metrology. They are as follows:

- set theory, including measures and metrics;
- theory of numbers, including the additive and metric ones;
- mathematical analysis, including the differential, integral, calculus of variations, operational, vector, and tensor calculus;
- higher algebra, including the algebra of sets, measures, functions; the theory of groups, rings, bodies, fields, grids, and other algebraic systems; the theory of mathematical models and algebraic invariants, etc.;
- higher geometry, including the Euclidean, affine, projective, Riemannian, Lobachevskian geometries, etc.;
- theory of functions and functional analysis, including the theory of metric spaces, norms, and representations;
- spectral and harmonic analysis, including the theory of orthogonal series and generalized functions;
- mathematical physics with models as well as with direct, inverse, and boundary problems;
- probability theory and mathematical statistics, including a statistical analysis: confluent, covariance, correlation, regression, dispersion, discriminant, factor, and cluster types, as well as a multivariate analysis and the theory of errors, observation treatment, and statistical evaluation; the theory of optimal experiment planning and others;
- game theory, including the utility theory and many-criterion problems;
- theory of systems, including dynamic ones; ergodic theory, perturbation and stability theory, optimal control theory, graph theory, theory of reliability and renewal theory, and others;
- mathematical logic, including theory of algorithms and programming, calculation of predicates, propositional calculus, mathematical linguistics, evidence theory, trainable system theory, and so on;
- computational mathematics.

An object of measurement, as the well-known Polish metrologist Ya. Piotrovsky [384] has remarked, is "the formation of some objective image of reality" in the form of a sign symbol, i.e. a number. Since we wish to get the results of measurement in the form of a number, particularly a named one, then the scale of a physical quantity has to correspond to axioms, postulates, and foundations [313] of a number system being in use.

At the same time, I. Newton [365] gave the following definition to the number: "Under a number we understand not so much a set of units, as an abstract ratio of some quantity to that of the same kind which is accepted as a unit". This definition, in an explicit form, corresponds to the proportional measurement scale or scale of ratios [268, 382].

The present volume sums up the work of many years done by the authors in the fields indicated above and includes the results of activities on investigation of some actual aspects and topics of measurement theory (printed in italics) over a period of more than 30 years [36, 37, 95-115, 347, 349-351, 353, 354, 451-457, 462-464, and others]. The priority of research is confirmed by the number of the authorship inventions certificates [42–44, 265, 272, 383, 448, 449, 465, 469–476, 480]. The basic part of the investigations was conducted at the D. I. Mendeleyev Institute for Metrology and the St. Petersburg State University of Aerocosmic Instrumentation, where the authors have been working during the course of nearly 40 years.

It should be noted that in the text of this book data accumulated for a long time period in Russian metrological institutes is presented.This data does not always coincide with the presemt-day and commonly accepted ideas and points of view of the global metrological community. However, they firstly do not conflict with the basic (fundamental) metrological statements and concepts, and secondly, they accentuate special features of metrology application in Russia and the member-states of the Commonwealth of Independent States (CIS).

The authors are grateful to T. N. Korzhakova for her help in translating the text into English.

Contents

Abbreviations

A	amplifier
AI	algebraic invariant
AC	amplitude characteristic
AD	adjustable divider; analysis of the documentation
ADC	analog-to-digital converter
AFC	amplitude-frequency characteristic
AFRIMET	Intra-Africa Metrology System
AG	adjustable generator
ALA	amplitude-limiting amplifiers
AM	amplitude modulation
AMMI	auxiliary MMI
AND	logic circuit "AND"
APM	amplitude-pulse modulation
APMP	Asia Pacific Metrology Programme
AS	amplitude selector
AT	attenuator
AX	auxiliary channel of MRI
BAM	balance-amplitude modulation
BCF	block of criterion formation
BIPM	Bureau International des Poids et Measures
BFD	block of forming the derivative
C	commutator
CA	computer aids
CC	complete centralization
CD	counter divider
CG	controllable gate
CIPM	Comité International des Poids et Measures
CIWT	code inspection and walk-through
CM	correlation meter
CMC	calibration and measurement capabilities
CO	comparison operator
COOMET	Euro-Asian Cooperation of National Metrological Institutions
D	differentiator
DAC	digital-to-analog converter
DC	complete decentralization
DDC	device for measurement of dynamic characteristic
DE	degree of equivalence
DFA	metrological data flow analysis
DMRE	device for measurement of magnetic registration error
DWPM	discrete width-pulse manipulation
EMF	electro-motive force
EMU	error measurement unit
EU	European Union
EURAMET	European Collaboration in Measurement Standards
EUT	equipment under test
FHF	filter of high frequencies
FLF	filter of low frequencies
FM	frequency modulation
FMP	frequency manipulation
FPM	frequency-pulse modulation
FPC	fundamental physical constant
FS	frequency selector
GCWM	General Conference on Weights and Measures
GRS	generator of reference signal
GSS	generator of sinusoidal signal

GTS	generator of test quasi-random signal
H	hardware
I	integrator
IC	investigated channel of MRI
IEC	International Electrotechnical Commission
IFCC	International Federation of Clinical Chemistry
IMS	information-measurement system; International Measurement System; indices of measurement software
ISO	International Organization for Standardization
IR	ionizing radiation
IRPM	ionizing radiation parameter measurement
IT	information technology
IUPAC	International Union of Pure and Applied Chemistry
IUPAP	International Union of Pure and Applied Physics
JCRB	Joint Committee of the Regional Metrology Organizations
KC	key comparison
KCRV	key comparison reference value
LC	local centralization
LSM	least-squares method
MC	multiple centralization
MF	median filter
MI	measurement instrument; measuring instrument
MM	measurement method
MMI	metrological measurement instrument
MMM	metrological measurement mean
MMWRZ	modified method without returning to zero
MP	Markov's parameters
MPE	maximum permissible error
MRA	Mutual Recognition Arrangement
MRI	magnetic recording instrument
MRV	method of redundant variables
MS	metrological system
MS-C	measurement standard-copy
MSR	mean square root
MVS	metrological validation of software
MWRZ	method without returning to zero
NB	notified body
ND	normative document
NMI	national metrology institute
NMC	normalized metrological characteristic
NSM	national system of measurement
OIML	Organisation Internationale de Métrologie Légal
LSM	least squares method
PAM	parasitical amplitude modulation
PBF	pass band filter
PC	phase changer
PDF	probability density function
PFC	phase-frequency characteristic
PM	phase modulation
PMU	phase measuring unit
PQ	physical quantity
PWF	pulse–weight function
QCA	quantitative chemical analysis
R	shift register
RF	rejection filter
RMMI	reference MMI
RPM	relative phase manipulation
RMO	Regional Metrology Organization
RTM	representative theory of measurement
RUTS systems	system of reproducing PQ units and transferring their sizes
RVM	redundant variables method
S	pulse shaper

S-C	standards-copies
SCMIQ	system of certification of measurement instruments quality
SCMM	system for certifying methods of measurements
SEME	system of ensuring measurement efficiency
SEMQ	system of ensuring measurement quality
SEMU	system of ensuring the measurements uniformity
SIM	Inter-American Metrology System
SMA	system of metrological assurance
SMB	system of metrological bodies
SMCS	system of metrological control and supervision
SME	system of metrological examination
SMMI	subordinate MMI
SMP	system of metrological personnel
SMS	state measurement standard
SMT	software module testing
SMU	system of measurement units
SND	system of normative documents
SNMD	system of normative metrological documents
SPMS	state primary measurement standards
SPQ	system of physical quantities
SPQU	system of physical quantity units
SRM	system of reproducing methods
SQ	unit for raise signal to the second power
SS	standard sample
SSM	state system of measurement
SSMS	state special measurement standard
SSS	system of standard samples
SSSRD	state system of standard reference data
SSTMI	state system for testing of MIs
SSVCMI	state system of verification and certification of MIs
STM	system of transfer methods
SU	subtraction unit
SVS	state verification scheme
SW	software
SSW	"standard" software
T	trigger
TCLE	temperature coefficient of linear expansion
TM	transfer method
TPM	time-pulse modulation
TTM	tape transporting mechanism
UC	unit of control
UCF	unit of coherent frequencies
VFTM	validation by functional testing of metrological functions
VFTSw	validation by functional testing of software functions
UPS	unit of power supply
UR	unit of registration
US	unit of synchronization
UTD	unit of time delay
UVD	unit of variable time delay
V	voltmeter
VR	voltage regulator
VS	verification setup
VSHA	verification setup of the highest accuracy
WCC	with continuous carrier
WMI	working measurement instrument
WMS	working measurement standard
WPM	width-pulse modulation
WRZ	with returning to zero
WSC	with square carrier
$\sum$	summator
"+1"	unit of generating command ("+1")

Chapter 1
International measurement system

1.1 Principles underlying the international measurement system

At a turn of the 20th to the 21st centuries, the methods and systems used to ensure the measurement uniformity in the world radically changed. This was caused by the need to significantly reduce the influence of international trade barriers, which put obstacles in the way of its further development and hampered the progress of business, health protection, preservation of the environment, cooperation in industry, etc.

At a meeting held in Paris on October 14, 1999, the directors of the national metrology institutes (NMIs) of 38 states signed a Mutual Recognition Arrangement (CIPM MRA) for the mutual recognition of national measurement standards and for calibration and measurement certificates issued by NMIs [119]. Later on a number of other institutes also signed this document. At present it is possible to get acquainted with information about all the MRA signatory countries at www.bipm.org. (The CIPM MRA has now been signed by the representatives of 90 organizations, from 51 member states, 35 associates of the General Conference on Weights and Measures (GCPM), and 4 international organizations, and covers a further 144 institutes designated by the signatory bodies.)

The MRA is a significant and important step towards further development of the International Measurement System (IMS). The IMS is based on the system of physical quantity units (SPQU) and decisions of the GCPM, which is the highest international body on issues of establishing the units of quantities, their definitions, and the methods of reproducing their size.

The Joint Committee of the Regional Metrology Organizations (JCRB) and the BIPM were created with the purpose of promoting the cooperation of metrological institutes and organizations on the basis of discussion on and solutions for mutual methodical problems connected with implementation of the MRA. The MRA is put into effect by the CIPM committees, five regional organizations (EURAMET, COOMET, APMP, SIM, AFRIMET), and national metrological institutes, signatories of the CIPM MRA.

The measurement uniformity assurance implies solving a complex of scientific-technical, legal, and organizational problems related to the development of measurement standards, enhancement of methods for transferring the size of physical quantities units, and the creation of modern advanced measuring instruments.

If the scientific-technical problems are first solved within the framework of national systems of measurement uniformity assurance, then the approaches to a solution of legal and organizational problems are developed, made to agree, and implemented within the framework of international organizations such as OIML, ISO, and others. Among the fundamental documents one ought first mention Recommendations: OIML D 1 Elements for a Law on Metrology, OIML D 5 Principles for establishment of hierarchy schemes for measuring instruments, OIML D 8 Measurement standards: choice, recognition, use, conservation, and documentation, and others.

Close cooperation in the field of metrology is impossible without a common terminology and unified methods applied to express the accuracy of measurements. The appearance of a new version of the International Vocabulary of Metrology (VIM) [246] has become an important step in the development of a "unified language". This version reflects the present-day trends of metrology development, which are connected with the dissemination of measurements to new fields such as medicine and biology. This has resulted in the expansion of the concepts "quantity", "measurement", and changes of the terms linked with the concept "metrological traceability". The appearance of the generalizing primary concept "reference", which, depending on the context, may be used as a measurement unit or standard, or as a reference procedure, deserves special attention.

The mutual recognition of measurement results rests on the assurance of the compatibility of measurement results, i.e., on such a property of measurement results which does not allow the difference of values measured to overstep limits corresponding to measurement uncertainties.

The reliability of accuracy estimates is provided by certifying measurement procedures and by using a hierarchy scheme for calibrating measuring instruments and standards which ensures the traceability of a measurement result to a reference measurement standard. Comparison of national standards leads to the establishment of their equivalence, which can be considered as the final stage of establishing the measurement result traceability to a measurement unit (Figure 1.1).

In 1993, under the auspices of the international organizations BIPM, IEC, ISO, OIML, the International Union of Pure and Applied Chemistry (IUPAC), the International Union of Pure and Applied Physics (IUPAP), and the International Federation of Clinical Chemistry (IFCC), a document, called the Guide to the Expression of Uncertainty in Measurement (GUM) [243] was developed and since then has begun to be implemented into everyday practice.

At present the Guide is the main international framework document on the assessment of the accuracy of measurements. The supplements to the GUM are worked out as the methods of accuracy assessment are developed and enhanced. By this time Supplement 1 [244], devoted to the application of the Monte Carlo method for transformation of probability density functions in evaluating the uncertainty of measurements, and Supplement 2, disseminating the evaluation of uncertainty to the models with an arbitrary number of output quantities, have been issued.

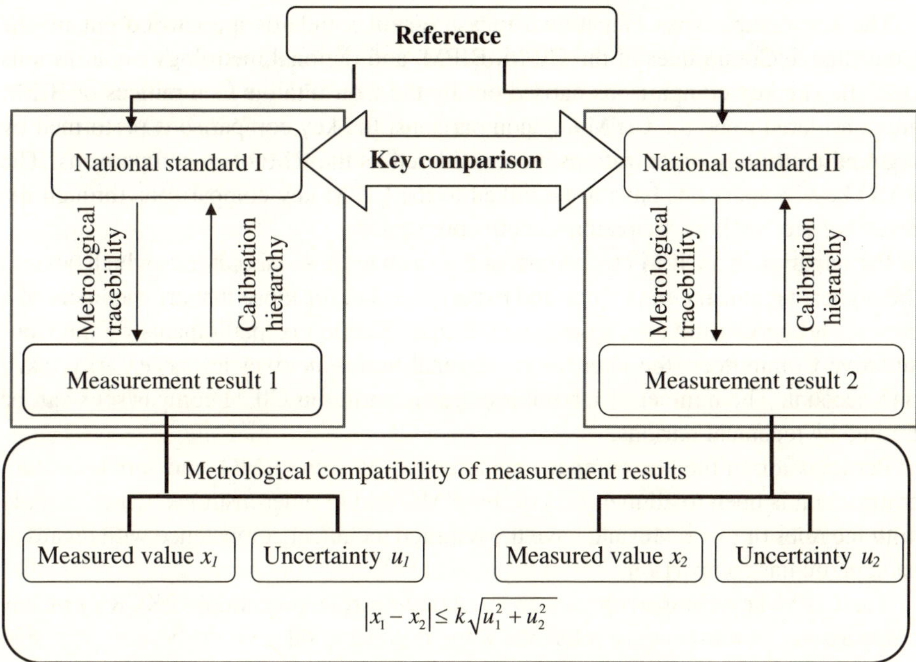

Figure 1.1. Assurance of the compatibility of measurement results.

This chapter deals with the methods of evaluating measurement results in key comparisons of national measurement standards and calibration of measuring instruments.

The main purpose of the MRA is to create an objective base for the mutual recognition of the results of measurements and measuring instrument calibrations obtained by different national laboratories. This main purpose is realized by solving the following problems:

- fulfillment of international comparisons of national measurement standards, called key comparisons, following well-defined procedures, which lead to a quantitative expression of the degree of equivalence of national measurement standards – participants of the key comparisons;
- successful participation of each NMI in corresponding supplementary comparisons of national measurement standards;
- creation of quality management systems in each NMI;
- demonstration of the competence of each NMI in the form of a declaration of calibration and measurement capabilities (CMC) of a given NMI, which is entered into a common database on key comparisons supported by the BIPM and open to general use through the internet (key comparison data base: www.bipm.org).

The key comparisons of national measurement standards are carried out by the Consultative Committees of the CIPM, BIPM, and regional metrology organizations (RMO). The key comparisons carried out by the Consultative Committees or BIPM are considered to be the CIPM key comparisons; the key comparisons performed by regional metrology organizations are considered as the RMO key comparisons. The RMO key comparisons have to be linked to the CIPM key comparisons through the results of the NMIs participating in both comparisons.

Participation in a CIPM key comparison is open to those laboratories which possess the highest technical competence and experience, i.e. for those that are members of a corresponding Consultative Committee. In the selection of participants by the Consultative Committees, the interests of regional representatives are necessarily taken into account. The number of laboratories participating in CIPM comparisons can be limited by technical reasons.

Participation in the key comparisons arranged by some RMO, and supplementary comparison is open to all members of this RMO and to other institutes which comply with the rules of this RMO and have the required technical competence with regard to each particular comparison.

The CIPM key comparison results in obtaining reference values (KCRV) of key comparisons. In most cases a reference value is understood to be the best estimate of a measurand obtained on the basis of results of all participants. The concept of a degree of equivalence of measurement standards means the degree of correspondence of the results obtained by the participants of a given comparison to a reference value of this key comparison.

The degree of equivalence of each national measurement standard is quantitatively expressed by two quantities: a deviation from a reference value of a key comparison, and the uncertainty of this deviation (at the level of confidence 95 %). The degree of equivalence between pairs of national measurement standards is expressed by a difference of their deviations from reference value and uncertainty of this difference (at the level of confidence 95 %).

The RMO key comparisons disseminate the metrology equivalence established in the process of CIPM key comparisons to a great number of the NMIs which for various reasons do not participate in the CIPM key comparisons. The results of the RMO key comparisons are related to the reference values established at the CIPM key comparisons on the basis of results obtained by the participating institutes, both in the CIPM and RMO comparisons (linking institutes).

Regional organizations are responsible for

- linking with CIPM key comparison through participation of a sufficient number of laboratories in both comparisons, in order to ensure that linking with the reference value of the key comparison is established with a very low level of uncertainty;
- procedures used in the regional comparisons – evaluation of measurement data has to be compatible with the procedures applied in CIPM key comparisons;

- coordination of time for carrying out RMO and CIPM key comparisons (at least their frequency);
- evaluation of results of RMO key comparisons and supplementary comparisons, application of appropriate procedures, presentation of results for publication in Supplement B of the key comparison database;
- postponing the linking of reference value of a RMO key comparison, carried out before the beginning of a corresponding CIPM key comparison to a reference value of this CIPM key comparison until both key comparisons are completed.

The CIPM and RMO key comparison results, reference values of key comparisons, deviations from reference values, and their uncertainties are published in the BIPM database (Appendix B).

The calibration and measurement capabilities of NMIs are represented in the form of quantities, ranges, and expanded uncertainty (usually at a level of confidence of 95%). The CMC are listed for each NMI in Appendix C of the BIPM database. They must be confirmed by the results obtained in the key comparisons indicated in Appendix B.

1.2 Classification of key comparisons of national measurement standards

The MRA establishes three types of national measurement standard comparisons of NMIs, which differ in the problems they solve and in the requirements for preparing, carrying out, and reporting the results of comparison:

- CIPM key comparisons (CIPM KC);
- RMO key comparisons (RMO KC): for example, key comparisons of the EURAMET, APMP, COOMET, and other regional organizations;
- supplementary comparisons.

CIPM key comparisons are performed by consultative committees. As a rule, the primary national measurement standards participate. The CIPM key comparisons solve two main problems:

(1) determination of a reference value of key comparisons (KCRV) and the degree of equivalence of national measurement standards (Appendix B of the BIPM database);

(2) confirmation of CMCs claimed (Appendix C of the BIPM database).

The number of participants in CIPM key comparisons is limited. Therefore, the problem of disseminating the metrological equivalence to a greater number of participants interested in the mutual recognition of measurement results and calibrations is urgent. This problem is solved within the framework of regional key comparisons carried out by regional metrology organizations. Within the framework of RMO KC two tasks are solved:

(1) determination of the degree of equivalence of national measurement standards;
(2) confirmation of the CMCs claimed.

It is important to stress that the reference value of key comparisons within the framework of a RMO KC is not determined, and the equivalence degree is established relative to a KCRV through results of a linking NMIs which take part in both comparisons.

Supplementary comparisons of national measurement standards solve only the task of confirming CMCs for which no degree of equivalence of national measurement standards have been established.

An analysis of reports on key comparisons has shown a variety of schemes used to perform the key comparisons. These schemes are characterized by properties of a traveling transfer measurement standard, number of participants, time limits, and so on. 'The tasks of planning and developing a schedule, preparing and investigating the traveling measurement standard, developing a protocol of comparisons, processing comparison results, and coordinating a report on the comparisons are fulfilled by a pilot laboratory. Sometimes pilot laboratories receive the assistance of other laboratories which are participants of comparisons.

The following classification features for selecting a method for the evaluation of comparison data are important.

Scheme of performing a comparison. Comparisons can be of a circular or radial type. When comparisons are circular, a traveling measurement standard leaves a particular pilot laboratory and, according to a schedule of comparison, measurements with this traveling measurement standard are carried out by all comparison participants. After that, the traveling measurement standard returns to the pilot laboratory. At the end of comparisons the pilot laboratory carries out measurements with the purpose of makeing sure that the traveling measurement standard is in an appropriate operational state. To reduce the time of comparisons it is possible to use a number of identical traveling transfer measurement standards. In this case a problem of linking different loops of one and the same key comparison arises. This problem should not be confused with the problem of linking results of RMO key comparisons with CIPM key comparisons. The scheme of circular comparisons implies a good stability of a traveling measurement standard and relatively short duration of comparisons. Provided that the comparisons require a long period of time and that a check of the traveling standard stability is needed in the process of comparison, then the radial scheme of comparisons is applied.

In the radial scheme of comparisons the traveling standard returns to the pilot laboratory several times in the comparison process, and the pilot laboratory carries out the comparisons while keeping track of a potential systematic drift or instability of a measurand.

Kind of measurand. In comparisons a measurand can be

- the value of a material measure or values of set of measures. The measures from a set can be of equal nominal values (in case of possible damages of these measures during transportation) or of significantly different nominal values;
- the value of a calibration coefficient of a measuring instrument;
- functional dependence – for example, the dependence of a calibration coefficient of a measuring instrument on a frequency. This implies the fulfillment of measurements at different frequencies.

A priori information about compared measurement standards. Participants of comparisons present to a pilot laboratory the measurement results, the uncertainty budget, and information about their measurement standard and measurement procedure. In particular, if secondary measurement standards participate in comparisons, then while processing it is important to take into account a possible covariance of the results of participants, which is caused by the traceability of these secondary measurement standards to the primary ones which are also participating in the given comparisons.

Form of reporting the results (the form of a set of reference values and degrees of equivalence). In most cases, at least at the initial stage of comparison data processing, the reference value is understood to be the best estimate of a measurand. Therefore, in processing one frequently obtains a set of reference values and, correspondingly, a sct of degrees of equivalence. Strictly speaking, only in specific cases are there grounds for calculating a mean of a number of reference values: for example, the case when a set of material measures with similar nominal values is used as a traveling measurement standard. A tendency to get a laconic presentation of the measurement results leads to the application of different procedures for averaging a number of reference values. In this case the direct connection of the reference value with a particular measurand disappears. When from the very beginning the reference value can be represented by certain dependence, the choice of a certain fixed point, for example, a frequency given for a joint presentation of results obtained by different participants, seems to be a more correct action.

As an example the scheme of key comparisons CCM.M-K1[1] is given in Figure 1.2. This scheme combines into one the participants of CIPM key comparisons and regional comparisons of the APMP, EUROMET and COOMET.

One practical result of national measurement standard comparisons is the confirmation of the calibration and measurement capabilities of a particular national laboratory. The laboratory carries out measurements in according to a routine procedure, presents to a pilot laboratory its measurement results, budget of uncertainty, information about the measurement standard, and measurement procedure.

[1] http://kcdb.bipm.org/AppendixB/KCDB_ApB_info.asp.

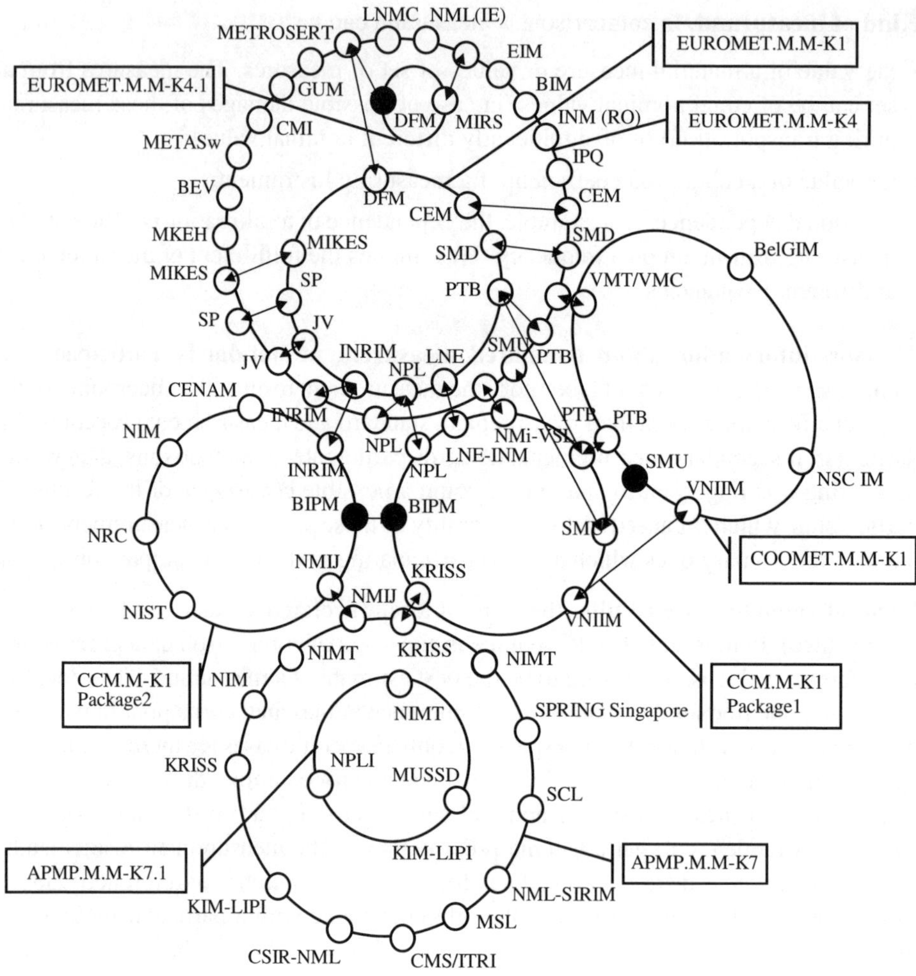

Figure 1.2. Scheme of the key comparisons CCM.M-K1. Here: ○ – laboratory participant of the key comparisons; • – pilot laboratory of the key comparisons.

Therefore, key comparisons are, as a matter of fact, the tool for a practical check of laboratory competence, the quality of metrological services rendered by a particular laboratory. The key comparisons can take more than a year. Sometimes discussions on the results and a search for a consensus on a method of data processing take even longer than the actual time required for performing the measurements. A high status of key comparisons is caused by the fact that the standards being compared are the reference national measurement standards in a chain of transferring the unit of a quantity.

1.3 Basic approaches to evaluating key comparison data

The MRA defines the degree of equivalence of national measurement standards as a deviation of a NMI measurement result from a reference value of key comparisons. Correspondingly, the supreme task is the evaluation of the reference value of key comparisons. In publications one may come across various estimates of the reference value, but the most wide-spread ones are weighted mean, mean, and median.

A weighted mean is the best one, provided the the uncertainties presented are reliable and that there is no correlation of the results obtained by different NMIs. In cases where these conditions are met, in [126, 127, 129] it is recommended using as the KCRV a weighted mean with weights inversely proportional to the squares of combined standard uncertainties.

In practice, judging by the reports on key comparisons contained in the BIPM database, a mean successfully competes with a weighted mean, since there exists a certain caution with respect to the measurement uncertainties declared by laboratories.

Let us consider the models used in evaluating key comparison data. The basic model corresponds to the situation of a traveling transfer measurement standard (of an invariable measurand) which is time stable and has the appearance

$$X_i = X, \tag{1.1}$$

where X is the measurand and X_i is the measurand obtained in the i-th laboratory. Participants in comparisons present the measured value and corresponding standard uncertainty x_i, u_i.

To process the key comparison data the least-squares method (LSM) is applied. The KCRV is calculated as a weighted mean of NMIs measurement results, using inverse squares of the corresponding values of the standard uncertainty:

$$x_w = \frac{x_1/u^2(x_1) + \cdots + x_n/u^2(x_n)}{1/u^2(x_1) + \cdots + 1/u^2(x_n)}.$$

The corresponding standard uncertainty is equal to

$$u^2(x_{\text{ref}}) = \left(\frac{1}{u^2(x_1)} + \cdots + \frac{1}{u^2(x_n)}\right)^{-1}.$$

The weighted mean can be taken as the KCRV only in the case where the data presented by the laboratories are consistent (agree with the model accepted), which can be checked using the criterion χ^2. For this an observed value of statistics χ^2 is calculated:

$$\chi^2_{obs} = \frac{(x_1 - y)^2}{u^2(x_1)} + \cdots + \frac{(x_n - y)^2}{u^2(x_n)}.$$

It is believed that a check of the data consistency does not "go through" (the data consistency is not confirmed), if $P\{\chi^2(n-1) > \chi^2_{\text{obs}}\} < 0.05$. Here the normal probability distributions are assumed.

In the case of positive check results the degree of equivalence is calculated as a pair of values – deviation of a measurement result from the reference value and uncertainty of this deviation:

$$d_i = x_i - x_{\text{ref}}$$
$$u^2(d_i) = u^2(x_i) - u^2(x_{\text{ref}})\,^{\cdot}$$

The degree of equivalence between the measurement standards of two NMIs, the results of which, as in the case given, have been obtained in the same comparison, is calculated by the formula

$$d_{ij} = d_i - d_j = x_i - x_{\text{ref}} - (x_j - x_{\text{ref}}) = x_i - x_j$$
$$u^2(d_{ij}) = u^2(x_i) + u^2(x_j).$$

There can be a number of reasons for a check by the criterion χ^2 not "going through". Among them there are two main reasons: failure of the transfer standard stability, and underestimation of the existing uncertainty of measurements by some of the comparison participants. In both cases we are dealing with inconsistent comparison data. There is no unified strictly grounded method for evaluating inconsistent data [158, 250, 309, 521, 548].

These methods of data evaluation can be conditionally divided into two groups. The first one realizes various procedures for removing the inconsistency of data. If the reason for data inconsistency is a traveling standard drift, then a correction can be introduced into the results obtained by the participants [116, 490, 559, 560]. After that, the traditional procedure of data evaluation can be applied.

If the reason for inconsistency is that discrepant data has turned up, then the form a consistent subset of measurement results is attempted [128, 159, 367, and 546]. There are different strategies for revealing and removing "outliers", i.e. the results that are not consistent with the remaining ones within the limits of the declared uncertainties. The criterion E_n, at which the normalized deviation of each result from the reference value is calculated, has become the most used:

$$E_n = \frac{|x_i - x_{\text{ref}}|}{2 \cdot u\,(x_i - x_{\text{ref}})}.$$

In case of data consistency the E_n values should not exceed the unit.

Another approach to forming a group of consistent results is based on the use of the criterion χ^2. The largest subset, the results of which are checked against the criterion χ^2, is selected [128]. If there are several subsets with a similar number of elements, then the most probable one is chosen, i.e. a subset for which a sample value of statistics χ^2_{obs} is the closest to the mathematical expectation $k-1$ of the distribution $\chi^2\,(k-1)$, where k is the number of subset elements.

On the basis of the subset formed of consistent measurement results, a reference value is determined as the weighted mean of these results. In case of several consistent subsets with a similar number of elements, the Bayesian model averaging procedure can be applied [158].

The second group of methods of inconsistent data evaluation is based on a more complicated model. As an alternative to model (1.1), the model including a bias of laboratory results is

$$X_i = X + B_i, \tag{1.2}$$

where B_i is the bias of a result in the i-th laboratory.

Various interpretations of model (1.2) connected with additional assumptions relative to B_i are possible [118, 308, 498, 542, 547, 548]. The introduction of additional assumptions in this case is needed to obtain a single solution, since the model parameter $(n + 1)$ is evaluated on the basis of n measurement results.

In the "hidden error" [548] and "random effect" [521] models it is supposed that B_i is the sample of one distribution with zero mathematical expectation and variance σ. An estimate of the measurand X and variance σ is obtained using the maximum likelihood method.

Model (1.2) can be applied, for example, when there are grounds to believe that, in the process of comparisons, insignificant, as compared to measurement, uncertainties $(\sigma < u_i)$, changes of the values of a traveling standard have taken place. In this case an additional source of the KCRV uncertainty arises, which is not related to the initial data uncertainties.

Moreover, model (1.2) is used in the situation when the data inconsistency is caused by underestimating the measurement uncertainty by comparison participants. In our view, this is possible only in the case when there is reason to believe that all participants have underestimated a certain common random factor when evaluating the uncertainty of measurements. For example, in [548] it is stated that an underestimate of the instrumental uncertainty can take place under the condition that all participants use measuring instruments of one and the same type.

In the "fix effect model" B_i describes systematic biases that have not been taken into account in the uncertainty budget. An additional assumption allowing the estimates B_i, X to be evaluated is required [309, 498, 521]. For example, it can be a condition under which the sum of systematic biases is equal to zero: $\sum B_i = 0$.

In Section 1.6 model (1.2) is treated for the case of consistent measurements, when information about uncertainty budget components, evaluated according to type A and type B, is used instead of the combined uncertainty of measurements [109, 110, 311]. This makes it possible to obtain estimates of the systematic biases of laboratory results without additional constrains, thanks to a joint analysis of all data obtained by comparison participants.

The base procedure of evaluating consistent comparison data can be extended to the case with several transfer measurement standards, using the generalized method of least squares [366]. In the general case the initial model for applying the least-squares method can be represented in the matrix form

$$\vec{Y} = X \cdot \vec{a},$$

where
$\vec{Y}$ is the vector of measurement results obtained in different laboratories for different traveling standards;
X is the experiment plan matrix the elements of which are the units and zeros depending on the fact, whether the measurements of a standard given have been performed by a certain particular laboratory or not;
$\vec{a}$ is the vector of the values of different measurement standards.

The least-squares method allows the following estimate for the vector $\vec{a}$ to be obtained:

$$\vec{a}_{\text{ref}} = (X^T \Sigma X)^{-1} X^T \Sigma^{-1} \vec{Y},$$

where Σ is the covariance matrix of measurement results: $(\Sigma)_{ij} = \text{cov}(y_i, y_j)$.

Correspondingly, we have for the vector of equivalence degrees

$$\vec{d} = \vec{Y} - X \cdot \vec{a}_{\text{ref}},$$

with the respective matrix

$$W = \Sigma - X(X^T \Sigma^{-1} X)^{-1} X^T.$$

An advantage of this approach is its generality and applicability to a great variety of schemes for performing key comparisons.

1.4 Expression of the degree of equivalence of measurement standards on the basis of a mixture of distributions

Since in the MRA there is no clear definition of the reference value and, correspondingly, the equivalence of measurement standards, then various interpretations of these concepts which on the whole do not contradict the MRA are possible [95, 96, 105, 117, 118, 546, 547].

The interpretation of the equivalence of measurement standards as the compatibility of measurement results obtained using these standards is the most common. Thus, if a result of the i-th laboratory is incompatible with a KCRV within the limits of the uncertainty claimed, then the standard given is not equivalent to the remaining group. The expansion of measurement uncertainties with the purpose of making key comparison measurement results consistent, as a rule, is not anticipated in evaluating key comparison data.

Within the framework of this subsection an alternative interpretation of the equivalence of measurement standards as the reproducibility of measurement results, obtained by a group of NMIs participating in key comparisons, is treated. Correspondingly, the degree of equivalence is understood as a quantitative measure of this reproducibility [95, 96, 105].

This interpretation can be useful for evaluating data of comparisons conducted for the first time, when from the very start there are doubts about taking into account all factors and, above all, systematic ones, in the budget of measurement uncertainty.

Let a measurement result be presented in the form of observation model that corresponds to model (1.2):

$$x_i = x + b_i + \varepsilon_i,$$

where x is the measurand value, b_i is the bias value of measurement results in a laboratory given, and ε_i is its error.

The measurement results obtained in laboratories can be interpreted as a sample of a general totality of measurements carried out by a group of laboratories participating in comparisons. It is important to note, that this distribution significantly depends on a membership of the group of participants. It is the group of definite laboratories that "generates" a new distribution, different from distributions inside each laboratory-participant, i.e., the mixture of the distributions:

$$F(x) = \frac{1}{N} \sum F_i(x),$$

where $F_i(x)$ is the frequency distribution function of the random quantity X_i, the realization of which provides results in the i-th laboratory, with the mathematical expectation $EX_i = x + b_i$.

It is suggested to choose the mathematical expectation of mixture distribution as the reference value:

$$EX = x + \frac{1}{N} \sum b_i.$$

Then the degree of equivalence of measurement standards can be defined as the difference of the mathematical expectations of two distributions by the formula

$$d_i = EX_i - EX = a + b_i - a - \bar{b} = b_i - \bar{b},$$

and may be interpreted as the difference between the "reference laboratory value" and reference value of key comparisons, or as the difference of a systematic bias of measurements in a particular laboratory and systematic bias of the reference value. It is necessary to note that in such an approach the degree of equivalence and the reference value are defined through model parameters.

Let a linear combination of measurement results be considered as an estimate of the reference value:

$$\hat{v} = \sum \omega_i x_i,$$

$$E\hat{v} = \sum \omega_i E x_i = a + \sum \omega_i b_i.$$

From this it follows that an unbiased estimate is obtained at the arithmetical mean:

$$\omega_i = \frac{1}{N}, \quad v = \frac{1}{N}\sum x_i, \quad D(v) = \frac{1}{N}\sum u_i^2.$$

Correspondingly, the estimate of degree of equivalence is

$$\hat{d}_i = x_i - \bar{x}, \quad u^2(\hat{d}_i) = u_i^2\left(1 - \frac{2}{N}\right) + \frac{1}{N^2}\sum u_i^2.$$

The distribution mix function $F(x) = \frac{1}{N}\sum F_i(x)$ is the most general form for describing the dispersion of measurements in the group of comparison participants. This function, on its own, its derivative or particular characteristics, for example, a tolerance interval, can serve as the measure of *mutual equivalence of a group of measurement standards.*

When introducing the quantitative equivalence measures on this basis of a mixture of probability distributions, it is necessary to distinguish different types of equivalence: *mutual equivalence of a group* of measurement standards, *pair-wise equivalence* inside of a group of measurement standards, and lastly, *equivalence of a particular measurement standard to a group* of measurement standards. The pair-wise equivalence of any pair of comparison participants are described by the difference of two independent random quantities X, Y. Each of these random quantities has the distribution $F(x)$. The density of difference distribution is the convolution of initial probability distribution densities: $p(z) = \int f(z-x) f(x) dx$.

In the same manner, the concept of equivalence *of each measurement standard to a group* is defined, which is quantitatively expressed by the distribution (or tolerance interval) of deviations of results obtained by this laboratory from any result of the group $X_i - X$, where $X_i \in F_i(x)$, $X \in F(x)$. Correspondingly, the density of difference distribution is the convolution $p_i(z) = \int f_i(z-x) f(x) dx$.

It should be noted that application of mixture of distribution to key comparison data analysis, which is advocated in this subsection is different from those used in [117, 546].

1.5 Evaluation of regional key comparison data

1.5.1 Two approaches to interpreting and evaluating data of regional key comparisons

The NMIs participate in the realization of the MRA through the regional metrological organizations. The basic problems of the RMO comparisons are

- the extension of metrological equivalence to the NMIs which have not participated in CIPM key comparisons by way of conducting RMO key comparisons;
- the confirmation of the calibration and measurement capabilities the NMIs have claimed by conducting supplementary comparisons.

In the MRA the procedure of linking RMO and CIPM key comparisons is treated as follows: *"The results of the RMO key comparisons are linked to key comparison reference values established by CIPM key comparisons by the common participation of some institutes in both CIPM and RMO comparisons. The uncertainty with which comparison data are propagated depends on the number of institutes taking part in both comparisons and on the quality of the results reported by these institutes"*.

The emphasis should be placed on the following aspects that are apparent from the quoted text:

- linking of RMO and CIPM key comparisons is understood as a linking of regional comparison results with the KCRV, the value of which is established in CIPM key comparisons;
- practically the linking is realized through the results of the participants of both comparisons, the so-called "linking institutes (laboratories)";
- the procedure of linking inevitably introduces an additional component of uncertainty which depends on the number of linking institutes and the quality of their measurement results.

Let us consider the two schemes of realizing the linking procedures which are most oftenrealized in practice.

Example 1. Comparisons of microphones within the frequency range 125 Hz to 8 kHz. First there were performed key comparisons CIPM – CCAUV.A-K1[2], in which twelve NMIs took part. The NPL (UK) was designated as pilot laboratory. The measurement procedure did not differ from routine calibration. All participants calibrated two traveling measurement standards in the points given. The KCRV was determined as a simple mean for a given frequency. Since the measurement results demonstrated the identical behavior for both microphones in the process of comparison, the results were averaged.

[2] http://kcdb.bipm.org/AppendixB/KCDB_ApB_info.asp.

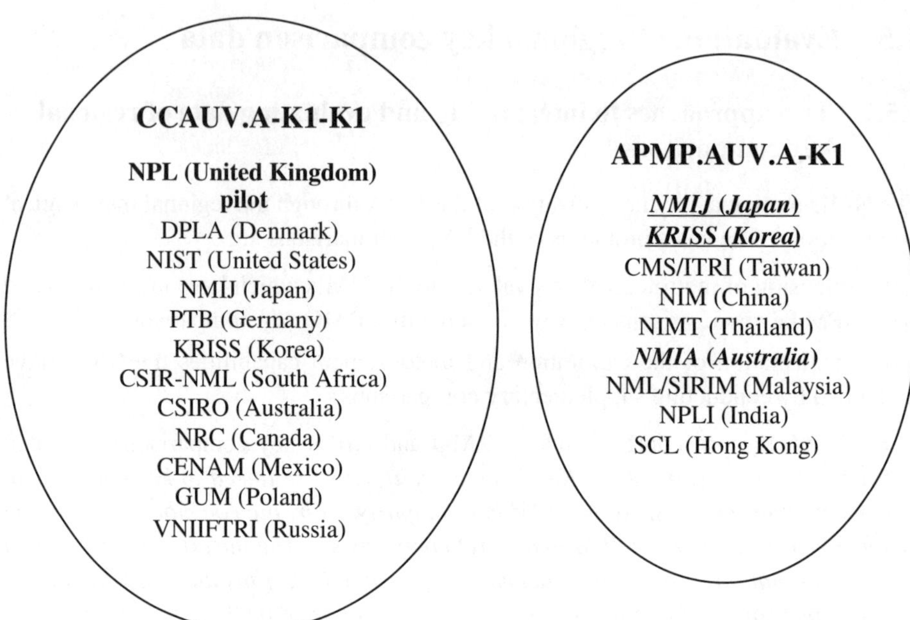

Figure 1.3. Participants of the CCAUV.A-K1-K1 – APMP.AUV.A-K1.comparisons.

Right after the CIPM comparisons, the regional comparisons APMP.AUV.A-K1[3], were carried out, in which nine NMIs took part. The comparison participants are presented in Figure 1.3. The regional comparisons were performed according to a similar scheme. Laboratories in Japan, Korea, and Australia participated in both comparisons. But only the laboratories in Japan and Korea were chosen as the linking ones, since the laboratory in Australia changed its equipment between the comparisons and, consequently, the systematic components also changed, and it was necessary to redefine its degrees of equivalence.

The linking between the comparisons was realized by adding a correction for the difference between the values of a measurand to the measurement results of the regional comparisons. This correction was calculated as a mean of the differences of the results which the linking institutes had obtained in the CIPM and RMO key comparisons. The deviations of the thusly transformed results of the regional comparisons from the KCRV did not exceed the corresponding uncertainties, which is the confirmation of their calibration capabilities according to the criterion E_n.

The scheme of linking the regional and the CIPM comparisons is rather common in practice. Conditionally, it may be called the scheme of "transformation" of regional comparison results to the level of CIPM comparisons. This transformation can be achieved by adding a term or multiplying factor [96, 121, 135, and 157]**.** Different principles of transformation will be considered below. It is important to note that a

[3] http://kcdb.bipm.org/AppendixB/KCDB_ApB_info.asp.

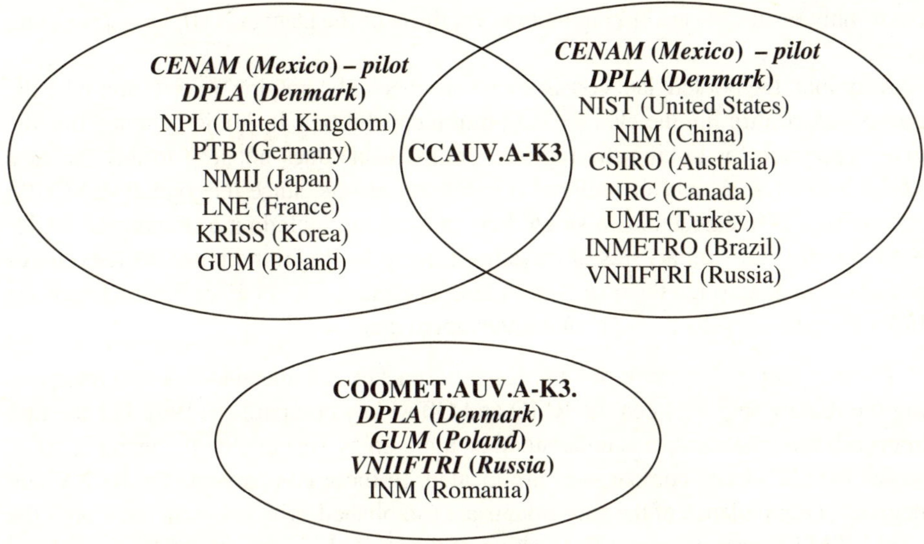

Figure 1.4. Participants of the comparisons CCAUV.A-K3 and COOMET.AUV.A-K3.

procedure such as this is always accompanied by an increase in the measurement uncertainty of regional comparisons.

Example 2. Comparisons CCAUV.A-K3, COOMET.AUV.A-K3[4]. The participants of comparisons are presented in Figure 1.4. These are also comparisons of microphones which a similar construction. But comparisons CCAUV.A-K3 were organized using a different scheme. Two complete sets of traveling measurement standards had been sent consisting of two microphones each to shorten the comparison time. Also, only two laboratories calibrated all four microphones during the comparison.

Thus the problem arose of providing a link of different circles of one and the same CIPM comparison. This problem was solved by applying the least-squares method [306]. The estimates of traveling standard values were obtained, and the degrees of equivalence were calculated. Then the criterion χ^2 was applied for checking the consistency of all data. This served as the confirmation of the calibration capabilities claimed by the participants of the whole group.

This approach was disseminated to the evaluation of data of the regional comparisons COOMET.AUV.A-K3, which were considered as an additional circle of key comparisons. The data of the regional comparisons was combined with the data of the CIPM comparisons, and the pooled data was evaluated by the least-squares method.

An approach such as this naturally leads to a recalculation of the reference values and degrees of equivalence established in the CIPM key comparisons. In order to reduce this effect to a minimum in processing the results of these particular compar-

[4] http://kcdb.bipm.org /AppendixB/KCDB_ApB_info.asp

isons, only results of one linking laboratory, those of the Danish NMI, were taken into account.

After that, the pooled data consistency was checked, applying χ^2 criterion with the purpose of confirming the claimed calibration capabilities. It should be noted that the given approach for linking the regional comparisons does not lead to any increase of the initial uncertainty of regional comparison results, since the joint RMO-CIPM comparison is interpreted as a single key comparison including a number of circles with various (but similar) sets of traveling standards. It is important for these circles to be "crossed" through the results of linking institutes. This enables the tracing of the link of any two results through an uninterrupted chain.

These examples illustrate two existing approaches to understanding and interpreting the relationship between the RMO and CIPM key comparisons [96]. For the first approach this relationship is understood as a hierarchy scheme, on the upper level of which the CIPM key comparisons are located. In these comparisons the KCRV and degrees of equivalence of the participants are established, which remain fixed until the next CIPM key comparisons. The linking of the RMO KC with the KCRV is realized through the results of the linking institutes.

In the second approach all subsequent regional comparisons are perceive as loops of a single uninterrupted key comparison. With the appearance of new information all perviously obtained estimates have to be continuously recalculated.

Below only the first approach to ensuring the linking of comparisons will be treated as more appropriate for the MRA methodology, and the mathematical approaches that will be presented are intended just for its realization.

1.5.2 Equation of linking RMO and CIPM KC. Optimization of the algorithm of evaluating degrees of equivalence

It is natural to begin solving the problem of evaluating the regional comparison data by formulating a measurement model. In the case given an output quantity is the degree of equivalence of the results obtained by a laboratory which is a RMO KC participant, or more precisely, the vector of the values of the degrees of equivalence, since in the general case in the regional comparisons several laboratories participate and have to be "linked".

The measurands determined in the process of the CIPM and RMO key comparisons and which influence the estimate of output quantities, i.e., the estimate of the degrees of equivalence of national measurement standards, are considered as input quantities.

In accordance with the MRA the pair-wise degree of equivalence of two measurement standards is expressed as the difference of their degrees of equivalence and its corresponding uncertainty:

$$D_{ij} = D_i - D_j. \tag{1.3}$$

Expression (1.3) is general and also valid in the case where the degrees of equivalence i and j are evaluated in different key comparisons [156]:

$$\begin{aligned} d_i &= x_i - x_{\text{ref1}}, \quad u(d_i), \\ d_j &= x_j - x_{\text{ref2}}, \quad u(d_j), \\ d_{ij} &= d_i - d_j, \quad u^2(d_{ij}) = u^2(d_i) + u^2(d_j) - 2\text{cov}(d_i, d_j). \end{aligned}$$

Provided that the degrees of equivalence are established in one comparison and for the same measurand, then the pair-wise degree of equivalence can be established without using the reference value:

$$\begin{aligned} d_{ij} &= x_i - x_j, \\ u^2(d_{ij}) &= u^2(x) + u^2(x_j) - 2\text{cov}(x_i, x_j) \end{aligned}$$

If we take the second term from the right part of equation (1.3) and place it in the left part, we will obtain

$$D_j = D_{ji} + D_i. \tag{1.4}$$

Expression (1.4) is the equation of linking, where in the left-hand side is the quantity looked for, i.e., the degree of equivalence D_j of the RMO KC laboratory participant, and in the right-hand side there are quantities through which are defined

- the degree of equivalence of the measurement standard of the linking laboratory D_i, the estimate of which has been obtained in CIPM KC - $\{d_i, u(d_i)\}_1^L$;
- the pair-wise degree of equivalence of measurement standards D_{ji}, the estimate of which has been obtained in RMO KC: $\{y_j - y_i, u(y_j - y_i) = \sqrt{u_j^2 + u_i^2}\}$.

From equation (1.4) the role of linking laboratories in the procedure of ensuring the traceability to the KCRV is evident. This traceability can be established only in comparisons with a linking laboratory, i.e., under the condition of establishing the pair-wise equivalence with the laboratory, the degree of equivalence of which has already been established.

Since the linking can be realized through any linking laboratory $1 \le i \le L$ and for any RMO KC participant, apart from linking laboratories $L + 1 \le j \le m$, then one obtains $L(m - L)$ equations

$$\begin{aligned} D_{L+1} &= D_{L+1,1} + D_1, \\ &\dots \\ D_{L+1} &= D_{L+1,L} + D_L, \\ &\dots \\ D_m &= D_{m1} + D_1, \\ &\dots \\ D_m &= D_{mL} + D_L. \end{aligned}$$

Application of the LSM gives a vector of estimates the degrees of equivalence $\vec{D}^T = (D_{L+1}, \dots, D_m)$ with the corresponding covariance matrix of these estimates $U(\vec{D})$ [109]:

$$\hat{\vec{D}} = (\Phi^T U^{-1} \Phi)^{-1} \Phi^T U^{-1} \vec{Z} \quad U(\vec{D}) = (\Phi^T U^{-1} \Phi)^{-1}, \tag{1.5}$$

where
$\vec{Z}^T = (y_{L+1} - y_1 + d_1, \dots, y_{L+!} - y_L + d_L, \dots, y_m - y_1 + d_1, \dots, y_m - y_L + d_L)$ is the vector of estimates of the input quantities of a dimension $L(m - L)$, formed of pair-wise differences of measurement results and estimates of the degrees of equivalence;
U is the covariance matrix of the input quantities estimates of dimension $L(m - L) \times L\,(m - L)$;
Φ is the matrix of dimension $L(m - L) \times (m - L)$, having a block structure and consisting of vector columns $\vec{j}\,(L)$ of the dimension L:

$$\Phi = \begin{pmatrix} j(L)\ 000...0000 \\ 0\ j\ (L)\ 0000 \\ .... \\ 00000 \quad j(L) \end{pmatrix}, \quad j(L) = \begin{pmatrix} 1 \\ \cdots \\ 1 \end{pmatrix}.$$

Despite the fact that the matrix Φ has a block structure, in the general case the system of equations does not fall to independent subsystems relative to elements of the vector $\vec{D}^T = (D_{L+1}, \dots, D_m)$, since covariance of the estimates of input data differ from zero.

Since in practice it is preferable to have an explicit solution, let us consider a certain simplification of the general problem. For this purpose some additional conditions will be introduced: (1) linking of the results of each laboratory is realized independently of the results of other laboratories; (2) the reference value is estimated as the weighted mean of the results of the laboratories participating in CIPM KC; (3) the degree of equivalence of linking institutes is estimated as $d_i = x_i - x_{\text{ref}}$.

Conditions (2) and (3) correspond to the case most common in practice and condition (1) means a transition from the dimension $L(m - L)$ to $(m - L)$ systems of the dimension L. Expression (1.5) for the degree of equivalence of the laboratory that has obtained the result y in RMO KC is rewritten in the form

$$\begin{aligned} \hat{d} &= (j^T U^{-1} j)^{-1} j^T U^{-1} \vec{Z} = (j^T U^{-1} j)^{-1} j^T U^{-1} \big((y - x_{\text{ref}}) j^T + \vec{X} - \vec{Y}\big) \\ &= y - x_{\text{ref}} + (j^T U^{-1} j)^{-1} j^T U^{-1} (\vec{X} - \vec{Y}), \qquad (1.6) \\ \vec{X}^T &= (x_1, \dots, x_L), \quad \vec{Y}^T = (y_1, \dots, y_L)\,. \end{aligned}$$

In the additional assumptions mentioned above and related to the stability of accuracy characteristics of the linking laboratories results, namely $\rho_i = \frac{\operatorname{cov}(x_i, y_i)}{u(x_i) u(y_i)}$,

$u(x_i) = u(y_i)$, the elements of the covariance U_{ij} are equal to

$$\operatorname{cov}\left(d_i, d_j\right) = \begin{cases} u^2(y) - (1 - \rho_i - \rho_j)\, u^2(x_{\text{ref}}) & i \neq j \\ u^2(y) - (1 - 2\rho_i)\, u^2(x_{\text{ref}}) + 2(1 - \rho_i)\, u^2(x_i) & i = j. \end{cases}$$

Expression (1.6) means that the estimate of the degree of equivalence can be presented in the form

$$\hat{d} = y - x_{\text{ref}} + \sum_{1}^{L} w_i\,(x_i - y_i) \tag{1.7}$$

In other words, in order to evaluate a degree of equivalence a correction has to be made to the result of RMO KC, which is equal to a weighted sum of the difference of linking laboratories results, and then the KCRV value is subtracted. Thus, the quote algorithm based on the LSM confirms the rule of transformation of the RMO KC results, which is natural at first sight.

Let the optimal weights w_i be determined, reasoning from the condition of an uncertainty minimum $u(\hat{d})$ and assuming the independence of results of the laboratory-participant of RMO KC as well as that of the results of the laboratory participants of CIPM KC:

$$u^2(\hat{d}) = u^2(y) + u^2(x_{\text{ref}}) + 2\sum_{1}^{L} w_i^2\,(1 - \rho_i)\, u_i^2 - 2\sum_{1}^{L} w_i\,(1 - \rho_i)\, u_{\text{ref}}^2.$$

In order to find the weights it is necessary to solve the following optimization problem under the condition where their sum is equal to 1:

$$\min_{w_i} \sum_{1}^{L} w_i^2\,(1 - \rho_i)\, u_i^2 - \sum_{1}^{L} w_i\,(1 - \rho_i)\, u_{\text{ref}}^2 + \lambda\left(\sum_{1}^{L} w_i - 1\right).$$

Hence the following expressions are obtained for the optimal weights:

$$w_i = \frac{1 - u_{\text{ref}}^2 \sum_1^L \frac{1}{2u_i^2}}{\left(2(1-\rho_i)\,u_i^2\right)\sum_1^L \frac{1}{2(1-\rho_i)u_i^2}} + \frac{u_{\text{ref}}^2}{2u_i^2}$$

$$= \frac{\frac{1}{(1-\rho_i)u_i^2}}{\sum_1^L \frac{1}{(1-\rho_i)u_i^2}} + u_{\text{ref}}^2\left(\frac{1}{2u_i^2} - \sum_{1}^{L}\frac{1}{2u_i^2} \times \frac{\frac{1}{2(1-\rho_i)u_i^2}}{\sum_1^L \frac{1}{2(1-\rho_i)u_i^2}}\right), \tag{1.8}$$

where $\rho_i = \dfrac{cov(x_i, y_i)}{u(x_i)u(y_i)}$ and $u(x_i) = u(y_i) = u_i$, $u(x_{\text{ref}}) = u_{\text{ref}}$.

The corresponding standard uncertainty of the degree of equivalence is given by the expression

$$u^2(\hat{d}) = u^2(y) + u^2_{\text{ref}} \sum_1^L w_i \rho_i + \frac{2}{\sum_1^L \frac{1}{(1-\rho_i)u_i^2}} \left(1 - u^2_{\text{ref}} \sum_1^L \frac{1}{2u_i^2}\right). \quad (1.9)$$

Provided that in (1.9) the uncertainty of the reference value is neglected regarding a first approximation, then it is obvious that at the transformation of RMO KC results their uncertainty increases by a quantity

$$\frac{2}{\sum_1^L \frac{1}{(1-\rho_i)u_i^2}}.$$

This uncertainty may be called the uncertainty of the linking algorithm. It is the smaller at the greater number of linking laboratories and higher coefficients of correlation between the results of these laboratories take place.

In conclusion it should be noted that equation (1.4) contains only the degrees of equivalence and does not directly include KCRV directly.

1.5.3 Different principles for transforming the results of regional comparisons

Expression (1.7) underlies many algorithms for the evaluation of the results of regional comparisons by transforming them to the level of CIPM key comparisons [109, 121, 135, 260, 261]. In fact, (1.7) can be rewritten in the form

$$\hat{d} = y + \sum_1^L w_i (x_i - y_i) - x_{\text{ref}} = y_{\text{transf}} - x_{\text{ref}},$$

$$y_{\text{transf}} = y + \Delta.$$

Some algorithms of the RMO KC results transformation are inferred from the condition of minimization of uncertainty associated with correction Δ, i.e., the uncertainty of the difference of measurands in the CIPM and RMO key comparisons. At the heart of an algorithm used in the COOMET Recommendation GM/RU/14:2006 [121], the condition of uncertainty minimization underlies: $u(\sum_1^L w_i(x_i - y_i)) = u(\Delta)$. This condition is provided at a choice of weights, inversely proportional to the squares of the uncertainty of linking results differences:

$$v_i = \frac{\frac{1}{(1-\rho_i)u_i^2}}{\sum_1^L \frac{1}{(1-\rho_i)u_i^2}}.$$

The comparison with the optimal weights given by expression (1.8) shows that the weights are close to the optimal ones in those cases where it is possible to neglect the uncertainty of KCRV.

Recommendation COOMET GM/RU/14:2006 [121] contains two data processing procedures which are conventionally called procedures C and D. These procedures correspond to the two methods of RMO KC data transformation. The transformation can be realized by introducing an additive or multiplicative correction (multiplication by a factor).

Procedure C calls for the introduction of an additive correction for the measurement results obtained in RMO KC, whereas procedure D requires a multiplicative correction. The multiplicative correction itself is calculated in a way similar to that of the additive one, where, instead of the differences of measurement results of the linking institutes, their ratios are taken.

An undoubted advantage of [121, 261] is the fact that it considers not only the situation where the RMO KC results are independent of each other and independent of CIPM KC results except the results of linking measurement standards, but also the situation where some of the RMO KC results are correlated among themselves or with the CIPM KC results. The main reason for such a correlation is the traceability of the laboratories participating in RMO KC to the laboratory participants of the CIPM KC, which is rather frequently met in practice. Then procedure D [261] is considered by comparing it with other multiplicative corrections.

The algorithms for transforming the RMO KC results by introducing a multiplicative correction are treated in [157, 261]. In [157] a correction is suggested in the form of the ratio of a KCRV to a weighted mean of the linking institutes results obtained in RMO KC:

$$r_1 = \frac{\dfrac{1}{\sum_1^n u^{-2}(x_i)} \sum_1^n \dfrac{x_i}{u^2(x_i)}}{\dfrac{1}{\sum_1^L u^{-2}(y_i)} \sum_1^L \dfrac{y_i}{u^2(y_i)}}. \tag{1.10}$$

In [261] the multiplicative correction is introduced as a weighted mean of linking the institute ratios obtained in CIPM and RMO key comparisons:

$$r_2 = \frac{\sum_1^L \dfrac{1}{u_{\text{rel}}^2(x_i)(1-\rho_i)} \times \dfrac{x_i}{y_i}}{\sum_1^L \dfrac{1}{u_{\text{rel}}^2(x_i)(1-\rho_i)}}. \tag{1.11}$$

Since a criterion of correction selection is the minimum of uncertainty associated with data transformation, it has been suggested to consider the multiplicative correction of the general type [112]:

$$r = \frac{\sum_1^n \omega_i x_i}{\sum_1^L \nu_j y_j}. \tag{1.12}$$

In addition, the condition of normalization $\sum_1^n \omega_i = 1$, $\sum_1^L \nu_i = 1$ has been introduced. Thus, the weighted mean of all the results of the CIPM key comparisons is located in the numerator and the weighted mean of the results which the linking institutes have obtained in RMO comparisons are in the denominator.

The weights are calculated reasoning from the condition of minimization of the relative uncertainty $u_{\mathrm{rel}}^2(r)$:

$$u_{\mathrm{rel}}^2(r) = \sum_1^n \omega_i^2 u_{\mathrm{rel}}^2(x_i) + \sum_1^L \nu_i^2 u_{\mathrm{rel}}^2(x_i) - 2\sum_1^L \omega_i \nu_i \rho_i u_{\mathrm{rel}}^2(x_i).$$

The solution of the problem gives the expression for the weights ω_i, ν_i through the constants π, λ:

$$\omega_i = -\frac{\pi\rho_i + \lambda}{u_{\mathrm{rel}}^2(x_i)(1-\rho_i^2)}, \qquad i = 1,\dots,L,$$

$$\omega_i = -\frac{\lambda}{u^2(x_i)}, \qquad i = L+1,\dots,n,$$

$$\nu_i = -\frac{\pi + \lambda\rho_i}{u^2(x_i)\left(1-\rho_i^2\right)}, \qquad i = 1,\dots,L.$$

These constants are determined from the condition of normalization $\sum_1^n \omega_i = 1$, $\sum_1^L \nu_i = 1$ by solving the linear system of equations:

$$\pi = \frac{\Delta(\pi)}{\Delta}, \quad \lambda = \frac{\Delta(\lambda)}{\Delta},$$

$$\Delta = -\sum_1^L \frac{1}{u^2(x_i)(1+\rho_i)} \times \sum_1^L \frac{1}{u^2(x_i)(1-\rho_i)} - \sum_1^L \frac{1}{u^2(x_i)\left(1-\rho_i^2\right)} \times \sum_{L+1}^n \frac{1}{u^2(x_i)},$$

$$\Delta(\lambda) = \sum_1^L \frac{1}{u^2(x_i)(1+\rho_i)},$$

$$\Delta(\pi) = \sum_1^L \frac{1}{u^2(x_i)(1+\rho_i)} + \sum_{L+1}^n \frac{1}{u^2(x_i)}.$$

It should be noted that the weights of the linking institutes exceed the weights of the remaining ones; this difference increases with the growth of the correlation coefficient between the results of the linking institutes. To make the results of the comparative analysis more obvious, let us assume that $\rho_i \approx \rho$, $i \le L$. This assumption allows the expression used for the weights to be simplified:

$$\nu_i = \frac{u^{-2}(x_i)}{\sum_1^L u^{-2}(x_i)}, \qquad i = 1, \ldots, L,$$

$$\omega_i = u^{-2}(x_i) \left(\frac{\rho}{\sum_1^L u^{-2}(x_i)} + \frac{1-\rho}{\sum_1^n u_{\text{rel}}^{-2}(x_i)} \right) \qquad i = 1, \ldots, L,$$

$$\omega_i = -\frac{\lambda}{u_{\text{rel}}^2(x_i)} = (1-\rho) \frac{u_{\text{rel}}^{-2}(x_i)}{\sum_1^n u_{\text{rel}}^{-2}(x_i)}, \qquad i > L.$$

In Table 1.1 the uncertainties of the multiplicative corrections considered and their comparison are illustrated. Two cases are treated: with several linking laboratories and with one linking laboratory.

Table 1.1. Uncertainties of the multiplicative corrections and their comparison.

General case	$L = 1$
$u_{\text{rel}}^2(r) = \frac{(1-\rho)^2}{\sum_1^n u_{\text{rel}}^{-2}(x_i)} + \frac{1-\rho^2}{\sum_1^L u_{\text{rel}}^{-2}(x_i)}$	$u_{\text{rel}}^2(r) = (1-\rho^2)\, u_{\text{rel}}^2(x_1) + \frac{(1-\rho)^2}{\sum_1^n u_{\text{rel}}^{-2}(x_i)}$
$u_{\text{rel}}^2(r_1) = \frac{1-2\rho}{\sum_1^n u_{\text{rel}}^{-2}(x_i)} + \frac{1}{\sum_1^L u_{\text{rel}}^{-2}(x_i)}$	$u_{\text{rel}}^2(r_1) = u_{\text{rel}}^2(x_1) + \frac{1-2\rho}{\sum_1^n u_{\text{rel}}^{-2}(x_i)}$
$u_{\text{rel}}^2(r_2) = 2(1-\rho) \frac{1}{\sum_1^L u_{\text{rel}}^{-2}(x_i)}$	$u_{\text{rel}}^2(r_2) = 2(1-\rho)\, u_{\text{rel}}^2(x_1)$
$u_{\text{rel}}^2(r_1) - u_{\text{rel}}^2(r) = \rho^2 \left(\frac{1}{\sum_1^L u_{rel.i}^{-2}} - \frac{1}{\sum_1^n u_{\text{rel}}^{-2}(x_i)} \right)$	$u_{\text{rel}}^2(r_1) - u_{\text{rel}}^2(r) = \rho^2 \left(u_{\text{rel}}^2(x_1) - \frac{1}{\sum_1^n u_{\text{rel}}^{-2}(x_i)} \right)$
$u_{\text{rel}}^2(r_2) - u_{\text{rel}}^2(r) = (1-\rho)^2 \left(\frac{1}{\sum_1^L u_{\text{rel}}^{-2}(x_i)} - \frac{1}{\sum_1^n u_{\text{rel}}^{-2}(x_i)} \right)$	$u_{\text{rel}}^2(r_2) - u_{\text{rel}}^2(r) = (1-\rho)^2 \left(u_{\text{rel}}^2(x_1) - \frac{1}{\sum_1^n u_{\text{rel}}^{-2}(x_i)} \right)$

The multiplicative correction r can be considered to be the generalization of corrections r_1 and r_2, since at $\rho = 0$ it coincides with r_1. Moreover, if the corrections r_1, r_2 are compared with each other, then at $\rho < 0.5$ the correction r_1 has a lesser uncertainty, and, vice versa, r_2 is more preferable at $\rho > 0.5$.

Differences in uncertainties associated with the corrections considered depend first of all on the coefficient of correlation among the results of linking institutes. It can be assumed that the CIPM KC participants have approximately similar uncertainties

$u(x_i) \approx u$, $i = 1, \ldots, n$, and then the difference between the generalized correction r and the corrections r_1, r_2 has a clear expression and is determined by the number of linking institutes and a coefficient of the correlation of their results:

$$u_{\text{rel}}^2(r_1) - u_{\text{rel}}^2(r) = \rho^2 u_{\text{rel}}^2 \frac{n-L}{nL},$$

$$u_{\text{rel}}^2(r_2) - u_{\text{rel}}^2(r) = (1-\rho)^2 u_{\text{rel}}^2 \frac{n-L}{nL}.$$

Further on, a significant correlation among linking laboratories results is assumed; therefore the correction r_2 is considered for linking comparisons [261]. The transformed results of the regional comparison $z_k = r_2 y_k$, $k > L$ can be added to the results of the corresponding CIPM comparison with the following uncertainties:

$$u_{\text{rel}}^2(z_k) = u_{\text{rel}}^2(y_k) + u_{\text{rel}}^2(r_2), \qquad u_{\text{rel}}^2(r_2) = \frac{2}{\sum_{k=1}^{L} \frac{1}{u_{\text{rel}}^2(y_k)(1-\rho_i)}}.$$

The degree of equivalence for the RMO KC ($k > L$) participants are given in the form

$$d_k = z_k - x_{\text{ref}}$$

with the corresponding uncertainty

$$u^2(d_k) = u^2(z_k) + u^2(x_{\text{ref}}) - 2u(z_k, x_{\text{ref}}).$$

Let us note that the correlation with the results of the institutes, which are the CIPM KC participants leads to a decrease in the uncertainty of degrees of equivalence. The following expressions for the uncertainties of degrees of equivalence have been obtained:

- $$u^2(d_i) = r_2^2 u^2(x_i) + u^2(x_{\text{ref}}) + 2\left(\sum_1^L \frac{1}{u^2(x_k)(1-\rho_k)}\right)^{-1} \times \left(1 - u^2(x_{\text{ref}}) \sum_1^L \frac{1}{u^2(x_k)}\right) \tag{1.13}$$

if there is no correlation between the regional comparison result and the results of the CIPM KC participants, and

- $$u^2(d_i) \cong r_2^2 u^2(x_i) - u^2(x_{\text{ref}}) + 2\left(\sum_1^L \frac{1}{u^2(x_k)(1-\rho_k)}\right)^{-1} \times \left(1 - u^2(x_{\text{ref}}) \sum_1^L \frac{1}{u^2(x_i)}\right)$$

if there is a correlation between the regional comparison results and the results of the linking institute (for example, the result of the regional comparison is traceable to the measurement standard of a CIPM KC participant).

Expression (1.13) for the uncertainty of the equivalence degrees of transformed data is the sum of three components: initial uncertainties, uncertainty of the reference value, and uncertainty caused by transformation.

1.6 Bayesian approach to the evaluation of systematic biases of measurement results in laboratories

As mentioned above, the MRA does not provide a univocal interpretation of the concept "equivalence of measurement standards", but introduces a quantitative measure of equivalence as a certain estimate and corresponding uncertainty, without any explanation of what kind of quantity is meant. Naturally, this provokes discussion as well as the wish to determine the degree of equivalence with the help of some particular quantity using an initial model of measurement result: $x_i = x + b_i + \varepsilon_i$.

In a number of works the degrees of equivalence are identified with the systematic biases of the measurement results obtained in a respective laboratory. Such considerations are based on the model (1.2). Hence, as was noted above, they demand additional assumptions at a simultaneous evaluation of a measurand and systematic biases.

We do not agree with the idea of identifying the degree of measurement standard equivalence with a bias of a measurement result. First of all, let us note that the equivalence of measurement standards is established on the basis of a particular set of standards. This is obviously shown above when using a mixture of distributions for a quantitative expression of the equivalence of standards. As to the bias of a measurement result of some laboratory, it is an intrinsic characteristic of the measurement results obtained in this particular laboratory.

However, it should be recognized that the problem of revealing and evaluating systematic biases in results of comparison participants is rather important. Key comparisons of national measurement standards provide getting new information about systematic biases of results obtained by laboratories, which is not taken into account in the traditional approach to evaluating comparison data. In this section the problem of evaluating the systematic biases of laboratory results on the basis of data obtained in key comparison of measurement standards is treated without reference to the calculation of the degrees of equivalence.

The application of the Bayesian theorem has allowed a posteriori estimates of systematic biases of laboratories to be obtained without any additional constrains related to the parameters evaluated. Bayesian methods were formerly also applied when evaluating key comparisons data, but there were difficulties connected with constructing a priori densities of distributions for the parameters being evaluated. These difficulties have been successfully overcome, using information contained in a budget of uncertainty, which is easy to access in each comparison [109, 310].

Let model (1.2) be considered over again:

$$X_i = X + B_i, \quad i = 1, \dots, n,$$

where X is the measurand that is measured by all comparison participants and remains to be unchangeable at a stable traveling standard, and B_i is the systematic bias of a measurement result of a laboratory.

Before carrying out a comparison, some a priori information is available:

- the characteristic (standard deviation) of the measurement precision in a laboratory σ_i, which is assumed to be known as an accurate one;
- the combined standard uncertainty caused by systematic effects $u_B(x_i)$. It is assumed that all known corrections have been already inserted and that the corresponding uncertainties of these corrections have been taken into account;
- the combined standard uncertainty, correspondingly, equal to

$$u(x_i) = \sqrt{u_B^2(x_i) + \sigma_i^2}.$$

Thus, before carrying out the comparisons there is no information about a measurand, and with respect to a systematic bias it is possible to tell only that its best interlaboratory estimate is equal to zero, with the corresponding uncertainty $u_B(x_i) = u_{Bi}$. This can be formalized in the form of a priori probability density functions (PDFs) of the parameters being evaluated in the following manner:

$$p(b_i) = p_{Bi}(t \mid u_{Bi}) = \frac{1}{\sqrt{2\pi} u_B(x_i)} \exp\left\{-\frac{t^2}{2u_B^2(x_i)}\right\}, \quad p(X) \propto 1.$$

On the basis of a priori information and measurement results x_i, obtained in a key comparison, in each laboratory a posteriori distribution density function of measurand values can be obtained. The application of the Bayesian theorem leads to the following expression for the a posteriori PDF of the measurand:

$$p(x|x_i) \propto \int l(x, b_i|x_i) p(x, b_i) \, \mathrm{d} b_i,$$

where $l(x, b_i|x_i) \propto \exp[-(x_i - (x + b_i))^2/2\sigma_i^2]$ is the likelihood function, and $p(x, b_i)$is the joint PDF of measurand and bias in the i-th laboratory, in the case being considered $p(x, b_i) \propto p(b_i)$.

Before joint handling of all data presented by the comparison participants, it is necessary to make sure that the data is consistent. To do this it is possible, for example, to use a known criterion, based on the analysis of the zone of overlapping the PDFs of measurand, obtained in different laboratories [498]. Let us pay attention once again to the fact that model (1.2) is applied in this case for an analysis of consistent data.

The application of the Bayesian theorem yields a joint a posteriori PDF of the measurand and biases of laboratory results:

$$p(x, b_1, \dots, b_N | x_1, \dots, x_N) \propto \\ l(x, b_1, \dots, b_N | x_1, \dots, x_N) p(x, b_1, \dots, b_N), \qquad (1.14)$$

where $l(x, b_1, \dots, b_N | x_1, \dots, x_N) = l(x, b_1 | x_1) \cdots l(x, b_N | x_N)$ is the likelihood function, and $p(x, b_1, \dots, b_N) \propto p(b_1) \cdots p(b_N)$is the PDF of measurand and systematic biases of laboratory results.

Integrating (1.14) yields a posteriori marginal PDFs of the measurand and of the systematic bias in a particular laboratory:

$$p(x | x_1, \dots, x_N) = \int p(x, b_1, \dots, b_N | x_1, \dots, x_N) \, \mathrm{d} b_1 \cdots \mathrm{d} b_N,$$

$$p(b_i | x_1, \dots, x_N) = \int p(x, b_1, \dots, b_N | x_1, \dots, x_N) \, \mathrm{d} x \, \mathrm{d} b_1 \cdots \mathrm{d} b_{i-1} \, \mathrm{d} b_{i+1} \cdots \mathrm{d} b_N.$$

Let us note that methods of statistic modeling can be used to obtain the a posteriori PDFs of the measurand and the systematic biases for the a priori PDEs of an arbitrary form and for an arbitrary likelihood function. For Gaussian a priori PDFs of systematic biases and the Gaussian likelihood function the joint a posteriori density of distribution is given by the expression

$$p(x, b_1, \dots, b_n | \dots, x_i, , u_{Bi} \dots . \sigma_i \dots .) \propto \\ \prod_i \frac{1}{\sqrt{2\pi} \sigma_i} \exp \left\{ -\frac{(x_i - x - b_i)^2}{2\sigma_i^2} \right\} \frac{1}{\sqrt{2\pi} u_{Bi}} \exp \left\{ -\frac{b_i^2}{2u_{Bi}^2} \right\}. \qquad (1.15)$$

The best estimates of a measurand, calculated as a mathematical expectation of the distribution $p(x | x_1, \dots, x_N)$, and of the corresponding standard uncertainty, calculated as a standard deviation, are equal to

$$x_w = u^2(x_w) \sum_{i=1}^{N} \frac{x_i}{(u_{Bi}^2 + \sigma_i^2)}, \qquad u^2(x_w) = \left(\sum_{i=1}^{N} \frac{1}{(u_{Bi}^2 + \sigma_i^2)} \right)^{-1}.$$

Thus, the Bayesian approach gives a weighted mean of measurement results as an estimate of the measurand. This coincides with the estimate obtained with the least-squares method.

In the same way estimates of systematic biases and corresponding uncertainties have been obtained:

$$b_i = \frac{u_{Bi}^2}{u_{Bi}^2 + \sigma_i^2} (x_i - x_w), \qquad u^2 (b_i) = u_{Bi}^2 \frac{\sigma_i^2 + u_{Bi}^2 \frac{u^2(x_w)}{\sigma_i^2 + u_{Bi}^2}}{\sigma_i^2 + u_{Bi}^2}.$$

An analysis of the estimates obtained for systematic biases shows that

- if the precision of measurements is high $\sigma_i \ll u_{Bi}$, then the estimate of a systematic bias is close to the difference between the measurement result and weighted mean $b_i \approx (x_i - x_w)$. At the same time, the corresponding standard uncertainty is equal to $u^2(b_i) \approx u^2(x_w)$, i.e., significantly less than the uncertainty obtained for the degree of equivalence;
- if a random component is significant, then the role of a priori estimate of a systematic bias in this case becomes more important and $|b_i| \ll |x_i - x_w|$.

The problem of the evaluation of a systematic difference between the results obtained in two laboratories is of independent interest [110, 309]. Based on the model for the i-th laboratory, it is easy to obtain a model of the systematic difference of the results of two laboratories which is equal to the difference of the systematic biases of the results of each laboratory:

$$X_i - X_j = B_i - B_j.$$

It is interesting to note that the estimate of the systematic difference between two results depends of the fact of whether the information from all the laboratories or only from those between which the systematic discrepancy of results is determined has been used.

The joint PDF of systematic biases of the i.th and j-th laboratories can be obtained by marginalization of (1.15):

$$p\left(b_i, b_j \,|\ldots, x_i, , u_{Bi} \ldots . \sigma_i \ldots .\right) \propto$$
$$\iiint \prod_i \frac{1}{\sqrt{2\pi} u_{Bi}} \exp\left\{-\frac{b_i^2}{2u_{Bi}^2}\right\} \frac{1}{\sqrt{2\pi}\sigma_i} \exp\left\{-\sum \frac{(x_i - x - b_i)^2}{2\sigma_i^2}\right\} dx\, db_1 \underset{i \neq j}{\cdots} db_n.$$

The mathematical expectation of the difference of quantities is equal to the difference of the mathematical expectations:

$$\Delta_{ij} = b_i - b_j = \frac{u_{Bi}^2}{u_{Bi}^2 + \sigma_i^2}(x_i - x_{\text{ref}}) - \frac{u_{Bj}^2}{u_{Bj}^2 + \sigma_j^2}(x_j - x_{\text{ref}}).$$

To calculate the corresponding uncertainty it is necessary to know the covariation $\text{cov}(b_i, b_j)$, which can be obtained from the joint probability density function (1.15). As a result the following is obtained:

$$u^2\left(\Delta_{ij}\right) = \frac{u_{Bi}^2 \sigma_i^2}{u_{Bi}^2 + \sigma_i^2} + \frac{u_{Bj}^2 \sigma_j^2}{u_{Bj}^2 + \sigma_j^2} + \left\{\frac{u_{Bi}^2}{u_{Bi}^2 + \sigma_i^2} - \frac{u_{Bj}^2}{u_{Bj}^2 + \sigma_j^2}\right\}^2 u^2(x_w).$$

It should be noted that estimates of the systematic difference and associated uncertainty obtained using information from all comparison participants differs from the corresponding estimates obtained on the basis of data from only two laboratories. The Bayesian approach on the basis of data from only two laboratories gives the following estimates of systematic differences:

$$\tilde{\Delta}_{ij} = (x_i - x_j)\frac{u_{Bi}^2 + u_{Bj}^2}{\sigma_i^2 + \sigma_j^2 + u_{Bi}^2 + u_{Bj}^2},$$

$$u^2(\tilde{\Delta}_{ij}) = (\sigma_i^2 + \sigma_j^2)\frac{u_{Bi}^2 + u_{Bj}^2}{\sigma_i^2 + \sigma_j^2 + u_{Bi}^2 + u_{Bj}^2}.$$

The comparison of two estimates of pair-wise systematic differences shows that

- estimates obtained on the basis of information from all participants always have a lesser uncertainty, and the following inequality is true:

$$u^2(\tilde{\Delta}_{ij}) - u^2(\Delta_{ij}) =$$

$$\left(\frac{u_{Bi}^2}{u^2(x_i)} - \frac{u_{Bj}^2}{u^2(x_j)}\right)^2 \left\{\frac{1}{\frac{1}{u^2(x_i)} + \frac{1}{u^2(x_j)}} - \frac{1}{\sum_i \frac{1}{u^2(x_i)}}\right\} > 0;$$

- estimates and the corresponding uncertainties coincide, provided that the laboratories have similar characteristics of precision and accuracy $\sigma_i = \sigma,\ u_{Bi} = u,\ \forall i$;
- estimates become close to each other at a high precision of measurement results in laboratories $\frac{\sigma_i}{u_{Bi}} \to 0$.

In Section 1.6 the problem of systematic bias estimation on the basis of the measurement results obtained in key comparisons has been considered. The approach applied implies that detailed information about uncertainty components is available. This approach is based on the Bayesian analysis of information from all participants, provided that the consistency of this information is present. As a result, the uncertainties associated with laboratory bias estimates can be reduced compared to the assessment of a particular laboratory.

1.7 Evaluation of measurement results in calibrating material measures and measuring instruments

The goal of calibrating measuring instruments is to provide the traceability of measurement results to a national or some other reference measurement standard. The relations between quantity values with associated uncertainties obtained using a measurement standard and the corresponding indications of the measuring instrument under calibration (which is inferior in the calibration hierarchy) and associated uncertainties are established during calibration. A measure indication implies the value of a quantity that is reproduced by this measure. When applying the measuring instrument, this relationship is used for getting a result according to measuring instrument indications.

When a single-value material measure is calibrated, a value or correction to a nominal value (or a value assigned to the measure at its previous calibration) and associated uncertainty are indicated. When multivalue measures are calibrated, a totality of new values or corrections (additive or multiplicative ones) is indicated for all points of a range which are calibrated. The values are reported with their associated uncertainties.

The relation (the calibration characteristic) determined for a measuring instrument in calibration can be expressed in one of the following basic forms:

- in the form of a table containing both the corrections to indications of a measuring instrument and the corresponding instrumental uncertainty values. The calibration characteristic can also be given in the form of corrections (an additive or/and multiplicative) to the (nominal) calibration characteristic assigned to the measuring instrument;
- in the form of a calibration function. The calibration function can be represented in an explicit form (analytically or graphically) with indication of the uncertainty at each point of a range or in the form of coefficients of the calibration function and also their uncertainties.

In needed, during the process of calibration some other metrological characteristics of measuring instruments can also be determined, for example,

- instability of the calibration characteristic of measuring instruments;
- standard deviation of indications of a measuring instrument under conditions of repeatability, which characterizes a random dispersion of indications under rated conditions when the calibration is performed;
- nonlinearity of the calibration function.

When evaluating the calibration characteristics, the data processing is performed in accordance with a general scheme proposed in the Guide for the expression of uncertainty in measurement [243] and includes two stages:

(1) formulation of a measurement model;

(2) evaluation of an uncertainty, which consists of establishing the uncertainties of the input quantities of a measurement model and transforming them into an uncertainty of an output quantity in accordance with a measurement model.

1.7.1 Formulating a measurement model

A measurement model is always a certain approximation of the dependence of an output quantity on input quantities, a particular form of which is determined by requirements for the accuracy of determining the metrological characteristic being calibrated.

The formation of a measurement model relates to the most complicated problems and are difficult to be formalized. A decision requires knowledge of a particular field of measurements. After the measurement model has been proposed, the process of calculating the uncertainty in many respects becomes a routine procedure, regulated by the GUM. It should be noted that one of the Supplements to the GUM is devoted to general issues of modeling measurements.

The measurement equation applied in calibration purposes expresses the dependence of a metrological characteristic of a measuring instrument (i.e., the dependence of an output quantity Y), which has to be determined, on an input quantity of the measuring instrument X, and of all other quantities X_i, influencing this metrological characteristic and its uncertainty:

$$Y = F(X, X_1 \dots, X_n). \tag{1.16}$$

In the measurement equation applied for the purpose of calibration, the output quantity can be represented as

- the value of a material measure under calibration or its deviation from a nominal value;
- the systematic error of a measuring instrument at a fixed point of a measurement scale;
- the (systematic) deviation of a measuring instrument indication from a nominal calibration characteristic;
- the calibration coefficient or parameters of a calibration function of a measuring instrument;
- other metrological characteristics of a measuring instrument.

Provided that in the process of calibration the stability of a material measure value is evaluated, then a change of the value for a certain time interval is taken as the output quantity.

Output quantities of the measurement model are the quantities influencing the result of determining the metrological characteristic of a measuring instrument and its uncertainty:

- X is the measurand at the input of a measuring instrument being calibrated, the value of which is determined using a reference measuring instrument or reference material measure, which are used for calibration purposes. X can be a vector quantity in the case where an indirect measurement method is applied in calibration;
- $X_1, \dots, X_n$ are the influencing quantities, the values of which are directly measured or used as reference data, established constants, etc.

Each kind of measurement and each type of measuring instrument to be calibrated are characterized by their own set of influencing quantities: $X_1, \dots, X_n$. However, there

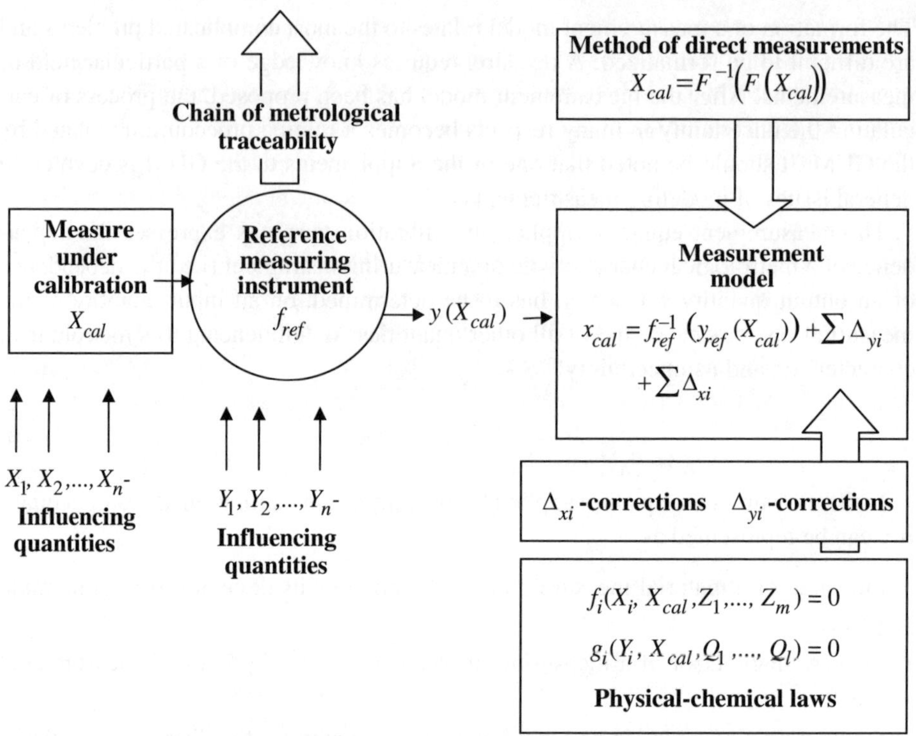

Figure 1.5. Method of direct measurements applied for measure calibration.

are some general measurement models to be suggested, depending on the method of measurement applied for calibration purposes. The process of forming a measurement equation is illustrated in Figure 1.5 for the case where a single-value measure is calibrated by the method of direct measurements with the help of a reference measuring instrument, when the relationship reflecting a measurement principle is trivial: $X_{\text{cal}} = F^{-1}(F(X_{\text{cal}}))$, where X_{cal} is the quantity maintained by the measure under calibration, and F is the true transformation function of the reference measuring instrument.

The calibration of single-value and multivalue material measures can be performed by the following methods [173, 427]:

- the method of direct measurement, according to which the values of a measure to be calibrated are evaluated with the help of a reference measuring instrument;
- the method of comparison with a reference material measure, which is realized with the help of a comparator;
- the method of indirect measurements, according to which the values of a material measure are determined on the basis of a known dependence of a quantity, reproduced by a material measure, on other directly measured quantities.

The method of comparison with a reference material measure, which is realized with the help of a comparator, has two modifications: the differential method of measurements and the method of substitution. In realizing the differential method of measurement, the difference of quantities maintained by a measure to be calibrated and the reference measure are evaluated with the help of a comparator. When realizing the method of substitution with the help a measuring instrument which performs the role of a comparator, the values of the measure to be calibrated and of the reference measure are subsequently determined, and then their relationship is found.

The calibration of a single-value material measure with the method of direct measurement consists of the multiple measurements of a quantity reproduced by the measure under calibration, which is carried out with a reference measuring instrument. In the general case the measurement equation is written in the form

$$x_{\text{cal}} = f_{\text{ref}}^{-1}\left(y_{\text{ref}}\left(X_{\text{cal}}\right) + \sum \Delta_{yi}\right) + \sum \Delta_{xi}, \tag{1.17}$$

where
x_{cal} is the value of a quantity reproduced by a measure to be calibrated;
f_{ref} is the calibration function of a reference measuring instrument;
$y_{\text{ref}}(X_{\text{cal}})$ is the indication of a reference measuring instrument, corresponding to a quantity that is reproduced by a measure being calibrated;
f_{ref}^{-1} is the function, which is inverse with respect to the calibration function established in the process of calibrating a reference measuring instrument;
Δ_{yi}, Δ_{xi} are the corrections introduced into indications of a reference measuring instrument and also into a final result of measurements, correspondingly.

A particular form of the measurement model depends on the method of representing a calibration characteristic of a reference measuring instrument. If the calibration characteristic of the reference measuring instrument is given with a table of corrections to its indications, then the measurement equation, as a rule, is represented in the following form:

$$x_{\text{cal}} = x_{\text{ref}} + \Delta\left(x_{\text{ref}}\right) + \sum_{1}^{n} \Delta\left(x_i\right), \tag{1.18}$$

where
x_{ref} is the indication of the reference measuring instrument;
$\Delta\left(x_{\text{ref}}\right)$ is the correction to the indications of the measuring instrument;
$\Delta\left(x_i\right)$ is the correction for other influencing quantities.

If the calibration characteristic of the reference measuring instrument is represented by calibration coefficient k_{ref}, then the measurement model, as a rule, can be represented in the form

$$y = \frac{y_{\text{ref}}\left(X_{\text{cal}}\right) + \sum_i \Delta_{yi}}{k_{\text{ref}}} + \sum_i \Delta_{xi}.$$

When calibrating a material measure with the method of comparison with a reference material measure, two measurement standards are applied: a reference material measure with a nominal value equal to a nominal value of a measure under calibration, and a reference measuring instrument applied as a comparator. The differential method consists in multiple measurements of the difference between the dimensions of a quantity, which are maintained by the reference measure and measure being calibrated, with the help of a comparator.

In the general case the measurement equation is written in the form

$$x_{\mathrm{cal}} = x_{\mathrm{ref}} + f_{\mathrm{ref}}^{-1}\left(y_{\mathrm{ref}}(X_{\mathrm{cal}} - X_{\mathrm{ref}}) + \sum \Delta_{yi}\right) + \sum \Delta_{xi},$$

where

x_{ref} is the value of a reference measure determined in the process of its calibration;
$X_{\mathrm{cal}}, X_{\mathrm{ref}}$ are the quantities reproduced by a reference measure and measure being calibrated, correspondingly;
$y_{\mathrm{ref}}\,(X_{\mathrm{cal}} - X_{\mathrm{ref}})$ are the indications of a reference measuring instrument, which correspond to the difference of quantities, reproduced by a reference measure and measure being calibrated;
Δ_{yi}, Δ_{xi} are the corrections introduced into indications of a reference measuring instrument and a final result of measurements, correspondingly.

Provided that the calibration characteristic of a reference measuring instrument is represented by a table of corrections to its indications, then the measurement model, as a rule, can be represented in the form

$$x_{\mathrm{cal}} = x_{\mathrm{ref}} + \Delta_{\mathrm{ref}}\,(X_{\mathrm{cal}} - X_{\mathrm{ref}}) + \Delta\,(X_{\mathrm{cal}} - X_{\mathrm{ref}}) + \sum_{1}^{n} \Delta\,(x_i),$$

where

$\Delta_{\mathrm{ref}}\,(X_{\mathrm{cal}} - X_{\mathrm{ref}})$ are the indications of a reference measuring instrument, corresponding to the difference of quantities reproduced by a measure being calibrated and reference measure;
$\Delta\,(X_{\mathrm{cal}} - X_{\mathrm{ref}})$ is the correction to indications of a reference measuring instrument;
$\Delta\,(x_i)$ is the correction for other influencing quantities.

If the calibration characteristic of a reference measuring instrument is represented by the calibration coefficient k_{ref}, then the measurement model, as a rule, can be represented in the form

$$x_{\mathrm{cal}} = x_{\mathrm{ref}} + \frac{\Delta_{\mathrm{ref}}(X_{\mathrm{cal}} - X_{\mathrm{ref}}) + \sum_i \Delta_{yi}}{k_{\mathrm{ref}}} + \sum_i \Delta_{xi}.$$

A particular case of the differential method of measurements is the zero method, according to which one tries to obtain the equality of dimensions of quantities reproduced by a reference measure and by the measure being calibrated. At the same time, a term in the right-hand side of equation which corresponds to the indications of a reference measuring instrument is equal to zero.

When realizing the method of substitution, the values of both reference and calibrated measures are subsequently determined using a reference measuring instrument, and then their relationship is found from

$$x_{cal} = x_{ref} \frac{y_{ref}(X_{cal})}{y_{ref}(X_{ref})} \cdot \prod \delta_{xi},$$

where X_{cal}, X_{ref} are the quantities reproduced by a measure being calibrated and the reference measure, respectively; $y_{ref}(X_{cal})$, $y_{ref}(X_{ref})$ are the indications of a reference measuring instrument, which correspond to a value reproduced by a measure being calibrated and reference measure, respectively; and δ_{xi} is the multiplicative correction.

Typical measurement models used in the case where the calibration of material measures is performed are summarized in Table 1.2

Table 1.2. Measurement equations applied for the case of calibrating measures.

Method of direct measurements with the help of a standard measuring instrument	
General case	$x_{cal} = f_{ref}^{-1}\left(y_{ref}(X_{cal}) + \sum \Delta_{yi}\right) + \sum \Delta_{xi}$
Particular cases of giving the calibration characteristic of a reference measuring instrument	$x_{cal} = x_{ref} + \Delta(x_{ref}) + \sum_{1}^{n} \Delta(x_i)$ $x_{cal} = \frac{y_{ref}(X_{cal}) + \sum_i \Delta_{yi}}{k_{ref}} + \sum_i \Delta_{xi}$
Calibration of measures with the method of comparison with a standard measure	
Differential method	
General case	$x_{cal} = x_{ref} + f_{ref}^{-1}\left(y_{ref}(X_{cal} - X_{ref}) + \sum \Delta_{yi}\right) + \sum \Delta_{xi}$
Particular cases of giving the calibration characteristic of a standard measuring instrument	$x_{cal} = x_{ref} + \Delta_{ref}(X_{cal} - X_{ref}) + \Delta(X_{cal} - X_{ref}) + \sum_{1}^{n} \Delta(x_i)$ $x_{cal} = x_{ref} + \frac{\Delta_{ref}(X_{cal} - X_{ref}) + \sum_i \Delta_{yi}}{k_{ref}} + \sum_i \Delta_{xi}$
Method of substitution	
	$x_{cal} = x_{ref} \frac{y_{ref}(X_{cal})}{y_{ref}(X_{ref})} \cdot \prod \delta_{xi}$

The calibration of measuring instruments may be carried out using the following methods:

- method of direct measurements, according to which a measuring instrument to be calibrated is used for measuring values of a multivalue referenced measure or a set of reference single-value measures;
- method of comparison with a reference measuring instrument;
- method of indirect (joint or combined) measurements.

The method of comparison with a reference measuring instrument has two options: (1) the method of comparison, which is realized with the help of a transfer measurement standard (a multivalue measure or set of single-value measures), and (2) the direct comparison of a measuring instrument under calibration with a reference measuring instrument.

When calibrating a measuring instrument using the method of direct measurement, the quantities reproduced by the reference measures, which correspond to different scale indications of the measuring instrument, are measured repeatedly with a measuring instrument under calibration.

Provided that in the process of calibrating a measuring instrument it is required to determine corrections to indications of a measuring instrument, or deviations from the nominal calibration characteristic at a point x_{ref}, then the measurement model, as a rule, can be represented in the form

- for the additive corrections:

$$\Delta(x_{\text{ref}}) = -\left(y_{\text{cal}}(X_{\text{ref}}) - x_{\text{ref}}\right) + \sum_{1}^{n} \Delta\,(x_i)$$

or

$$\Delta(x_{\text{ref}}) = -\left(y_{\text{cal}}(X_{\text{ref}}) - f_{\text{nominal}}(x_{\text{ref}})\right) + \sum_{1}^{n} \Delta(x_i),$$

- for the multiplicative corrections:

$$\delta(X_{\text{ref}}) = \left(\frac{y_{\text{cal}}(X_{\text{ref}})}{x_{\text{ref}}}\right)^{-1} \prod \delta\,(x_i)$$

or

$$\delta\,(X_{\text{ref}}) = \left(\frac{y_{\text{cal}}\,(X_{\text{ref}})}{f_{\text{nominal}}\,(x_{\text{ref}})}\right)^{-1} \prod \delta\,(x_i),$$

where

$y_{\text{cal}}(X_{\text{ref}})$ are the indications of a measuring instrument under calibration at specified value of a quantity reproduced by the reference measure X_{ref};

$x_{\text{ref}}$is the value of a reference measure;

$f_{\text{nominal}}(x_{\text{ref}})$ is the value obtained using a nominal calibration characteristic of a measuring instrument under calibration at specified value of a quantity x_{ref};
$\Delta(x_i)$, $\delta(x_i)$are the corrections for instability of a reference measure and other influencing quantities.

If in the process of the calibration of a measuring instrument it is required to determine its calibration coefficient k, then the measurement model is represented in the form

$$k = \frac{y_{\text{cal}}(X_{\text{ref}})}{x_{\text{ref}}} \prod \delta(x_i),$$

where
$y_{\text{cal}}(X_{\text{ref}})$ is the indication of a measuring instrument being calibrated at a specified value of a quantity reproduced by the reference measure X_{ref};
x_{ref} is the value of a reference measure;
$\delta(x_i)$ is the corrections for the instability of a standard measure and other influencing quantities.

When the method of comparison is applied to calibration of a measuring instrument, the latter, together with a reference measuring instrument, are connected to one and the same source of an input measurement signal, and efforts are made to provide the equality of the input or output signals of the measuring instruments, which are compared. After that, the calibration characteristic of the measuring instrument being calibrated is established by changing the value of the input quantity. If the equality of values of the input quantities of both measuring instruments is achieved, then, first using the indications of the reference measuring instrument $y_{\text{ref}}(X)$ and its calibration characteristic, an estimate of the input quantity value is found as $x_{\text{ref}} = f_{\text{ref}}^{-1}(y_{\text{ref}}(X))$, and then the associated uncertainty is calculated.

As a rule, in the process of calibration a series of repeated indications of the measuring instrument being calibrated and of the reference measuring instrument is obtained: $y_{\text{cal}}(X_{\text{ref}})$, $y_{\text{ref}}(X_{\text{ref}})$. In this case, when handling the data it is necessary to take into account a possible correlation of indications of the standard measuring instrument and instrument under calibration, which is caused by fluctuations of the measurand.

Typical measurement models, used in the case where the calibration of the measuring instrument is performed, are summarized in Table 1.3.

1.7.2 Evaluation of measurement uncertainty

The uncertainty of a calibration characteristic is caused by several sources including properties of the reference measuring instrument used and the measuring instrument to be calibrated, as well as measurement procedure applied for calibration. The basic components of the uncertainty are

- instrumental uncertainty of a reference measuring instrument;

Table 1.3. Measurement models applied in calibrating measuring instruments.

Method of direct measurements using reference measure (a set of measures)	
Evaluation of additive corrections	$\Delta(x_{\text{ref}}) = -(y_{\text{cal}}(X_{\text{ref}}) - x_{\text{ref}}) + \sum_{1}^{n} \Delta(x_i)$ or $\Delta(x_{\text{ref}}) = -(y_{\text{cal}}(X_{\text{ref}}) - f_{nominal}(x_{\text{ref}})) + \sum_{1}^{n} \Delta(x_i)$
Evaluation of multiplicative corrections	$\delta(X_{\text{ref}}) = \left(\frac{y_{\text{cal}}(X_{\text{ref}})}{x_{\text{ref}}}\right)^{-1} \prod \delta(x_i)$ or $\delta(X_{\text{ref}}) = \left(\frac{y_{\text{cal}}(X_{\text{ref}})}{f_{\text{nominal}}(x_{\text{ref}})}\right)^{-1} \prod \delta(x_i)$
Evaluation of a calibration coefficient	$k = \frac{y_{\text{cal}}(X_{\text{ref}})}{x_{\text{ref}}} \prod \delta(x_i)$

- drift of the calibration characteristic of a reference measuring instrument over a period after its previous calibration;
- methodical errors of the calibration method;
- random factors in performing the calibration;
- short-term instability of the measuring instrument being calibrated.

Initial data for evaluating measurement uncertainty according to type A are the repeated measurements carried out in the process of calibration. The uncertainty component caused by a combined influence of a number of factors, such as the reproducibility of indications of a reference and calibrated measuring instruments, variation of measurement conditions within the limits specified by the measurement procedure, short-term instability of the reference measuring instrument, etc.

Initial information for evaluating measurement uncertainty according to type B can be obtained from the calibration certificates of a reference measuring instrument, the precision and trueness characteristics of a measurement procedure used for calibration, previous calibration certificates of the measuring instrument to be calibrated, the accuracy class of the measuring instrument to be calibrated, the limits of permissible errors indicated in a measurement method specification, and engineering documentation for the measuring instrument.

The evaluation of uncertainty is regulated by GUM [243]. It is based on a measurement model, and in a condensed form it is represented as a Table 1.4, illustrating an uncertainty budget.

In column 1 are the input quantities of the measurement model. In column 2 are estimates of the input quantities or ranges of input quantity values. In column 3 are

Table 1.4. Budget of uncertainty.

1	2	3	4	5	6
Input quantity	Estimate/ range of values	Standard uncertainty	Type of evaluation	Coefficient of sensitivity	Contribution to the combined standard uncertainty
X_i	x_i	$u(x_i)$	A (B)	$c_i = \frac{\partial F}{\partial x_i}$	$u_i(y_q) = \left\| \frac{\partial F}{\partial x_i} \right\| \cdot u(x_i)$
...	...	...	...	...	...
					$u(y) = \sqrt{\sum_{i=1}^{n} u_i^2(y)}$

values of the standard uncertainty calculated in accordance with the GUM on the basis of available a priori information or a number of repeated measurements. In column 4 the type of uncertainty evaluation is indicated (in case it is needed, an assumed PDF is indicated). In column 5 are the coefficients of the sensitivity of input quantities $c_i = \frac{\partial F}{\partial x_i}$. In column 6 are the values of contributions of the input quantities $u_i(y_q) = |\frac{\partial F}{\partial x_i}| \cdot u(x_i)$ to the combined standard uncertainty $u(y)$ (the product of the values from column 3 and a module of a value from column 5).

When the estimates of input quantities are not correlated, the combined standard uncertainty of a measurement result is calculated by the formula:

$$u(y) = \sqrt{\sum_{i=1}^{n} u_i^2(y)}.$$

Provided the estimates of input quantities are correlated, then the combined standard uncertainty is calculated by the formula

$$u(y) = \sqrt{\sum_{i=1}^{n} u_i^2(y) + \sum_{i,j=1,\, i \neq j}^{n} c_{ij} r(x_i, x_j) u(x_i) u(x_j)},$$

where $c_{ij} = \frac{\partial^2 F(x_1, \dots, x_n)}{\partial x_i \partial x_j} |_{(x_1, \dots, x_n)}$, $r(x_i, x_j) = \frac{u(x_i, x_j)}{u(x_i) u(x_j)}$ is the correlation coefficient of the estimates x_i and x_j, and $u(x_i,\ x_j)$ is the covariance of the estimates x_i and x_j.

If the input quantities of the same measurement procedures, where measuring instruments or reference data are used, is to be evaluated, then there is a logical correlation of the corresponding estimates. In particular, if the input quantities X_1 and X_2 depend on the mutually independent quantities Q_l $(l = 1, \dots, L)$, then they are correlated, and the corresponding covariance is equal to $u(x_1, x_2) = \sum_{l=1}^{L} c_{1l} c_{2l} u^2(q_l)$, where c_{1l}, c_{2l} are the coefficients of sensitivity of the quantities X_1 and X_2 to the variables Q_l $(l = 1, \ldots, L)$.

The combined uncertainty $u(y)$ is given In the last line of the Table 1.4. The expanded uncertainty $U(y)$ is equal to the product of the combined uncertainty $u(y)$ of the output quantity y by the coverage factor k, usually assumed to be equal to 2 for confidence level $P = 0.95$:

$$U(y) = ku(y).$$

The above procedure for calculating uncertainty is based on the GUM and well founded for the case of a linear model of measurements (or a model permitting linearization) and with assumption that the distributions of input quantities are normal.

If these conditions are violated, then it is possible to apply methods of transforming of PDFs in accordance with [244], or the evaluation of a a posteriori density of output quantity probabilities using the Bayesian analysis.

1.7.3 Calculation of measurement uncertainty associated with a value of a material measure using Baycsian analysis

Let the application of the Bayesian theorem be considered in order to obtain an estimate and the corresponding uncertainty of a measure when the latter is calibrated by the method of direct measurements (1.18). For the sake of simplicity, the corrections for influencing quantities are not given. Then equation (1.18) takes the form

$$x_{\text{cal}} = x_{\text{ref}} + \Delta(x_{\text{ref}}) = x_{\text{ref}} + \Delta, \tag{1.19}$$

where x_{ref} is an indication of the reference measuring instrument, and Δ is an error of the reference measuring instrument.

The Bayesian analysis is a tool which allows the whole available a priori information about the measurand, the applied measuring instruments, and the measurement procedure to be taken into account for obtaining the measured value and its associated uncertainty [155]. The a posteriori PDF of the output quantity is calculated as the product of the a priori PDF of the input quantities and the likelihood function of the data obtained in the process of calibration.

The calibration of measuring instruments relates to those measurement problems where significant a priori information is available. This information contains

- instrumental uncertainty of a reference measuring instrument: $u_{\text{ref}}(x_{\text{ref}}) = u_{\text{ref}}$;
- precision and trueness of a measurement procedure, in particular, the repeatability variance σ_R;
- results of previous calibrations of a given material measure $(x_0,\ u(x_0))$.

The methods of assigning the PDFs to the measurement model quantities on the basis of available information are discussed, for example, in [550]. In particular, if only the estimate of the quantity value and its associated standard uncertainty are known, then in accordance with the principle of the maximum entropy a normal PDF is assigned. In the case given this is true for the a priori PDF of material measure value:

$$p_X\left(x \,|\, x_0,\ u\left(x_0\right)\right) = \frac{1}{\sqrt{2\pi} u\left(x_0\right)} \exp\left\{-\frac{\left(x - x_0\right)^2}{2u^2\left(x_0\right)}\right\}, \tag{1.20}$$

as well as for an error of the measuring instrument in use (it is assumed that all known corrections have been introduced, as prescribed by the measurement procedure):

$$p_\Delta\left(\Delta \,|u_{\text{ref}}\right) = \frac{1}{\sqrt{2\pi} u_{\text{ref}}} \exp\left\{-\frac{\Delta^2}{2u_{\text{ref}}^2}\right\}.$$

It should be noted that the information about the value of material measure obtained in the process of the previous calibration can be taken into account only in the case where it is consistent with the data of the present calibration. Especially if within the period between the calibrations there was a drift of the measure value, then the results of the previous calibration can only be used for establishing the parameters of this drift rather than for reducing the uncertainty of a current measurement. Consequently, before using the data of the previous calibration it is necessary to check the consistency of the two PDFs for the measure values, a priori PDF (1.20) and the PDF obtained on the basis of the results of the current calibration without using the data of the previous calibration. For this purpose it is possible, for example, to use [546].

The a posteriori PDF of measure value without taking into account the data of the previous calibration is calculated as

$$p\left(x_{\text{cal}}\right) = \int p\left(x_{\text{cal}}, \Delta \,|x_i, u_{\text{ref}}, \sigma_R \ldots.\right) d\Delta, \tag{1.21}$$

where

$$p\left(x_{\text{cal}}, \Delta \,|x_i, u_{\text{ref}}, \sigma_R \ldots.\right) \propto \prod_i \frac{1}{\sqrt{2\pi}\sigma_R} \exp\left\{-\frac{\left(x_{\text{cal}} - x_{\text{ref},i} - \Delta\right)^2}{2\sigma_R^2}\right\} \frac{1}{\sqrt{2\pi} u_{\text{ref}}} \exp\left\{-\frac{\Delta^2}{2u_{\text{ref}}^2}\right\}$$

is a joint a posteriori PDF of the measure value and measuring instrument error,

$$L\left(x_{\text{ref},i} \,|\sigma_R\right) = \prod_i \frac{1}{\sqrt{2\pi}\sigma_R} \exp\left\{-\frac{\left(x_{\text{cal}} - x_{\text{ref},i} - \Delta\right)^2}{2\sigma_R^2}\right\}$$

is a likelihood function, and $x_{\text{ref},i}$ is the indication of the reference measuring instrument, $i = 1, \ldots, n$.

It is well known that under the assumption made the a posteriori PDF of the measure value is the Gaussian one with the parameters

$$\left\{\frac{1}{n}\sum x_{\mathrm{ref},i},\ \sqrt{u_{\mathrm{ref}}^2+\frac{\sigma_R^2}{n}}\right\}.$$

Provided that checking has shown that the PDFs (1.20) and (1.21) are consistent, then PDF (1.20) can be used as the a priori PDF of the output quantity, i.e., the value of the measure under calibration [135]. Such situation is possible when the measure is calibrated rather often. If for this only the same standard measuring instrument is used, then a correlation arises between the error of the measuring instrument and the measure value, which is taken into account by considering the joint PDF: $p_0(x_{\mathrm{cal}}, \Delta)$. Then expression (1.21) is rewritten and takes the form

$$p\left(x_{\mathrm{cal}}\right)=\int\prod_i\frac{1}{\sqrt{2\pi}\sigma_R}\exp\left\{-\frac{\left(x_{\mathrm{cal}}-x_{\mathrm{ref},i}-\Delta\right)^2}{2\sigma_R^2}\right\}p_0\left(x_{\mathrm{cal}},\Delta\right)d\Delta.$$

The availability of the a posteriori PDF allows the measure value and associated standard uncertainty to be obtained as a mathematical expectation and standard deviation of this PDF, respectively, as well for a coverage interval to be constructed for any given coverage probability.

1.7.4 Determination of the linear calibration functions of measuring instruments

In the most general case the problem of determining the calibration functions of measuring instruments can be formulated in the following manner. It is known that between the measurand X at the input of a measuring instrument and output quantity Y of this instrument there is a functional dependence $Y = F(X)$. In the process of calibration the values of X and Y are measured or given with some uncertainties. It is required to determine a parametric form of the dependence $F(X)$, to evaluate its parameters, and to calculate the uncertainty of the curve fitted to the experimental data.

The form of $F(X)$ is by no means always known in advance, for example, from certain chemical regularities. Usually, in practice the matter concerns some kind of approximation of $F(X)$ with functions of the kind given $f(x)$, which satisfies the requirements for the target uncertainty associated with fitting.

The accuracy of determining the calibration functions of a measuring instrument is caused by the uncertainty of the experimental data (x_i, y_i), by the algorithm for evaluating the parameters of the $f(X)$ dependence, and by the correctness of the approximation of the theoretical calibration dependence of the function selected. These factors make a joint impact on the uncertainty of the calibration function estimate, which can be reduced at the expense of the optimal design of the measurement experiment when evaluating the calibration functions of measuring instruments.

The importance of measurement design increases when developing the measurement procedures intended for the routine calibration of measuring instruments, where it is necessary to provide the target uncertainty of the calibration function. The measurement design consists of selecting a number of values x_i, $i = 1, \dots, N$, at which measurements of the dependent quantity y_i are made, and the location of these values x_i, $i = 1, \dots, N$ within the measurement range and the number of repeated measurements y_{ij}, $j = 1, \dots, n_i$ at the specified value x_i are found, and, and finally, the requirements for the uncertainties associated with the values x_i, $i = 1, \dots, N$ themselves (values of the independent variable X) are determined.

In view of this the problem of determining the calibration curves of the measuring instruments is divided into the following stages [100, 106, 144, 430, 463]:

(1) selection of a model of the calibration function $f(X)$;

(2) design of a measurement experiment;

(3) evaluation of calibration curve parameters and associated uncertainties.

The error of determining the calibration function can be represented in the form of two summands:

$$\widehat{f}(x) - F(x) = \widehat{f}(x) - f(x) + f(x) - F(x),$$

where $F(x)$ is the theoretical (true) calibration dependence, $f(x)$ is the accepted model of the calibration curve, and $\widehat{f}(x)$ is the estimate of the model on the basis of experimental data.

The summand $\widehat{f}(x) - f(x)$ is the transformed error that is completely determined with the uncertainties of the initial experimental data and the data processing algorithm. The second summand $f(x) - F(x)$ is the methodical error that has a systematic character and characterizes the inadequacy of the model selected.

Provided that the extreme cases are being considered, when one of the summands is converted into zero, then the equality of the second summand to zero leads to a statistical problem of evaluating the parameters of a regression, the classical statement of which corresponds to the absence of errors of an independent variable x_i. The equality of the first summand to zero corresponds to the case of the absence of the experimental data errors, which leads to the classical problem of the function interpolation on the basis of its values y_i, $i = 1, \dots, N$ given at specified points x_i, $i = 1, \dots, N$.

There is always a certain "threshold" inconsistency of the accepted model with real data. Therefore, the second summand always differs from zero. It is important to take into account this inconsistency when evaluating the uncertainties of determining the functional dependence or when ensuring that this inconsistency does not significantly influence the calculation of the measurement uncertainty.

Usually, in practice the second way is selected. A certain form of the calibration function is assumed, reasoning from the experience of a researcher, the simplicity requirements, and its suitability for control at repeated applications. Then the parameters of the accepted model are evaluated. After that, the adequacy of the model for the real data is checked using the criteria of agreement – for example, the criterion χ^2. If the model and data are consistent, then the data on the uncertainty of the model is neglected.

As previously mentioned, the selection of the model defies formalized description. It is only possible to formulate the general principles, the main ones of which are

- the rational relationship between the model complexity and the uncertainty of the model evaluation. It should be taken into account that with a limited volume of experimental data the increase in the number of parameters being evaluated results in an increase in the combined uncertainty of the calibration function (with a probable decrease in the systematic error due to the model inadequacy);
- ensuring the uncertainty required for determining the functional dependence reasoning for the purposes of its further application.

Based on these principles, in practice the linear calibration functions are rather often used. Provided that the accuracy required in the range being considered is still achieved, then the range is divided into subranges, and a piecewise-linear approximation of the calibration curve is used.

The initial data for evaluating the calibration function are the pairs of values with the associated measurement uncertainties: $\{(y_i, u(y_i)), (x_i, u(x_i)), i = 1, \ldots, N\}$. As a rule, the uncertainties of measuring instrument indications $u(y_i)$ are calculated according to the type A evaluation on the basis of repeated indications of the measuring instrument or given variance of indication repeatability of the measuring instrument. The uncertainty of the values of the independent variable $u(x_i)$ is calculated on the basis of the available information about the uncertainty of giving the points x_i – for example, about the procedure of preparing calibration solutions based on certified materials.

Then after evaluating the parameters, the linear curve can be easily represented in the form $Y = a_0 + b(X - \bar{X})$ [430]. For such representation the least-squares method makes it possible to obtain unbiased and uncorrelated estimates(a_0, b), in case of the equal measurement uncertainties and uncorrelated data $\{(y_i, u(y)), (x_i, u(x)), i = 1, \ldots, N\}$ they have

$$a_0 = \frac{\sum_{i=1}^{N} y_i}{N}, \qquad b = \frac{\sum_{i=1}^{N} y_i (x_i - \bar{x})}{\sum_{i=1}^{N} (x_i - \bar{x})^2}.$$

The standard uncertainty of the calibration curve at the point x is calculated by the formula

$$u^2(x) = \left(u^2(y) + b^2 u^2(x)\right) \cdot \left(\frac{1}{N} + \frac{(x - \bar{x})^2}{\sum_{i=1}^{N} (x_i - \bar{x})^2}\right). \tag{1.22}$$

The expanded uncertainty is calculated by the formula $U_P(x) = k \times u(x)$; the coverage factor is assumed to be equal to $k = 2$ for the confidence level $P = 0.95$ and $k = 3$ for the confidence level $P = 0.99$.

For the arbitrary covariance matrix of initial data $\{(x_i, y_i), i = 1, \ldots, N\}$ the generalized LSM is used. The corresponding estimates of parameters (a_0, b) are obtained using the numerical methods. In [311] an approach to evaluating parameters of the linear calibration curve on the basis of the calculation of a joint probability density of the parameters (a_0, b), taking into account the available a priori information, is considered.

The design of a measurement experiment for evaluating the calibration curves is realized taking into account the accepted algorithm for processing thd emeasurement results, since a particular dependence of the uncertainty associated with (a_0, b) on the parameters of the experiment plan is determined by the applied algorithm for data processing.

The distinction of defining a problem of designing a measurement experiment in metrology from the similar procedure in the theory of an optimal experiment is revealed in the following.

(1) Usually the matter concerns the availability of providing the target uncertainty of the calibration curve, whereas in the theory of an optimal experiment there is a minimum or maximum of the accepted criterion of optimality.

(2) The uncertainty of the calibration curve is defined not only by random factors the influence of which can be decreased at the expense of increasing the number of experimental data, but also by uncertainties associated with values x_i, which are usually caused by the systematic factors. It is not always possible to neglect the uncertainties of independent quantities x_i. Consequently, in the general case the requirements for the uncertainty of the calibration curve are finally transformed into the requirements for the uncertainty associated with valuesx_i. Therefore, the design of a measurement experiment includes not only the determination of a number of values x_i and their location within the range, but also the required uncertainty of these values.

(3) Usually when determining the calibration curve the coefficients of a model of the given form, which can be different from a "true" one, are evaluated. Therefore, there is always a systematic error caused by the model inadequacy that has to be taken into account in the criterion of optimality.

It is well known that the optimal design spectrum of the measurement experiment for determining the linear regression is concentrated at the ends of the range. However, this theoretical result cannot be applied in practice, in view of the presence of systematic errors and deviations of the calibration curve from the linear regression.

An experiment plan $\varepsilon(M)$ is identified as the following set of quantities:

$$\begin{Bmatrix} x_1, x_2, \ldots, x_N \\ n_1, n_2, \ldots, n_N \end{Bmatrix}, \quad \sum n_i = M.$$

A set of points $\{x_1, \ldots, x_N\}$ is called the design spectrum.

There are a great number of different criteria regarding the optimality of experiment plans, which may be divided into two groups: criteria connected with the covariance matrix of estimates of the model parameters, and criteria related to the uncertainty associated with the curve fitted. The basic criteria in the groups listed are as follow.

(A) Criteria related to the covariance matrix $D(\varepsilon)$ of the estimates of the calibration curve parameters:

- $\min\limits_{\varepsilon} |D(\varepsilon)|$ is D-optimality;
- $\min\limits_{\varepsilon} Sp\,[D(\varepsilon)]$, $(Sp\,[D(\varepsilon)] = \sum_{i=1}^{m} D_{ii})$ is A-optimality;
- $\min\limits_{\varepsilon} \max\limits_{i} \lambda_{ii}(\varepsilon)$ (where λ_{ii} are the proper numbers of the covariance matrix) is E-optimality.

(B) criteria related to the uncertainty associated with the curve fitted:

- $\min\limits_{\varepsilon} \max\limits_{x} Ed^2(x, \varepsilon)$ is G-optimality;
- $\min\limits_{\varepsilon} \int Ed^2(x, \varepsilon)dx$ is Q-optimality,

where $Ed^2(x, \varepsilon)$ is the mathematical expectation of the square of an error of the functional dependence at the point x: $d(x, \varepsilon) = |\hat{f}(x) - f_{\text{true}}(x)|$.

For metrology the criteria of optimality in the form of a functional are the more natural. Therefore, later on we shall consider the$G$-optimal plans.

The problem of finding an optimal plan of the measurement experiment while fitting the linear calibration curves will be considered below, taking into account deviations of the real calibration curve from the linear one. Let the G-optimal plan for evaluating the parameters of the linear regression $y = a + bx$ be determined for the case when the model error is evaluated by a deviation from the parabolic regression $y = c_0 + c_1x + c_2x^2$. In addition it is assumed that the uncertainty of the independent variable is negligible, the experimental data are uncorrelated, and their uncertainties are constant within the range.

The parameters are evaluated using the LSM:

$$\hat{a}_0 = \sum y_i/N, \quad \hat{b} = \sum y_i(x_i - \bar{x})/\sum(x_i - \bar{x})^2,$$
$$\bar{x} = \sum x_i/N, \quad \hat{a} = \hat{a}_0 - \hat{b}\bar{x}.$$

The optimal plan for the experiment is defined from the condition of minimizing a squared maximum value of the error of the calibration curve within the interval of the independent variable values:

$$d = \max_{x \in X} E\{\hat{a} + \hat{b}x - c_0 - c_1 x - c_2 x^2\}^2.$$

The problem is solved in the following manner. The optimal plan is defined for some N, and then the value of N is selected from the condition of providing the required uncertainty of the calibration curve. If only the symmetrical plans are considered, then the expression for the criterion d after simple calculations will be reduced to the form

$$d = \max_{x \in X} \left[c_2^2 \left\{ (x - \bar{x})^2 - \frac{\sum (x_i - \bar{x})^2}{N} \right\}^2 + u^2(y) \left\{ \frac{(x - \bar{x})^2}{\sum (x_i - \bar{x})^2} + \frac{1}{N} \right\} \right].$$

In finding the maximum it is convenient to pass on to a new variable $z = (\frac{x - \bar{x}}{T})^2$ where T is half the measurement range. In new designations d is written in the form

$$d = \max_{0<z<1} \left[c_2^2 T^4 \left\{ z^2 - \frac{\sum z_i{}^2}{N} \right\}^2 + u^2(y) \left\{ \frac{z^2}{\sum z_i^2} + \frac{1}{N} \right\} \right].$$

If $c_2 = 0$ (i.e., the linear curve is true), then the maximum is reached at $z = 1$ (and is equal to $u^2(y)\{\frac{1}{N} + \frac{1}{\sum z_i}\}$). The minimum d of the criterion, which is equal to $\frac{2u^2(y)}{N}$, is reached at z_i=1 (x_i=$\bar{x}$?T). This is a well known result. If $u(y) = 0$ (i.e., there is not any measurement error, and the basic error is caused by the inadequacy of the model), then the maximum is achieved at $z = 0$, and the minimum of the criterion is achieved at $z_i = 1/2(x_i = \bar{x} \pm \frac{\sqrt{2}}{2}T)$ and is equal to $\frac{c_2^2 T^4}{4}$. In the general case the design spectrum is concentrated at the points

$$x_i = \bar{x} \pm \sqrt{z^*} T, \tag{1.23}$$

where

$$z^* = \begin{cases} \frac{1}{4} \left\{ 1 + \sqrt{1 + \frac{8u^2(y)}{c_2^2 T^4 N}} \right\}, & \frac{8u^2(y)}{c_2^2 T^4 N} \le 1 \\ 1, & \frac{8u^2(y)}{c_2^2 T^4 N} \ge 1. \end{cases}$$

The z^* values are within the range $(0.5, 1)$ depending on the relationship of the measurement uncertainty associated with experimental data and deviation of calibration curve from the linear one. The d criterion value for the optimal plan is equal to ${c_2}^2 T^4 {z^*}^2 + \frac{u^2(y)}{N}$.

To provide the target uncertainty u^2_{target} of the calibration curve it is necessary to meet the condition

$${c_2}^2 T^4 {z^*}^2 + \frac{u^2(y)}{N} \le u^2_{\text{target}}. \tag{1.24}$$

The second summand in the left side of inequality (1.24) can be decreased at the expense of increasing the number of repeated measurements at the points of design spectrum (1.23). The first summand reflects the uncertainty component caused by the model nonlinearity and can be decreased only at the expense of reducing the range T or dividing it into subranges.

Thus, if the calibration curve for a particular measuring instrument is studied rather carefully and the correctness of approximating it with the linear function is grounded, then, performing the calibrations repeatedly, it is useful to conduct measurements at the calibration points corresponding to design spectrum (1.23).

In practice a uniform design spectrum is often applied. Such a design spectrum is in a certain sense a compromise and is used in two ways: on the one hand it is used for evaluating the parameters of the linear curve, on the other hand for checking the compliance of the model and experimental data. These two tasks are different; the second one relates to the design of discriminating experiments and it is not considered in this chapter.

When applying the uniform design spectrum, the uncertainty of the calibration curve can be evaluated on the basis of deviations of the experimental data from the curve being fitted. In contrast to the calculation uncertainty, according to (1.22) this way of uncertainty evaluation may be seen as experimental:

$$u^2(x) = u_A^2(y) \cdot \left(\frac{1}{N} + \frac{(x - \bar{x})^2}{\sum_{i=1}^{N} (x_i - \bar{x})^2} \right), \quad u_A(y) = \sqrt{\frac{1}{N-2} \sum_{i=1}^{N} (y_i - \hat{y}_i)^2}, \tag{1.25}$$

where $\hat{y}_i$ is the estimate obtained on the basis of the curve beging fitted.

Using the estimates of form (1.25) it is necessary to take into account that the dispersion of the experimental data around the calibration curve is due to a number of reasons (but it is not possible to say that these reasons are fully taken into account in equation (1.25)): the repeatability of measuring instrument indications, the deviation from the linearity, and the uncertainty of values of independent quantity.

The design of the measurements for evaluating calibration curves is the methodical base for developing a procedure of calibration which will be able to provide the target uncertainty of the calibration function evaluation. The target uncertainty of the calibration function is dictated by its further application. In particular this can be the use of a given measuring instrument for carrying out measurements with a required accuracy, or the attribute of a given measuring instrument to a definite accuracy class of instruments by results of its calibration, etc.

1.8 Summary

The compatibility of measurement results is the most desirable property of measurement results with regard to their practical use. Two experimental measurement procedures serve as an objective foundation for the compatibility of measurement results: comparisons of the primary measurement standards and calibration of the measuring instruments.

The goal of calibration is to provide the metrological traceability of measurement results to a primary measurement standard. The metrological traceability is based on a chain of subsequent calibrations where each link of the chain introduces an additional uncertainty into the final measurement uncertainty. It should be emphasized that the assurance of the traceability to a primary standard, in other words the use of a calibrated measuring instrument in measurements, is a necessary but insufficient condition of the compatibility of measurement results. The application of validated measurement procedures is also required. Validation assumes the calculation of measurement uncertainty under rated conditions in accordance with a recognized approach, which for the present is the GUM [243]. The rated conditions correspond to the conditions of intended use of measuring instruments. The comparisons of primary measurement standards with the goal of establishing their equivalence can be regarded as a final step in providing metrological traceability of measurement results to a reference.

In this chapter some problems in evaluating measurement results when calibrating measuring instruments or comparing measurement standards have been considered. Comparisons of the measurement standards and the calibration of the measuring instruments are the basic experimental procedures confirming the measurement uncertainties declared by measurement and calibration laboratories. These procedures, accompanied by quality management systems which are supported by measurement and calibration laboratories give an objective basis for the mutual recognition of the measurement results obtained at these laboratories.

The problem of evaluating the data of the key comparisons of national measurement standards is comparatively new in the field of handling measurement results and evaluating the accuracy, and has arisen after the signing of the MRA by the national metrology institutes in 1999. Publications in the international journal *Metrologia* show that the interest in the problem has not decreased.

Its current importance is confirmed by the regular presentation of reports and papers on the evaluation of key comparison data at seminars and conferences on methods of evaluating measurement results. As before, the discussion concerning the content of the initial concepts "reference value of key comparisons" and "degree of equivalence of national measurement standards" is continued.

In this chapter the models used in evaluatng key comparison data were analyzed, and various approaches for expressing the degree of equivalence of measurement standards were discussed, including those which apply mixtures of distributions, and the problem of linking regional key comparisons with CIPM key comparisons were treated in more detail.

In measurement standards comparisons the routine calibration procedures are used. Working out the calibration procedure of measuring instruments implies formulating the measurement model and a design of the plan of measurement experiment. Some general aspects of these issues were also treated in this chapter.

Chapter 2

Systems of reproducing physical quantities units and transferring their sizes

2.1 Classification of reproducing physical quantities units and systems for transferring their sizes (RUTS)

2.1.1 General ideas

2.1.1.1 Aim, object, and tasks of the investigation

The aim of this chapter is to develop a generalized type of system for reproducing units of physical quantity units and transferring their sizes (further – RUTS systems) into various fields of measurements.

The RUTS systems, the graphical display of which is provided with so-called "verification schemes" (hierarchy schemes for verification of measuring instruments) or metrological traceability chains, are the material and technical basis of measurement assurance, or metrological traceability of measurement, in a country. The number of national verification schemes which have been constructed and are currently in use corresponds to the number of approved national measurement standards which are in force today.

However, in spite of the great number of practically realized systems intended for reproducing particular physical quantity units and transferring their sizes, until now there is still no generalizing theory for constructing such systems. There are a relatively small number of theoretical works, the most important ideas of which were used as a foundation [194] for the procedure for determining the parameters of verification schemes. However, all these works are are devoted to only some of the problems of the unit size transfer, mainly to those which refer to accuracy correlations in the unit transfer system. At the same time, only one method for constructing such systems has been obviously or not obviously assumed everywhere, i.e., an accuracy hierarchy of measurement standards and working measuring instruments with an upper link in the form of a national measurement standard.

At the same time, the fact that today in practice the number of quantities and parameters to be measured is, according to different literature sources, somewhere between 250 [372] and 700 [503] (there is an undoubted tendency to its increasing), and all available verification schemes provide the uniformity of measurements only for a little more than 30 quantities (see Section 2.1.5.1), indicating that there are some other ways to provide the uniformity of measurements and, correspondingly, RUTS systems of other types.

On the other hand, the number of available verification schemes designed to provide the uniformity of measurements significantly exceeds the number of quantities to be measured, and this discrepancy seems to be increasing, a fact which is connected to a weak clearness on the necessity and sufficiency criteria used in developing such systems. Provided that both tendencies will remain in the future (without any deliberate interference by metrologists), this could result in an unlimited increase in the number of verification schemes, corresponding measurement standards, material, labor and financial costs, and manpower reserves. Thus, just now the realization of the necessity of *optimizing the entire RUTS system* (for the whole totality of quantities and parameters to be measured) is about to happen.

This naturally requires a reliable mechanism, i.e., a scientific study of the problem of optimizing the RUTS systems. Therefore, the importance of generalizing and systematizing the results of theoretical studies of the problems inherent in both the reproduction of physical quantity units and the transfer of their sizes, the optimization of the system providing the assurance of the measurement uniformity and, finally, the development of the theory of constructing systems of such a type, is evident.

The first stage of creating such a generalization theory is the development of the classification issues, since classification serves as a basis for systematization and is an important moment in the development of any domain of knowledge. According to M. Planck, "the correct classification is a already high kind of knowledge".

A great number of facts have been already accumulated about the RUTS systems construction in different types of measurements , and this is the objective prerequisite for posing the problem of the classification of these systems.

However, it should be noted that classification must not be considered as a final separate topic. It only defines more exactly (intensifies, deepens) the differences of classification groups, but by no means makes impossible the existence of groups which belong simultaneously belong to several classes. This is particularly true for the systems of a complicated type, to which the RUTS systems being considered can be related.

2.1.1.2 Method and structure of the investigation

Classification being a "logical procedure", the essence of which consists in dividing the whole set of objects according to detected similarities and differences into separate groups (classes, subordinate sets), is by itself the method of investigation.

It is known that any set of objects can be classified into separate groups using two methods: by either listing all objects entering a given group, or by indicating an attribute (attributes) peculiar to each member of the group but which is lacking (due to notation) in objects which are not members of the given group. The more effective is the second method, namely the method of *logical classification* accepted for a given study.

If the signs (attributes) of classification are sufficiently clear, then the classification itself as the division of the set does not present any significant difficulties – this is, truthfully speaking, correct for the sets with elements having very pronounced signs, but in real situations it is by no means always the case.

Therefore, the main difficulty of classification is simply the establishment of sufficiently clear *classification signs*. This is in turn determined by the extent of the clearness with which these signs become apparent in various elements of the set being studied. At the same time, the efficiency criterion of a classification being developed is the possibility of further profound study of selected classes (establishment of their *inherent* properties and interrelation of classes) and the construction of an integral interconnected and consistent system from elements of the set under study.

All above shows that the value of any classification is determined by the extent with which the singled-out signs (grounds of classification) are *essential* for the set under study. With all this going on, not only is the essentiality of the sign important, but also the fact that this sign is initial. The signs singled out should be, as far as possible, *initial* (primary, determinant) with regard to other possible substantial signs.

In connection with this, in the development of the RUTS systems classification, the problem inevitably lay in a detailed analysis of this set as an integer and as a set consisting of parts, as well as in the identification of the entity of this set by studying its place and the role they play among other sets and systems both connected with it and more general ones. This determined the structure of studying.

Development of the RUTS systems classification was carried out under two aspects: *interspecific* classification (i.e., classification which does not depend on the type of measurements or physical quantity under measurement) and *specific* classification (i.e., classification based on physical quantities under measurement). In accordance with this, there two groups of signs were used (see Sections 2.1.4 and 2.1.5).

However, in connection with the fact that the problems and results of the specific classification of RUTS systems are in many respects similar to those of the problems and results of the classification carried out before, here the authors consider only some of the problems, namely those which were not dealt with in [144].

Since during the further development of the RUTS systems theory (and, perhaps, of some other metrological systems) which consists of a more complete formalization of the detected entities and interrelations, at the stage of classification an apparatus of the mathematical theory of systems and the set theory is used where possible.

2.1.2 Analysis of the RUTS systems

In order to carry out an analysis of the working systems of reproducing units of physical quantities and transferring their sizes and approach to the choice of classification signs of such systems and to classification itself more soundly, it is necessary to understand what is meant by the term "RUTS system". To do this, it is necessary to show the substantial signs of these systems, both the external (with respect to other systems) and internal signs (those which are substantial with respect to the RUTS systems themselves).

2.1.2.1 Concept of a metrological system

In the modern literature the concept "system" seems to be very widespread and is used for various studied objects, although it remains, to a certain extent, intuitive. This is not surprising, since it is one of the most generalized concepta (a metanotion similar to philosophical categories such as "substance", "common" and "particular", "similarity" and "difference", etc.), and determining such metanotions is rather difficult, since for this purpose one needs some other metanotions which are clear enough and not less common.

The analysis of the various approaches for determining the concept "system" allows the system to be determined as a "complex (totality) of interrelated elements forming a certain integrity" [37]. At the same time the two most important properties of any system, i.e., its integrity and its ability to be divided, are determined. According to this point of view, it would be possible to assume one more definition which is, in our point of view, more general: "The system is a nonempty set of singled out (i.e., detectable) essentialities (elements, objects) united by a certain more general substantiality relatively stable with respect to time".

In publications on metrology the term "system" have also become widely used, e.g., in word-combinations such as "system of units", "system of standards", "system of transferring unit sizes", "system of measurement uniformity assurance" ("system of providing measurement traceability"), etc. Unfortunately, in most cases these terms (word-combinations) were introduced without appropriate definitions, which does not allow for them a sufficiently monosemantic interpretation and (which is very important) nor demarcate them in a very clear way. At the same time, the use of the achievements of the general theory of systems permits this to be done, provided that the objects of metrological study be subjected to a deep analysis and clear identification. Under such a point of view it would be useful to introduce a generalized concept "metrological system" as a system of a (definite) category of objects being studied in metrology.

The objects under study in metrology are extremely diverse and can be both material (measuring instruments, standards, technical and engineering constructions and structures, scientist as keeper of measurement standards, etc.) and nonmaterial (terms and definitions, measurement unit symbols and their definitions, the context of normative documents, actions of specialists, the order of work at metrological organizations, etc.), as well as of a mixed type (methods and means of verification, definitions of units and their reproduction with the help of measurement standards, metrological institution as a whole, etc.).

In accordance with the above the metrological systems can be divided (as is done with the systems in other areas of knowledge) into formal (elements or objects of which are the essence of concepts, symbols, descriptions), informal (involving material elements), and mixed (including all socioeconomic systems).

2.1.2.2 "Elementary" metrological systems

As is known, in metrology the main (final) object of study is *measurement*. Therefore, any objects essential for measurement can be the object of metrological systems, particularly all objects which are the *components* (elements) *of measurement* [390, 437] such as a physical quantity to be measured (PQ), unit for measuring a given physical quantity, measuring instrument (MI), measurement method, measurement conditions and observer (i.e., an observer carrying out measurements and possessing knowledge needed to do them). A totality of such objects in different combinations forms various metrological systems.

For convenience (and a greater clearness) of further reasoning let us introduce the following symbols for *"elementary" metrological systems*:

- $\Phi \equiv \{\varphi_i\}$ is the set of all measured physical quantities, $\varphi_i \in \Phi$;
- $[\Phi] = \{[\varphi_i]\}$ is the set of all measurement units of PQs, $[\varphi_i] \in [\Phi]$;
- $\Psi \equiv \{\psi_i\}$ is the set of all possible measurement conditions [under "measurement conditions" in a generalized form we will assume both external influencing factors and features of an object, i.e., a carrier of a given PQ, as well as features of its realization (including a measurement range)], $\psi_i \in \Psi$;
- $S \equiv \{s_i\}$ is the set of all available measuring instruments, $s_i \in S$;
- $M \equiv \{m_i\}$ is the set of all measurement methods, $m_i \in M$;
- $K \equiv \{k_i\}$ is the set of all observers, $k_i \in K$.

At the same time, the sets Φ, $[\Phi]$, and M are the formal systems (consisting only of names and descriptions) and the sets Ψ, S, and K are the informal systems. We also consider that all sets are put in order, i.e., they are the vectors. As the metrological systems become more complicated, the formal systems, as it will be seen later, can "pass" (reflected) into informal realizations.

Each of these elementary sets and their various combinations can serve as the object of studying by metrology, the result of which will be the establishment of their properties and interdependencies of properties of corresponding metrological systems. Some of these properties are evident.

Thus, the power $|[\Phi]|$ of the measurement unit set $[\Phi]$ in a general form is equal to the power $|\Phi|$ of the measured PQs set $|[\Phi]| \geq |\Phi|$; each element of the set $[\Phi]$ is similar to a corresponding element of the set Φ]. The set S of all available measuring means is the combination $n \equiv |\Phi|$ of MI sets for various PQs (provided the universal MI's are excluded from consideration), i.e.,

$$S = S(\varphi_1) \cup S(\varphi_2) \cup \cdots \cup S(\varphi_{|\Phi|}) \equiv \bigcup_{i=1}^{n=|\Phi|} S(\varphi_i).$$

Here, the nomenclature of MI types for every $S(\varphi_i)$ will be determined by the subset $\Psi(\psi_i)$ of all measurement conditions of a given PQ.

The set M is determined by the intersection n of the subsets $M(\varphi_i)$:

$$M = M(\varphi_1) \cap M(\varphi_2) \cap \cdots \cap M(\varphi_{|\Phi|}) \equiv \bigcap_{i=1}^{n=|M|} M(\varphi_i).$$

In a similar manner it would be possible to determine the set Ψ. However, Ψ is the complex uncountable set, and we will not formalize it here.

It is evident that the set K is a certain function of the sets Φ, S, and M; the set K, in its turn, determines the system of manpower training for all observers, performing measurements.

It is also evident that the totality (in the general form as the vector product) of all indicated "elementary sets" determines the whole set of measurements (*system of measurement*).

Here it is important to emphasize that from the point of view of the integrity property in the definition accepted for the concept "system", any set of both the indicated ones and their totalities forms a corresponding system. This requires indicating a "linking entity". In particular, in order that the *system of measurements* corresponding to the set of all measurements may really be the system, it is necessary to indicate a concrete public formation to which it applies. Thus, it is possible to speak of the international or national systems of measurements as well as of an industry-branch system of measurements, etc.

It should also be noted that the "elementary nature" of the sets indicated is relative, since their division is also relative. Properties of practically every set depend on the properties of other sets. This is seen from the example for the set K. But here the issue concerning the "elementary nature" ("primary nature") of metrological sets is only mentioned. It is the subject of independent deep study (in our view, only on the way of selecting determinants, i.e., primary properties of the systems, it is possible to construct their scientific classification and the theory of their construction).

2.1.2.3 Measurement as a trivial metrological system

Let some formal properties of the elements of elementary metrological sets be considered, reasoning from their essence. Without going into detail, and reasoning from a method of forming these sets, it is possible with a commonality degree sufficient for practice (the element Φ – physical quantity as an abstraction – determining the formal set Φ, does not depend on (t, p), but as a concrete realization has to be localized in a concrete measurement. A similar remark can be made with respect to $[\varphi]$):

$$\begin{aligned}&[\varphi_i] = f(\varphi_i); \Psi_i = \Psi(\varphi_i, t, p); S_i = S(\varphi_i, \Psi_i); m_i = m(\varphi_i, \Psi_i, s_i);\\ &K_i = K(t_i, p_i, S_i, m_i, \Psi_i); \varphi_i = \varphi(\varphi_1, \varphi_2, \ldots, \varphi_j, \ldots, \varphi_{|\Phi|-1}),\end{aligned} \tag{2.1}$$

where $\varphi \in \Phi$, $[\varphi] \in [\Phi]$, $\psi \in \Psi$, $s \in S$, $k \in K$, their indices are identical within the limits of each dependency, but, generally speaking, different for different dependencies; t and p are the space-temporal coordinates of the metrological system in which appropriate elements of sets are included; $p = \{x, y, z\}$.

The space-temporal *localization* of elements is important in any system being studied, since as it is seen from equation (2.1) that their properties depend on place and time where and when they are located in the given system, which in its turn (as already emphasized for the measurement system) has to be determined in the space–temporal continuum, i.e., the system boundaries (T_c, P_c) have to be given.

Here, $t \in T_c$ and $p \in P_c$. This, in particular, is especially evident for Ψ (concrete conditions of measurements) and k (a concrete observer). At the same time we suppose that we deal with the metrological systems (including the measurement systems which are the most common) in which the scales T and P have to be of such a kind, so that within them the sets Φ, $[\Phi]$, Ψ, S, M and K, as well as their elements φ_i, $[\varphi_i]$, S_i, and m_i will not change. Without this assumption relative to the nomenclature stability (that of definitions of PQs, their units, MIs and measurement methods) the system study would be impossible.

Then any measurement (here the measurement means a totality of its inevitable components) in the system considered $t \in T$ and $p \in P$ can be represented as the set (a vector) $\{t_i, p_i, \varphi_i, [\varphi_i], \Psi_i, S_i, m_i, K_i\}$ in which all indices i are fixed and equal.

In this respect it seems important to introduce a *correctness condition* of measurement which is formulated in [29] at fixed p_i, $[\varphi_i]$, S_i, m_i, and K_i, but under the stipulation that the most essential temporal changes of φ_i and Ψ_i, as well as of parameters S_i over the measurement time τ_{meas} are taken into account. According to this condition and due to temporal changes in the measurement system $\{\varphi_i, [\varphi_i], S_i, m_i, \Psi_i, K_i\}$, the measurement error is caused by three components:

$$\Delta(\tau_{\text{meas}}) = \Delta\varphi + \Delta s + \Delta\psi,$$

where in the first approximation

$$\Delta\varphi = \int_t^{t+\tau_{\text{meas}}} \frac{\partial\varphi}{\partial t} dt, \qquad \Delta s = \sum_i \int_t^{t+\tau_{\text{meas}}} \frac{\partial\varphi}{\partial p_{s_i}} \frac{\partial\varphi_{s_i}}{\partial t} dt,$$

$$\Delta\psi = \sum_j \int_t^{t+\tau_{\text{meas}}} \frac{\partial\varphi}{\partial\psi_j} \frac{\partial\psi_j}{\partial t_j} dt. \tag{2.2}$$

Here $P_{s_i} \in P_s$ are the parameters (characteristics) of MIs in the static mode, which give the normalized MI error Δ_{norm}, e.g., an accuracy class.

Then the goal of measurement (i.e., getting a value of a PQ within the limits of an error given Δ_{giv} can be only under the condition

$$\int_t^{t+\tau_{\text{meas}}} \left\{ \frac{\partial \varphi}{\partial t} + \sum_i \frac{\partial \varphi}{\partial p s_i} \cdot \frac{\partial p s_i}{\partial t} + \sum_i \frac{\partial \varphi}{\partial \psi_j} \cdot \frac{\partial \psi_j}{\partial t} \right\} dt \leq \Delta_{\text{giv}} - \Delta_{\text{norm}}. \quad (2.3)$$

This condition of correctness can be generalized taking into account errors caused by space changes within the limits of the measurement procedure.

2.1.2.4 Metrological system of itself

The elementary metrological systems introduced above and some of their more or less complicated combinations are primary (external) with respect to metrology in the sense that, as a matter of principle, they can exist irrespective of metrological activity, though in the presence of the latter they become objects under study.

However metrological activity consists not only of studying the systems (objects) which are external with respect to it, but also of developing proper metrological systems (as a result of productive activity). The specific character of "metrological production" (and that of corresponding proper metrological systems) is determined by the specific character (by the essence) of practical metrology problems.

We think that at present it is possible to mark out three relatively independent general problems of practical metrology which determine the content and volume of the concept "metrological assurance" as a totality of all kinds of metrological activities and their results (in the documentary standard the term "metrological assurance" has already been defined as the establishment and use of various (listed there) grounds which are required for solution of two problems, i.e., achievement of the uniformity and accuracy of measurements).

As will be seen later, the above corresponds only to the system of measurement quality assurance. Moreover, enumeration of the grounds determining kinds of metrological activity also restricts the concept in an artificial manner, since it does not allow the main problems to be solved by some other way:

(1) the problem of measurement uniformity assurance;

(2) the problem of the technical improvement of measuring instruments and measurement methods;

(3) the problem of measurement efficiency assurance.

The solution to the first two problems leads to the formation of a *system of ensuring the measurement quality* (SEMQ) within the framework of the given measurement system, and the solution of the third problem, in its turn, leads to formation of *a system for ensuring the measurement efficiency* (SEME) of this given measurement system.

Considering some national system of measurements (NSM) as the measurement system, it is possible to determine the most common proper metrological system, i.e.,

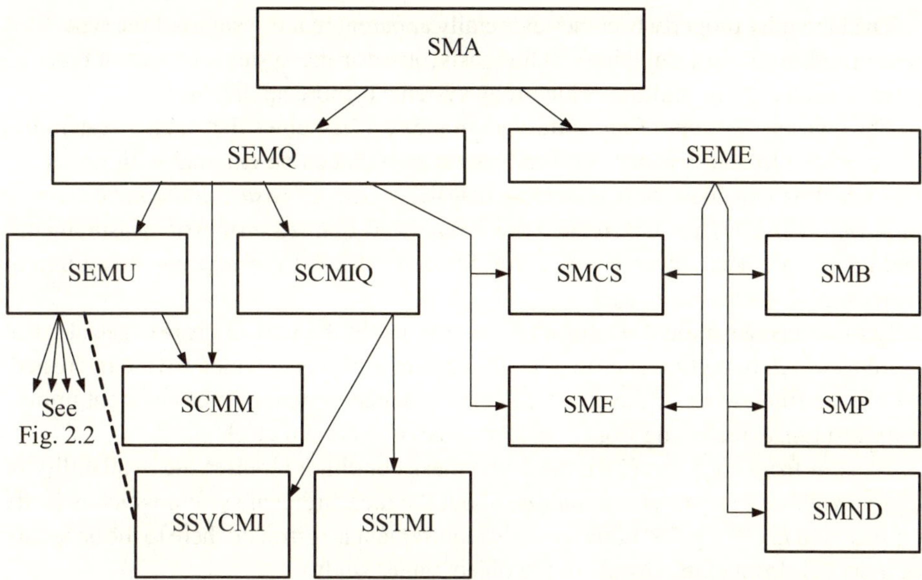

Figure 2.1. Interrelation of "proper metrological systems" on the upper levels of the metrological assurance system. Here: SMA is the system of metrological assurance; SEMQ is the system of ensuring the measurement quality; SEME is the system of ensuring the measurement efficiency; SEMU is the system of ensuring the measurements uniformity; SCMM is the system for certifying methods of measurements; SCMIQ is the system of certification of measurement instruments quality; SSVCMI is the state system of verification and certification of measurement instruments; SSTMI is system for the state testing of measurement instruments; SMCS is the system of metrological control and supervision; SMB is the system of metrological bodies; SME is the system of metrological examination; SMP is the system of metrological personnel; SMND is the system of metrological normative documents.

the "*system of metrological assurance of national economy*" as a totality of all proper metrological systems which appear as a result of solution of the above main problems of practical metrology.

Figure 2.1 shows an attempt to systematize proper metrological systems in an implicit or explicit form (without detailing them), which have been formed for today. Some grounds of the interrelations indicated in Figure 2.1 will be given below.

2.1.2.5 Hierarchy of metrological systems

The hierarchy property of a structure is inherent to all existent systems, particularly to complicated ones. This is seen from the possibility of describing on any level (to abstract) any system, various features for joining the system elements into groups, and others.

The hierarchy property becomes especially apparent in the systems of the type "system of uniformity (of anything)". Obviously, just for the systems of such a type the formalized theory of multilevel hierarchy systems is more applicable.

The concept "system of uniformity assurance ..." assumes that such a system has to be related to another more general system (which can be external with respect to that which is considered). It is evident that the *system of ensuring the measurement uniformity* (SEMU) has to be related to a certain well-defined system of measurements, and taking into account what is noted in Section 2.1.2.4, to an appropriate system of metrological assurance as well.

Let us consider mutual relations of the systems in the part of proper metrological systems which directly relates to a RUTS class of systems in which we are interested, within the framework of the above selected national system of measurement uniformity assurance, reasoning from considerations of general metrology.

As seen from equation (2.3), the correctness condition requires the availability of a great volume of *a priori information* about the measurement system which is justly emphasized in [390, p. 79] (although this requirement is restricted there to the necessity of a priori information only about the object under study).

Even if we suppose that the space–temporal changes in the measurement procedure do not take place, and even in the case where the assumption is made that the space–temporal changes take place and direct single point (spot) measurements are possible (obviously, those which are of the most interest from the point of view of mass measurements in the state system of measurement uniformity assurance), a priori information meeting a simpler correctness condition is also required:

$$\Delta_{\text{meas}} \leq \Delta_{\text{norm}}. \tag{2.4}$$

This information is present in the implicit form in Δ_{meas}, since $\Delta_{\text{meas}} = (\Delta\varphi'_{\text{meas}}$, where $(\Delta\varphi'_{\text{meas}}$ is the real measured value of the PQ (of a concrete dimension), i.e., $\varphi' = \varphi'(s_i, m_i, \psi_i, k_i)$, some of the parameters of this dependence are connected by relations (2.1) with the exception of those which contain t and p fixed in the case given.

In our view this circumstance (an inevitable necessity that a priori information be present and any necessity of this kind, to put it more precisely, for a different volume of this information under conditions given) is rather significant for understanding the specific character of metrological activity and therefore can be used for analyszing metrological systems. In this connection the use of this circumstance in the classification of measurements according to accuracy (by the method of error evaluation) see in [144].

We can formulate the following *postulates*:

(1) any measurement requires a certain a priori measurement information;

(2) the volume of a priori information has to be determined by the accuracy required: the higher the accuracy required, the greater the volume of a priori information needed;

(3) at a given accuracy of measurements the volume of required a posteriori measurement information will be determined by the volume of a priori measurement information: the less the volume of a priori measurement information is, the greater the volume of required a posteriori information.

To all appearance, these postulates can be useful for further developing information theory of measurements. However, in the case given they allow some forms of metrological activity (and corresponding metrological systems) to be combined into a *system of ensuring the measurement quality* as a constituent part of the system of metrological assurance (SMA) (see Figure 2.1).

Since it is the availability of certain a priori information which determines to a significant extent the quality of measurements (i.e., the correctness of solving measurement problems according to condition of information correctness), providing this a priori information for solving all the measurement problems of importance for a national economy is the main essence of the *system of ensuring the measurement quality* in a country as the metrological system of the latter.

And, since according to the third postulate, at a small volume of known a priori information the volume of necessary a posteriori increases (i.e., the labour-intensiveness of the measurements themselves), the task of metrology is to provide the whole national system of measurements with *maximal* a priori information. Hence the need of the system of metrological assurance to have a subsystem of measurement efficiency assurance and its interconnection with the system of measurement quality assurance can be seen.

What is the distinction of the terms "*ensuring the measurement quality*" and "measurement uniformity assurance? According to [195] under *measurement uniformity* is meant "a state of measurements where the measurement results are expressed in terms of legal units and the measurement errors are known with the given probability" (the last part of the phrase, obviously, should be understood simply as the availability of not only an estimate of the physical quantity value but also of *an estimate of its error*).

Formally this can be written in the following way (*the comparability condition*). There is Δ such that at any $j \neq i$ it is possible to have

$$\varphi_j - \varphi_i \leq \Delta \quad \text{at} \quad \varphi_j = \varphi_i \quad \text{and} \quad \psi_j = \psi_i, \tag{2.5}$$

where $\varphi_j = \varphi(t_j, P_j, s_j, m_j, k_j, [\varphi_j])$ and $\varphi_i = \varphi(t_i, P_i, s_i, m_i, k_i, [\varphi_i])$ are the really measured values of the physical quantity φ, and φ_i are its true values. Let us note that here the indices "i" and "j" mean that we are dealing not with different physical quantities (as the set Φ has) but with different results of measurements of one and the same physical quantity.

Obviously, the comparability of measurements (measurement uniformity) can also be reached with a measurement accuracy which is unsatisfactory for a given national system of measurements. Since the condition of measurement correctness (2.3) or (2.4) was related with the achievement in obtaining the physical quantity value within the given accuracy characterized by Δ_{norm}, it can be considered as a condition for getting the accuracy required.

For the simplest (and from the point of view of the national system of measurements interesting) case expressed by inequality (2.4), taking into account comments to it, the *condition of achievement the accuracy required* can be formulated in the following manner. There exists i (on the sets $S, M, \ldots$ in the given NSM) such that at any i (in the given NSM) we have

$$\Delta_{\varphi} \leq \Delta_{\text{norm}}, \tag{2.6}$$

where $\varphi = \varphi(s_i, m_i, \psi_i, k_i)$ if equation (2.1) is taken into account.

Now it is possible to determine in a formal way *the system of ensuring measurement quality* as the metrological system that provides fulfillment of conditions (2.5) and (2.6), i.e., conditions of comparability and correctness of measurements, and *the system of ensuring the measurement uniformity* (SEMU) as the system providing the fulfilment of the condition of measurement comparability (2.5).

An attempt to use the quality indices of measurements (accuracy, correctness, repeatability, reproducibility) determined in [195] has not given good results, since these concepts to a certain extent overlap. They can be useful in further detailing of the systems under the condition of their more precise specification.

From the informal point of view, the condition of measurement result comparability means the possibility of "speaking in one language", i.e., the *unification* of measurement information, mainly of a priori information, because it determines the unification of a posteriori information (if for example, the same units are used, then the results will be expressed in similar units).

On the other hand the assurance of the comparability of measurement results is the most efficient way to reveal *systematic errors* of measurements (see [304]). This is exactly the main task of the system of measurement uniformity assurance (and of metrology as a whole).

2.1.2.6 RUTS system as a subsystem

In Section 2.1.2.3 a formal definition of measurement (i.e., from the point of view of its structure) is given. At the same time the structure of any measurement can be represented as the vector $U_i = \{\varphi_i, [\varphi_i], \psi_i, S_i, m_i, k_i\}$ of a state of the i-th considered measurement system $\{\Phi, [\Phi], \Psi, S, M, K\}$.

Now let the *measurement* be considered as a *process* of determining (finding) the PQ value (by means of an experiment as in [195]).

The PQ value is usually written in the form of the equality

$$\varphi = \{\varphi\}[\varphi], \tag{2.7}$$

where $\{\varphi\}$ is the numerical value of the physical quantity and $[\varphi]$ is its unit. At the same time under φ in the left-hand side of the equation they understand the physical quantity itself to be both a concrete realization and the quantity in general [437, p. 16]. Within the framework of the assumed formalism this is adequate. The expression $\varphi' \in \Phi$ is the designation of the physical quantity (a symbol, name, or definition irrespective of its numerical value). The same also relates to $[\varphi]$, i.e., to its unit. Therefore the physical quantity value does not coincide by implication with the PQ itself. To distinguish them, let us introduce a symbol $\varphi^{(\mathrm{meas})}$ for the value φ. Then

$$\varphi^{(\mathrm{meas})} = \{\varphi\}[\varphi], \tag{2.7a}$$

can be considered to be the element of a formal set $\Phi^{(\mathrm{meas})} = [\Phi] \times R$, where R is generally (in the general case) the set of all real numbers, and $[\Phi]$ is the set of PQ units introduced earlier.

Since the time of Prof. M. F. Malikov [323] equation (2.7) has been considered to be the basic equation of measurements (also see [390, p. 10]) but at that the equation of measurements for the simplest case, i.e., for direct measurements, is written in the form $\varphi = X$, where φ is the physical quantity value *in question* and X is the value *obtained* experimentally. Here one can also see an explicit inexactitude.

Let us consider the simplest but most important case of measurements, i.e., a *direct single measurement*, since any more complicated measurement is in the long run reduced to a direct single measurement (just in this case the PQ value is directly obtained *by an experiment*), the transfer to the result of the more complicated measurements is realized on the basis of methods of measurement result processing, the latter being the most developed issue in metrology.

In case of such measurements, the PQ value, as a formal result of measurements coinciding with expression (2.7a), has to correspond to a rather determined *indication* of a measuring instrument (as to the informal state of the measuring instrument) which can be conditionally called *an informal value* of the physical quantity under measurements:

$$\varphi^{(\mathrm{meas})} = N_s \varphi_0', \tag{2.8}$$

where N_s is the dial readouts according to any numerical scale of the given MI and φ_0' is the constant of the MI for this scale.

Since according to the definition $\varphi_0' = N_0[\varphi']$, where $[\varphi]$ is the symbol of the PQ unit as before, and $[\varphi]$ is the PQ unit as its certain size (the concrete realization that was "memorized" by a given measuring instrument), then

$$\varphi^{(\mathrm{meas})'} = N_s N_0[\varphi'] = N\{[\varphi']\}[\varphi] = N[\varphi'].$$

Elements of the informal set (of the system) corresponding to the informal set (to the system) will be accompanied with a mark ($'$) for the purpose of distinction. Let us note that the sets R and $\{\Phi\} \equiv \{\{\varphi\}\}$ are formal, since their elements are numbers.

Hence it is seen that it is always possible to choose such a scale of the given measuring instrument, such that its constant φ_0 will coincide with the informal (inherent in a given concrete measuring instrument) unit of measurements, i.e., $\varphi_0' = [\varphi']$. Therefore, taking into account the fact that the formal measurement result is taken in the form of a PQ value indicated by the measuring instrument, the *equation of the direct single measurement* (to which all other kinds of measurements are reduced) can always be presented in the form

$$\varphi^{(\text{meas})} = N[\varphi']. \tag{2.8a}$$

However, it should be kept in mind that $\varphi_0' = [\varphi']$ (as a specific constant of the measuring instrument) is not only the unit symbol, it is *a priori* (embedded into properties of the measuring instrument) *information about the size of the physical quantity chosen as the unit*, i.e.,

$$\varphi_0' = [\varphi'] = \{[\varphi]\}. \tag{2.9}$$

Just in this a significant difference exists between (2.8a) and (2.7a), which in appearance seem to be the same.

From this point of view it is interesting to classify the basic types of measuring instruments: *measures*, measuring *instruments*, and measuring *transducers*.

- *Measures* can be presented as MI for which both N_s and $\varphi_0' = [\varphi']$ are a priori measurement information.
- *Instruments* can be considered to be MI for which there is a priori information and N_s is a posteriori information.
- For *transducers* the equation of measurements can be written as

$$\varphi_j^{(\text{meas})'} = K_{\text{tr}}\, \varphi_i^{(\text{meas})'}, \tag{A}$$

  where φ_i and φ_j are the different values, and K_{tr} is the coefficient of transformation (in a general case it is a nonlinear operator).

For the MI described by such an equation, taking into account equations (2.7a) and (2.8a), a priori information will be

$$K_{\text{tr}} = \frac{\varphi_j^{(\text{meas})'}}{\varphi_i^{(\text{meas})'}} = \frac{N_j\{[\varphi_j']\}}{N_i\{[\varphi_i']\}} \cdot \frac{[\varphi_j]}{[\varphi_i]} = N_k \frac{[\varphi_j']}{[\varphi_i']}, \tag{B}$$

i.e., the ratio of the sizes of different physical quantities (the input and output ones).

Let us return to the condition of comparability (2.5), having noticed that, as a matter of fact, it should contain not simply φ_i and φ_j, but their values $\varphi_i^{(\text{meas})}$ and $\varphi_j^{(\text{meas})}$. Then taking into account the above,

$$\begin{aligned}\varphi_j^{(\text{meas})'} - \varphi_i^{(\text{meas})'} &= N_j\,\{[\varphi_j]\}\,[\varphi_j] - N_i\,\{[\varphi_i]\}\,[\varphi_i] \\ &= N_j\,\{[\varphi_j]\}\,[\varphi_j]\left(1 - \frac{N_i\,\{[\varphi_i]\}}{N_j\,\{[\varphi_j]\}} \cdot \frac{[\varphi_i]}{[\varphi_j]}\right).\end{aligned} \tag{2.10}$$

Hence it is seen that if $[\varphi_i] \neq [\varphi_j]$, i.e., "the symbols" of the units do not coincide in measurements of i and j, then condition (2.5) cannot be fulfilled at any Δ, since a final number (dimensionless) has to be obtained in equation (2.10) in parenthesis, and the ratio $\frac{[\varphi_i]}{[\varphi_j]}$ is nonsense that has no quantitative distinctness.

Thus, to provide the uniformity of measurements (comparability of their results) it is necessary first of all that $[\varphi_j] = [\varphi_i]$, i.e., that the unit of each physical value have *only one designation* (symbol, definition or name, which is formally the same). Consequently, the system of measurement uniformity assurance has to include the *system of PQ units* as its subsystem (as a set of their designations, symbols, and definitions).

In Section 2.1.2.2 this system is introduced outside the framework of the proper metrological systems. It inevitably becomes such a system when the social need in measurement uniformity assurance increases.

The comparability condition according to equation (2.10) also requires the comparison of PQ unit sizes realized in various measurement experiments (see the ratio of numerical values of units in equation (2.10) in parenthesis). Hence, it follows that the system of measurement uniformity assurance also has to include the system that allows the unit sizes realized in each concrete measuring instrument in the form of a priori information to be compared. This system in its general form should be called the *RUTS system*.

Up to now we have restricted ourselves to the case of direct measurements. Moving on to indirect measurements we will inevitably come to the necessity of providing the uniformity of values of different physical constants entering equations of indirect measurements. We will come back to this in Section 2.1.5.3. But here it is important that this is also the problem of one additional proper metrological subsystem, i.e., *the state system of standard reference data* (SSRD).

It is quite possible that the structure of the system of measurement uniformity assurance, which includes these subsystems, extends further. In particular, the system of measurement uniformity assurance to all appearance should include (at least in part) the *system of certifying methods of measurements* SCMM (certification is a rather new trend of metrological activity).

In Figure 2.2, taking into account the above and Section 2.1.2.8, an interaction of various subsystems of the SEMU is schematically shown.

2.1.2.7 Analysis of the concepts "reproduction of a unit of a physical quantity" and "transfer of the size of a physical quantity unit"

The concept "RUTS system" is usually determined with its full name, i.e., with enumeration of the main functions and their objects. In this connection it is important to analyze what is meant by the concepts "reproduction of a unit size" and "transfer of the unit size" of a physical quantity. Unfortunately, there is no clarity on this issue, especially with regard to the interpretation of the concept "reproduction of the unit" (compare with [323, 390, 437, p. 12]).

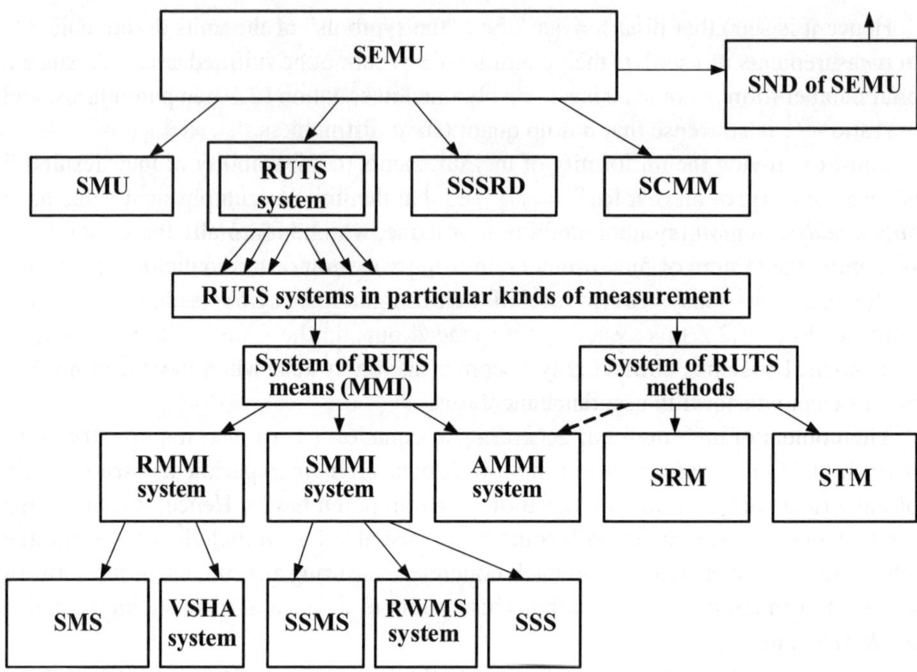

Figure 2.2. Interrelation of subsystems of the system of ensuring the measurement uniformity. Here: SEMU is the system of ensuring the measurements uniformity; SND is the subsystem of normative documents of SEMU; SMU is the subsystem of measurement units; RUTS system is the subsystem of reproducing PQ units and transferring their sizes; SSSRD is the state system of standard reference data; SCMM is the subsystem for certifying methods of measurements; RUTS subsystems are the subsystems of reproducing PQ units and transferring their sizes in particular kinds of measurement; Subsystem of RUTS means is variety of metrological measurement instruments (MMI); Subsystem of RUTS methods is variety of methods of reproducing PQ units and transferring their sizes; RMMI subsystem is variety of reference MMIs; SMMI subsystem is variety of subordinate MMIs; AMMI subsystem is variety of auxiliary MMIs; SMS is the subsystem of the state measurement standards; VSHA is the subsystem of the verification setups of the highest accuracy; SSMS is the subsystem of the secondary measurement standards; WMS is the subsystem of the working measurement standards; SSS is the subsystem of the standard samples; SRM is the subsystem of the reproducing methods; STM is the subsystem of the transfer methods.

In the Reference Supplement to GOST 8.057–80 (a documentary standard) the reproduction of a physical quantity unit is defined as the "totality of metrological operations aimed at determination of a PQ value reproduced by a *state measurement standard* at an accuracy at the level of recent achievements of metrology".

Thus, here the reproduction of the unit is connected with a measuring instrument which is the most accurate in the given system of measurements. However "reproduction of a unit" is defined in terms of "reproduction of physical quantity" and does not expose the concept essence (let us note that a unit is one of the realizations of the same physical quantity).

However such a point of view does not conflict with another official document [195], according to which *any measure* by its own definition reproduces the physical quantity of dimensions given (including the unit size). In this connection let us note that the largest part of measurement standards (also including national ones) reproduce not strictly the unit, but frequently the dimensions of the physical quantity which are too far from the unit. That is one of the difficulties in interpreting the term "reproduction of a unit".

From the above it is obvious that there is a need to study more deeply the essence of this concept as one of the fundamental concepts of practical metrology, in particular of the problem of measurement uniformity assurance.

A single work specially devoted to this problem is the paper "Definition and reproduction of physical quantity units" by Prof. S. V. Gorbatcevich[1]. In that work the author analyses the essence of the concept "reproduction of the units" and gives a more or less successful definition: "reproduction of the unit is the creation of a such an object (body or instrument), the properties or parameters of which are expressed by the quantity in terms of units corresponding their definition".

A deep understanding is examined (unfortunately without giving an unambiguous formulation) in a monograph by M. F. Malikov [323, pp. 315–325]. The author also connects the "reproduction of the unit" with the practical realization of its theoretical definition. However, here, as distinct from the previous work, the author applies "reproduction of the unit" not only to all measurement standards (including secondary ones) but also to working measurement standards, though on p. 285 he ascribes this property only to measures.

However there is one *common* thing which joins various points of view at the concept "reproduction of the unit". This is the relation of the unit to a concrete measuring instrument, its *materialization* (realization, embodiment).

It was shown above (Section 2.1.2.6) that a certain realization of the PQ unit in a concrete measuring instrument (a real content of measurement information about the unit size) with the help of which the measurement is carried out, is, according to expression (2.8a), a necessary condition of any measurement. In other words: *any MI has to realize* (actualize, materialize) *in itself the unit size*.

The vailability of a numerical factor for converting the unit size into its multiple or submultiple part does not play any principle role.

From this it follows that to apply the concept "reproduction of the unit size" in the sense of its materialization (realization) only to measurement standards is meaningless. It is also applicable to working measuring instruments in this sense.

---

[1] *Metrologiya* **12** (1972), 3.

Let us note, however, that information about the unit size really contained in every particular measuring instruments is a priori measurement information (in any measurement with the help of the given measuring instrument) obtained by an earlier experiment (as a posteriori measurement information in measurements with another measuring instrument).

A unique *exception*, obviously, is *the most accurate* measuring instrument in the given system, for which *there is no a priori information* about the PQ dimension. The unique a priori information about the unit of the given PQ for such a measuring instrument is *formal information contained in the unit definition.*

Thus, a *unique possible version* of separating the concept "reproduction of the unit size" simply from its "materialization" can be the correlation of the first concept only with the most accurate measuring instrument in the given system of measurements.

So, *the reproduction of the unit is its materialization (realization) of such a type, that the unique a priori information about the unit is its theoretical definition.*

Such a definition in addition to its logical strictness allows us for the first time to approach the *classification of RUTS systems*, as well as to *unify their type structures.*

Formalization of the concept "reproduction" needs to be made in further studies. In the simplest case of one measuring instrument used for the purpose of reproducing, which corresponds to measurement equation (2.8), then taking into account (2.9) and believing that according to the definition for this measuring instrument $\{\varphi]\} = 1$, we obtain the equation of unit reproduction

$$\varphi_{\text{st}}^{(\text{meas})'} = N_{\text{st}}[\varphi]. \tag{C}$$

"Reproduction means" is also a measuring means, only without experimental a priori information about the unit.

The concept "transfer of the unit size" is less disputable, since it contains in its name (term) rather full information about its essence.

In GOST 8.057–80 [195] the following definition of "transfer of the unit size" is given: the totality of metrological procedures aimed at defining on the basis of calibration or verification the PQ value which has to be ascribed to a secondary or working standard at the time of either their calibration or metrological certification, or to a working measuring instrument in the process of its verification.

In [390, p. 173] the "transfer of the unit size" (under which the transfer of the unit size as a particular case of PQ realization can be understood) is defined as "founding (confirmation, ascription) metrological characteristics of a verified or certified measuring instrument with the help of a more accurate measuring instrument".

In all other cases (without definitions) the "transfer of the unit size" is understood to be simply the totality of hierarchy comparison (within the limits of the given verification scheme) of unit sizes "embedded" in subordinate measuring instruments by their verification or calibration (see, e.g., [437, p. 83]).

We think that further detailing is unnecessary (in the same way as in the preceding case), since a nongraduated measuring instrument, strictly speaking, is not a measuring instrument (nothing can be measured with it). Moreover, the concepts "verification", "certification", "comparison", "calibration", and "graduation" reflect varieties of metrological procedures on transferring the unit size and are secondary with regard to the concept "transfer of the unit size " presently being considered.

From this it follows that the main content of the concept "transfer of the unit size" is constituted by *comparison of unit sizes, "frozen" (embedded in the form of a priori measurement information) in (subordinate) measuring instruments varying in accuracy.*

To make this concept formal, let us use the same equation (2.8). Since in transferring (comparison) they try to reach the situation when $\varphi_2^{(\text{meas})\prime} = \varphi_1^{(\text{meas})\prime}$, where $\varphi_1^{(\text{meas})\prime}$ is the PQ value obtained with a more accurate measuring instrument, and $\varphi_2^{(\text{meas})\prime}$ is the value of the same PQ obtained with a subordinate measuring instrument, there we have $N_2[\varphi_2'] = N_1[\varphi_1']$, from which we get the *equation of unit size transfer*:

$$[\varphi_2'] = \frac{N_1}{N_2}[\varphi_1']. \tag{D}$$

This case covers all the basic methods of transfer. Methods of indirect measurements which are frequently included into varieties of transfer methods, strictly speaking, cannot be related to the transfer methods (see Section 2.1.3.4).

An analysis of the main concepts of RUTS systems would not be complete without considering concept "*maintaining the PQ unit*", which necessarily enters the legal definition of the concept "standard" [195].

For the concept "maintaining the PQ unit" there is only one explanation given in the reference supplement to GOST 8.057–80 [195]: "unit storage is the totality of metrological procedures aimed to keep the PQ, the value of which has been ascribed to the primary or working measurement standards at their metrological certification, invariable with time.

However such a definition should rather be referred to the concept "maintaining the measurement standard" (compare with [323, p. 326] and [331]). Needless to say, at proper maintenance (which is absolutely mandatory, according to GOST 8.057–80) of the measurement standard, the latter, as well as the size of the unit it reproduces, will be kept. But generally speaking, the *unit is maintained by any measuring instrument* since the latter materializes (memorizes) it.

Therefore, *maintaining the unit* is simply *its realization* with a given measuring instrument *at any time*. In particular, under *maintaining the unit with the measurement standard* one should understand *its reproduction at any time*. However long the measurement standard reproduces the unit, so long will it maintains this unit (certainly, the longer the better).

In this connection it should be noted that the reproduction of the unit with the measurement standard (with the state measurement standard as the most accurate one in the SSM) in a definite time can be performed by two methods:

(1) the *discrete* method, according to which the unit is reproduced only at any time interval where the standard setup is “switched on”; in the remaining time the standard is the reproducer of the unit only *potentially*;

(2) the *analogous* method, according to which the standard reproduces the unit at any moment of time until it exists as the standard.

From this, inter alia, a very important requirement becomes obvious which concerns the nomenclature of the metrological characteristics of national measurement standards connected with their ability to maintain corresponding units. Whereas according to the analogous method of reproduction it is sufficient to characterize the error of maintaining the unit by *instability* (with a stability index), when using the discrete method of reproduction (as will be seen in Section 2.1.4.2, this method is the most widespread one of the national measurement standards in force) it is also necessary to indicate together with the stability index the reproducibility of this standard (as an error of the standard due to its different “inclusions”.

2.1.2.8 Structure of the RUTS system

First of all it we should notice that frequently the concept “RUTS system” is used for a separate kind of measurements, i.e., for a PQ, the unit of which is reproduced and transferred by a given system. At the same time it is no less important whether or not the concept of the *full RUTS system* complies with the system covering all physical quantities which can be measured.

Apparently it is useful to introduce a specific name “*particular RUTS system*” for a RUTS system related to a particular physical quantity.

As to the structure of the *full RUTS system*, it is possible to mark out two subsystems: *a full system of reproducing PQ units* and *a full system of transferring unit sizes*. At present no doubts about the usefulness of introducing the first subsystem seem to arise. At least it agrees with the system of all national measurement standards which constitute the most important part of the measurement standard base of the country.

The independence and importance of the problem of establishing the interrelationship of reproductions of the units of various (different) quantities became quite apparent in the development of the system of national measurement standards of the most important units of electrical quantities [501].

The second subsystem was introduced rather as an addition to the first one. At present the category “systems of transferring the unit size” is more important for particular RUTS systems.

However from the point of view of classifying the RUTS systems as their set, the *structure of a particular RUTS system* is of much greater interest. For this purpose let us introduce *generalized elements* of a particular RUTS system, which in the general case can be both elements of the simplest metrological set (Section 2.1.2.2.) and elements of more complicated metrological systems. As elements of the particular RUTS system let us choose only elements of two systems: the *means* of reproducing units and transferring their sizes as well as the *methods* of reproducing units and transferring their sizes.

The necessary facts, at least in the form of verification schemes, exist only for these sets. Moreover, these sets are the most essential for the RUTS systems (systems of normative documents, metrological specialists, metrological bodies, etc.). They are also to a great extent common to all metrological systems of their own, as well as in the structure of the national system of metrological assurance (Figure 2.1).

Let us notice that the means of reproducing the PQ units and transferring their sizes are the single, particularly *metrological* technical means. Their distinction from the technical means intended for measurements outside the metrological practice, i.e., working measuring instruments (WMI), follows from the preceding point and does not transfer its size (in definitions of Section 2.1.2.6). At the same time any means of reproducing the unit or transferring its size is simultaneously the undoubted means of measurements or *measuring instruments*.

Taking into account these two circumstances it is useful to name a generalized element of a particular RUTS system, which is the means of reproducing the unit and (or) transferring its size, by metrological measuring means (MMM): "*metrological measuring means are the measuring instruments intended for and (or) applied in a certain system of measurements* (for example, in the SSM) *for reproducing the unit and (or) transferring its size*".

There were some other attempts to introduce a generalized concept for measuring instruments used in metrological practice which distinguishes them from the category working measuring instruments. Thus, in [390, p. 172] all measuring instruments are divided into working measuring instruments and measurement standards (similar to M. F. Malikov's work [323, p. 285]. However such terminology is not very successful.

From the essence of the concepts "reproducing the unit" and "transferring the unit size" established above inevitably follows the hierarchical nature of the structure of the MMI system for a given particular RUTS system (see Figure 2.2).

From the point of view of the hierarchy of different MMIs *within one of a particular RUTS system* it is possible to distinguish *three independent classes of metrological measuring instruments*, which are suitable for any variety of RUTS system:

- *reference MMI of the system*: the MMI whichallows a given particular RUTS system to reproduce and transfer the size of a PQ unit;
- *subordinate MMI of the system*: the MMI which allows a given particular RUTS system to transfer (but not to reproduce) the unit size;
- *auxiliary MMI of the system*: the MMI providing the necessary functioning of an initial and (or) a number of subordinate MMIs of a given particular RUTS system (another version: providing normal operation of the hierarchial system of the initial and subordinate MMIs).

The ntroduction of the class of “auxiliary MMIs of the system” does not follow from the logics of preceding constructions. It was dictated by the practice of applying initial and subordinate MMIs (see Section 2.1.3.3).

2.1.3 Varieties of the elements of a particular RUTS system

2.1.3.1 Reference metrological measuring instruments (RMMI)

Within the framework of the state system of measurements being considered, it is possible to refer to reference MMIs for a particular RUTS system, the following categories of MMIs:

- state measurement standards (SMS) defined in GOST 16263; among them are distinguished (see also GOST 8.057 [195]) state primary measurement standards (SPMS) and state special measurement standards (SSMS);
- verification setups for reproducing the units by the indirect method. In practice these setups obtained the name “verification setups of the highest accuracy” (VSHA);
- complex of working measurement standards (WMS) borrowed from other verification schemes (is admitted too by GOST 8.057–80).

However the definitions of each category are fuzzy to such an extent that they do not permit the MMI given to refer specifically to one of them.

In fact, all PQ units (except the basic ones) can be reproduced by the method of indirect measurements (through the direct measurements of other PQs, the units of which are known). Therefore, if a verification setup of a high accuracy means a complex fixed MIs necessary for indirect measurements, then there will be no principle distinctions from the national measurement standard.

On the other hand if one assumes that in the third category, too (of borrowed metrological measurement means: measuring instruments or working measurement standards), all standards included into the complex are fixed within the framework of one setup, then the latter will not differ at all from the first two categories (since MIs of other PQs can doubtlessly reproduce the PQ given only by an indirect method.

Since the assumptions made are not regulated by anything, then there seems to be some difficulty in referring MMI to some category.

Evidently the nomenclature of the varieties of reference MMIs has to follow from the distinction of the RUTS systems types on the basis of different essential signs of their reproduction, i.e., it has to be based on the classification of RUTS systems, which will be considered in Section 2.1.4.

Here we will only notice that this issue requires an independent terminology development taking into account the matter stated in this item, as well as in Section 2.1.4.

2.1.3.2 Subordinate MMI

The category of subordinate metrological measuring instruments intended for metrological application includes

- standards-copies (S-C);
- varieties of secondary measurement standards;
- working measurement standards which can be related to some grade of standards;
- verification setups (VS).

All of these varieties are indicated in GOST 16263 and included into GOST 8.057 [195]. Here, too, the interpretation of these varieties requires further improvement.

Thus, the working standard is for some reason defined not generally by a measuring instrument (it would be more logical to ascribe to it the corresponding metrological functions), but by three concrete varieties of measuring instruments (measures, instruments, transducers), and as a result it is necessary to artificially introduce the concept "verification setup" (VS).

No distinctions (including those of metrological kind) are seen between the categories of secondary standards intended for transferring the unit size: S-C and WMS. In accordance with Section 2.1.2.8 (and the definitions in GOST 16263–70) they all can serve only in transferring the unit size; and the introduction of the S-C as an additional transfer link with a new name but the same functions filled by the VS and WMS appears even less convincing.

Moreover, the name S-C is itself not really adequate. The idea springs to mind that it would be better to use this term for a measuring instrument which really fulfils functions of the copy (as a reserve version, see Section 2.1.3.3). It should be noted that because of this a number of real verification schemes and standard-copies have been introduced with various functions.

In our opinion, from a metrological point of view for subordinate metrological measuring instruments, it would be quite enough to have only one category of working standards of various grades.

2.1.3.3 Auxiliary MMI

This class of metrological measuring instruments, as mentioned above, was formed historically on the basis of experience gained in applying reference and subordinate metrological measuring instruments. This experience has shown that in some cases it is useful and necessary to have, together with reference and subordinate metrological measuring instruments, a number of metrological measuring instruments of an auxiliary destination, which fulfill one of the *functions* indicated below, which result from the *metrological application* of reference and subordinate metrological measuring instruments:

- performing *comparison* of reference and subordinate metrological measuring instruments of a given particular RUTS system, or corresponding basic (as a rule, the reference ones) metrological measuring instruments of the RUTS systems of one and the same physical quantity but for various measurement systems (national and international ones);
- checking of the *safety* (invariability) of metrological characteristics of a reference metrological measuring instrument;
- replacement of a reference metrological measuring instrument in case of its failure.

To perform the *first* function the following varieties of measuring instruments are required: *reference measurement standards* (according GOST 16263–70 and 8.057–80 [195]) and *comparators* (they are not regulated anywhere, but provided for in GOST 8.061–80 [194]). To perform the *second* and *third* functions only one category is provided for. It is a *standard–witness* (according to GOST 16263–70).

According to GOST 8.057–80 the measuring instruments which control the invariability of the size of a unit reproduced by a state measurement standard have to be included into it, i.e., to perform the second function there is no any independent category of metrological measuring instruments. Thus, it turns out that the standard-witness serves only for replacing the state measurement standard.

Taking this into account it is possible to say that the term "standard-witness" used for the function of replacing the state measurement standard cannot be considered to be very accurate. For this purpose the name "standard-copy" seems more appropriate.

It is also not satisfactory that until now the important category of comparators has not been regulated though any transfer of the unit size performed through comparison (see Section 2.1.2.7); and if it is necessary for this purpose to use additional measuring instruments, which undoubtedly influence the accuracy and method of transferring the unit size, then they are worthy of the same attention as is the case with other metrological measuring instruments.

This "nonsense" is mostly observable in the case of a verification scheme for mass measuring instruments (weights and balances). Only weights are the metrological measuring instruments, whereas the balances, although they have grades, are not regulated by the national verification scheme, i.e., they are considered to be versions of working measuring instruments.

2.1.3.4 Methods of transferring a measurement unit size

In accordance with GOST 8.061–80 [194], four categories are provided for methods of transferring the unit sizes (in the GOST they are called "verification methods", which, in our opinion, is not entirely correct):

(1) direct collation (without comparison means, i.e., not using auxiliary MMM);

(2) collation with a comparator or other comparison means;

(3) the method of direct measurements;

(4) the method of indirect measurements.

Disregarding the first three varieties, which are interpreted more or less distinctly (see, for example [394, p. 179]), let us dwell only on the method of indirect measurements.

In indirect measurements, as is known, the value of a physical quantity (a PQ being sought) *being measured* is determined on the basis of direct measurements of *other* physical quantities, i.e.,

$$\varphi_i^{(\mathrm{meas})'} = a\big(\varphi_1^{(\mathrm{meas})'} \ \varphi_2^{(\mathrm{meas})'}, \ldots, \varphi_j^{(\mathrm{meas})'}, \ldots, \varphi_i^{(\mathrm{meas})'}, C_k\big), \tag{2.11}$$

where $i \neq j$, and C_k is the totality of some constants which appear when determining units in a certain system of units.

Taking into account the equation of direct measurements (2.8a), as well as equation (2.11), it is possible to see that when carrying out such measurements *a priori information* concerns only the units of *other* physical quantities, Therefore, in view of the aforesaid (Section 2.1.2.7), equation (2.11) can be only the equation of the *reproduction* of the unit given.

However, if the value $\varphi_i^{(\mathrm{meas})'}$ of the quantity sought is known from some other more accurate measurements (perhaps also obtained by an indirect method), then the transfer of the unit size becomes possible:

$$\varphi_{i2}^{(\mathrm{meas})} = \varphi_{i1}^{(\mathrm{meas})} \quad \text{and} \quad \{\varphi_{i2}\}\,[\varphi'_{i2}] = \{\varphi_{i1}\}\,[\varphi'_{i1}],$$

hence

$$[\varphi'_{i2}] = \frac{\{\varphi_{i1}\}}{\{\varphi_{i2}\}}[\varphi'_{i1}].$$

But such a comparison is possible when either a more accurate measuring instrument is the measure, or both measuring instruments measure one and the same concrete realization of the physical quantity by an indirect method, i.e., they are transportable.

2.1.4 Interspecific classification of particular RUTS systems

When classifying particular RUTS systems, as the object of analyzing they used national verification schemes which were in force at that time. Moreover, information was obtained from literature and specialists on those physical quantities for which at present there is no legal verification schemes.

In Section 2.1.4 we have an interspecific classification (i.e., the typical one for various types of measurements). In Section 2.1.5 we some aspects of particular classification of RUTS systems will be considered.

The choice of *classification signs* was made on the basis of the analysis of basic concepts and ideas connected with the RUTS system as a whole, which was made in the preceding parts of the chapter. Since the classification given has as its final goal a development of a theory of RUTS system construction (perhaps, of other metrological systems as well), then the essentiality of a sign was determined by its significance in the view of *constructing* such systems.

It is enough, for example, to say that such an important sign of the RUTS systems as the unit size reproduction and transfer accuracy in itself is less essential from this point of view, since it depends first of all on the concrete realization of system elements, though some parameters of the system on the whole are its derivatives.

Two classification signs chosen as the basic ones are

(1) the degree of the unit size reproduction centralization of a given particular RUTS system within the framework of the state system of measurements;
(2) the method of the unit size reproduction by the reference metrological measuring instrument of the system.

Moreover, an additional analysis of the RUTS systems was done according the following criteria:

- the relationship of the element accuracy of adjacent levels of a particular RUTS system (also with respect to working measuring instruments);
- the number of steps of the unit size transfer;
- the "population" of each level of the RUTS system with subordinate and auxiliary metrological measuring instruments.

2.1.4.1 Classification on the centralization degree of unit size reproduction

Every particular RUTS system applies to one physical quantity. That is assumption of the definition given earlier. Both the concept itself and which of the parameters should be considered as one physical quantity are not so trivial issues (see further on in Section 2.1.5). They all function within the framework of one system of measurements, i.e., the state system of measurements. Because of it the classification of all the RUTS systems available for analysis revealed accordingly four categories (type groups) of particular RUTS systems.

The first group includes the RUTS systems with the *complete centralization* of the unit size reproduction (type CC). In each of these systems there is only one reference (initial) metrological measuring instrument for the whole country, i.e., the RUTS system applies to *all* the measuring instruments of a given physical value in the country.

Although the formal sign of such systems can be the presence of only one national verification scheme for a given physical quantity, the actual detection of such systems shows certain difficulties – formost, from the point of view of interpreting the physical quantities (see Section 2.1.5.1), such as similar ones with regard to names and units, but different with respect to their physical sense (plane angle and phase shift between electrical voltages, mass and mass of Ra), as well as particularly different in names but similar in sense and units (magnetic field intensity and magnetization, radiant power and radiant flux, concentration and pH, etc.).

However, if such cases are left out (the list will consist of 17 PQs; see Table 2.1), then a deeper analysis of the remaining 17 state verification schemes shows that in some of them (for force, velocity at vibration motion, inductance, and others) other degrees of unit reproduction centralization are present.

Table 2.1. Number of RUTS systems with a given multiplicity degree.

Multiplicity degree	2	3	4	5	6	7	8	≥ 9
Number of RUTS systems with a given multiplicity degree	8	4	3	2	–	–	1	–

Taking into account that in a given state measurement system (in the country) there are always working measuring instruments not covered by the national verification scheme (in particular, measuring instruments for scientific investigations which use indirect methods of measurements), we can say that there are practically no "pure" RUTS systems of the complete centralization type at all. The RUTS systems for mass, density, time, light intensity, and some other quantities most closely correspond to them.

The second group consists of RUTS systems with a *multiple centralization* of the unit reproduction (type MC). In each of these systems there are some *different* reference metrological measuring instruments of the country, each of them reproducing the unit under various (not overlapping) conditions of measurements ($\Psi_i \neq \Psi_k$). In this case the RUTS system is, as it were, subdivided into a number of subsystems (within the limits of one physical quantity), i.e., into *divisions* of a particular RUTS system.

At present this type of RUTS system corresponds to the availability of a number of national measurement standards of one PQ (and correspondingly, to a number of state verification schemes). This type of system is naturally correct only in the case where every state measurement standard (a reference metrological measuring instrument) reproduces the unit under given conditions at accuracy higher than that of *all* other measuring instruments of a given PQ under the same conditions (within the framework of the state system of measurements); otherwise such a reference metrological measuring instrument has to change its rank and become a "subordinate" metrological measuring instrument.

Here it should be noted that in this respect GOST 8.061–80 and GOST 8.057–80 [194, 195] are contradictory. In Section 1.2.1 of the first document it is stated that every state verification scheme applies *to all* measuring instruments of a given PQ used in the country. At the same time every state verification scheme is headed by one state measurement standard, but according to the second GOST it is possible to have some measurement standards for one PQ.

If a *multiplicity degree* is introduced for the unit reproduction, which corresponds to a number of independent reference metrological measuring instruments of such a type, then from an analysis of the functioning state verification schemes the following data can be obtained.

The RUTS system applied for pressure has the greatest degree of multiplicity. There are a number of functioning state standards and state verification schemes provided for different measurement conditions, for example:

- excessive pressure (up to 250 MPa) according to GOST 8.017–79;
- pressure within the range of $10 \cdot 10^8$ to $40 \cdot 10^8$ Pa according to GOST 8.094–73;
- absolute pressure (from 10^- to 10^3 Pa) according to GOST 8.107–81;
- absolute pressure (from $2,7 \cdot 10^2$ to 4.10^5 Pa) according to GOST 8.223–76;
- variable pressure according to GOST 8.433–81;
- periodic pressure (up to $4 \cdot 104$ Pa) according to GOST 8.501–84;
- pulse excessive pressure according to MI 1710–87;
- sound pressure in an aquatic environment according to MI 2098–90.

The multiplicity degree, equal to 8, belongs to the RUTS system for one of the basic quantity, i.e., temperature. Fivefold RUTS systems exist only for specific heat capacity and length.

The third group includes RUTS systems with a *local centralization* of the unit reproduction (type LC). In each system of this type there are some *similar* reference metrological measuring instruments reproducing the unit under similar conditions ($\Psi_i = \Psi_k$).

The possibility of the existence of RUTS systems of this was noticed by E. F. Dolinsky [141, p. 43].

The availability of RUTS systems of this type is indirectly admitted to by GOST 8.061–80 (points 2.6 and 2.7) [194] by way of introducing a category of borrowed working measurement standards for reproducing the unit by the method of indirect measurements, which is supported by a documentary standard according to which *verification* by the method described above are made legal.

In the first case there is no reference metrological measuring instrument. The measuring instrument complex is not set up with individual copies, and the reference metrological measuring instrument is a kind of potential (virtual) one. In this case it would be better to speak of *a reference method of reproduction* rather than of the reference metrological measuring instrument. Prof. M. F. Malikov meant something like this under the concept used earlier for the reproduction of the unit with the help of the *standard method* [323, p. 331].

In the second case this complex of measuring instruments has been set up within one (official) setup, but it is possible that some complexes of such a type exist (N_0), and quite possible that $2 \leq N_0 \leq N$ (WMI).

No such RUTS systems exist officially, but they actually do exist – in an explicit form: volume capacity (occupancy), constant linear acceleration, sound velocity, and others; and for a number of quantities (angle, force, inductance) as the form of an "admixture" (branch) in the state verification scheme for flow.

The fourth group includes RUTS systems with the *complete decentralization* of the unit reproduction (type DC). There are no initial metrological measuring instruments in such systems. In all cases the measurements of a given PQ are carried out by the indirect method, i.e., the unit of this PQ is reproduced in each concrete measurement.

As examples of PQs belonging to such a type of RUTS systems it is possible to give the following ones: area, inertia moments, force pulse, mechanical energy, heat energy in heat engineering.

As it is seen from an analysis of functioning systems, in practice the RUTS system types listed above are met in a pure form extremely seldom. As a rule, there are systems of a hybrid type with a different degree of coverage of measuring instruments for a given PQ. Moreover, there is a possibility of combining RUTS systems for various quantities. This is rather well seen from Table 2.2, where it is shown that a number of PQs (according to GOST 8.023–03 four PQs: light intensity, brightness, luminous flux, luminance) belong to one state verification scheme.

However, in the presence of one initial metrological measuring instrument, such a combination of RUTS systems can, strictly speaking, only take place in one case: when the physical quantities combined are connected by a purely *mathematical* dependence (square or some other degree, logarithm, etc.). A RUTS system for time and frequency can serve as an example.

The purely *mathematical* dependence of the PQ, by definition, are not be confused with a mathematical model of the PQ being measured (on a concrete object). For example, the fact that the area of a square is equal to the square of its side does not mean that, having precisely measured a side of a square stool, we will obtain a value of the sitting place area of the stool with the same accuracy. To check this it is necessary to verify the model adequacy to the object and introduce the corrections required into their dependence.

2.1.4.2 Classification of the RUTS systems on a unit reproduction method

Among the many possible interpretations of the term "method of unit reproduction" we have chosen the following two:

(1) from the point of view of the behavior *of the reproduction in the course of time* ("maintenance" of the unit size);

(2) from the point of view of the *reproduction-dependence on the process* of the unit size transfer.

The first aspect was discussed in Section 2.1.2.7. Here it is only possible to ascertain that at present only RUTS systems for mass, density, viscosity, plane angle, EMF, electrical capacitance, inductance, and radium mass belong to the continuous method of unit reproduction (in the course of time).

Table 2.2. RUTS systems with the complete centralization of the unit reproduction (selectively).

No	Physical quantity	Unit of PQ	Number of a normative document corresponding to the state verification scheme
1	Time	s	GOST 8.129–99
2	Mass	kg	GOST 8.021–05
3	Force	N	GOST 8.065–85
4	Line speed (at oscillatory movement of a solid body)	m/s	MI 2070–90
5	Angular velocity	rad/s	GOST 8.288–78
6	Angular acceleration	rad/s^2	GOST 8.289–78
7	Luminous intencity	cd	GOST 8.023–03
8	Radiance	W/(sr · m^2)	GOST 8.106–01
9	Brightness	W/m^2	GOST 8.195-89
10	Inductance	H	GOST 8.029–80
11	Differential resonance paramagnetic susceptibility	T	GOST 8.182–76
12	Magnetic flux	Wb	GOST 8.030–91
13	Magnetic moment	A · m^2	GOST 8.231–84
14	Activity concentration	Bq/m^3	GOST 8.090–79
15	Exposure rate	A/kg	GOST 8.034–82
16	Liquid density	kg/m^3	GOST 8.024–02
17	Kinematic viscosity (of liquid)	m^2/s	GOST 8.025–96

Notes:

1. The conclusion that a physical quantity belongs to a RUTS system of the first group (with the complete centralization of the unit size reproduction) has only been made on a formal basis, i.e., on the availability of a single state verification scheme for the physical quantity (with elimination of uniform physical quantities different in names but similar in sense and units).
2. Energetic quantities of the different nature of physical events are given in this table.

In all the remaining cases we are dealing with a *discrete* method of the unit reproduction. From this it follows that the introduction of an additional metrological characteristic of reference metrological measuring instruments (in particular, national measurement standards), i.e., the reproducibility, becomes urgent.

From an analysis of the RUTS system from the point of view of the relation of the unit reproduction to the process of its size transfer, it is possible to establish two situations:

(1) when a reference metrological measuring instrument reproduces the unit independent of the fact whether or not its transfer is realized;

(2) when a reference metrological measuring instrument reproduces the unit *only at the moment* of its transfer.

The difference between these situations is caused by the difference in the composition of the reference metrological measuring instruments. If the reference metrological measuring instrument contains in its structure a "source" of the PQ itself, then the reproduction of the unit can naturally be realized without any transfer. If the reference metrological measuring instrument does not contain any "source" which is at the same times located in a metrological measuring instrument by directly subordinated to the reference metrological measuring instrument, then it is possible to reproduce the unit only at the moment of transfer.

The second situation takes place, for example, in operating RUTS systems for temperature coefficient of linear expansion (GOST 8.018–07, and others), specific heat capacity (GOST 8.141–75, and others), heat conductivity, hardness by various scales, relative distribution of power density in a cross section of a beam of laser continuous radiation, and others.

To all appearance, it is unlikely that it is useful to call such reference metrological measuring instruments by national measurement standards. More likely they are *verification* intended, as a rule, for certifying standard samples (SS).

2.1.4.3 Other aspects of the interspecific classification

The remaining classification criteria according to which the analysis of RUTS systems was carried out relate mainly to the system of the unit size transfer. The classification of RUTS systems on the basis of the relationship between the measurement ranges of the reference MMI and the lowest subordinate (or of the working measuring instruments finally connected to them) can be of the greatest interest. Since at present such an analysis is possible only for the state verification schemes with a state measurement standard at the head, therefore only state measurement standards will be considered further on in the form of a reference metrological measuring instrument.

For this purpose three categories of RUTS systems have been selected:

(1) $\mathrm{D_{SMS}} \geq \mathrm{D_{WMI}}$;

(2) $\mathrm{D_{SMS}} = X_{\mathrm{nom}} \ll \mathrm{D_{WMI}}$;

(3) $\mathrm{D_{SMS}} < \mathrm{D_{WMI}}$.

Here D marks the range of MI.

As a matter of fact, in such an analysis the consideration was given not to the RUTS systems on the whole, but to their subdivisions (taking into account the degree of multiplicity of the systems). At the same time, it is possible to attribute 33 cases to the first category, 18 cases to the second one, and 53 to the third one.

The first case means the unit reproduction and transfer of its size within the whole range of measurements, i.e., the reproduction and transfer of the PQ *scale*. This is rather reasonable for those quantities which are determined up to ratios of the order

Table 2.3. Comparison of the number of physical quantities in various divisions of physics according to various literature sources.

Division (section, part) of physics	A	B	C	Results of summation		
				I	**II**	**III**
1	*2*	*3*	*4*	*5*	*6*	*7*
Space and time (I)	10	10	15	15	13	13
Periodical phenomena (II)	2	7	4	9	9	8
Mechanics (III)	23	24	41	45	44	42
Heat (IV)	18	22	15	24	21	19
Electricity and magnetism (V)	38	45	38	54	52	38
Light and optics (VI)	16	38	15	42	42	38
Acoustics (VII)	9	17	18	21	19	4
Physical-chemical and molecular phenomena (VIII)	1	37	6	39	40	26
Ionizing radiations (IX)	14	10	7	14	13	11
Atomic and nuclear physics (X)	0	54	17	60	51	36
Total	131	264	176	323	304	**235**

and equivalence of the sizes or intervals between them [100]: hardness, temperature, time, and others.

The second case (which can be considered as a limit one for the third case, when $D_{SMS} \to 0$) and the third case can be realized only for additive quantities, for which the ratios of arithmetical operations of summation, etc., are appropriate.

It is obvious that in these cases a question of the accuracy of *scale transformation* concerning the transfer of the unit size from the reference MMI to WMI arises. This question needs to be the object of an independent analysis from the position of a generalizing theoretical consideration.

On the basis of results the analysis carried out at the time of classification and taking into account Sections 2.1.4 and 2.1.5, one can make the following conclusions on the interspecific classification.

1) A total number of particular RUTS systems (i.e., the measured physical quantities) operating in the state system of measurements is not known accurately.

If the nonoverlapping list of physical quantities is of interest, then according to the estimate made in Section 2.1.5 (see column 7 in Table 2.3) their number is 235. If it is assumed that the particular RUTS systems can have divisions (under the conditions Ψ of measurements (see also Section 2.1.6), then the number of measured physical quantities and RUTS systems (taking into account their divisions) will be equal to 304 (column 6 of Table 2.3).

2) According the data obtained, the number of RUTS systems with the complete centralization of the unit reproduction (type MC) is equal to 27. The number of RUTS

systems with the multiple centralization of the reproduction (type MC) is 31, and the total number of divisions these systems have is 95.

Some RUTS systems are combined ones. This is why the sum of the RUTS systems of the CC type and the sum of the divisions of systems of the MC type do not coincide with the number of operating state verification schemes.

3) The number of RUTS systems of the LC and MC types can only be approximately evaluated. According to the examples given above, the number of systems of the LC type is equal to ~ 10. Then the remaining systems (minus the CC, MC, and LC types) will be the systems with the complete centralization of unit reproduction. Their total number is $235 - (27 + 31 + 10) \cong 167$.

4) Although all operating state verification schemes are considered to be the systems with the complete centralization of the unit reproduction, according to Section 2.1.2 they are first of all subdivided into systems of the CC and MC types.

Moreover, a deeper analysis shows that some of the state verification schemes do not correspond even to a division of the system of the MC type, i.e., they cannot independently reproduce the unit, since they actually "receive" it from some other state verification scheme and, thus, have to be included into the last one. The state verification schemes for high pressures, for temperature related to infrared and ultraviolet radiations, for alternating electric voltage and others can be used as examples, i.e., subordinate metrological measuring instruments from some other state verification scheme of the same physical quantity belong to corresponding state measurement standards.

Furthermore, as noted earlier, in many of them there are "admixtures" of systems of the LC and DC types.

5) The nomenclature of reference MMI varieties which we dispose of (Section 2.1.3.1) is evidently insufficient for the systems and their reference MMIs to be distinguished either by a degree of the unit reproduction centralization or by a method of reproduction. This question requires an independent terminology analysis.

2.1.5 Some problems of the specific classification of particular RUTS systems

In [456] the classification of measurements with respect to physical quantities and various conditions of their realization is considered. This classification, particularly with regard to dividing *types of measurements* into subtypes etc., is based to a considerable degree on available information contained in the state verification scheme. Therefore, in a certain sense it reflects the structure of the RUTS system and its separate (particular) subsystems.

For example, the availability of a number of state verification schemes for measuring instruments of a given physical quantity under different specific conditions connected with singularities of realizing this PQ, range of its values, behavior in the course of time, etc., finds its reflection both in the available subtypes (varieties) of measurements and in the availability of n-multiple partly centralized RUTS systems (see Section 2.1.4.1).

The classifier described in [456] was used in developing proposals on improvement the specialization of metrological bodies of the Gosstandart of the USSR. The main idea of these suggestions consisted of imparting the hierarchical character to the structure of metrological bodies, as well as to the corresponding structure of the system for assurance of the uniformity of measurements and full RUTS system consisting of subsystems and levels. In connection with this, the developed proposals to a certain extent reflect the classifier of particular RUTS systems according to the fields and types (varieties) of measurements.

Here only some principle aspects of the specific classification of the RUTS systems will be considered which were not considered in [456].

2.1.5.1 About formal and actual number of physical quantities

An analysis was performed with regard to the lists of physical quantities which can be found in the literature on metrology:

(A) G. D. Burdun, B. N. Osnovi, metrologii. M.: Izd-vo standartov, 1985. 256 p.;

(B) Metodicheskiye ukazaniya RD 50-169-79 Vnedreniye i primeneniye ST SEV 1052-78 “Metrologiya. Edinitci fizicheskih velichin”. M.: Izd-vo standartov, 1979;

(C) L. A. Sena, Edinitci fizicheskih velichin i ih razmernosti. M.: Nauka, 1977. 336 p. (all of them are in Russian).

First of all let us consider the quantitative characteristics of the lists. If we locate all physical quantities defined in this literature in an order according to divisions of physics as in (B), then the picture shown in Table 2.3 (p. 84) is obtained.

Putting aside the question concerning the completeness of the PQ system on each source (undoubtedly the real list of physical quantities used in physics is much wider), we have compared the data trying to make up a *uniform list* of physical quantities.

At the *first stage* (designated as I; see column 5 in Table 2.3) within the limits of each division (section, part) of physics the quantities which are present at least in one work, were summed up, i.e., a formal operation of combining the sets

$$\mathrm{A}_i = \mathrm{L}_{1i} \cup \mathrm{L}_{2i} \cup \mathrm{L}_{3i},$$

where i is the number of a division of physics, and $\mathrm{L}_{1,2,3}$ are the lists of physical quantities, containing at separate works.

Thus, the list A constitutes mutually supplemental lists of different works, only the *name of a physical quantity* being a distinction sign.

At the *second stage* (designated as II; see column 6 in Table 2.3) the lists A_i were composed from various divisions of physics, and the elements which were repeated in various divisions were removed. And again the formal sign was the name of a physical value, or, to put it more precisely, its significant (fundamental) part.

Thus, for example, from division I there were excluded various "moments of plane figures", since they are met (in a substantial form) in division III; from division II the "wave number" was removed, since it was in division VI; from all divisions (except division III) the "energy" was removed too without looking at various adjectives such as: acoustic, electromagnetic, and so on; from division IV the "heat of a chemical reaction" and "molar heat capacity" which are applied in division VIII were removed, and so on.

Simultaneously a derivative of the "chemical reaction heat", namely the "specific heat of a chemical reaction" was transferred into division VIII.

However, if one tries to gain a real, rather than formal understanding, then it appears that to make up a *uniform* and *consistent list of physical quantities* is not very simple.

The fact is that in each division of physics together with quantities specific only to this division, other quantities, the nature of which is close (uniform) to quantities of other divisions of physics, were introduced. For example: *energy flux – radiant flux – radiation power.*

The situation of such a type is especially vivid in division VII ("Acoustics"), where together with the adjective "acoustical" (sound) other quantities such as *period* and *frequency* (of oscillations) (division II), *velocity, pressure, energy, power,* and some others appear. If an attempt is made to combine all the cognate quantities into one division, then in the division "Acoustics", for example, almost nothing will remain (see the result "C" in column 7 of Table 2.3): for the most part specific quantities connected with the acoustics of manufacturing facilities (acoustic penetrability of a partition wall, etc.).

To all appearances, if one "digs" still deeper, then the whole division "Acoustics" has to be transferred into divisions I–III (mainly divisions such as mechanics of gaseous and liquid media). In any case, such an analysis results in a significant decrease of the number of quantities (the result "C" in column 7 of Table 2.3).

2.1.5.2 Uniformity of PQs and RUTS systems

The analysis carried out essentially reveals one more circumstance: there are *uniform* quantities which not only have *different names* but also *different constitutive equations* even within one division of physics; for example:

- pressure, normal stress, modulus of longitudinal elasticity, shear modulus, etc.;
- power (of electrical circuits) – active and reactive;
- electrical voltage, potential, electro-motive force (EMF), and others.

Physical quantities frequently have different semantic content (such as, i.e., in optics, "surface density of a radiant flux", "radiancy", "energy illumination", which, although they formally have one and the same constitutive equation, their sense differs. The first quantity refers to the radiation itself, the second one to its source and the third one to an object of irradiation).

In part this issue has been solved. However there are many quantities for which such a task has not been solved. This refers, first of all, to such general physics quantities as energy, power, length, and time. Suffice it to say that length as a general physics quantity obtains the following specific names (and partly the content) in various divisions of physics: *length of a average particle track*, *thickness of a layer of half attenuation*, *length of neutron moderation*, *elementary particle radius*, *length of an oscillation wave* (sound, electromagnetic, and others), *diffusion length*, *focal distance* (for a lens), etc.

It is obvious that from the point of view of measurement uniformity assurance (and RUTS systems) it is important that uniform quantities have *the same units* and that they therefore have to belong to *one* RUTS system. From the point of view of the formal construction of the RUTS systems, such a fullness of concrete realizations of physical quantities has to reflect on the set $\{\Psi_i\}$ of measurement conditions.

At the same time, it is at present unlikely to combine such values as energy and power within the framework of one RUTS system, since the natures of the phenomena to which they apply are too diverse. This circumstance was taken into account in analyzing and classifying the RUTS systems in Section 2.1.4.

From the aforesaid it is possible to make the following conclusions:

(1) the system of physical quantities (SPQ) also has a certain hierarchical structure (there are more general and less general quantities);

(2) at present the system of physical quantities breaks up into a number of *subsystems* (relating to physical phenomena different in nature and level) which are not coordinated with each other from the point of view of PQ terminology;

(3) it is necessary to *deepen the concept* "physical quantity" in metrology and to develop an optimal (from the point of view of a contemporary level of the physics development) *system* of physical quantities.

2.1.5.3 The problem of "physical constants" in the RUTS systems

Physical constants are also physical quantities and also have to be measured. Naturally, the question arises: What is their relation to the RUTS systems? Or, to be more precise: What is the type of the RUTS system they can belong to?

An analysis of operating national verification schemes shows that there are already about 15 RUTS systems and their subdivisions for the quantities of such a kind which are, as a matter of fact, the physical constants (wavelengths for optical radiation, density and viscosity of liquids, coefficient of linear heat expansion, specific heat and heat conductivity of solid bodies, heat of various phase transfer of substances, rotation angle of the polarization plane of optical radiation in a substance, dielectric permittivity, etc.).

However, these systems, constructed in the form of systems of the CC type (with a national measurement standard at the head), were created under the influence and by analogy with the RUTS systems which had been already created for "nonconstant" physical quantities without any proper reasoning.

First of all, let us note that the hierarchical property of physical quantities, which was discussed above, is especially obvious when one deals with physical constants. There are constants of a *fundamental* character (the so-called *universal constants* or *fundamental physical constants* or *fundamental physical constants* (FPC)) and *less universal* (local) constants, i.e., *constants of concrete substances* (objects).

Among other things, it is impossible to draw a distinction between them. So, in many investigations and calculations the acceleration of gravity for the earth is assumed to be a universal constant, while in other cases (requiring a greater accuracy) it is assumed to be a constant of especially local significance (at a given point on the earth and under a constant environment).

Another important moment is the *permanency of constants* (even the universal ones) which in its absolute sense is only *hypothetical*. Experiment (absolute measurements of their values) gives the actual information about how much they are permanent (in time, space as well as under the influence of some fields). At the same time at present a number of constants (more or less universal) are used for reproducing units of *other* physical quantities, their values being considered as *fixed*.

This is caused, first of all, by the fact that within the space and time limits of the measurement system, where these constants are used for reproducing the units, their permanency has been really determined with high accuracy.

However, the (a priori) fixation of a dimensional constant value means the introduction of a new unit into the system of units [432, p. 31]. Therefore, if the constant C_k is used for reproducing a unit of some other physical quantity φ_i according to the equation of indirect measurements (2.11), then this PQ thereby becomes one of the basic ones. Thus, for example, the value of the constant $\frac{h}{2e}$ and that of the constant μ_0 are fixed when reproducing the EMF (Volt) and electrical current units, respectively.

The physical constants, however, are frequently used when reproducing physical quantity of the same nature, i.e., the nature uniform with respect to the constant itself. In exactly the same way the physical constants are used in national measurement standards of frequency and time units (where the frequency of transition between certain levels of a ^{137}Cs atom is fixed; generally speaking this frequency is an especially "local" constant having a high stability), length (where the wavelength of a certain monochromatic electromagnetic radiation is fixed), temperature (where the temperature of one of the phase transitions of a certain substance is fixed), density and viscosity (where the values of corresponding quantities for water under certain conditions are fixed), and others.

It is of the essence that the fixation in these cases can differ. If the size of a constant is fixed *a priori* and *formally* (only through the definition of the unit), then in this case the corresponding unit becomes one of the basic units (the cases with frequency, length, temperature).

It is clear that the physical constants of such a type (leading to the basic units) can have only RUTS systems of the CC type and coinciding with the RUTS system corresponding to the *basic* physical quantity (uniform in respect to the constant).

For example, the RUTS systems for length (meter) and wavelength in spectroscopy have to coincide. If the constant *value* is fixed (i.e., a priori *measurement* information about its unit is used), then in this case according to the above the constant given can, strictly speaking, not be used for reproducing the unit of the given physical quantity. This will not be the *unit reproduction*.

However the reproduction of the density and viscosity units with the help of corresponding national measurement standards does not contradict the above, since these national standards function within the frames of one system of measurements, and the "standardized" values of water density and viscosity are obtained from measurements carried out within the framework of the other system of measurements, i.e., the international one (although in these cases the choice of the RUTS system of only the CC type is not well-founded).

This is the very essence of the work connected with the State Service of Standard Reference Data (SSSRD) which is an additional subsystem with respect to the RUTS system within the framework of the system of enshuring uniformity measurement in the country (Figure 2.2).

At the same time if the C_k value is fixed within the frames of the national system of measurements, then the improvement of the C_k value, which can be realized by indirect measurements of other physical quantities, can only take place within the framework of the international RUTS system. This is what the *reversibility* of a measurement standard setup consists of; either the reproduction of a unit given (with a system of the CC or LC) on the basis of an international value of the constant, or the improvement of the constant on the basis of its international reproduction (with a system of the LC type).

From this it follows that for constants and other PQs, *any type of RUTS system* (CC, MC, LC, DC) is possible. The choice will be determined only under considerations of necessity and possibility (except the case leading to the basic PQ units).

However, if for physical constants leading to the basic units such systems can be only the systems of the CC or MC type, then for the remaining cases (density, viscosity, TCLE, etc.) such a type of the system is most improbable. This is seen from the fact that the physical constants are stable in time, i.e., measurements are not numerous and they are carried out infrequently (the number of working measuring instruments is small). Further, practically no new instruments are created for its direct measurements (with the exception of density and viscosity).

Systems of the MC type can arise if the physical constant has to be measured under different conditions (for example, the TCLE, specific heat capacity, thermal conductivity of solid bodies at different temperatures).

Finally, the most important thing is that the ability of the "standardized" international value of the physical constant allows this value to be used for reproducing the unit in any setup suitable for measuring the PQ by an indirect method (i.e., the realization of a RUTS system of the LC or even DC type is economically simple and easy to do).

2.1.5.4 The RUTS systems for dimensionless PQs (factors)

Until now, on the analogy of the usual RUTS systems (in the form of a national verification scheme with a state measurement standard at the head) a significant number of RUTS systems have been built for dimensionless physical quantities – coefficients of various types (Table 2.4).

Table 2.4. Dimensionless (relative) quantities in the system of operating state verification schemes (selectively).

N°	Quantity name	No. GOST & SVS
1	Spectral coefficient transmission and reflection (for visible radiation)	8.557–07
2	Relative dielectrical permittivity	8.412–81
3	Magnetic susceptibility	8.231–84
4	Volumetric specific humidity in oil and petroleum products	8.190–76

At the same time any dimensionless quantity (coefficient) is the ratio of uniform dimension quantities. Some dimensionless quantities are determined with the help of a combination of some quantities, which has as a result the zero dimension (for example, the fine structure constant in electrodynamics). Their measurements can be realized in the form of an indirect measurement; however, this case is not considered here, taking into account that in Table 2.4 the real measured coefficients are just the ratios of uniform quantities.

Therefore, using equation (2.8) and considering that RUTS systems for corresponding dimension quantities have been realized (i.e., the units have been coordinated for different measuring instruments and within the limits of a necessary range of measurements), we have

$$X_k^{(\text{meas})} = \frac{\varphi_2^{(\text{meas})}}{\varphi_1^{(\text{meas})}} = \frac{\{\varphi_2\}}{\{\varphi_1\}} \cdot \frac{[\varphi_2']}{[\varphi_1']} = \frac{\{\varphi_2\}}{\{\varphi_1\}} = \{X_k\} \cdot [1],$$

where X_k is the dimensionless quantity, φ_2 and φ_1 are two concrete realizations of one physical quantity (different dimensions), and [1] is the dimensionless unit.

From this it is seen that the problem of unit reproduction and its transfer as an independent problem for dimensionless quantities (coefficients) does not arise. It is reduced to the problem of a RUTS for the initial physical quantity φ irrespective of the fact whether or not there is a need to establish a unit for φ (i.e., to use the unit from the accepted system of units) or to take it as an arbitrary one (as in the case of relative measurements), but identical within the whole required range of φ values (dimensions).

In connection with this, the existence of the RUTS systems for dimensionless quantities (all the more of the CC type) appears insufficiently grounded. Most likely instead of them there could be setups of the highest accuracy for verifying measuring instruments for relative measurements.

2.1.6 The technical-economic efficiency of various RUTS systems

Problems of the RUTS systems efficiency from a purely economical point of view have to be undoubtedly solved in the form of an indivisible (unified, common) complex of problems concerning the economical efficiency of not only the metrological systems itself but also of the system of measurements as a whole. The hierarchical structure of proper metrological systems (Figures 2.1 and 2.2) and the fact that these systems arise in the form of a "superstructure" over the current functioning system of measurements in the country indicate this.

This is indeed the reason why the influence of metrology (and of its metrological systems) on the efficiency of all sectors of the national economy is only indirectly apparent, i.e., through the increase in the general level of the measurement quality in the country.

In connection with this, let us consider merely metrological aspects of the technical and economical *efficiency* of the RUTS systems which will be conditionally determined as

$$\varepsilon(C_r) = \frac{E(C_r)}{E(C_r)}, \tag{2.12}$$

where $E(C_r)$ is the reduced expenses connected with the creation and operation of the C_r system, and $\varepsilon(C_r)$ is the overall index of a metrological effect of applying this system.

The metrological effect can be determined reasoning from the degree a final aim of the RUTS system has been achieved, which according to Section 2.1.2.6 is reduced to the implementation of the condition of measurements comparability (2.5) if it is considered that within the state system of measurements a unified system of PQ units is operating.

The *degree of compliance* with condition (2.5) can be determined, certainly, as the ratio of the sets of all $j \neq i;\ j \in \{j\}$, for which condition (2.5) is fulfilled by the given (C_r) RUTS system, to the set of all $j \neq i;\ j \in \{j\}$, which are possible within the given national system of measurements. Then

$$\varepsilon(C_r) = \frac{\{j\}_r}{C(C_r) \cdot \{j\}} = \frac{1}{C(C_r)} \cdot \frac{\{j\}_r}{\{j\}}. \tag{2.12a}$$

In order to *compare* the possibilities of various RUTS system types it is sufficient to analyze the relative efficiency of the RUTS systems (for example, C_1 and C_2):

$$\varepsilon_{\text{rel}}\left(\frac{C_1}{C_2}\right) \equiv \frac{\varepsilon(C_1)}{\varepsilon(C_2)} = \frac{C(C_2)}{C(C_1)} \cdot \frac{\{j\}_1}{\{j\}_2}, \tag{2.13}$$

since the set $\{j\}$ is common with respect to them.

First of all let us consider the RUTS systems of the CC type, for which $\{j\}_r = \{j\}$ evidently takes place, and then discuss their relative efficiency from the point of view of the difference between RUTS system structures. For such systems the efficiency will be determined only by expenditures connected with the creation and operation of the system, which consist of both the costs for creating and operating a national system of measurements and the costs for developing and realizing a transfer method (TM) for the unit size:

$$\varepsilon^{-1}(C_{\text{cc}}) = \sum_i C_i(\text{MMI}) + \sum_i C_i(\text{TM}).$$

Consequently, the task of finding the most efficient structure of a CC type RUTS systems is reduced to minimizing the joint costs connected with creation and operation of such a system. From the metrological point of view it is evident that in a general case any complication of the RUTS system structure (increase of a number of transfer steps and, in the long run, the number of MMI) leads to an increase in the general cost (a decrease in efficiency), since the increase in the number of MMIs is followed by an increase in the number of transfer methods (the number of elements connections) as well as an increase in the number of the common staff of specialists in metrology. On the other hand, the decrease in the number of MMI in the RUTS system structure is connected with a rise in price of each metrological measuring instrument due to the need to provide them with a required productivity.

Actually, the productivity of a given metrological measuring instrument (i.e., a maximum number of subordinate measuring instruments verified or calibrated with its help) is determined by the expression

$$P_0 = \frac{\overline{T}_1}{\bar{t}_1},$$

where $\overline{T}_j$ is the average verification (calibration) interval for measuring instruments $\{S_i\}_0$, subordinated to a given metrological measuring instrument (S_0), and $\bar{t}_1$ is the average time duration of the subordinate MI verification or calibration procedures.

Then the "transfer capability" of the RUTS systems at n steps of verification (n-levels of MMI) is

$$P(C_{cc}) = \frac{\overline{T}_1}{\bar{t}_1} \cdot \frac{\overline{T}_2 - \overline{t}'_1}{\bar{t}_2} \cdots \frac{\overline{T}_n - \overline{t}'_{n-1}}{\bar{t}_n} \cong \prod_{j=1}^{n} \frac{\overline{T}_j}{\bar{t}_j}, \tag{2.14}$$

where $\overline{t'_j}$ is the average time of the MMI disuse (S_j), that includes the time of its verification and transportation (for the verification); usually we have $\overline{t'_j} \ll \overline{T}_j$, therefore here a transition has been made to an approximate equality.

It should be noted that, as a rule, $T_0 > T_1 > T_2 > \cdots > T_n$ and $t_1 > t_2 > \cdots > t_n$. Moreover, usually the calibration (verification) interval of a given metrological measuring instrument is less, the greater its productivity is, i.e., in the first approximation:

$$T_j \approx \frac{1}{P} = \frac{\bar{t}_{j+1}}{\overline{T}_{j+1}} \quad \text{or} \quad \frac{\overline{T}_j \cdot \overline{T} : j+1}{\bar{t}_{j+1}} \cong \text{const}\,.$$

Therefore,

$$P(C_{cc}) \cong \prod_{j=1}^{n} \frac{\overline{T}_j}{\bar{t}_j} = \begin{cases} \dfrac{(\text{const})^{\frac{n}{2}}}{t_i} & \text{at } n \text{ even,} \\[2ex] \dfrac{(\text{const})^{\frac{n-1}{2}}}{t^2} \cdot \dfrac{\overline{T}_n}{\bar{t}_n} & \text{at } n \text{ odd.} \end{cases} \tag{2.15}$$

The maximum productivity value is achieved at $T_0 = \text{const}$ (an intercertification period of an RMMI):

$$P_{\max}(C_{cc}) = \frac{(T_0)^{n/2}}{\bar{t}_1}.$$

Comparing the C_1 system with n-levels and the C_2 with one level of the MMI (an ideal, extreme case of the RUTS), one obtains that the equal productivity $P(C_1) = P(C_2)$ (i.e., the same degree of fulfilling condition (2.5) on condition that the verification time of a WMI is the same in both cases) can be achieved only when

$$(T_{01})^{n/2} = T_{02}.$$

This means that the RMMI of a one-level system should have a greater metrological reliability (to $n/2$ power) than that in a system with n levels. It is known that the increase in the system reliability by a factor of k leads to an increase in its price on average by a factor of k^2 (i.e., in the case considered by a factor of 4). Moreover, there is a limit, i.e., a minimal time of verification of a subordinate MMI. To get more accurate estimates, special investigation and corresponding initial data are required.

It is possible to show that at a sufficient detailed elaboration of the measurement comparability condition (2.5), the RUTS system of the MC type can be reduced to systems of the CC type (from the point of view of the metrological effect, i.e., the degree of fulfilment of this condition), since, at a k-multiple system,

$$\{j\} = \sum_{r=1}^{k} \{j\}_r.$$

As regards the RUTS systems of the LC and DC types, their metrological effect, as a rule, is significantly less than that of the systems of the CC or MC type, since the powers of the sets greatly differ: $|\{j\}_r| \ll |\{j\}| = |\Psi|$.

The fulfillment of the RUTS systems functions in full volume (i.e., the condition of comparability of measurements) is possible, strictly speaking, only when "constant" physical quantities are available which allow $\varphi_{2u} = \varphi_{1u}$ to be realized in equation (2.5) and, consequently, provide the possibility of an indirect comparison of unit sizes in the presence of many RMMI. However the costs for creation and operation of RUTS systems of the LC and DC types are much less that those for systems of the CC and MC types. In connection with this, their technical and economical efficiency can be compared, and the creation of systems of the LC and DC is justified.

The use of other classification signs for comparing the technical and economical efficiency of the RUTS systems unlikely makes sense since this efficiency weakly depends on them.

In conclusion let us formulate the necessary and sufficient conditions (signs of grounding) which are common for creating centralized RUTS systems of one or another type (i.e., and for creating RMMI).

Necessary conditions (i.e., in the aggregate the obligatory ones) are:

- the availability of WMI in the SSM which are intended for direct measurements of a given PQ;
- the technical possibility of realizing comparisons of various types of MI for a given PQ (i.e., the availability of corresponding methods of the unit size transfer) within the framework of a chosen degree of centralization;
- the availability of a unified system of units, accepted in the SSM;
- the availability of a necessary resource of accuracy between an RMMI and WMI.

The *sufficient* conditions (fulfillment of at least one of which in meeting all necessary conditions serves as the basis for creating a system) are:

- the unit of a given PQ belonging to the basic units of an accepted system of units (the adequacy condition for CC type systems);
- dependence of the accuracy of reproducing a derivative unit mainly on properties of a given RMMI (as a measuring setup) further than on the accuracy of direct measurements of other PQs determining the given one;
- availability of a primary RMMI, which at the same time cannot provide the accuracy required for transferring the unit size of a given PQ for the considered measurement conditions (the adequacy condition for the MC type systems);

- localization (for example, within the limits of one industry branch or economical region of the country) of a significant park of WMI of a given PQ in SSM (the adequacy condition for the LC type systems);
- maximum economical efficiency of a given type of RUTS systems for a given PQ.

2.2 Physical-metrological fundamentals of constructing the RUTS systems

2.2.1 General ideas

The system for ensuring the uniformity of measurements is the most important one of the measurement assurance systems in a country. Its aim is to "serve" providing the prominent economical complex, particularly important under conditions of scientific and technical progress, with a required quality of measurements, i.e., the "national system of measurements", which "runs through" all sections of the economy and, above all, the sections which determine the rate of its development.

In order to realize a required reconstruction of the system for ensuring the uniformity of measurements in a well-considered and purposeful manner with the purpose of increasing its efficiency and quality, naturally a corresponding reliable instrument is required, and above all a developed theoretical basis.

It should be noted that at present the theory of constructing a system for ensuring the uniformity of measurements, including a system for reproducing physical quantity units and transferring their sizes, is in fact lacking, a fact that has been noted many times, in particular in [37].

There are only separate theoretical works devoted to the solution of some problems of measurement uniformity assurance (see details in Section 2.2.2.1). Suffice it to say that one of the most important problems of the present is *optimization* in accordance with the criteria of intensification, economy, increase in the quality of production and services. This problem also has to touch on the system of ensuring the measurements uniformity. It has not been applied to normative documents (ND) which determine the construction of such a system.

Initial concepts of a long-term development of the system of ensuring measurement uniformity in a country were formulated in the Forecast (see the reference in [37]). Later on, in developing a long-term program for the development of fundamental investigations on metrology, works were planned for creating a fundamental theoretical basis for constructing various metrological systems, including systems of physical quantity unit reproduction and the transfer of their sizes, which is the main scientific and engineering base of the SEMU.

In Section 2.1 the authors give the analysis of the concept "RUTS system", reveal the relationship between this system and other metrological systems, classify the RUTS systems, and state the fundamentals of a formalized description of various metrological systems.

The purpose of the investigation in Section 2.2 is the development of theoretical fundamentals for constructing a typical RUTS system in a separate kind of measurement, as well as of a complete RUTS system, i.e., the creation of the fundamentals of a theory which will be able, where possible, to take into account all the problems connected with studying, constructing, optimizing, and using such systems.

2.2.2 Analysis of the state of the issue and the choice of the direction for basic research

2.2.2.1 Standpoint for the analysis

The subject of this investigation is the systems for reproducing the units of physical quantities and transferring their sizes. The theory of systems, due to the complexity and uncertainty of this subject, requires a more precise definition to a circle of questions, according to which a brief review of works published and their analysis will be made.

We will proceed from the concept, formulated in Section 2.1, that the RUTS system is a *real* complicated system of the means and methods for reproducing the physical quantity units and transferring their sizes, which is the main component of the system for ensuring measurement uniformity.

Unfortunately, frequently in the literature a real RUTS system is identified with a so-called "verification scheme" which is only its graphical (schematic) reflection, and the "theory of verification schemes", which to our opinion is unsuccessful, is discussed.

As early as in the first half of the 20th century, Prof. L. V. Zalutskiy, who first introduced the concept "verification scheme" into metrology, interpreted this concept as a *document* stating an order of transferring unit sizes "from one measure to another". Such an interpretation of the verification scheme has been fixed with the official documents GOST 16263–70 and GOST 8.061–80 [194, 195].

Taking into account the above, it is important to give an overview of theoretical works which consider the issues of constructing real RUTS systems is called for. At the same time, since the practical construction of the real RUTS systems is based on basic normative documents (GOSTs and others) it is useful to give a brief analysis and estimation of these documents.

One more category of works to be considered concerns the theory of constructing systems of physical quantities and their units, since, firstly, the system of units of quantities measured sides most tightly with the RUTS system (see Figure 2.2) and secondly, because the choice of a system of units (particularly the basic ones) determines the structure and properties of RUTS systems. The analysis of and references for this problem are given in Section 2.2.5.1.

Finally, as long as the case in point is the construction of the theory fundamentals, then it is useful to examine the ideas of what the theory is in general and what attributes it has, which are described in the available literature.

2.2.2.2 Brief review of papers on the theory of the construction of RUTS systems

The first attempt of a theoretical consideration of the RUTS system construction problem was made by Prof. M. F. Malikov in his classical monograph [323] in the 1940s. In this monograph Prof. Malikov tried to link a number of working standard ranks between primary measurement standards and working measuring instruments with a ratio of the accuracies of a verified and verifying measuring instruments, reasoning from the condition that an error of a verifying measuring instrument does not influence the accuracy of a verified measuring instrument ("lack of influence" condition), i.e., on the basis of a so-called "criterion of a negligible error".

In the same work an issue concerning the error accumulation in transferring the unit size from a measurement standard is examined. Among other requirements for constructing RUTS systems, only those are indicated according to which these systems havc to be "on the level of contemporary achievements of measurement technique" and "useful", i.e., they should not complicate the verification procedure when there is no need to do this; moreover the elements and measures which these system contain have to provide a required accuracy in verifying working measuring instruments.

Further on, it is also noted that the construction of verification schemes is realized top-down, i.e., from a primary measurement standard. An analysis of the ratio of working measurement standard and verified measuring instrument errors are examined in the first foreign works [153, 551] on the influence of verification schemes.

An interesting approach in the examination of RUTS systems (the authors call them "metrological circuits") is described by V. N. Sretenskiy and others. This approach is based on the investigation of the influence of metrological circuits on external systems (measurement technique, science, industry, nonmanufacturing business) and links generalized parameters of these external circuits to the parameters of metrological circuits through general equations of external system operation efficiency. Unfortunately, did not receive proper attention, and the suggested approach was not developed.

Another approach to grounding and choosing a reserve of accuracy between the levels of a verification scheme was first suggested by N. A. Borodachev in his monograph [63].

In this monograph the author proceeds from considering the verification defect, i.e., he considers verification to be one of various kinds of technical controls for product parameters (compliance of product parameters with their requirements).

Between 1958 and 1974 a great number of theoretical works were devoted to the examination of this approach. Among them are E. F. Dolinskiy [140, 141], A. N. Kartasheva [254], and K. A. Reznik [400–402].

These works deal with only one form of verification, i.e., the control of metrological success of a measuring instrument (fit–unfit) and, strictly speaking, are not related to the problem of transferring the unit size. However these works stimulated the solution of the problem of constructing verification schemes. In practice this trend of investigations corresponded to the solution of the task of selecting reference measuring instruments for the verification of working measuring instruments. The first review on works of this type was collected in 1975 by A. V. Kramov and A. L. Semeniuk [279].

Subsequently this trend evolved into the problem of the "quality of verification": N. N. Vostroknutov, M. A. Zemel'man, and V. M. Kashlakov [255, 538].

The first attempts to introduce the "collective" parameters of verification schemes into consideration, with the purpose of optimizing these systems, were made by N. A. Rubichev and V. D. Frumkin [414], and E. L. Crow [132].

The task of optimizing the structure of a verification system was solved reasoning from the criterion of its minimum cost by the method of undefined Lagrange multipliers. In the first of these, the following initial parameters of calculation were chosen: the number of working measuring instruments to be verified, verification time and the working measurement standard of the lowest category, the verification (calibration) interval, the error of a working standard of the lowest category, and the costs of realization and maintenance of a verification system.

The results in question for solving the problem were: the total number of categories, the number of working standards in each category and error ratio for working standards between the categories. However, calculations have shown that the function of verification system costs turns out to be weakly sensitive to the ratios of measuring instrument errors between the categories of the verification system as well as to the law of accumulating errors (an arithmetic or quadratic error), i.e., the minimum of the cost function appears to be rather "blunt".

Moreover, it was noticed that there was no sufficiently reliable input data, especially of an economical character. Similar tasks were solved in other works: L. G. Rosengren [409]; V. E. Wiener and F. M. Cretu [544].

In the paper written by S. A. Kravchenko [282] a calculation of a concrete verification system in the field of phase measurements is realized. The author, using information about the costs of developing and manufacturing the phase measuring setups of different accuracies, makes an attempt to optimize both the number of verification scheme levels and the ratio of working measurement standard accuracies between separate levels.

A further development in investigation of an optimal construction of the system of transferring the unit sizes followed the path either by detailing separate aspects of the system structure and the process of its operation or by using new methods of problem solution.

K. A. Reznik in [401] sums up the results of his long-term investigations on determination of a number of verification system steps on the basis of the productivity and verification intervals of measuring instruments on each level, as well as a joint park of working measuring instruments (the maximum possible number of steps) with a detailed study of the influence of a kind of distributions of the corresponding errors. The results of these works were used as a basis of the normative document MI 83–76 [331].

In A. M. Shilov's work [435] the problems of accumulating in a verification scheme the errors which change with time as well as those of a relationship between errors of a reference measure and a verified one (a working measure) are taken into account. It is claimed that the approach suggested demands significantly less a priori information than the approach based on the probability of a verification defect.

V. P. Petrov and Yu. V. Riasniy in [380] analyze a method of constructing optimal verification schemes according to the criterion of a minimum of economical costs, however taking into account the probabilities of verification defects. As the initial parameters of the task, the operation costs dependencies of measuring instruments, disposable for service and unworkable, on their errors, as well as the losses due to operation of unworkable measuring instruments, are superinduced.

In a series of works by A. I. Vasil'ev, B. P. Zelentsov, A. A. Tsibina, A. M. Shilov N. G. Nizovkina [435, 525–527, 532, 533] from the city of Novosibirsk, great attention is given to investigating the role of and determining the verification interval values for working and reference measuring instruments, as well investigating the influence of verification defects on metrological reliability of measuring instruments, the calculation of parameters and modeling of a system of transferring the unit size and, finally, the problems of a system approach to the assurance of measurement uniformity.

It is also proposed to use verification intervals as variable parameters, with the help of which it is possible to control the uniformity of measurement (or, more precisely, to control the uniformity of measuring instruments). Moreover, the system for ensuring the uniformity of measurements in a country (which the authors identify in its sense with the system of transferring the unit sizes) is represented in the form of two interconnected subsystems, one which supplies metrological service bodies with verification equipment, and another which provides the uniformity of applied measuring instruments. Factors influencing the structure and operation of the system for ensuring the uniformity of measurements are determined:

(1) errors of measuring instruments and methods of verification on all levels of a verification scheme;
(2) normalized metrological characteristics of measuring instruments;
(3) availability of standardized procedures of carrying out measurements and verifications;
(4) quality of verifications (on distribution of errors and probabilities of verification defects);
(5) number of verification scheme steps;

(6) nomenclature and park of measuring instruments throughout the steps of a verification scheme;

(7) metrological state of measurement standards and working measuring instruments;

(8) coefficients of their usage;

(9) productivity of measurement standards;

(10) distribution of measuring instruments over the territory and a need of its transportation for verification;

(11) need for verifications in situ;

(12) full strength of metrological bodies with verification equipment;

(13) periodicity of measuring instrument verification;

(14) prices of measurement standards, verifications, operation costs;

(15) losses due to disfunction of measurement uniformity;

(16) losses due to the retirement of measurement standards and measuring instruments for verification and repair;

(17) limitation of resources for metrological assurance.

In these works an attempt is made to give definitions for the concepts "measurement uniformity assurance" and "uniformity of measuring instruments", as well as some properties of a system for ensuring measurement uniformity as a complicated cybernetic system are pointed out.

L. A. Semenov and N. P. Ushakov's work [431] is devoted to the problem of reducing the costs connected with the arrangement and function of a network of verification bodies at the expense of their rational disposition. The authors carry out an analysis of an economic-mathematical model of the problem concerning integer-valued linear programming for given sets of the they types of working measuring instruments and their parks, points of working measuring instruments and working measurement standards dislocation, verification intervals, taking into account the costs of verification and transportation of measuring instruments.

A brief review of the methods and theory of mass service in determining the need for verification means (including problems of optimal disposition of verification equipment) is given by V. V. Belyakov and L. N. Zakashanskiy [50].

In Ya. A. Krimshtein's work [283] the problem of the synthesis of an optimal structure of a metrological network (according to the criterion of a total cost minimum) for transferring the unit size to the park of working measuring instruments of one group. In contrast to [431], here there were chosen not only versions of disposals of the reference measuring instruments realizing the verification of working measuring instruments, but the structure of the whole verification network. Optimal parameters of the problem are determined by methods of integer-valued measurement programming.

In a series of works by specialists from the Tomsk Politechnical Institute, which are summarized by A. I. Kanunov, T. V. Kondakova, E. P. Ruzaev, and E. I. Tsimbalist

[252], it is suggested to solve the synthesis problem of optimal structures of verification schemes by way of a formalized description using graph theory. At the same time sets of working measurement standards and working measuring instruments are represented in the form of peaks, and links between them are shown in the form of graph ribs. Due to a lack of links between the elements of one level in a considered multilevel hierarchical system, its optimization can be performed step by step rising from the bottom levels to the upper ones and having an algorithm of optimization within the adjacent levels.

In the work of G. V. Isakov [229] attention is paid to the complexity of a real system of transferring the unit sizes and to the need for solving the problems relating not only to the structure of methods and means for transferring the unit sizes, but to executive routine support components (including organizational, trained, financial, informationa,l and lawful ones).

In L. A. Semenov's work [428] an idea for a new approach to optimization of the unit size transfer system is expressed for the first time, according to which it is necessary to consider the interconnection of three systems: the system of the unit reproduction (throughout the totality of quantities measured), proper systems of the unit size transfer (for separate quantities), and a system of information consumption (all sections of the economy of a country). The first and third systems impose limitations from the "top" and "bottom" onto the second one.

Taking this into account, the optimization of the unit size transfer system is suggested to be carried out stepwise: at the first stage it is necessary to determine a number of steps and characteristics of the unit size transfer system (construction of a verification scheme); at the second stage the optimization of the disposal of reference measuring instruments (the system structure) is realized on the basis of economical criteria. Moreover, some issues of the increase in error in transferring the unit size from a measurement standard to a working measuring instrument are considered, particularly in finding new graduation characteristics of a verified measuring instrument.

In T. N. Siraya's work [442] errors of unit size transfer from a group measurement standard, as well as errors of unit reproduction and maintenance by a group standard are considered.

One more approach to the problem of optimizing a system of the unit size transfer by an economical criterion is given in [499] by V. S. Svintsov This work is a total of all his previous studies. It is based on the fact that the periodicity of verification (verification interval) simultaneously influences the quality of verification and the amount of losses because of the use of damaged working measuring instruments (with a latent metrological failure), i.e., a variable parameter is the verification interval. Thereby the above-mentioned idea in L. A. Semenov's work [428] is partially realized.

However the author acknowledges that to use the approach he suggests is difficult in practice, because a great amount of initial data of a technical and economical character is required. In the next work by the same author [500] a model of an operation process is described which was designed to perform, in principle, "optimization of metrological maintenance of measuring instruments not only in accordance with the periodicity of verification but also simultaneously in accordance with the duration of their previous no-failure operation".

The paper written by V. A. Dolgov, V. A. Krivov, A. N. Ol'hovskiy, and A. A. Grishanov [139] and that by A. A. Avakyan [28] are also of interest, although they are not directly related to the problem of optimizing the park and nomenclature of working measuring instruments and the methods of measurements, which, certainly, influences input parameters of the system considered.

An original approach to the analysis and synthesis of verification schemes, which is based on applying the theory-group methods, was proposed by V. A. Ivanov in [239]. But in fact, the practical use of this approach now seems rather unclear.

Finally, in the work by O. A. Kudriavtsev, L. A. Semenov, and A. E. Fridman [286] an attempt is made to classify the main problems of constructing a system for ensuring the uniformity of measurements, which yield to mathematic modeling (including the problems concerning the RUTS systems). However, due to the fuzziness of the signs chosen, the classification given cannot be seen as a successful one.

The classification of the efficiency indices of the separate elements of the system appears to be insufficiently complete, clear, and grounded. More interesting is an attempt to develop a general model of a system based on a synthesis of the existing ideas about it. However the requirements for the model stated in the work are also insufficiently grounded, and the problem of formalization itself in a general form (through a minimum of the efficiency criterion, which appears to be proportional to a number of working measuring instruments) is simply wrong.

Thus, even this brief review of published works (we tried to select the most "central" ones) shows that at present there have been a significantly large number of theoretical investigations on various problems connected with constructing the RUTS systems in separate kinds of measurements (see, for example [501]). Unfortunately, among all these works there is nothing on constructing a system for reproducing the units (i.e., the interspecific construction of a general RUTS system). References concerning this problem are given in the respective paragraphs of this chapter.

2.2.2.3 Current normative documents for the RUTS systems

In the USSR, previous to 1981 the following basic normative documents regulating general problems of the unit reproduction, transfer of their sizes and construction of corresponding systems were in force.

1) GOST 16263 "National system of measurements. Metrology. Terms and definitions" [195], where the terms related to the problems considered were determined: a *standard, reference measuring instruments, verification* and others. Unfortunately, in this GOST no definitions of the terms "reproduction of a physical quantity" and "transfer of the unit size" were given.

2) GOST 8.061 "National system of measurements. Verification schemes. Scope and layout" [194], where the classification of verification schemes (national and local) was given and general requirements for the maintenance of verification schemes were formulated. The GOST also indicated the possible methods for verification and metrological requirements and requirements for text and graphical representation of verification schemes in normative documents. It was foreseen that the national verification schemes were "headed" either by a national measurement standard or by measuring instruments borrowed from other verification schemes.

3) GOST 8.057 "National system of measurements. Standards of physical quantities units. Basic provisions" [193], where the ideas of centralized and decentralized reproduction of the unit were given and grounds for centralizing the reproduction of the unit were formulated. Moreover, here the classification of standards (according to their composition and designation) and general requirements for the order of certification, maintenance and application of standards were presented.

The possibility of the existence of verification setups of the highest accuracy was also provided for. These verification setups were "the head" of the verification schemes when the decentralized method of the unit reproduction was used, and replaced the measurement standards, but for some reason they were legally equated with the status of working standards. In the Supplement to the this GOST, for the first time definitions of terms "reproduction of the unit" and "transfer of the unit" were given.

4) MI 83 "Procedure for determining parameters of verification schemes" [331], which developed the regulations of GOST 8.061 and established methods for determining general parameters of RUTS systems: the relationship between the normalized errors of measurement standard and verified measuring instruments and the estimation of a number of accuracy steps (maximum and minimum). For a more accurate determination of the number of steps it was recommended to take into account the following: the necessity of a WMS reserve in case of a failure of the basic ones; the availability of the working measuring instruments of limited occurrence for which a deliberately underloaded WMS was created; geographical "atomism" of working measuring instruments; and special features of the kind of measurements and applied verification means.

The remaining fundamental normative documents (GOST 8.009 [191], 8.010, 8.011, 8.042 and others) concerned only indirectly the problem considered.

All the normative documents on RUTS systems indicated above are based on theoretical studies and mostly on the works performed at the VNIIM (by Prof. K. P. Shirokov [436, 437], K. A. Reznik, and others) before 1973–1975. Unfortunately, they poorly synthesize the large amount of other research.

The experience of applying these normative documents has shown that, together with their undoubtedly organizing effect, very frequently certain difficulties arose connected to an insufficient clarity of the regulations they contained and to some self-contradictions. In particular, in the work by M. S. Pedan and M. N. Selivanov [377] the difficulties connected with an inadequate definition of the status of primary and special state measurement standards, uncertainty of their composition, and others are shown. Some other defects of the these normative documents are stated above.

Since 1981 instead of GOST GOCT 8.057 and 8.061 a new complex of fundamental documentary standards are in force: GOSTs 8.057 [193], 8.061 [194], 8.372–80, and 8.525–85. In these documents an attempt is made to take into account some defects of previous GOSTs. Specifically, in GOST 8.057–80 the composition of national and secondary standards is stipulated, and in GOST 8.525–85 the problems concerning the legal position of setups of the highest accuracy are considered in detail.

However the new complex has new defects.

(1) Disjunction of two GOSTs into a number of GOSTs, the authors of which are different bodies (groups of specialists), has complicated firstly the work with normative documents (although judging by the amount of data they are equivalent to the two previous GOSTs) and, secondly, has resulted in a number of internal contradictions within this complex. For example, GOST 8.061–80 states that verification schemes have at the head a national measurement standard, while GOST 8.525–85 provides for the possibility of reproducing the unit with a "verification setup of the highest accuracy".

(2) The status itself and the reasons for VSHA creation remain very dim. In GOST 8.525–85 it is assumed that they are designated to be used either for decentralized reproduction of the unit or for dimensionless quantities and quantities having a "strictly specialized range or field of use". At the same time it is not clear what the difference in essence is between a setup of the highest accuracy and a national measurement standard, especially taking into consideration that the transfer of unit sizes from the VSHA has to be realized also in accordance with a national verification scheme.

(3) Introduction of an additional category of verification schemes into GOST 8.061–80, which are departmental. At the same time there is an existing category of local verification schemes (which can be of any level of commonness except the national one).

(4) As before the role, place, and general metrological requirements for comparators are not determined.

(5) In GOST 8.057–80 the definitions of terms such as "reproduction of the unit" and "transfer of the unit size" are not given, i.e., at present they are left without any regulation anywhere.

It should be noted that during a long perior after the development of the fundamental GOSTs 8.057–80 and 8.061–80, a significant number of works on theoretical problems of constructing the RUTS systems (see Section 2.2.2.2) have appeared, but in the new complex of GOSTs the results of these works were practically not taken into account and the ideology of constructing the RUTS systems, as regulated by the GOSTs remained practically unchanged.

The aforesaid, however, testifies not only to the lack of attention by the authors of the GOSTs to theoretical work, but also to the fact that the results of these theoretical studies are of little use and are difficult to translate into a simple language for their practical application. In any case the measurement procedure, MI 83–76, has not been revised, although GOST 8.061–80 has established the obligation for quantitative grounds of the optimal decision of national verification schemes.

2.2.2.4 Choice of the main directions of investigation

In general a scientific theory implies the whole totality of scientific (theoretical) knowledge about an object under study. Scientific theories can be extremely diverse, both in form and inner structure. At present, there are no distinct ideas of what theories should be considered as the most perfect for expressing knowledge. Evidently, this depends on the level of development of a corresponding field of knowledge, the specific character of objects under consideration, and the degree of abstraction, which is essentially determined by the goals of the theoretical investigations.

The estimate characteristics of a theory can be such features as accuracy, reliability, completeness, insularity, depth, consistency, and simplicity. Since the features indicated do not always correlate among themselves (for example, the reliability with the completeness, the completeness with the consistency, etc.) then the construction of a theory can be considered, in a sense, to be a problem of multicriterion optimization. Taking into account that our initial information and comprehension of a goal are limited, this problem is fuzzy with regard to its setting.

Apparently, the most general characteristic of the quality of a scientific theory can be a *degree of the order* of all kinds of knowledge on which the theory is based. In I. Kant's opinion, any science is the system, i.e., a totality of knowledge put in order on the basis of known principles.

An estimate of the quality of metrology theory should evidently be based on setting an interrelation of gnoseological and metrical accuracies at a selected totality of estimation parameters.

The selection of the optimal initial (basic, not deduced) concepts is particularly important.

Thus, since generally the theory in itself is the system, and in the case considered the matter concerns a "theory of systems …", then it will be useful to dwell on some principal conceptions of general systems theory.

- The *features* are the properties (characteristics) of the system on the whole, its parts or separate elements. In the capacity of features, in particular, some estimation characteristics of the theory (accuracy, reliability, completeness, insularity, depth, etc.) mentioned above can be used. These features can be quantitative (objective) and qualitative (subjective).
- The *system state* is characterized by a set of system values at a certain instant.

- *System behavior* is the variation of its state with time.
- The *system structure* is determined by a totality of relations (links) of system elements and depends on a number and type of interrelations of these elements. A hierarchical structure is inherent in complex systems, i.e., the putting in a definite order of the levels of structure subsystems and elements.
- *Environment* means the external (with respect to a system given) systems and elements which do not belong to it.
- *System function* is the function that determines its interrelations with the environment (with other systems).
- Elements entering a system are called *input elements* and those leaving it are called *output elements*, the latter being the *results* of the transformation processes in the system.
- For *controlled* (organizational) *systems* such concepts as *goals and tasks of the system*, which define its designation, are very important.
- The *system operation* is the realization of the goals and problems the system poses.

When using the system approach for analyzing and synthesizing the system under study the following problems are usually subdivided into:

- determination of the system boundaries on the whole as well as of environmental boundaries (interacting systems);
- establishment of a function of the system, its goals, and tasks;
- determination of the structure and detection features of the system and its parts;
- construction of the software matrix for transforming input elements into output ones;
- description of the system control.

Thus, the description and optimal construction of the system is finally reduced to composing an equation that allows the results (output) to be calculated and their efficiency to be evaluated on the basis of comparatively reasonable number of controlled parameters, i.e., providing the possibility of controlling the system.

Taking this into account, let us sum up the main results of the analysis of works on issues of the RUTS system theory (in Section 2.2.2.4). The least studied (or not at all studied) are:

- determination of the system of basic initial concepts of the theory, its boundaries, and attributes (special features);
- formulation of the basic initial concepts of the theory while improving its object, and the selection of a rational theory language for describing the object;
- selection of the most significant features of the RUTS systems and its components (subsystems and elements);

- determination of the environment and the study of interrelations with external systems;
- determination of the goal and designation of the RUTS system on the whole, particularly reasoning from interrelations with external systems;
- determination of inputs and outputs of a RUTS system, and its resources;
- study of the (self-dependent) system of unit reproduction as an essential subsystem of the RUTS system;
- versions of the goal achievement (in particular, the decentralized method of reproducing unit dimensions) and establishment of an efficiency measure for evaluating the degree of the goal achievement;
- study of the structure of RUTS system elements;
- modeling of the structure and process of RUTS system operation on the whole;
- formulation of the quality criteria (of functioning) of a RUTS system and selection of optimization models.

It is clear that the problems indicated which are "white spots" in the present-day state of the theory of RUTS systems construction can serve as the main direction of investigations for further development of this theory (actually for its construction). However the directions indicated are a long-term program.

In the present work the task has been posed to lay the basis for such a theory only reasoning above all from consideration of the RUTS system on the whole and its interrelations with the main external systems, i.e., with the system of ensuring the measurement uniformity and the system of measurable physical quantities.

Since from a theoretical point of view the external systems themselves have also been insufficiently studied, an attempt has been made to express our opinion about these systems. The directions indicated correspond to the current important tasks facing theoretical metrology [501, 503].

2.2.3 Foundations of the description of RUTS systems

2.2.3.1 Initial concepts

Let us indicate the terms and their definitions for those initial concepts of metrology which will be needed to describing RUTS systems. The majority of terms were taken from GOST 16263 [195]; for some of them there were new definitions [246, and others] which, from our point of view, better reflect the concept content and, chiefly, put fundamental terms of metrology into a logically more harmonious and consistent system. Terms (and corresponding concepts) introduced anew are marked by an asterisk (*).

- *Physical quantity* is the property, general from the point of view of quality with regard to various real objects, but individual from the point of view of quantity for each of them (by objects the bodies, processes, and fields).
- *Physical quantity gender** is the qualitative characteristic of the property identified with a given physical quantity (see [390, p. 9]).
- *Physical quantity dimension* is the quantitative content of the property, corresponding to a concept of the given PQ in a concrete object.
- *Physical quantity unit* (unit of measurement) is the PQ that is uniform with regard to a quantity under measurement and has a definite dimension adopted under agreement for some object for establishing dimensions of the given PQ for other objects.
- *Physical quantity value* is information about a PQ dimension which is expressed in the form of a denominate number in terms of accepted units.
- *Measurement problem** is the problem of founding the value of some PQ under definite (given) conditions.
- *Measurement* is the process of solving some specific *measurement problem.*
- *Measuring instrument* is the technical mean designated to perform measurements of a definite PQ and for keeping information about the unit of this PQ.
- *Metrological characteristics of a measuring instrument* are the characteristics of properties of this MI which influence measurement results.
- *Measurement method* is the totality of procedures for using principles and measuring instruments in a given measurement.
- *Metrological system** is the system of objects used and studied in metrology.

Measurement presupposes a description of the quantity commensurate with the intended use of a measurement result, a measurement procedure, and a calibrated measuring system operating according to the specified measurement procedure, including the measurement conditions.

Here we have not given definitions for some of the necessary concepts, obvious from the point of view of their names, such as "physical value under measurement (measurand)", "measurement object" (measurand carrier), "measurement result" (result of solving a posed measurement problem). The main concepts of the RUTS systems theory itself will be also introduced further on.

2.2.3.2 Measurement as an elementary metrological system

A rather fruitful approach for studying and describing various metrological systems is the use of measurement components as an elementary metrological system. An approach of this kind is sufficiently "natural", since it uses the main object of studying by metrology, i.e., a measurement, and is rather efficient for studying various metrological systems from the single positions, which is very important in systemic investigations.

Let us consider a general description of measurement to be the process of measurement problem solution outlined in [390]. Since a measurement in the most general form is understood to be one of a variety of perceptions at which information is always transformed, then an initial equation for describing the measurement can be written in the form

$$J : I_a \to I_p, \tag{2.2.1}$$

where J is the operator of transformation, I_a is a priori information, and I_p is a posteriori information.

Specification of this rather abstract expression takes place in the process of introducing the concept "measurement problem" (see Section 2.2.3.1), i.e., the problem of finding the value of a PQ under definite conditions. It should be noted that this is an important moment for carrying out any measurement. Without the formulation of a specific measurement problem the measurement becomes meaningless. To meet this demand it is necessary to indicate the kind of the PQ, the object, and the conditions of measurement, time (and so on) within which the measurement should be carried out, i.e., to assign parameters (components) of a specific measurement problem z_i:

$$z_i = (\varphi_i, o_i, \psi_i, \delta\varphi_i, g_i, t_i, \Delta t_i, p_i, \ldots), \tag{2.2.2}$$

where

φ_i is the measurand as the quality (by definition);
o_i is the object of the study (the carrier of the measurand);
ψ_i are the measurement conditions (the totality of the given influencing factors);
$\delta\varphi_i$ is the given measurement error;
g_i is the given form of presenting a measurement result;
t_i is the moment of time at which the measurement is realized;
Δt_i is the time interval within which it is necessary to perform the measurement;
p_i are the coordinates of a place (of space) where the measurement is performed.

This collection of components forms a set of given (i.e., uncontrollable in the process of a given measurement) parameters of measurement as a system. It should be noted that the collection indicated is sufficiently general (universal) and can be applied for any measurement problem.

After formulating the measurement problem, it is naturally possible to consider any measurement as a *process of solving a measurement problem,* which can be divided into three stages.

At the first stage, according to a measurement problem (reflecting the question "what has to be done?") a plan for a measurement experiment is developed (answering the question "in what way must it be done?"). At this stage a method and the required measuring instruments are selected, an observer (an operator who is able to carry out the measurement experiment) is chosen, a procedure (algorithm) for using selected measuring instruments and the methods and means of processing experimental data are defined more accurately (and so on). The development of this plan is realized on the basis of a priori information (accumulated before posing the given measurement problem), reasoning from the content of the problem itself.

This first stage of the measurement process can be represented by the following equation of transformation:

$$J_v(z) : (I_a, z_i) \to I_z, \tag{2.2.3}$$

where
$J_v(z)$ is the corresponding transformation function realized by an operator (a subject) v, processing initial information (in a general case it may differ from an observer);
I_z is information corresponding to the received plan of measurement experiment (here the space-time parameters are omitted for simplicity):

$$I_z = \left(\varphi_i, o_i, \psi_i, \delta\varphi_i, g_i, \cdots \middle| [\varphi_i], m_i, s_i, v_i, w_i, \ldots\right), \tag{2.2.4}$$

where
$[\varphi_i]$ is the selected measurand unit;
m_i is the chosen measurement method;
$s_i \equiv \{s_k\}$ is the totality of measuring instruments used to solve a given measurement problem;
v_i is the observer (operator) realizing the plan of measurement experiment;
w_i is the means for processing measurement experiment results.

All these parameters located on the right of a vertical line in equation (2.2.4) are the *controllable components of measurement* (as of the system), which can be varied within the time of developing the measurement experiment plan. In Figure 2.3a the first stage of the measurement process is shown in diagram form.

At the second stage a process of real (physical) transformations takes place. These transformations are connected with a physical interaction of the selected measuring instruments with the object, external conditions and observer (Figure 2.3b). Usually, this process is compared with the concept "measurement"; however such an interpretation appears to be insufficient for revealing all THE components of measurement as the system and also of the process of the measurement problem solution, it also presents difficulties for describing the measurements different from the direct and single ones. In terms of the designations accepted here this stage in the general form can be expressed by the equation

$$\hat{J}_s(I_z): \quad (o, \psi) \to h_{s_i}, \tag{2.16}$$

where h_{s_i} is the indications of measuring instruments s_i obtained as a result of the measurement experiment.

At the third stage the processing of the obtained measurement information h_{s_i} is performed on the basis of the measurement experiment plan using in the general case

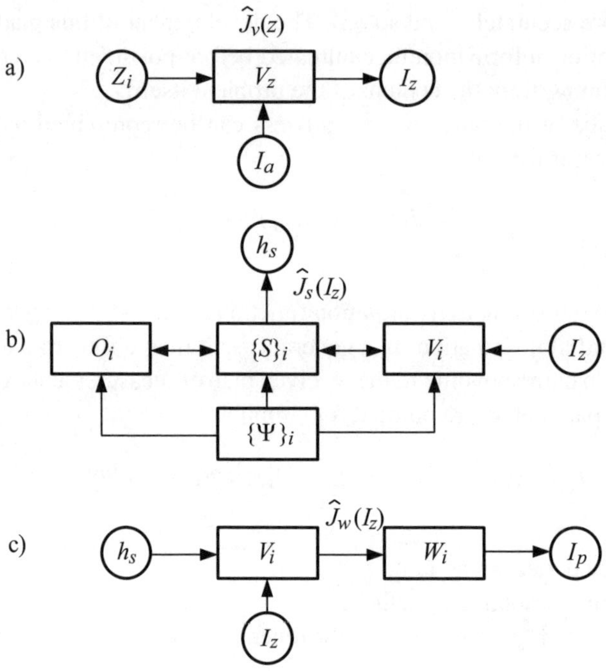

Figure 2.3. Stages of measurement as the process of the measurement problem solution.

an auxiliary computer technique w_i (Figure 2.3c).This stage, by analogy to the preceding ones, can be described by the equation

$$J_w(I_z)\colon\ (h_s, I_z) \to I_p. \tag{2.2.6}$$

The a posteriori information I_p is the measurement result Q in accordance with the given form of its presentation. These forms are indicated in GOST 8.011. An algorithm of transition from h_s to $I_p = Q$ depends not only on the form of the presentation of the result but also on the measurement method. So, for direct single measurements the result is a value of the measurand $\varphi^{(\text{meas})}$, according to a readout h_s of the measuring instrument:

$$\varphi^{(\text{meas})} = h \cdot \varphi_{os} = n_s[\varphi]_s, \tag{2.2.7}$$

where φ_{os} is the constant of the given measuring instrument for a selected digital scale (scale-division value) n_s, and φ_s are the numerical value of the measurand and "true" (or assigned) value of the unit realized in the given measuring instrument.

In a more general case the measurand value obtained as a result of a specific measurement has to be presented in the form

$$\varphi^{(\text{meas})}(z_i) = n_{\varphi_r}(I_{z_i})[\varphi_r]_{I_{z_i}} \tag{2.2.7}$$

This means that both the numerical value and the measurand value φ of the dimension r are determined by the whole totality of parameters of the measurement system I_{z_i}, with the help of which the given measurement problem z_i is solved. In other words, this means that for the measurement system on the whole there are some "metrological characteristics" (much as now they are traditionally introduced for measuring instruments).

2.2.3.3 A priori information in measurements

The important role of the a priori information has become increasingly better understood in recent years [32, 33, 201, 201, 438, and others]. The approach to the general measurement description set out above accentuates this role. For each component of the measurement like that of the system (2.2.4) it is necessary to have some a priori information about dependencies

(1) of external influencing quantities on space and time coordinates within the framework of the measurement system I_{z_i}:

$$\psi_{l_i} = \psi_l(t_i, p_i); \tag{2.2.9.1}$$

(2) of metrological characteristics (q_s) of a measuring instrument on external conditions and the same space and time coordinates (time instability q_{s_i}, their dependence on the space orientation of a measuring instrument, etc.):

$$q_{s_i} = q_s(\psi_{l_i}, t_i, p_i); \tag{2.2.9.2}$$

(3) of the true value (dimension) of the measurand in an object on the state of this object, on the point of location (coordinates) of the object; on the time (principal instability of the measurand), and on external conditions:

$$\varphi_{0_i} = \varphi_0(t_i, p_i, \psi_i); \tag{2.2.9.3}$$

(4) of the measurement method on a selected measuring instrument, representation of results, and measurement error:

$$m_i = m(s_i, g_i, \delta\varphi_i); \tag{2.2.9.4}$$

(5) of the unit of the measurand on a definition of the measurand itself (the most trivial and unambiguous relationship):

$$[\varphi_i] = k \cdot \varphi_i; \tag{2.2.9.5}$$

(6) of measuring instrument types (and, consequently, the totality of their metrological characteristics) on the measurand, given measurement error and external

conditions:

$$s_i = s(\varphi_i, \delta\varphi_i, \psi_i); \tag{2.2.9.6}$$

(7) of the operator for performing his/her functions on external conditions as well as on a place and time of measurement:

$$v = v(\psi_i, t_i, p_i). \tag{2.2.9.7}$$

Apart from these dependencies it is necessary to know a priori the following: the approximate value of the measurand in objects of a given type; the existing types of measuring instruments and their characteristics; the corresponding constants and dependence parameters (2.2.9.1)–(2.2.9.3); the normative documents in the field of metrology which are in force (fundamental and relating to a given type of measurement); the skill of operators (observers); etc.

Only when the necessary volume of a priori information (one measurement part of which is from the preceding experiments, the second is from calculations and the last is of a qualitative nature) is available, then is it possible to obtain the *right measurement result* (provided that the plan of the measurement experiment was skillfully developed). In this connection it will be useful to generalize *the condition of measurement correctness* (given in [37] for the case of temporal changes in this system within the framework of the measurement experiment:

$$\begin{aligned}
\delta\varphi_i = &\int_{t_i}^{t_i+\Delta t_i} \left(\left(\frac{\partial\varphi(0)}{\partial t} \right)_i + \sum_k \frac{\partial\varphi}{\partial q_k} \cdot \frac{\partial q_k}{\partial t} + \sum_l \left(\frac{\partial\varphi}{\partial\psi_l} \cdot \frac{\partial\psi_l}{\partial t} + \frac{\partial\varphi}{\partial v} \cdot \frac{\partial v}{\partial t} \right) \right) dt \\
&+ \int_{p_i-\Delta p_i}^{p_i+\Delta p_i} \left(\left(\frac{\partial\varphi(0)}{\partial p} \right)_i + \sum_k \frac{\partial\varphi}{\partial q_k} \cdot \frac{\partial q_k}{\partial p} + \sum_l \left(\frac{\partial\varphi}{\partial\psi_l} \cdot \frac{\partial\psi_l}{\partial p} + \frac{\partial\varphi}{\partial v} \cdot \frac{\partial v}{\partial p} \right) \right) dp \\
&+ \sum_l \int_{\psi_{l_i}-\Delta\psi_{l_i}}^{\psi_{l_i}+\Delta\psi_{l_i}} \left(\frac{\partial\varphi(0)}{\partial\psi} + \frac{\partial\varphi}{\partial v} \cdot \frac{\partial v}{\partial\psi_l} + \sum_k \frac{\partial\varphi}{\partial q_k} \cdot \frac{\partial q_k}{\partial\psi_l} \right) d\psi_l \\
&\leq (\delta\varphi_i)_{\text{given}} - (\delta\varphi_i)_{\text{norm}},
\end{aligned} \tag{2.2.10}$$

where

$p_i \pm \Delta p_i$ are the space boundaries of the measurement system;

$\pm\Delta\psi_{l_i}$ are the errors of setting the values of external influencing quantities within the measurement experiment;

$(\delta\varphi_i)_{\text{given}}$ is the given error of measurement;

$\delta\varphi_i)_{\text{norm}}$ is the normalized error of measurement which is specified in a normative document for corresponding measuring instruments and measurement methods within the limits of working conditions of their application.

From this it follows that if the influence of all components of measurement system (2.2.4) except those which are provided for in a normative document for metrological characteristics of the measuring instrument (see 2.2.9.6) is either taken into account or is negligible (which is practically the same thing), then expression (2.2.8) for the measured physical value will have the form

$$\varphi^{\text{meas}}(z_i) = \varphi^{\text{meas}}(s_i) = n_{\varphi_r}(s_i)[\varphi]_{s_i}, \tag{2.2.8a}$$

where the index r, as previously, denotes that the matter concerns a definite dimension of the measurand.

On the basis of the above it is possible to interpret the existing varieties (types) of measuring instruments in the following way:

- *measures* (s_m) are instruments for which both $n_\varphi(s)$ and $[\varphi]_s$ are a priori information.
- *measuring instruments* (s_n) are instruments for which $[\varphi]_s$ is a priori information and $n_\varphi(s)$ is a posteriori measurement information.
- *Transformers* (s_t) are measuring instruments for which the transformation coefficient (K_t), i.e., the ratio of the output and input values measurands, is the main a priori information.

Thus, a priori information is an essential condition for performing any measurement. When this information is absent, it is impossible to carry out any measurement. On the other hand, if everything about the measurand is known, i.e., the measurement result is known a priori, then the measurement is not needed.

In [141] three postulates are formulated. These postulates establish a qualitative connection of the volume of a priori and a posteriori information in measurement with a given (required) accuracy. Condition (2.2.10) confirms the correctness of these postulates.

In this connection it is useful in further investigations to consider *two extreme cases* of measurements with different required accuracies and different limitations imposed on other components of the measurement system (2.2.4):

(1) the class of research standard measurements, i.e., the measurements of the highest accuracy performed when measurement standards are studied and the values of fundamental constants are improved;

(2) the class of technical (working) measurements, i.e. mass measurements for production and maintenance purposes, in those cases when measurements are not the goal but a means to reach the goal.

Therefore, when determining the cost of measurements in problems of metrological system optimization it is necessary to take into account not only the cost of the measuring instruments (depending on their accuracy) but also the cost of the other components of an elementary metrological system, i.e., a separate measurement, taking into account the interrelation of these components.

2.2.3.4 Structure of an RUTS system

The essence of an RUTS system is well reflected in its name (only in the case of the clear definition of the concepts studied in detail which this name includes). Let us introduce the concepts which are *fundamental* for the RUTS systems.

Reproduction of a unit is a unit embodiment (realization) at which a single a priori piece of information about the dimension of this unit is its definition.

If it is correct to say that the definition of the unit is "absolutely accurate", i.e., ideal; then taking into account (2.2.8a) it is possible to write the following *equation of the unit reproduction*:

$$\varphi^{(\text{meas})}(z_{\text{st}}) = n_{\varphi_r}(s_{\text{st}}) \cdot [\varphi_r]_{\text{st}} = n_{\varphi_r}(s_{\text{st}}) \cdot n[\varphi]_{\text{st}} \cdot s_{\text{st}} \cdot [\varphi_0], \tag{2.2.11}$$

where

$[\varphi_0]$ is the unit of a physical quantity by its definition, i.e., the unit having an ideal dimension accepted as the unit precisely by definition;

$[\varphi_r]_{\text{st}}$ is the unit realized in an initial measurement standard (basic reference measurement standard) in the range of physical quantity values which correspond to its dimension r and differ from $[\varphi_0]$ due to an error of realization.

Transfer of the unit is the comparison of unit dimensions realized in measuring instruments which are intersubordinate in accuracy and rank.

According to routine practice (which is justified by economical expediency) the process of the unit dimension transfer is realized by two methods:

(1) *certification* (calibration) of a subordinate measuring instrument with respect to a higher one (by accuracy and rank) when as a result of their readouts, corrections are introduced into readouts of the subordinate measuring instrument;

(2) *verification* (control) of a subordinate measuring instrument with respect to a higher one when the difference of readouts of the verified and verifying measuring instruments is compared with a limit of the allowable error of the verified measuring instrument and a conclusion is made whether or not the measuring instrument is suitable for application.

The second method is applied predominantly for verifying working measuring instruments, whereas the first one is used for higher measuring instruments.

(3) Finally, *the storage of a unit* is the unit realization in the process of maintenance of the higher measuring instrument, i.e., within the whole service life period of this measuring instrument. It should be noted that any measuring instrument stores the unit by its definition, since otherwise it cannot be used for its direct destination, i.e., for measurements.

Taking into account the given definitions, it can be said that the *RUTS system elements* are the methods and means of unit reproduction and transfer of their dimensions, and a *total system* is given by the expression

$$\sum_{0} \equiv \sum_{\text{RUTS}} = \{s_{i_1 b}, s_{i_2 n}, m_{i_3 b}, m_{i_4 n}\},$$
$$s_{i_b} \in \sum_{0}, \quad s_{i_n} \in \sum_{0}, \quad m_{i_b} \in \sum_{0}, \quad m_{i_n} \in \sum_{0}, \tag{2.2.12}$$

where the designations are evident from indices and designations accepted earlier in Section 2.2.3.2:

$$i_1 = \overline{(1, n_1)}, \quad i_2 = \overline{(1, n_2)}, \quad i_3 = \overline{(1, n_3)}, \quad i_4 = \overline{(1, n_4)}$$

In this *total* system it is possible to mark out various subsystems (which are its "sections"):

- *Subsystem of reproducing* units:

$$\sum_{\text{pru}} = \{s_{i_1 b}, m_{i_{3,}b}\}, \quad \sum_{\text{pru}} \subset \sum_{0}. \tag{2.2.12.1}$$

- *Subsystem of transferring* units:

$$\sum_{\text{ptu}} = \{s_{i_2 n}, m_{i_4 n}\}, \quad \sum_{\text{ptu}} \subset \sum_{0}. \tag{2.2.12.2}$$

- *Subsystem of RUTS for a given measurand φ:*

$$\sum_{\varphi} \equiv \sum_{\text{RUTS}} (\varphi) = \{s_{i_1 b}(\varphi),\ s_{i_2 n}(\varphi),\ m_{i_3 b}(\varphi),\ m_{i_4 b}(\varphi)\},$$
$$\sum_{\text{RUTS}} (\varphi) \subset \sum_{0}. \tag{2.2.12.3}$$

- *Subsystem of means* for reproducing units and transferring their dimensions:

$$\sum_{s} = \{s_{i_1 b},\ s_{i_2 n}\}, \quad \sum_{s} \subset \sum_{0}. \tag{2.2.12.4}$$

- *Subsystem of methods* of reproducing units and transferring their dimensions:

$$\sum_{m} = \{m_{i_3 b},\ m_{i_4 n}\}, \quad \sum_{m} \subset \sum_{0}. \tag{2.2.12.5}$$

- *Subsystem of means* for reproducing a unit and transferring its dimension for a given measurand φ:

$$\sum_{s}(\varphi) = \{s_{i_1 b}(\varphi),\ s_{i_2 n}(\varphi)\}, \quad \sum_{s}(\varphi) \subset \sum_{\varphi}. \tag{2.2.12.6}$$

- *Subsystem of RUTS methods* for a given measurand φ:

$$\sum_{m}(\varphi) = \{m_{i_3 b}(\varphi), m_{i_4 n}(\varphi)\} \subset \sum_{\varphi}. \tag{2.2.12.7}$$

- *Subsystem of transferring* the dimension of a unit for a given measurand φ:

$$\sum_{s_n}(\varphi) = \{s_{i_2 n}(\varphi), m_{i_4 n}(\varphi)\} \subset \sum_{s}(\varphi). \tag{2.2.12.8}$$

The following relations are obvious:

$$\sum_{0} = \sum_{\text{pru}} \cup \sum_{n} = \sum_{s} \cup \sum_{m} = \bigcup_{i \in |\{\varphi_i\}|} \sum \varphi_i; \quad \sum_{\varphi} = \sum_{s}(\varphi) = \cup \sum_{m}(\varphi);$$

$$\sum_{m} \bigcap_{i \in \{\varphi_i\}} \sum \varphi_i = \bigcup_{i \in \{\varphi_i\}} \bigcap_{i,j} \sum_{m}(\varphi_i) \sum_{m}(\varphi_j); \quad \sum_{s} \bigcup_{i \in |\{\varphi_i|\}} \sum_{s}(\varphi_i), \quad \text{and others.}$$

The review of the technical literature in Section 2.2.2.2 shows that theoretical studies have been related for the most part to systems of the form described by equation (2.2.12.8). Usually the measurement standards of a certain country are implied by a system, and by subsystem (2.2.12.3) they imply a verification system (a slangy analogue is a verification scheme).

Earlier the system $\sum_o$ was called the *total* system of reproducing units and transferring their sizes, and the subsystem $\sum_\varphi$ was called the *particular* RUTS system (for a definite measurand).

Let us consider systems of the form (2.2.12.3), (2.2.12.5), and (2.2.12.6) in greater detail.

Let us introduce a *generalized element* of systems of the form described by equations (2.2.12.4) and (2.2.12.6), which we will call a *metrological measuring instrument* (a measuring instrument intended and used specially to reproduce a unit and/or to transfer its dimension). This especially metrological destination is a principle distinction of a metrological measuring instrument from a working measuring instrument, since a working measuring instrument does not reproduce a unit, but only stores it.

Elements of a measuring instrument of any RUTS system (containing it) are located in the system in a hierarchical manner as follows from the content of the concepts "reproduction" and "transfer", i.e., they preset the hierarchical structure of a particular RUTS system Σ_φ. At the head of this system there are used means of reproducing the unit, i.e., *reference measuring instruments* of the system Σ_φ, which allow the unit dimension to be reproduced, stored, and transferred *to subordinate measuring instruments* of the system. Sometimes for necessary (normal) operation of the initial and (or) subordinate measuring instruments of the given system Σ_φ *auxiliary metrological measuring instruments* are introduced. A more detailed analysis of the MMI classification and their variety is given in Section 2.1.3.

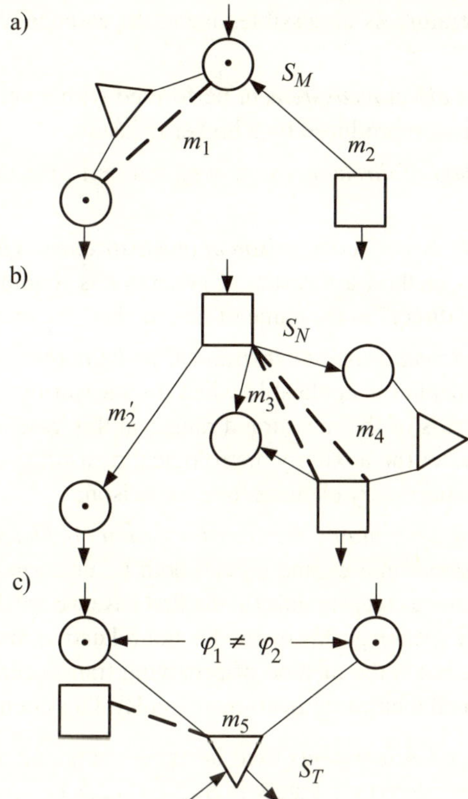

Figure 2.4. Interrelations (transfer methods) between subordinate metrological measuring instruments of different types. Here: ○ is the measurand generator; ⊙ is the measure (the certified generator); □ is the measuring instrument; △ is the measurement transformer; m_i – see text.

Any MMI, in spite of its exceptional metrological role, remains a measuring instrument (sometimes it is called a complex). Therefore, as any measuring instrument the MMIs can be realized in the form of measures, devices, or transformers. Figure 2.4 illustrates the interrelation of the MMIs as the subordinate elements of a RUTS system of a particular form (for a particular measurand) depending on their belonging to one or another type of measuring instruments.

This scheme allows possible *methods for transferring* unit dimensions among subordinate MMIs to be established:-

- m_1 is *the method of comparing two measures with the help of a comparator* (transformer); in this case either a scale transformer of the type bridge circuits (in case of different ratings) or a device of the null detector type can serve as a comparator. In our opinion to apply the method of direct measure comparison, as is frequently called for in the literature, is impossible, due to the essence of the measures themselves;

- m_2 is *the method of direct measurement* performed with a verified instrument of a measurand dimension, reproduced by a higher measure;
- m_2' is also *the method of direct measurement*, but performed with a higher instrument;
- m_3 *is the method of "direct" comparison of one instrument with another one*; however, to realize this method a measurand generator is required; this is clear, and therefore the word "direct" in the name of this method are in quotation marks;
- m_4 is the *method of comparing one instrument with another one with the help of a comparator*. This method is applicable when the measuring instruments are compared in different parts of the measurand range; in this case a scaling transformer is used as a comparator (an auxiliary metrological measuring instrument). Here a a measurand is also a necessary element for comparison;
- m_5 is *the method of determining the transformation coefficient* of a measurement transducer; at the same time φ_1 and φ_2 can both be uniform (of the same dimensions) and heterogeneous measurands (in the first case we are dealing with a scaling transducer). Strictly speaking, this method is not related to methods of transferring the unit dimension, but is the method of certifying transducers which can serve as auxiliary metrological measuring instruments in RUTS systems.

The method of indirect measurements used in reproducing and transferring units and their dimensions is considered separately in Section 2.2.5.1.

Let us consider the general equation of the unit dimension transfer between two steps of the RUTS system for the main method applied between metrological measuring instruments, i.e., the certification method.

Let s_2 be a verified metrological measuring instrument and s_1 be a verifying metrological measuring instrument.

When carrying out the comparisons s_1 and s_2 by any of the main transfer methods (m_1–m_3) the MI readouts of s_1 and s_2 are compared at the same dimension of the φ_r measurand. Let c be the readout difference between the s_1 and s_2 readouts, which was established as a result of comparison.

In this case according to the approach considered in Section 2.2.3.2 we have two measurement problems, z_1 and z_2, which practically coincide in all components except s_i. Therefore, taking into account the above and equation (2.2.8a) we have

$$\begin{aligned}\varphi^{\text{meas}}(z_1) &= n_\varphi(s_1)\cdot[\varphi]_{s_1},\\ \varphi^{\text{meas}}(z_2) &= n_\varphi(s_2)\cdot[\varphi]_{s_2},\\ c \equiv n_\varphi(s_1) - n_\varphi(s_2) &\equiv \Delta n_\varphi(s_1 - s_2)\end{aligned}$$

under the condition that

$$\varphi_2(z_1) = \varphi_2(z_2). \tag{2.2.13}$$

Comparing readouts s_1 and s_2 and considering values of the measurand φ, we think (just so we have to think) that dimensions of the units realized in s_1 and s_2, are similar and equal to the unit in s_1 as a more accurate measuring instrument:

$$\varphi_{\text{com}}^{\text{meas}}(z_1) = n_\varphi(s_1) \cdot [\varphi]_{s_1},$$

$$\varphi_{\text{com}}^{\text{meas}}(z_2) = n_\varphi(s_2) \cdot [\varphi]_{s_2}.$$

Therefore, when introducing the correction c into the readout s_2, a corrected (real) value φ in the system z_2 is obtained for the $[\varphi]_{s_2}$ unit:

$$\varphi_D^{\text{meas}}(z_2) = [n_\varphi(s_2) + \Delta n_\varphi(s_1 - s_2)] \cdot [\varphi]_{s_1} = n_{\varphi D}(s_{2D}) \cdot [\varphi]_{s_1}. \tag{2.2.14}$$

Let us see what a real dimension of the unit is in s_2. Using equations (2.2.13) and (2.2.14), we obtain

$$[\varphi]_{s_2} = \left[\frac{n_{\varphi D}(s_1)}{n_\varphi(s_2)}\right] \cdot [\varphi]_{s_1} = \left[1 + \frac{c}{n_\varphi(s_2)}\right] \cdot [\varphi]_{s_1} \tag{2.2.15}$$

or

$$\Delta[\varphi]_{s_2} = [\varphi]_{s_1} - [\varphi]_{s_2} = \left[\frac{c}{n_\varphi(s_2)}\right] \cdot [\varphi]_{s_1}. \tag{2.2.15a}$$

Thus, the introduction of the correction $c = \Delta n_\varphi(s_1 - s_2)$ into the readout of s_2 on the basis of the results of its comparison with the more accurate s_1 is equivalent to a change of the unit dimension realized in s_2, by

$$\Delta[\varphi]_{s_2} = \left[\frac{c}{n_\varphi(s_2)}\right] \cdot [\varphi]_{s_1}. \tag{2.2.15b}$$

Questions concerning other features characterizing the structure of the RUTS system and its individual elements are considered in Section 2.2.3.5.

2.2.3.5 Environment and boundaries of a RUTS system

As is clear from merits of case, a RUTS system is the main constituent of the system of ensuring the uniformity of measurements, which, in its turn, is the main component of the system of metrological measurement assurance, etc. However, for a more detailed (formalized) description let us consider the issue from another point of view.

Let the concept *"generalized system of measurements"* be the set of all measurements performed within some finite time interval (T) in space (P), i.e., within the limits of a definite space–time continuum (P,T). Using the presentation of a separate measurement in the form of measuring system (2.2.4), it is possible to present the generalized system in the following form:

$$\sum_{0}(P,T) = \bigcup_{k}\{I_{z_i}(\varphi_k)\},$$

$$I_{z_i}(\varphi_k) = (\varphi_{k_i}, o_i, \psi_i, \delta_i, g_i, t_i, \Delta t_i, p_i \ldots |[\varphi_k]_i|, m_i, s_i, v_i, w_i, \ldots),$$
$$1 \le i \le |\{z_i(\varphi_k)|\} \quad t_i \in T, \quad p_i \in P, \quad 1 \le k \le |\{(\varphi_k)\}|. \tag{2.2.16}$$

Localizing the space–time continuum (P, T) we will pass on to the quite definite systems of measurements. So, confining P within the framework of a country, we have a *national system of measurements* (*NSM*). The latter is a concept which has already been put into practice by metrology:

$$\mathrm{NSM} = \sum_{\mathrm{NSM}}(T). \tag{2.2.16.1}$$

Let us try to formulate the conditions which would reflect the role of metrology with regard to the state of NSM, i.e., the influence that proper metrological systems (2.2.2.4) upon the NSM. Obviously, the set of measurement problems

$$Z_i(\varphi_k) = Z(\varphi_{k_i}, o_i, \psi_i, \delta_i, g_i, t_i, \Delta t_i, p_i, \ldots) \tag{2.2.17}$$

is determined by the needs of all spheres of societal activities (science, industry, maintenance, etc.), and it is possible to consider that practically this does not depend on parameters of proper metrological systems, although as was shown below the problem of optimizing the set of measurement problems (for both the given measurand and their totality) is of current importance and can have an appreciable economic effect.

In their work (where such systems are called metrological chains or networks) the authors V. N. Sretenskiy and others consider the general character of the influence of the generalized system of measurements and proper metrological systems on the spheres indicated. For the sphere of science the availability of a strong positive feedback is characteristic. It causes the acceleration of its development in connection with the fact that "science moves forward proportionally to the mass of knowledge inherited from the preceding generation" (F. Engels).

Metrological systems have both positive components (of development) and negative (components of stability and quality) feedbacks. Therefore, here there should be an optimal relationship between the costs for metrological systems and industrial losses due to inefficiency and poor quality. Linking metrological systems with the sphere of maintenance has a pronounced negative feedback (absorption of measurement information), since the problem of supporting the stability of consumer properties (parameters) of operated objects is solved. Here the problem of optimization is connected with parameters of the proper metrological systems.

Thus, in all cases the influence of the proper metrological systems is reduced to the influence on the quality and efficiency of measurements performed in the NSM (and of theirs results). This is natural for metrology as the science of measurements and corresponding practical activity (see 2.2.4).

At the same time, there is no doubt that the main "responsibility" for proper metrological systems is above all to ensure the quality of measurements, since their efficiency has a great influence on other spheres of activity (primarily on the production of the measurement technique itself). Therefore, let us choose the quality criterion of measurements performed in the NSM as the main index of the efficiency of proper metrological operation.

- *Indices of measurement quality* [438] are: accuracy, trustworthiness, measurement trueness, measurement repeatability and reproducibility.
- *Accuracy of measurements* characterizes the closeness of a measurement result to a true value of a measurand.
- *Trueness of measurements* is determined by the closeness of a systematic error to zero as a consequence of measurements.
- *Trustworthiness of a measurement* is determined by the degree of confidence in its result and is characterized by the probability of the fact that the true value of a measurand is in real value neighborhoods with indicated boundaries.
- *Repeatability of measurements* reflects the closeness of results of measurement of one and the same measurand (of the same dimension), carried out under similar conditions.
- *Reproducibilty of measurements* reflects the closeness of results of measurement of one and the same measurand carried out under different conditions (according to a method, measuring instruments being applied, conditions and observer).

The first two definitions are given in accordance with GOST 16263, the third and last two definitions were taken from [230] and [438] respectively. Let us note that the evaluation of measurement reliability is a merely mathematical maneuver using the sufficiently developed probability theory and always has to be performed in evaluating the results of measurements due to the inevitably probabilistic (random) character of measurements (measurement results and their errors). Therefore, this index cannot be applied to the number of those which are controllable from "within" a metrological system, although it is important from the point of view of customer of measurement information.

From the remaining four indices the first three (accuracy, trueness, and repeatability) are completely determined by one index, i.e., by the measurement accuracy as an integral index characterizing the closeness to zero of systematic and random components of a separate measurement error (in single and repeated observations).

From the point of view of the issue under consideration the last index of measurement quality, i.e., the reproducibility, from all those listed above, is much more interesting, since it characterizes the "collective properties" of a generalized system of measurements (including the NSM). This index can be called the comparability of measurements (or more precisely, the comparability of measurement results) which is in our opinion a better term than the "reproducibility of measurements", since the latter has a hint of the repeatability of a measurement problem. This contradicts the content of the concept. Hereinafter we will use the short term "comparability of measurements" (with the meaning given above for the term "reproducibility").

Taking into account the above, as the basic indices of the measurement quality which is the main criterion of efficiency of the proper metrological system influence on the NSM, we chose two: accuracy and comparability of measurements. A high degree of measurement comparability can be provided at a significant systematic error Δ_c (a deviation of measurement result, i.e., a real measurand value from a true measurand one). At the same time the value Δ_c can known or unknown, but it must be the same in all measurements. It goes without saying that the comparability will "automatically" increase when the accuracy of all the measurements increases. However in practice both problems are of current importance.

In Figure 2.5 an attempt is made to illustrate the relationship between the different quality indices.

Let us consider NSM (2.2.16.1) to be a closed system (without inputs and outputs). This corresponds to the choice of a comparably short time interval T when a set of measurement problems z_i, as well as components needed to solve them [on the right of a vertical line in the expression for z in equation (2.2.16)] remain unchanged (constant). Obviously, for the first stage of the description of such a large system this assumption is quite defensible.

Now it is possible to consider the formulation of conditions for providing a required level of measurement quality in the NSM for both indices of quality: accuracy and comparability.

1) Since the condition of separate measurement correctness (2.2.10), as indicated in Section 2.2.3.3, means the condition of the maximum likely closeness to a given accuracy of measurement by way of taking into account all the measurement components of the system, then condition (2.2.10) can be used to formulate a *condition* of reaching the *given* (*required*) *measurement accuracy* within the framework of the NSM.

In a given measurement system of the NSM (see equation (2.2.16) and equation (2.2.16.1))

- for any measurement problem $z_i(\varphi_k)$ (see equation (2.2.17)) there exists a set of controllable parameters of the system

$$U_i(\varphi_k) = ([\varphi_k]_i, m_i, s_i, v_i, w_i, \ldots), \tag{2.2.18}$$

as well as a volume of a priori information I_a at which the performance of condition (2.2.10) is provided.

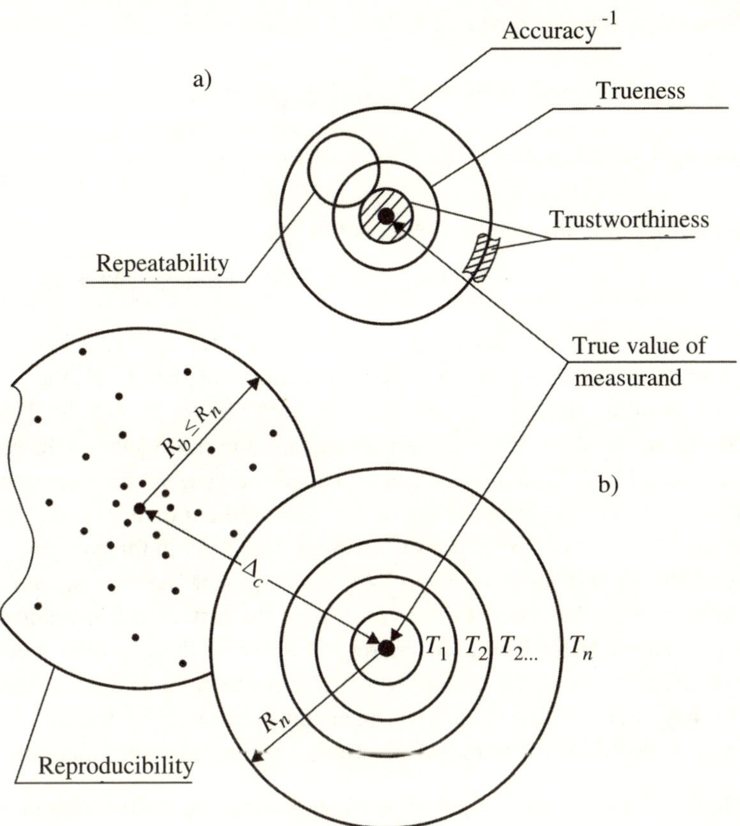

Figure 2.5. Illustration of the relationships of various indices of measurement quality: (a) accuracy, trueness, repeatability, and trustworthiness of a measurement results of one measurement problem solution; (b) reproducibility and accuracy of results of a solution of various measurement problems, performed at a different "visible" accuracy $T_1 > T_2 > T_3 > \cdots > T_n$.

In formal logic language this condition is expressed as

$$\forall z_i(\varphi_k)\exists\ i \in U_i(\varphi_k)\ \ \&\ \ I_a \to (2.2.10) \tag{2.2.18a}$$

It is clear that a priori information has to be related first of all to knowledge about the kinds and parameters of dependencies (2.2.9.1)–(2.2.9.7), linking parameter values of a given measurement system.

2) *The measurement comparability condition* within the framework of the NSM is formulated in the following way. On a given set $\{z_i(\varphi_k)\}$ of measurement problems in a given system of measurement at any $i \neq j$ but when $\varphi_{kr}(o_i) = \varphi_{kr}(o_j)$, there is a group of controllable system parameters such as

$$\langle [\varphi_k]_i, [\varphi_k]_j, m_i, m_j, s_i, s_j, v_i, v_j, w_i, w_j, \ldots \rangle, \tag{2.2.19}$$

which allows the condition to be met

$$\varphi_k^{\text{meas}}(z_j) - \varphi_k^{\text{meas}}(z_i) \leq \sqrt{\delta_i^2 + \delta_j^2}.$$

In formal logic language this condition is expressed as

$$\exists_i \in z_i(\varphi_k)A_i^N \neq j \;\;\&\;\; [\varphi_{kr}(o_i) = \varphi_{kr}(o_j)] \to \forall N[\varphi_k^{\text{meas}}(z_j) - \varphi_k^{\text{meas}}(z_i)] \leq \sqrt{\delta_i^2 + \delta_j^2}. \quad (2.2.19a)$$

Thus, a simultaneous fulfillment of conditions (2.2.18) and (2.2.19) within the framework of the general measurement systems being considered, i.e., the NSM, provides a corresponding quality measurement of the NSM on the whole. This makes it necessary to speak about the necessity of having one more subsystem within the framework of the NSM. This system can be called the *system of ensuring measurement quality* (*SEMQ*). It is used to control the measurement quality in the NSM, i.e., to achieve the fulfillment of conditions (2.2.18) and (2.2.19) in this system.

Unfortunately, at this stage of metrology development the formalization of the SEMQ has not succeeded, but this does not affect the task of our investigation. An attempt has been made to determine (merely intuitively) those proper metrological systems (or more precisely, the problems) which have to be included into the SEMQ. Here, on the basis of condition (2.2.18) it can be more definitely be said that the SEMQ has to provide the solution of the following problems (general requirements to SEMQ):

- research, development and output of working measuring instruments of a needed nomenclature and accuracy, as well as of computer engineering products (the task of the instrument-making industry on the basis of an analysis of "blanks" (unsolved problems) in equation (2.2.18) for components s_i and w_i);
- manpower training (operators, observers) of a corresponding qualification (component v_i);
- development of corresponding methods of measurements (m_i);
- introduction of required measurand units (components $[\varphi_k]_i$);

Undoubtedly the last three problems concern first of all the proper metrological systems (to all appearance, to the system of metrological assurance SMA).

The concept "comparability of measurements" closely correlates with the concept "uniformity of measurements". This is confirmed by an analysis of literature data where understanding the uniformity of measurements is identified with a state of the system where a given accuracy is provided in different places, time, conditions, methods, operators, and measuring instruments. In other words, it is possible to give the following initial definition.

Measurement is the state of a general system of measurements at which any two measurements of PQ, having the same dimension, carried out within the framework of this system, give results which do not overstep the limits of the evaluated errors of these measurements. It is therefore naturally possible to define a system of ensuring the measurements uniformity as a system that provides fulfillment of condition (2.2.19) in the system (2.2.16.1).

It is possible to say a bit more about this system (SEMU) than about the SEMQ.

First of all, measurement comparability condition (2.2.19) is not really a part of any measurement problem in the NSM. There whould be still be a need of a problem situation on ensuring the uniformity of measurements (or comparability of measurement results, which is the same thing). A typical (and maybe simply characteristic, i.e., determining) practical case of such a problem situation consists of the following.

Let a *customer* A be situated in a space–time continuum similar to that where the NSM is, but, having his own coordinates $(\Delta P_a, \Delta T_a)$, needs an object (product) "a" characterized by a set of consumption properties (measurable indices of quality) $\varphi_k(a)$, $k \in \overline{(1, n)}$. At the same time the product (object) "a" satisfies desires (problems) of the customer A only in the case where the values of each consumption property do not overstep the definite limits within the tolerance values $\Delta_{\text{tol}}\varphi_k(a)$ at the confidence probability $f_{\text{tol}}(\varphi_k)$.

A *supplier* B (manufacturer) with his own coordinates $(\Delta P_b, \Delta T_b)$, should set up the manufacturing of product "a" with the values of indices $\varphi_k(a)$ within the limits of the indicated (given) tolerance and confidence probability values.

Since both customer A and supplier B have their own interests and are able to use their own set of methods and means for determining the consumption properties $\varphi_k(a)$ of the product, then the main problem for settling their interpersonal relations (and for the whole economy of their country) consists of achieving a guarantee which will allows the supplier and customer to obtain comparable measurement results relative to the corresponding indices φ_k in spite of the different ways (methods) they use for one and the same measurand φ_k. To achieve this means meeting condition (2.2.19).

It is clear that the system providing the measurement uniformity between the systems B (the supplier) and A (the customer) has to be "external" with respect to both of them, but to belong to a common space–time continuum. It should be remarked that the described problem situation "supplier–customer" can take place between enterprises of a region or country, as well as between different countries.

From this it follows that for constructing a system of ensuring the uniformity of measurements not so much the quality of measurements carried out or the solution of measurement problems in the common system of measurements is important, as the number of interrelations "supplier–customer" for each measurable property φ_k, i.e., the number of problem situations in a system considered (for example, in some NSM). From the point of view of economics this is determined by the degree of specialization and cooperation of social manufacturing.

Now let us consider the interpretation of the measurement result comparability condition with regard to problems z_i and z_j:

$$\varphi^{\text{meas}}(z_j) - \varphi^{\text{meas}}(z_i) \le \sqrt{\delta_i^2 + \delta_j^2} \quad (2.2.19b)$$

under the condition $\varphi_r(z_i) = \varphi_r(z_j)$.

In accordance with equation (2.2.8) the results of solving these problems are

$$\varphi^{\text{meas}}(z_j) = n_\varphi(z_j)[\varphi]_{s_j}; \quad \varphi^{\text{meas}}(z_i) = n_\varphi(z_j)[\varphi]_{s_i}. \quad (2.2.20)$$

They correspond to the true dimensions of the units $[\varphi]_{s_j}$ and $[\varphi]_{s_i}$ which are realized in s_j and s_i, but which are unknown.

Observers in z_i and z_j believe that both of these express the results in terms of “accepted units” $[\varphi]_0$, i.e., they have the “seeming” results

$$\varphi_{\text{seem}}^{\text{meas}}(z_j) = n_\varphi(z_j) \cdot [\varphi]_0; \quad \varphi_{\text{seem}}^{\text{meas}}(z_i) = n_\varphi(z_i) \cdot [\varphi]_0, \quad (2.2.20a)$$

the difference between which is

$$\varphi_{\text{seem}}^{\text{meas}}(z_j) - \varphi_{\text{seem}}^{\text{meas}}(z_i) = \{n_\varphi(z_j) - n_\varphi(z_i)\} \cdot [\varphi]_0 = \Delta n_\varphi(z_j, z_i) \cdot [\varphi]_0,$$

and it is compared with $\sqrt{\delta_i^2 + \delta_j^2}$, i.e., the comparison is performed according to indications of the corresponding measuring instruments. As to a real difference of the values, it can differ from the seeming one due to the difference between the real unit dimensions realized in s_i and s_j and the accepted dimension:

$$\varphi_{\text{real}}^{\text{meas}}(z_j) - \varphi_{\text{real}}^{\text{meas}}(z_i) = \Delta n_\varphi(z_j, z_i) \cdot [\varphi]_{\text{real}}.$$

We can see this at $[\varphi]_{\text{real}} \ne [\varphi]_0$, $\varphi_{\text{real}}^{\text{meas}} \ne \varphi_{\text{seem}}^{\text{meas}}$. In other words, the reproducibility (the comparability) of measurements depends on the dimension of the unit realized in measuring instruments. From this it follows that the validity of meeting condition (2.2.19) can be secured only in the case when the units in the measurement systems z_i and z_j being compared are similar and equal to (close to) the dimension of the unit *accepted* in the given system of measurements.

This means that SEMU has to include a subsystem which would not only provide the uniformity of units (from the point of view of the dimensions realized in the measuring instruments of the generalized system of measurements) but also their compliance with conventionally accepted units (with their definition). *The RUTS system plays this role* in accordance with the essence that was prescribed to it earlier in Section 2.2.3.4. Thus, the following interrelation of systems has been established:

$$\text{NSM} \leftrightarrow \text{SEMQ} \subset \text{SMA} \subset \text{SEUM} \subset \text{RUTS system.} \quad (2.2.21)$$

2.2.4 Fundamentals of constructing a RUTS system

2.2.4.1 Basic properties of a RUTS system

Let us sum up what we know about the RUTS system. According to Section 2.2.3.5 the *goal of the RUTS system* is to provide an objective estimate of the condition of measurement comparability (2.2.19), which is reduced to ensuring the compliance of these units with those conventionally accepted (their closeness to the definition) within the framework of the same generalized measurement system to whcih the given RUTS system is related.

As the generalized system we choose, as before, a closed NSM, i.e., a system of the type

$$\sum_{\text{NSM}}(T) = \bigcup_k \{I_{z_i}(\varphi_k)\}, \tag{2.2.21.1}$$

where

$$I_{z_i}(\varphi_k) \equiv \langle \varphi_{k_i}, o_i, \psi_i, \delta_i, g_i, t_i, \Delta t_i, p_i, \ldots \,|\, [\varphi_k]_i, m_i, s_i, v_i, w_i, \ldots \rangle;$$
$$i \in \{z_i(\varphi_k)\}, \quad k \in \{\varphi_k\}, \quad t_i \in T, \quad p_i \in P_{\text{NSM}}.$$

The closedness of the NSM means the stability of the sets

$$\Phi\{\varphi_k\} = \{\varphi_1, \ldots, \varphi_k, \ldots, \varphi_N\}, \tag{2.2.22}$$
$$Z(\varphi_k) \equiv \{Z_i(\varphi_k)\} = \{\varphi_{k_i}, o_i, \psi_i, \delta_i, g_i, t_i, \Delta t_i, p_i, \ldots\}, \tag{2.2.23}$$
$$U(\varphi_k) \equiv \{U_i(\varphi_k)\} = \{[\varphi_k]_i, m_i, s_i, v_i, w_i, \ldots\}. \tag{2.2.24}$$

Set (2.2.24) is the set of all controllable elements of system (2.2.16.1), i.e., elements chosen for solving a particular measurement problem $Z_i(\varphi_k)$ when a measurement system I_{z_i} is constructed for this problem to be solved.

Let the general concept of total RUTS system (2.2.12) be rewritten as

$$\sum_{\text{RUTS}} = \bigcup_k \sum_{\text{RUTS}}(\varphi_k) = \bigcup_k \{s_{i_1 b}(\varphi_k), s_{i_2 n}(\varphi_k), m_{i_3 b}(\varphi_k), m_{i_4 n}(\varphi_k)\},$$
$$k \in \{\varphi_k\}, i_1 = (\overline{1, n_1}), i_2 = (\overline{1, n_2}), i_3 = (\overline{1, n_3}), i_4 = (\overline{1, n_4}). \tag{2.2.25}$$

The structure of a particular RUTS system (for a given measurand) is hierarchal. The general form of this structure is shown in Figure 2.6.

The *inputs* of the total RUTS system are the following attributes:

- definitions of the units (the main input) including the conception about each measurand in the system of physical quantities;
- a priori information about the basic dependencies of measurands which characterize physical phenomena for reproducing the unit;

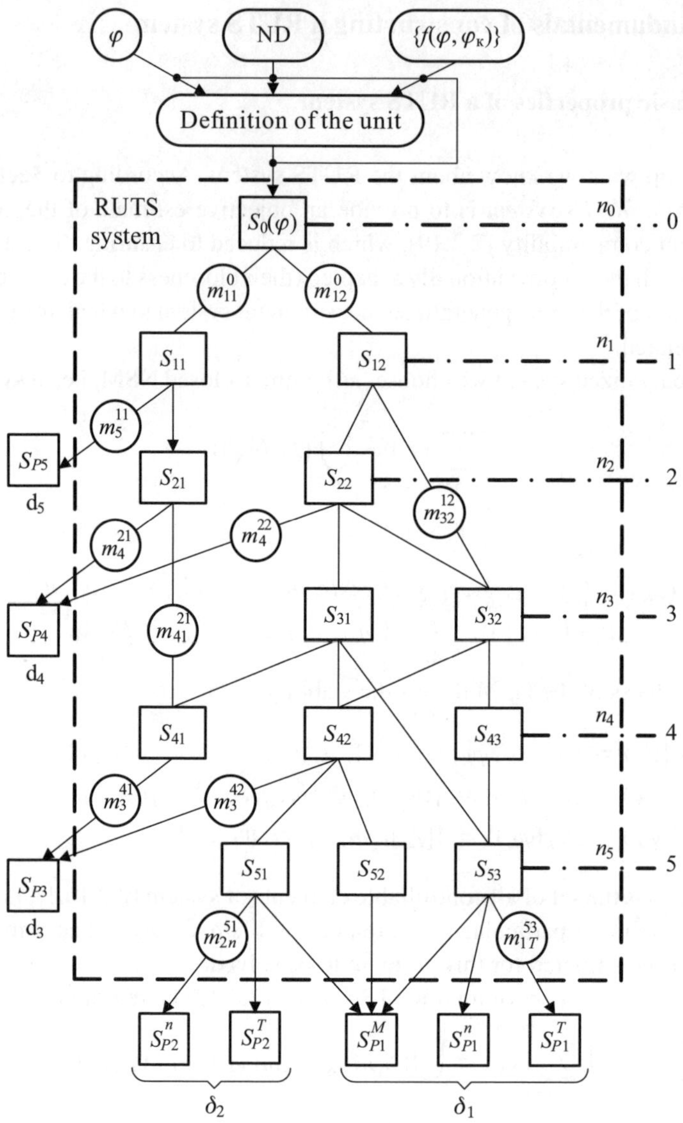

Figure 2.6. General structure of a particular (for a given φ RUTS system).

- data about physical constants which are used in definitions of units, as well as about constants of substances, materials, processes and events, all of which are used for realizing the units (standard reference data (SRD));
- general requirements for reference and subordinate metrological measuring instruments.

The *outputs* of the total RUTS system are the dimensions of units, realized in subordinate metrological measuring instruments and then used in working measuring instruments of the generalized system of measurements, as well as the methods of verifying these working measuring instruments.

For a particular RUTS system everything noted refers to a given measurand.

However, for the theoretical construction of an RUTS system it is necessary to know all the parameters (features) of the measurement system of the NSM, entering its description (2.2.16.1).

2.2.4.2 Volume of initial data for constructing an RUTS system

Let us attempt to evaluate the power of uniform sets for each parameter (feature), characterizing the NSM according to description (2.2.16.1).

1) *Number of measurable measurands* in a NSM. An analysis shows that their number is ~250:

$$|\Phi| \cong |\{\varphi_k\}| \sim 250. \qquad (2.2.25.1)$$

2) *Number of units of measurands*. This number, strictly speaking, has to correspond to the number of measurands. As a matter of fact, some measurands have several units (not counting multiple, submultiple, and other units, linked unambiguously by an accurate number): time (on macroscale), area, hardness, pressure, and others. It is much more complicated when one and the same unit is used for heterogeneous measurands (for more detail see Section 2.2.5.2). Let us consider that

$$|[\Phi]| \equiv |\{[\varphi_k]\}| \approx 250. \qquad (2.2.25.2)$$

Further estimates will be obtained on the basis of making calculation for one measurand (on average).

3) *The number of PQ realizations* on a set of objects $\{o_i\}$. The number of objects on which the measurements of even one measurand are made is infinitely great. Let us make an evaluation of the maximum of gradations of perceptible dimensions of a measurand, reasoning from the logarithmic law of measurand sizes distribution [368], the maximum ranges of their changes and the middle range of measurand dimension perception (which corresponds to a conventional gradation of objects on the basis of dimensions of the measurand they reproduce.

Considering that $\varphi_{\max} : \varphi_{\min} \approx 10^{17} : 10^{-17} = 10^{34}$, and the range of perceptible values corresponds to ~10, we obtain

$$|\{\varphi(o_i)\}| \sim 30 \div 40. \qquad (2.2.25.3)$$

It should be kept in mind that the objects are characterized not only by the dimension of a measurand realized in them, but also by other parameters which influence the solution of the measurement problem (aggregative state, informative parameters, substance composition, and others). Therefore, estimate (2.2.25.3) can be increased approximately by one order.

4) *Number of realizations of measurement conditions*. Obviously, it is possible to consider that the number has the same order as compared to the number of various (according to perception) gradations of measurand dimensions (at the same time the coefficient is usually taken as 3–4, i.e., according to the number of influencing quantities). Therefore

$$|\Psi| \equiv |\{\psi_i\}| \sim 100 = 10^2. \qquad (2.2.25.4)$$

5) *Number of given error values*. This number can be estimated on the basis of a number of accuracy classes as applied to measuring instruments in practice. It contains $\cong$10 values. Thus,

$$|\Delta| \equiv |\{\delta_i\}| \sim 10. \qquad (2.2.25.5)$$

6) *Number of presentation forms of measurement results*. This number is regulated by GOST 8.011 and is $\leq$10.

$$|G| \equiv |\{g_i\}| \sim 10. \qquad (2.2.25.6)$$

7) *Number of measurement time intervals*. It is also useful in the same way as in Section 2.2.2.3 to evaluate (estimate) this by the number of gradations in a real practical scale of intervals Δt from 10^{-9} to 10^7 s.

$$|\{\Delta t_i\}| \sim 20. \qquad (2.2.25.7)$$

8) *Number of measurement methods*. From the point of view of the ways to use measuring instruments it is the units. However, taking the principle applied (physical ones) into account, it is the number that is very difficult to evaluate. Neglecting this fact, i.e., considering that a measurement method means only general ways of using measuring instruments (the interpretation most often applied) we will obtain the estimate

$$|M| \equiv |\{m_i\}| \sim 10. \qquad (2.2.25.8)$$

9) *Number (nomenclature) of applied measuring instruments*. This number depends on the range of PQ dimensions being measured, the accuracy and conditions of measurements, i.e., on estimates (2.2.25.3)–(2.2.25.6). On the average, it is possible to consider that for each measurand

$$|S| \equiv |\{s_i\}| \sim 10^3. \qquad (2.2.25.9)$$

For many problems it is important to know the nomenclature of measuring instrument properties influencing a measurement result, i.e., *metrological characteristics* of measuring instruments. However from the point of view of the RUTS systems construction in most cases this can be neglected.

10) *Number of professional qualifications of an operator* (an observer). In the absence of normative data (if at all they exist) we consider that it will be useful to divide the operators into three classes: (III) for technical (mass) measurements; (II) for verification works and for engineering and laboratory measurements; and (I) for measurement standard works and high-accuracy physical experiments. Since this matter concerns the parameters of the NSM, then only three classes of operators is the most typical; thus,

$$|V| \equiv |\{v_i\}| \leq 3. \tag{2.2.25.10}$$

11) *Number (nomenclature) of computer engineering means* used in measurements. At present this number is difficult to evaluate. The general nomenclature is ~ 100 but similar to all measurement problems. From the point of view of constructing RUTS systems at a given stage this estimate seems to be insignificant.

Moreover, it is unlikely that it is useful to evaluate the number of space–time coordinates (t_i, p_i) in which measurements are realized. These parameters become essential when considering the issues of the RUTS system *operation*. Here it is important that $t_i \in T$, $p_i \in P$ (in the NSM).

Finally, the *number of measurement problems* (nomenclature) in system (2.2.16.1) will be determined by the totality of estimates for all the parameters indicated above. At the same time it is necessary to take into account that the estimate values for some parameters appear to be interconnected (correlated): for example, the form of presenting a result, measurement error, operator skill, volume of a priori information in use, etc.

It is therefore very difficult to give a real estimate of the measurement problem nomenclature (obviously, there are tens of thousands of varieties for even one measurand). Of one thing there is no doubt: even a particular NSM as a totality of all measurements for one measurand is a quite complicated system – at the least a 10-dimensional one, which has sets of values for each component of a 10-dimensional vector space.

2.2.4.3 General principles and an algorithm for constructing RUTS systems

In view of the huge volume of initial data and the complexity of the general structure of the RUTS system, a solution to the problem of constructing this system in a general form is not possible. It should be taken into account that a large part of the initial data, as a rule, is either entirely absent or known to have a significant degree of uncertainty (inaccuracy), which for a multifactor problem can reduce to zero the efforts spent to solve it.

Therefore, it is useful to choose principles and methods which allow a real effect from the fairly general approach suggested here.

The principle of successive approximation is the most natural. Some assumptions, i.e., restrictions, which simplify the problem and decrease its dimension, are successively introduced.

In the case given the *method of ranking and gradation* has to be efficient. In the system there are selected separate blocks (subsystems) or classes (groups) of problems which are ranked according to both their degree of generality and algorithmic sequence, and then the *principle of stage solution* is realized, i.e., the sequence of the problem solution "by parts" is determined.

At last, for solving the problems of such a kind it is efficient to use the "*shuttle method*", i.e., a repeated solution of direct and inverse problems, verification of solutions by several "cuts".

On the basis of these principles we offer a scheme and algorithm of problem solution for theoretical constructing the total RUTS system.

At the first stage the problem of constructing a system in a "bottom-up" manner is solved. At the same time the system is disjoined into subsystems connected with the definite measurand φ, and the problems are solved for each particular subsystem $\sum_{\text{RUTS}}(\varphi)$, i.e., the "$\varphi_i$ problems".

In these problems the concept "*chain of transferring*" a unit dimension, i.e., the chain of subsequent elements of the system (means of transferring the dimension S_n), connecting reference metrological measuring instrument of the system with some definite group (some type) of working measuring instruments (see Figure 2.4). The use of the graph theory apparatus for formalizing the connections between the adjacent levels can appear to be useful here.

The algorithm of the "problem φ_i" solution consists of the following.

1) Initial data (see Section 2.2.4.2) for the given measurand φ_i are grouped into blocks and "tied to" the WMI type S_{p_i}. At the same time it is assumed that the nomenclature (types) of working measuring instruments has been optimized earlier according to measurement problems.

2) Types of working measuring instruments are grouped in accordance with the accuracy $S_p(\delta_1), S_p(\delta_2), \ldots, S_p(\delta_m)$, and

$$\delta_1 > \delta_2 > \cdots > \delta_m. \tag{2.2.26}$$

Ranking the working measuring instruments according to their accuracy is performed for two reasons: firstly, because the goal of the RUTS system is the maximum approximation of a real unit dimension of working measuring instruments to an ideal one (according to the definition). This is achieved above all at the expense of measurement accuracy. Secondly, among the remaining parameters of WMI only the range of measurements is essential for constructing a system. It is quite correct to assume that the influence of other parameters (above all the conditions of measurements) on the accuracy of working measuring instruments is foreseen by a normative document. As to the measurement range of a particular working measuring instrument, its ranking takes place "automatically", while the working measuring instruments are subjected to the accuracy ranking procedure. This follows from the most general ideas about information (or resolution) capability of these instruments. The measuring instruments having the widest measurement range appear to have the least accuracy.

3) The longest chain of transfer is chosen. Paradoxical as it may seem at first glance, the chain of such a type can be determined a priori before constructing the system. It is the transfer chain aimed at the least accurate group of working measuring instruments $S_p(\delta_1)$. It is clear first of all from the fact that transferring the unit dimension to a more accurate working measuring instrument will require fewer transfer steps (levels) at a reference MMI accuracy equal for all working measuring instruments. Moreover, the most numerous group of working measuring instruments consists of less accurate ones, i.e., the expression given below is true almost without exception:

$$N_p(\delta_1) > N_p(\delta_2) > \cdots > N(\delta_m), \tag{2.2.27}$$

where $N_p(\delta_1)$ is the number (park) of working measuring instruments of the given accuracy group. So, from the point of view of the productivity (passing capacity), the chain of transferring the unit to the less accurate working measuring instruments has to be the longest one.

4) *Constructing the longest transfer chain* to the first approximation is performed. The construction methond depends on the choice (the availability) of initial data and vice versa. In any case it is necessary to know the park of working measuring instruments N_{p_1}. The following versions are possible:

Version 4a. Given

- t_{i+1}: the verification time for one measuring instrument at the $(i + 1)$-th level performed with a metrological measuring instrument of the i-th level;
- l_i: the number of measuring instruments of the $(i + 1)$-th level that can be verified with a metrological measuring instrument of the i-th level;
- q_{i+1}: the part of measuring instruments of the $(i + 1)$-th level, which are recognized in accordance with verification results as unsuitable, restored, and verified again;
- T_{i+1}: the verification interval (an average time of faultless operation with regard to the metrological reliability) for a measuring instrument of the $(i + 1)$-th level;
- τ_i: the part of time within which a metrological measuring instrument of the i-th level is used for verification (within the frames of a given operation period, i.e., within the frames of T_i).

The number of metrological measuring instruments of the i-th level necessary for verifying n_{i+1} measuring instruments at the $(i + 1)$-th level is determined from the continuity condition of the unit transfer process within their verification interval:

$$T_{i+1}\tau_i n_i = \frac{t_{i+1}N_{i+1}(1+q_{i+1})}{l_i}. \qquad (2.2.28)$$

Knowing that $N_{p_1} = n_{m+1}$ we first find n_m and then, using the recurrent method, the population of each level (n_i) and the number of levels m, are determined.

Version 4b. Given

- n_1: the number of subordinate metrological measuring instruments of the 1-st level in the given chain, which are verified with S_0 (with the reference metrological measuring instrument) within its verification interval T_0;
- c_Σ: the ratio between an error of the reference metrological measuring instrument to an error of the working measuring instrument;
- g_0 and g_p: the student coefficients for confidential errors of the reference metrological measuring instrument and working measuring instrument;
- the permissible probability of verification defects (preferable ones from the point of view of the aim of the RUTS system with regard to the defect of the second kind).

Using the procedure of the MI 83-76 National System of Measurement Assurance. Method of Determination of Verification Sceme Parameters, a maximum possible and minimal necessary (under these conditions) number of levels m is found. At the same time while calculating according to MI 83-76, it is necessary to introduce a correction that is the inverse one with regard to the coefficient of the population level occupied by subordinated MMI, since in the considered case of the linear transfer chain the "side" flows are absent (see condition (2.2.28)).

Version 4c. Given

- n_1: the number of MMI of the 1-st level;
- t_{m+1}: the time of verifying one working measuring instrument against an MMI of the lowest level m;
- l_i: the same as in version 4b;
- h: the ratio of the verification time of one MMI on the second level against an MMI of the first level to the verification time of one working measuring instrument: $h = \frac{t_2}{t_m}$;
- τ_i: the same as in version 4a;
- $\mu = (1 + q_{i+1}) = 1.25$: the average estimate of metrologically serviceable (fit according to verification results) measuring instruments on all levels;
- it is also assumed that verification intervals for all MMI are of a lower level than the first one, and also for working measuring instruments they are similar and equal to $T_m = 1$ year and $l_i = 1$.

Then by the procedure outlined in [526], the maximum number of MMI at each level and the minimum number of levels are found.

5) Similar construction procedures are made for the remaining chains relating to the verification of working measuring instruments of the other accuracy groups. On the basis of the results obtained an improved *structure of the RUTS system of the first approximation* is constructed.

At the second stage the nomenclature of MMI and methods of transferring the unit dimension on each step of the system which satisfy the conditions shown below are constructed:

(a) "compatibility" of the transfer method and types of MMI at the adjacent levels connected by this method m_{ij}^{kl} (see Figure 2.4);

(b) problems for which the system construction was performed at the preceding stage (versions 4a–4c).

If it is not possible to satisfy some conditions with an "available" set of means and methods suitable for MMI, then appropriate corrections are introduced into the conditions of versions 4a–4c, and the process of constructing is performed anew. This results in an improved *structure and composition of a particular concrete specific RUTS system of the first approximation.*

An analogous structure according to the first two stages is made for all other measurands $\varphi_k \in \{\varphi_k\}$ measured within the frames of the general measurement system, i.e., NSM. As a result we obtain a *total RUTS system of the first approximation* which satisfies the goals of this system to a first approximation.

The queue of performing *two subsequent stages* of constructing the RUTS system can change, depending on the availability of corresponding data, particular aims of further improvement or of the level of the consideration commonness, of available resources, etc.

At one of these stages (for example, at the *third* one) the compatibility of different particular systems is checked from the point of view of the general properties of the system of unit reproduction (see Section 2.2.5.1). At another stage it is possible to solve a number of optimization problems using economic criteria, registration of performance parameters of the RUTS system (or its subsystems), and optimization by the efficiency criterion of system performance.

At the final, *fifth stage* some inconsistency (discrepancy) of solutions obtained to separate conditions at all preceding stages are revealed, the necessary corrections are introduced into the initial data, and corrections are determined for a final solution of the first approximation.

After that we have a complete RUTS system, optimal with regard to the composition and structure and at most corresponding to the goal and quality of this system at an available knowledge level.

In Section 2.2.4.4 some problems of the operation and optimization of the RUTS systems are considered. Then a number of issues connected with the system of unit dimension reproduction as well as the influence of this system on the complete RUTS system construction are analyzed.

2.2.4.4 Efficiency of the RUTS systems operation and their optimization

Efficiency of RUTS systems

Let us now introduce the general concepts of "efficiency" and "quality" systems. The system efficiency (ε) in a general form is determined by the ratio of the effect obtained from the system to the costs of its construction:

$$\varepsilon_\Sigma \equiv \text{E (effect)}/\text{C (costs)}, \tag{2.2.29}$$

where the effect is determined as the extent of reaching the goal (G) of the system under consideration

$$\text{E} \equiv q\text{G}. \tag{2.2.30}$$

The multiplier q is the system quality, i.e., the extent of system compliance to the goal achievement. This interpretation eliminates the confusion that can constantly arise when using these concepts (frequently simply identifying them) and permits them to be given a specific content.

Indeed, definition (2.2.29) becomes current and means

$$\text{Efficiency} = \text{effect/costs} = \text{what does it give/what does it take}$$

(for the system with regard to some external system).

If the concept "project efficiency", (ε_0), is introduced as follows:

$$\varepsilon_0 = \text{what has it to give/what does it take}$$

(for the system with regard to an external one), then $\varepsilon = q/\varepsilon_0$ and q (the quality) takes on the sense

$$q = \varepsilon/\varepsilon_0 = \text{what does it give/what does it have to give}$$

(to the system).

Thus, *the quality is an index of the internal properties of the system which are determined by the goal of the external system, and the efficiency is an index of its external properties caused by its quality.*

The goal (G) of the system of ensuring measurement uniformity has been formulated earlier as fulfilling two conditions simultaneously for NSM:

(a) closeness of the units which various measuring instruments of a given measurand have in NSM;

(b) closeness of these units to the ideal ones (on determination).

Let us formalize these conditions in the language of the system approach used here:

$$[\varphi]_{s_j} - [\varphi]_{s_i} \leq \sqrt{\delta_i^2 + \delta_j^2} \equiv \delta_{ij}; \tag{2.2.31a}$$

$$[\varphi]_{s_r} - [\varphi]_0 = \delta_r \ll \delta_{ij}. \tag{2.2.31b}$$

Condition (goal) (2.2.31a) is provided in manufacturing and verifying working measuring instruments against standard measuring instruments (S_m as adopted in Section 2.2.4.3). For the most part, it is a problem of the instrument-producing industry and SEUM in the part relating to departmental metrological services.

Condition (2.2.31b) means that the error of the standard measuring instrument S_m, against which the working measuring instruments S_i and S_j were verified and which are used in solving the problems z_i and z_j under conditions of measurement compatibility (2.2.19), is negligible with regard to the errors S_i and S_j, but not smaller than the difference between the unit dimension realized in this standard measuring instrument and the ideal one. This is exactly the problem that the particular RUTS system solves.

Thus, *the generalized goal of a particular RUTS system* (for a measurand φ given in SEUM) is meeting condition (2.2.31b) for any measurement problems satisfying condition (2.2.19), or

$$\mathrm{G}\Big(\sum_{\varphi}\Big): \frac{\delta(s_m)}{\sqrt{\delta_i^2 + \delta_j^2}} = \frac{[\varphi]_{s_m} - [\varphi]_0}{\sqrt{\delta_i^2 + \delta_j^2}} \ll 1$$

for any $[i \neq j]$ in system (2.2.16.1), which satisfy the condition

$$\varphi^{\text{meas}}(z_i) - \varphi^{\text{meas}}(z_j) = \sqrt{\delta_i^2 + \delta_j^2} \quad \text{and} \quad \varphi(z_i) = \varphi(z_j). \tag{2.2.32}$$

Strictly speaking, it is necessary to impose *problem situations limits* on this condition, as well as on condition (2.2.19) concerning SEUM, i.e., equations (2.2.19) and (2.2.32) should be applied not to any pair (z_i, z_j) but mainly to those pairs of problems that are solved in the systems "supplier–customer".

In practice it is very difficult to evaluate this. However the limitation indicated also plays another role that is more essential: it allows only those pairs of measurement problems to be considered which are predominantly related to *one type of applied working measuring instruments* (by both the range and accuracy). This significantly reduces the dimension of problems connected with determining the efficiency of the RUTS system (and SEUM), but, to be more precise, results in dividing the variables and bringing an additive form of summation of the efficiency of working measuring instruments of different groups.

Here one more note should be added, which concerns an effect of the systems RUTS → SEUM → SMI → NSM. Since the working measuring instruments are used for evaluating the product quality by both a manufacturer and a customer and as a rule tolerances for corresponding parameters being measured, φ_k, of a product are similar ($\Delta_{\text{pm}} = \Delta_{\text{pc}} = \Delta_{\text{perm}}$), in both cases, it becomes important for a customer and manufacturer to meet not only condition (2.2.19), but also the *error relationship of the working measuring instruments for both a manufacturer and a customer*.

Provided, for example, $\delta_{\text{pc}} > \delta_{\text{pm}}$, then at $\Delta_{\text{perm}} = \delta_{\text{pc}}$ the customer risk (connected with decreasing of product quality due to such a measurement) will exceed the manufacturer risk (connected with the increasing costs of the output of the products), which creates a situation where the product quality in fulfilling tasks of a production plan decreases; and vice versa, at $\Delta_{\text{perm}} = \delta_{\text{pc}} > \delta_{\text{pm}}$ the quality of products obtained by the customer will increase due to the increase of manufacturer expenses (failure to execute the plan).

These important conclusions lead to the following *practical recommendations*:

(1) it is useful to equip manufacturers and customers with working measuring instruments of similar accuracy ($\delta_{im} = \delta_{jc}$);

(2) when working measuring instruments are used by manufacturers and customers it is necessary to select the relationship $\frac{\delta}{\Delta_{\text{d}}}$ such, that the probability of a defect of the verification of the first or second kind would be similar: $p(1) = p(2)$.

The quality q of RUTS system (φ) may be naturally determined by increasing the coefficient of accuracy reserve between WMI (s_i and s_j) and MMI (s_m) in condition (2.2.32), i.e.,

$$q\Big(\sum_{\varphi}\Big) = \frac{\sqrt{\delta_i^2 + \delta_j^2}}{\delta_m^{\text{lim}}} \cdot \frac{\delta_m}{\sqrt{\delta_i^2 + \delta_j^2}} = \frac{\delta_m}{\delta_m^{\text{lim}}}, \tag{2.2.33}$$

where δ_m is the really obtained in the RUTS system error of the MMI, (φ), against which the given WMI group characterized by $\delta_p = \sqrt{\delta_{ipt}^2 + \delta_{jpt}^2} = \delta_{pt}$ ($\delta_i = \delta_j$) taking into account the recommendations obtained) is verified, and δ_m^{lim} is the maximum value of the error of these MMIs meeting condition (2.2.32).

Thus, in expression (2.2.29) for efficiency only the costs of the system, C, remain unfixed. Considering the RUTS system as a system that *really* allows condition (2.2.32) to be met, the *total costs*, C_Σ, of realizing the total RUTS system (2.2.12a) in such a sense should contain: costs C_1 of research works on the development of its elements (MMI and methods of measurements), costs C_2 of manufacturing a needed park of MMI, costs C_3 of their allocation, costs C_{ex} of the operation of system elements, transport costs C_{tr} of WMIs and costs C_{con} of realizing the control of the system:

$$C_\Sigma = (C_1 + C_2 + C_3) + E_n + C_{ex} + C_{tr} + C_{\text{con}}, \tag{2.2.34}$$

where E_n is the normative efficiency coefficient. (Here all costs should be brought to an identical interval: to T, total time of NSM at the invariability of its indices, or to $T_{ivi} \in T$, the verification interval of WMIs, or to $T_s \in T$, the selected unit of time, usually a year; then the costs will be as they are indicated.)

The costs $(C_1 + C_2 + C_e)$ are the amount spent on pure research and engineering problems, and the costs $(C_e + C_{tr} + C_{\text{con}})$ are connected with solving and realizing organizational and legal problems (including the development of normative documents concerning the RUTS system and its elements, creation of control bodies, system of metrological control and inspection into C_{con}, etc.).

Since the costs of the second kind relate to elements of more generalized systems such as SEMU, SMA, and SEMQ, they will not be considered here. Then the costs relating to only to the RUTS system of the kind presented by (2.2.12a) are expressed in the form

$$C(\text{RUTS}) = (C_1 + C_2)E_n C_{ex} = C(\varphi). \tag{2.2.34a}$$

Then the *efficiency of the RUTS system* (φ), satisfying goal (2.2.32), taking into account equations (2.2.33) and (2.2.34a), can be written in the form

$$\varepsilon_{\Sigma(\varphi)} \equiv \frac{E(\varphi)}{C(\varphi)} = \frac{q(\varphi)\cdot \mathrm{G}(\varphi)}{C(\varphi)} = \frac{\delta_m}{\delta_m^{\lim}} \quad \text{at}$$

$$\left(\delta_m^{\lim} \leq \delta_p \frac{1}{C(\varphi)}\right) \geq \frac{\delta_m}{\delta_p[(C_1 + C_2)E_{\text{n}} + C_{\text{e}}]_\varphi} \quad \text{or}$$

$$\varepsilon_{\Sigma(\varphi,t)} \geq \frac{\delta_{mt}}{\delta_{p_t}[C_{\text{e}}(\varphi_k,\ t) + C_{\text{c}}(\varphi_n, t)E_{\text{n}}]} \tag{2.2.35}$$

for each group of WMIs of the given accuracy s_{p_i}; $t \in (\overline{1,m})$; $C_c = C_1 + < C_2$.

Optimization problems

A general statement of any problem on the optimization of any system Σ consists of founding a value of the functional:

$$\varepsilon_\Sigma = F(x, u), \tag{2.2.36}$$

where x is the uncontrollable parameters of the system and u is its controllable parameters satisfying the maximum efficiency of the system ε_Σ, i.e., determination of $u_{i(max)}$, corresponding to

$$\frac{\partial\varepsilon}{\partial u_i} = 0 \quad \text{and} \quad u_{il} \leq u_{i(\max)} \leq u_{ih}, \tag{2.2.37}$$

where $i \in I$ is the number of controllable parameters.

At the same time the following statements of the problem are possible:

(1) maximization of efficiency function at given resource limitations (and also of other types), which corresponds to finding the maximum effect in equation (2.2.29);

(2) minimization of costs (resources) at a given effect (level of a goal function and quality of the system).

In the case considered the problem of optimizing the RUTS system is formulated as follows: Find the parameters u_i on which the indices of the efficiency ε_Σ (RUTS) depends:

$$\frac{\delta_m}{\delta_p} = f_\delta(u_i); C_e = f_e(u_i) \quad \text{and} \quad C_c = f_c(u_i), \tag{2.2.38}$$

at which the system efficiency reaches the maximum:

$$\varepsilon_\Sigma^0 = \frac{\delta_{(m)}}{\delta_p(C_{\text{ec}} + C_{\text{n}}E)} \to \max_{u_i}. \tag{2.2.39}$$

At the same time, the limitations can be imposed on all indices of the system including those which follow from the construction of its structure and are described in Section 2.2.4.3:

$$u_{il} \le u_i(\max) \le u_{ih}, \quad C \equiv C_{\text{e}} + C_{\text{c}}E_{\text{n}} \le C_{\text{max}}. \tag{2.2.40}$$

Thus, the *choice of changeable (controllable) parameters* of the system considered is determinant. This choice depends on the level of consideration, the specific problem posed when optimizing (a choice of some parameter u_i which is of interest), the intuition of the researcher, since the efficiency of optimization (the payback of costs for the optimization as a result of its effect) depends on how sensitive the system efficiency becomes due to choosing a parameter of interest).

Here various ways of posing the *optimization problems* are possible. In each of them it is necessary to choose its own set of controllable parameters which should be solved in a definite order. Let us formulate some of them.

Problem 1. *Determination of an optimal accuracy relationship between the levels of RUTS system (φ) in one chain of the unit dimension transfer.* In this case the variable (controllable) parameters are the efficiency indices $\frac{\delta_m}{\delta_p}$ and $\frac{\delta_i}{\delta_m}$. Cost of the creation and operation of each element of the system (MMI) is expressed as a function of its accuracy. As a rule the dependence of the form given below is chosen:

$$C = \frac{C_0}{\delta_{\text{rel}}^\alpha}. \tag{2.2.41}$$

The coefficients α and C_0 are determined empirically from the available data on the cost of measuring instruments (of a given measurand) of different accuracy. Usually we have $1 \le \alpha \le 2$. The parameter C_0 is different for the cost of creation and cost of operation of a MMI, and strongly depends on the kind of measurements (the measurand being measured).

In the simplest case the parameters m and n_i of the RUTS system (φ) are given, i.e., are evaluated in advance on the basis of algorithms (see Section 2.2.4.3). Moreover, the efficiency functional is constructed:

$$\varepsilon^0_{\Sigma(\varphi)} = \frac{\delta_m}{\delta_p}\left(\sum_{i=0}^{m}\frac{n_i C_{0e}}{\delta_i^\alpha} + E_\text{n}\sum_{i=0}^{m}\frac{n_i C_{0c}}{\delta_i^\alpha}\right)^{-1}$$

$$= \frac{\delta_m}{\delta_p}\left((C_{0e} + E_\text{n}C_{0c})\sum_{i=0}^{m}\frac{n_i}{\delta_i^\alpha}\right)^{-1} \tag{2.2.42}$$

and its maximum is found under the condition (limitation) resulting from the law of error accumulation at the time of transferring the unit dimension:

$$\delta_p = \sqrt{\sum_{i=0}^{m}\delta_i^2}. \tag{2.2.42.1}$$

Further, the problem can be complicated for the case of all transfer chains.

Problem 2. *Determination of the optimal parameters of the RUTS system structure, (m and n_k), and MMI errors at given WMI parameters, (N_p, T_p, δ_p), as well as the means of reproducing the unit, $(n_0 = 1, T_0, \delta_0)$, i.e., of a reference MMI, (S_0).* In this case the controllable parameters are m, n_k and δ_k $(k = 1, \dots, m-1)$. At the same time a limitation is imposed:

$$\prod_{k=0}^{m-1}\frac{\delta_k}{\delta_{k+1}} \le \frac{\delta_0}{\delta_p} \quad \text{and} \quad \delta_p = \sqrt{\delta_0^2 + \sum_{i=k+1}^{m}\delta_i^2}, \tag{2.2.43}$$

and a form of empirical dependence connecting the transfer capacity of MMI and their cost with the accuracy are chosen.

Problem 3. *Determination of the optimal parameters of the RUTS system structure, (m and n_k), and the MMI errors, taking into account the probability of the defects of verification both when transferring the unit dimension within the system limits and when verifying WMIs.* The problem is similar to the preceding one, but is more complicated and requires additional initial data:

- dependence on the accuracy of a cost of creating and operating both suitable and unsuitable MMIs;
- cost of losses due to operating unsuitable WMIs;
- laws of error distribution of all MMIs and WMIs (this is the main difficulty).

Problem 4. *Optimization of the parameters of the RUTS system on the structure of interactions among its elements (dependence of methods of transferring unit dimensions on a MMI type).* This problem may be formulated on the language of graph theory but is difficult for practical realization.

Problems of the types indicated have already been considered in the literature: the statement of the problem of the RUTS system optimization on the *extent of unit reproduction centralization* (see Section 2.2.5.1). Actually, the problem is reduced to determining such a value for n_0 (the number of MMI at a top level of the RUTS system) at which the corresponding functional (2.2.39) becomes the maximum. For that it is necessary to somehow express n_0 through other parameters of the system. For example, among them there can be recurrent relationships of type (2.2.28), to which the empirical dependence of type (2.2.41) of one of controllable parameters $(\tau_i, T_i, t_{i1}, q_{i1}, l_i)$ on the error of the corresponding MMI is added, as well as taking into account the probability of verification defects, etc.

It is clear that at $n_0 = 1$ we have *complete centralization* of the unit reproduction, and at $n_0 \to N_p$ the *total decentralization.*

In this connection it is convenient to describe the *extent of centralization* by the coefficient which is invers to n_0:

$$\chi \equiv n_0^{-1}; \quad 1 \geq \chi \geq N_p^{-1}. \tag{2.2.44}$$

Problem 5. *Determination of the optimal extent of centralization (χ) of reproducing the unit in the RUTS system (φ) while varying the remaining parameters of the system.*

$$\begin{aligned}
n_i &= n_{i+1}\frac{t_{i+1}}{T_{i+1}} \cdot \frac{1+q_{i+1}}{\tau_i l_i}, \qquad i = \overline{0,m};\\
0 \leq n_i &- n_{i+1}\frac{t_{i+1}}{T_{i+1}} \cdot \frac{1+q_{i+1}}{\tau_i l_i} < 1;\\
\sigma_i &= f_0(t_{i+1},\ T_{i+1},\ \tau_i,\ q_{i+1},\ l_i);\\
p_i &= f_p\left(\frac{\Delta_i}{\sigma_i}, \sigma_{i-1}\right), \qquad \rho_i = f_p(\sigma_i);\\
C_{C_i} &= f_C(\sigma_i), \quad C_{E_i} = f_E(\sigma_i);\\
\sigma_p &= \sqrt{\sum_{i=0}^{m} \sigma_i^2},
\end{aligned} \tag{2.2.45}$$

where σ is the mean square (standard) deviation of the unit transfer; Δ is the permissible error of MMI; p is the probability of a verification defect; ρ is the distribution of the mean square deviation of the given MMI; the remaining designations have been introduced earlier.

The solution in the general form of such a complicated problem is practically impossible. Therefore, it is necessary to make some allowance based on the earlier obtained empirical relationships that are sufficiently general. Some of them can be

(1) the choice of empirical dependencies of costs on errors in the form (2.2.41) at $1 \leq \alpha \leq 2$;

(2) the limitation of the interval of a relationship of adjacent levels with the values obtained from the previously solved optimization problems: $2 < (\delta_{i+1}/\delta_i) < 3.5$.

It is known that the influence of errors of the top elements of the system $(\zeta_0, \zeta_1, \ldots)$ will be less, the more the number of levels separating them from WMI. Consequently, possible and useful are

- decreasing the relationship of errors of the adjacent categories transfer in moving towards the top levels;
- not taking into account the verification defects at the upper levels on the verification defects at the bottom level, and the probability of revealing unsuitable WMIs (at $m \gg 1$), and others.

Certainly, these issues require additional independent investigations.

Here it is important to stress that the known solution for problem 5 (by n_0 a number of working standards was meant) has shown that there exists an optimal relationship between n_0, m, and ζ_i at $n_0 \neq 1$, i.e., at a definite extent of decentralization of the reproduction. In other words, decentralization can be economically substantiated.

2.2.5 System of reproduction of physical quantity units

2.2.5.1 Main properties of a system

Let the system of reproducing units (2.2.12.1) be presented by way of

$$\sum_b \{sb_{i_k}(\varphi_k), mb_{i_k}(\varphi_k)\}; \quad k = 1, 2, \ldots, |\Phi|; \quad i_k = 1, 2, \ldots, n_{0_k}. \tag{2.2.46}$$

1) *Input parameters* (inputs) of the system are:

- $\Phi = \{\varphi_k\}$ is the set of measurand definitions, i.e., information I_{a1} about properties of every measurand within the framework of the NSM;
- $[\Phi] = \{[\varphi_k]\}$ is the set of PQ units definitions (a priori information I_{a2});
- $\{\mu_i\}$ is information I_{a3} about values of fundamental physical constants and other constants of substances, materials, events and processes used for reproducing units;
- $\{f(\varphi_i, \varphi_j)\}$ is information I_{a4} about dependences among measurands participating in the measurement system realizing the unit reproduction (for all φ_k);
- information I_{a5} on general requirements for the measurement system realizing the unit reproduction (from basic normative documents for initial MMI.

Output parameters of the system are the actual unit dimensions transferred to subordinate MMI with the corresponding transfer methods.

Thus, inputs of the system are *only a priori information*, which corresponds to the concept of the unit reproduction that has been formulated earlier.

2) From Section 2.2.4.4 it follows that already from the economical point of view the RUTS systems (φ), having a certain extent of the *decentralization of unit reproduction*, are possible and useful. These systems can possess a number of reference MMI with similar properties. Let us analyze the limitations of a *metrological nature*, arising in this case.

The equation of unit reproduction is given in Section 2.2.11:

$$\varphi^{\text{val}}(z_{st}) = n_{\varphi}(S_{st}) \cdot [\varphi_r]_{st} = n_{\varphi r}(S_{st}) \cdot n_{[\varphi]}(S_{st}) \cdot [\varphi_0], \tag{2.2.47}$$

from which it follows that the unit dimension φ_{st} realized in the reference MMI differs from the "ideal dimension" of the unit, $[\varphi_0]$, which strictly corresponds to its definition, by

$$\Delta[\varphi]_{st} = [\varphi]_0 - [\varphi]_k = \{1 - n_{[\varphi]}(S_{st})\} \cdot [\varphi_0]. \tag{2.2.48}$$

It is clear that, in the presence of a number of reference MMI located in different places, the condition of measurement result comparability (2.2.19) should be met in the same way as in the case of any measurement systems in the NSM. This condition, as shown at the end of Section 2.2.3.5, is equivalent to conditions (2.2.31a) and (2.2.31b). However in the case given (for reference MMI) there is no other MMI from which it would be possible to get a unit dimension. Therefore, it remains only to strictly meet the condition of measurement correctness (2.2.10), which requires

(a) a sufficiently large volume of qualitative information about dependencies in the form of equations (2.2.9.1)–(2.2.9.7);

(b) the creation of at most identical conditions in the measurement systems $z(S_{st.i})$ and $z(S_{st.j})$, realizing the "parallel" unit reproduction.

Thus, there are no principal limitations of the presence of some reference MMI (i.e., reproduction means) of a given measurand. At the same time these reference MMI can be, in principle, at any number of levels of the RUTS system, right up to the first one ($m = 1$). Evidently, the construction of the RUTS system from bottom to top should be performed up to the level where the differences in methods and means, knowledge and qualification of operators (and other components) of measurement systems located in different places and under different conditions can be made negligible.

Such a situation, as the practice suggests, is in tachometry. As the equipment and qualification level of metrological laboratories increases, the extent of decentralization has to rise and spread over other kinds of measurements.

3) Let us take two apparent axioms of reproduction as the basis:

(a) reproduction of the unit within the framework of a NSM is always required if only there is a need to measure a certain measurand (since any measurement needs a unit);

(b) at any measurement with an indirect method, the unit of a measurand φ, independent of the availability of other measuring instruments of the given PQ is produced.

The second axiom, in particular, argues that all derivative units can be reproduced through indirect methods with the help of measuring instruments of other PQ on which this given measurand depends (through which it is determined), i.e., with the help of so-called borrowed measuring instruments. If they are the measuring instruments of the basic quantities (see Section 2.2.5.2), then for the case of determining φ with the help of three basic quantities:

$$\begin{aligned} \varphi &\equiv f(\varphi_a, \varphi_b, \varphi_c), [\varphi] \\ &\equiv [\varphi_a]^{\alpha}, [\varphi_b]^{\beta}, [\varphi_c]^{\gamma}, \end{aligned} \tag{2.2.49}$$

$$\begin{aligned} \varphi^{\text{val}}(z_\varphi) &\equiv f(n_{\varphi a}(s_a), n_{\varphi b}(s_b), n_{\varphi c}(s_c)) \cdot [\varphi_a]^{\alpha}[\varphi_b]^{\beta}[\varphi_c]^{\gamma} \\ &= n_\varphi[z_\varphi] \cdot [\varphi]. \end{aligned} \tag{2.2.50}$$

Here it is important to note that although the result of measuring with the help of a borrowed measuring instruments is expressed in terms of the unit of the measurand, the totality of the borrowed measuring instruments (s_a, s_b, s_c), by itself has not yet any real dimension $[\varphi]$. It is quite another matter if we join these measuring instruments into one setup (into one measuring system s_φ) *certified only for the measurand* φ. In that case this setup already has an analogue "readout of the instrument" on φ and realizes a definite dimension $[\varphi]_s$. From this it follows that if it is possible to reproduce the unit with an indirect method, then *it is not in principle possible to transfer its dimension with the indirect method.* Unfortunately, some verification schemes, particularly in the field of thermo-physical measurements (for example, in GOST 8.140, 8.026 and others) contain a similar error.

Thus, in the case of *derivative units* the latter can be reproduced either with certified setups (reference MMI) gathered from determining (basic) quantities borrowed by measuring instruments, or with a *reference (standard) method*, when the totality of the φ borrowed measuring instruments are not certified, but the reproduction method using the borrowed measuring instruments is certified.

For basic units the obligation of the available reference MMI in RUTS systems is clear for the following two reasons:

- arbitrariness of choosing a dimension of their units ("according to an agreement");
- need to use basic units in measurements of derivative quantities.

An additional metrological requirement which determines the need to introduce an MMI at some level of the RUTS system (even for derivative units) is connected with the presence among WMIs of a park of measuring instruments which can directly evaluate φ (of instruments), which necessarily requires the certification for φ, i.e., transferring the dimension of the corresponding unit.

4) The reference MMI of any RUTS system has to provide for

(a) reproducing an unit;
(b) keeping it within a time interval exceeding the verification intervals of all subordinate metrological instruments (theoretically within the limits of operation of the whole RUTS system);
(c) transferring a unit dimension to all directly subordinate MMIs, which determines a minimal *needed composition of a reference MMI*:
   - *generator* (a source) of a physical quantity the unit of which has to be reproduced;
   - *certifying instrument* of a physical quantity dimension which, in fact, establishes the unit $[\varphi]_{st}$;
   - *comparator* (comparators) providing the unit transfer to all directly subordinate MMIs.

Because of the fact that in reference MMIs a maximum constancy of conditions (and other components of the system) is achieved and the influence of external factors is taken into account as much as possible, auxiliary devices are introduced into the composition of a complex of the reference MMI to reach the indicated goals.

5) In practice the reference MMI, even when corresponding to the structural requirements of the RUTS system, cannot always provide the unit reproduction and its dimension transfer at a required accuracy to the whole nomenclature of WMIs of a measurand given, for example, within the range of this measurand.

If that is the case, one of following two ways can be chosen:

(a) the use of one more (or a number of) type of reference MMIs for the required range of φ;
(a) an attempt at extending the range of every stage of the verification scheme from top to bottom (the process of scale transformation). The latter in particular is frequently applied when the reference MMI reproduces the unit at one point of the scale of φ values.

In both cases one problem situation arises, namely how the unit dimension should be compared for different parts of the scale of measurand values. The condition of dimension comparability in the form of (2.2.19) is not suitable for this case, since here $\varphi_r(z_i) \neq \varphi_r(z_j)$. In practice various methods of calibrating measures with the properties of additivity are used. Without going into detail about these methods, let us attempt to formulate the condition of measurement results comparability for this case.

Let us introduce the concept "relative measurement result" using equation (2.2.8):

$$\varphi_{\text{rel}}^{\text{meas}}(z_i) = \frac{\varphi^{\text{meas}}(z_i)}{n_\varphi(z_i)} = [\varphi_r]_{s_i} = n_{[\varphi_r]}(s_i) \cdot [\varphi]_0.$$

Then the condition of measurement comparability for the case $\varphi_r(z_i) \in \varphi_r(z_j)$ can be formulated in the following manner.

For every measured PQ, $\varphi \in \Phi$, in the considered general system of measurements there exist such sets of controllable system parameters (2.2.26),

$$U_i = U([\varphi]_i, m_i, s_i, v_i, w_i), \quad U_j = U([\varphi]_j, m_j, s_j, v_j, w_j),$$

that at any $j \neq i$ and $\varphi_r(z_i) \neq \varphi_r(z_j)$ it is possible to provide the fulfilment of

$$\left|\varphi_{\text{rel}}^{\text{meas}}(z_j) - \varphi_{\text{rel}}^{\text{meas}}(z_i)\right| = \left|[\varphi_{rj}]_{s_j} - [\varphi_{ri}]_{s_i}\right| \leq \sqrt{\delta_{i_{rel}}^2 + \delta_{j_{rel}}^2} \cdot [\varphi]_0$$

or

$$\left|n_{[\varphi]}(z_j) - n_{[\varphi]}(z_i)\right| \leq \sqrt{\delta_{i_{rel}}^2 + \delta_{j_{rel}}^2}. \tag{2.2.51}$$

Introduction of designations $\Delta_{[\varphi]}(z_i) \equiv 1 - n_{[\varphi]}(z_i)$ and $\Delta_{[\varphi]}(z_j) \equiv 1 - n_{[\varphi]}(z_j)$ makes it possible to write equation (2.2.51) in the form

$$\left|\Delta_{[\varphi]}(z_i) - \Delta_{[\varphi]}(z_j)\right| \leq \sqrt{\delta_{i_{\text{rel}}}^2 + \delta_{j_{\text{rel}}}^2}. \tag{2.2.51a}$$

This means that the difference in the deviations of the unit dimensions realized in the measuring systems I_{z_i} and I_{z_j} from the ideal dimension $[\varphi_0]$ should not exceed a mean square sum of the given relative errors of measurements in these systems.

Condition (2.2.51a) is true not only for all measuring instruments of NSM but also for reference and subordinate MMIs.

Such are the principal requirements of a physical and metrological character for the unit reproduction. All the remaining ones are connected with particular engineering and economical arguments, a part of which is described in Section 2.2.4.4.

2.2.5.2 The choice of base quantities and base units

Issues of constructing unit systems (SI) in metrological theory and practice have thoroughly discussed for a long time (for example, in [323, 432, 439]). The concept "system of physical quantities" (SPQ) can be found in the literature on metrology far more seldomly. Some authors deny the usefulness of introducing and considering such a system. Here, in connection with the problems of constructing a RUTS system and analysing properties of the unit reproduction system, only the separate general issues concerning the interrelation of these systems (quantities and units) will be considered.

1) Although physical quantities and their units can be introduced *principally at will* and a system of physical quantities (as well as a system of units, corresponding to it) can also be constructed principally at will, this is true only for the unlikely situation where someone decides to describe the material world and our surroundings completely disregarding all previous knowledge which mankind has accumulated. In reality, the physical world view (natural science ideology) being the reflection of ob-

jective reality properties has chosen for itself and approved for centuries an *entirely definite* system of ideas concerning this reality. These ideas are based on concepts and quantities reflecting general and stable properties of the environment.

The most general idea consists of the following: a substance exists in time and space, the form of its existence is movement. The base quantities must reflect these fundamental attributes of this substance. The fundamental laws of conservation (of energy, pulse, linear momentum) are connected with the properties of time and space. Quantities such as *time*, *length*, and *angle* are introduced as characteristics of the material objects placed in the space–time continuum.

As to length and time, no one has doubted or doubts that they should be considered as base quantities of SPQ. This also applies for their units.

Unfortunately, up to now there is no any consensus of opinion with regard to the angle and its unit. This is reflected in the International system SI, where not so long ago a specific (and poor-understandable) category of additional units was introduced for angular units. As to the unit of a plane angle, an absurd situation has arisen. On the one hand, it is a dimensionless unit, and on the other hand it has its name, i.e., it reveals properties of a denominate number. The majority of authors continue to consider the plane angle to be a derivative quantity.

However, even from the physical point of view all possible space relations cannot be described only in length. One cannot imagine a real rather than formal combination of linear quantities replacing the angle. Rotational movements are not reduced to forward ones, etc. A detailed review of the angle status and its unit is given in [147]. The results of an analysis contained in this review, as a whole, also testify to the usefulness of recognizing it as a base unit of the SPQ. Let us consider that the space–time properties of the material world required the introduction of three base (independent) quantities:

$$L \text{ – length, } T \text{ – time, } \Phi \text{ – angle (plane).}$$

2) The next, not less general attribute of substance is movement. It is comprehensible for our perception through an interaction, the measure of which is the *energy* E which is one of the basic general concepts of the physical world view.

It would be possible to consider only the set of quantities (L, T, Φ, E) as the main (initial) concepts for describing external objects, if there were not a variety of forms in which energy demonstrates itself (mechanical, electromagnetic, thermal, nuclear, etc. Although all these forms of energy are related to *one generic* concept, they are so specific that frequently it is very difficult to connect them. The inaccuracy of corresponding equivalents causes a specific quantity, characterizing a corresponding type of interactions and a sphere of events connected with energy to be introduced into each specific field where energy is revealed.

In mechanics where *gravitational* interaction is the determinant factor, it is body (object) mass (m). Here, we will not dwell on all the minute details of the difference between gravitational and inertial mass, the more so as the general theory of relativity shows (postulates) their equivalents.

In electrodynamics the main type of interaction is *electromagnetic* interaction. The main source of an electromagnetic field is an electric *charge*, q, and as the base quantity either a charge or a speed of its changing with time, i.e., *electric current* (I), should be chosen.

In thermodynamics the characteristic features are the interactions of a statistic nature connected with an energy exchange between macrosystems. The extent of macrosystem energy variation in the processes of heat exchange is characterized by *temperature* (Θ) that is also chosen as the base quantity.

Optical phenomena are a variety of electromagnetic interactions. Therefore, to introduce into the SI a unit of light (candela) as a base unit is caused not so much by any objective necessity connected with the specificity of the phenomena in optics, as by the need to measure some light characteristics (visible region of the optical spectrum) connected with a subjective perception of light fields by the eye. They are, as a matter of fact, off-system quantities and units.

The interactions which at present are the most complicated and difficult to understand are in the field of nuclear phenomena (weak and strong), particularly, strong interactions, which are difficult to characterize using physical quantities. For weak interactions it is possible in the capacity of the base characteristic to choose the *activity* of radionuclides which characterize the speed of a radioactive decay (a result of weak interaction).

In the fields of physics and chemistry a quantity of substance is considered as the base one; its unit is the mole. However their role in the system of physical quantities is uncertain and debatable to an extent that a great number of specialists are disposed to remove them from the category of base quantities.

Let us note that the mole, according its present definition, is simply a scale coefficient linking the macromass unit (kilogram) with the unit of atomic (micro) mass.

Thus, in the light of today's physical world view, as the base quantities of the system of physical quantities it is possible to accept the following: L (length), Φ (angle), T (time), M (mass), I (electrical current), Θ (temperature), A (radionuclide activity).

3) It is possible to show that the concept of dimension is related both to the units and quantities. Dimension serves as it were as a qualitative characteristic of a physical quantity, determining its gender. In this connection two aspects are important.

Firstly, the removal of base quantities (or the ungrounded decrease of their numbers) from the system can lead to the situation of heterogeneous quantities having similar dimensions. This is not permissable in an exact science. This is exactly what happens, when the plane angle is excluded from the category of base quantities, which is evidence of its base quantity and unit status.

Secondly, since the quantity and its unit are homogeneous concepts and must be expressed in terms of the denominate number of one type, an ambiguous link between the physical quantity and its unit in the system of the SPQ $\leftrightarrow$ SU type cannot be permittted. At the same time, such is indeed the case of many speed (rotation, flow) and counting quantities, where quantities of a quite different physical nature are expressed in terms of similar units. This was noted more than once, for example in [190].

4) The practice of implementing the measurement standards of base unit reproduction shows that frequently as the base unit adopted by convention for the system of units, a completely different unit is reproduced. This can be accepted, as a rule, with the following explanation. The base units are chosen on the basis of expediency, simplicity, and commonness of the material world description, while the base units are adopted due to practical considerations, primarily when achieving the highest accuracy is necessary, i.e., on the basis of somewhat different criteria. At the same time, transfers from the reproduction of the base units adopted by convention to the most accurate ones of the other type are always realized on the basis of their tight interconnection through fundamental physical constants (FPC) (comp.: Ampere, Volt, Ohm and μ; length, frequency and c_0).

At the same time, the *number of base units* (the base ones in a given field) *remains unchanged.* Therefore, it is useful to formulate the following rule: *the number of base units must be equal to the number of base quantities of a system; the number of base units adopted by convention has to be equal to the number of units reproduced in the real RUTS system.*

2.2.5.3 System of unit reproduction in the field of ionizing radiation parameter measurements (an example)

The system of unit reproduction in the field of *ionizing radiation parameter measurements* (IRPM) includes measurements (widely used in science and practice) of the radionuclide activity, dosimetry of photon, beta- and neutron radiations, measurements of the field parameters and sources of electron, photon and neutron radiations. The RUTS system in the field of ionizing radiation parameter measurements is based on seven national primary measurement standards and seven special national measurement standards. Usually the measurement standard consists of a number of setups. Each of these setups reproduces the dimension of one and the same unit, but within different ranges of particles or photons or for different kinds of radiation, etc.

An analysis of the existent system of measurement standards reveals a number of significant drawbacks caused by the lack of a unified approach in the formation of the RUTS system in such a complicated field, characterized by the diversity of different kinds of radiation (and the corresponding radiation sources), as well as conditions of their measurements (energetic ranges, environment, specificity of radiation (bremsstrahlung, pulse radiation, etc.)). Only one RUTS system designed for measuring the activity of various kinds of radiations and various RUTS systems used in dosimetry (photon, beta- and neutron radiation) functions in such a manner.

The special standard of the unit of neutron flux density has a lesser error than the primary standard of the same unit (at least on a board of the range of reproduction).

Only one special national measurement standard functions for reproducing the unit of volume activity. There are some other examples of drawbacks.

A system approach to constructing a system of unit reproduction in a given field of measurements has to be based on the general physical ideas of both the ionizing radiations and the field and its sources, the choice of initial (base) quantities in terms of which of the remaining quantities (completeness of the system description, interrelation of elements, hierarchical dependence) can be expressed. An attempt of this kind was made in 1974 by V. V. Skotnikov et al. [445].

Let us try to systemize and develop the ideas of this work in the light of a modern overview. First, we should defined more exactly physical objects and phenomena participating in the process of measurement in this field:

- proper "ionizing radiation", i.e., all radiation types connected with the radioactive decay of nuclei (α-, β-, γ- and conversion radiation) or with nuclear reactions (fluxes of charged and neutral elementary particles and photons), as well as X-radiation and bremsstrahlung (i.e., hard electromagnetic) radiation caused by the interaction of radiation of the types indicated with the atomic systems;
- the sources of this radiation (mainly the sources of radioactive radiation);
- the environment (of substance) with which the radiation interacts;
- the phenomena taking place in this environment as a result of radiation passing through it.

By the ionizing radiation (IR) we mean, strictly speaking, any electromagnetic or corpuscular radiation that is able in an appreciable manner to produce an ionization of environmental atoms through which it penetrates ("in an appreciable manner" means here that ionization is one of the main phenomena when the given radiation passes through the environment).

Since ionization requires a definite energy consumption for overcoming the forces of attracting electrons in atoms (or in molecules), it is possible to say, in other words, that the ionizing radiation is an electromagnetic or corpuscular radiation for which ionizing losses of energy are the main losses when this radiation passes through the substance.

In fact this field includes measurements of neutron flux parameters, i.e., of neutral (in the sense of a charge) radiation, the ionizing action of which is negligible as compared to the types of its interaction (nuclear reactions of various kinds). Moreover, the ionizing radiation measurements cover not only parameter measurements of the ionizing radiation itself, but measurements of the sources of this radiation, as well as the phenomena caused by this radiation in the environment.

In a wider plan it may be necessary to single out a field of "nuclear measurements" covering the whole specific field of measurements in nuclear physics and nuclear engineering (including its application fields).

Let the generalized field of radiation [445] be determined as a state in the space, the physical properties of which in a point given and at a given time moment are caused by the availability of particles or quanta at this point. The field of a definite type of particle in a given point and at a given moment in time are characterized by pulse values and energy distributions. Consequently, the main characteristic of the field of radiation has to be a value differential for all the properties of the generalized field. The most complete and universal (both from a theoretical and practical point of view) characteristic of the fields and sources of IR is the spectral characteristic of radiation or (in a wide sense) spectrum of radiation.

At the given (chosen) space and time coordinates for the system the radiation spectrum in the quite general form is described by the dependence of the form

$$S = \sum_i n_i(\vec{r}, t, E, \vec{\Omega}) = f(\vec{r}, t, E, \vec{\Omega}, i), \tag{2.2.52}$$

where n_i is the number of particles of a given radiation sort i in the point with the coordinates $\vec{r}$ having the energy E and moving in the direction $\vec{\Omega}$ (unit angular vector) at the moment in time t. At fixed values of the arguments the physical quantity n, corresponding to dependence (2.2.52), has the sense of a differential (space time, energetic-angular) density of radiation.

The knowledge of this quantity at the points in space interest gives practically everything needed from the point of view of measurement quantities and parameters of fields and the sources of ionizing radiations obtained (in most cases with a simple mathematic integration).

(1) *Differential radiation flux density* φ (the number of particles at the point with coordinates $\vec{r}$, having an energy E, which move in a direction $\vec{\Omega}$ and cross the plane surface of the 1 cm^2 area at a right angle to the vector $\vec{\Omega}$ for 1 s at the time moment t):

$$\varphi_i(\vec{r}, E, \vec{\Omega}, t) = n_i(\vec{r}, E, \vec{\Omega}, , t) \cdot U_i, \tag{2.2.53}$$

where U is the movement speed of particles under the same conditions.

(2) *Differential radiation flux through the plane* ΔS*:*

$$P_i(\vec{r}, E, \vec{\Omega}, t) = \int_{s=0}^{s+\Delta s} \varphi_i(\vec{r}, E, \vec{\Omega}, t) dS. \tag{2.2.54}$$

(3) *Differential intensity* (energy flux density) of radiation:

$$I_i(\vec{r}, \vec{\Omega}, t) = \int_{E=0}^{\infty} \varphi_i(\vec{r}, E, \vec{\Omega}, t) E_i dE. \tag{2.2.55}$$

(4) *Differential radiation absorbed dose* for the given substance (B) characterized by the dependence $\sigma_i(E_i)$ of the absorption cross-section of the i type on the energy:

$$D(\vec{r}, b) = \frac{d\bar{\varepsilon}}{dm}; \tag{2.2.56}$$

$$d\bar{\varepsilon} = \int_t \sum_t \frac{\int_E \int_\Omega \varphi_i(\vec{r}, E, \vec{\Omega}, t) E_i dE d\Omega}{\int_E \int_\Omega \varphi_i(\vec{r}, E, \vec{\Omega}, t) dE d\Omega} = \int_t \sum_t \frac{\int_{\Omega=0}^{4\pi} I_i(\vec{r}, \vec{\Omega}, t) d\Omega}{\varphi_i(\vec{r}, t)}.$$

(5) *Activity of radionuclides*:

$$A(t) = \frac{dN}{dt} = \frac{1}{P_i,\ E_0} \int_{\Omega=0}^{4\pi} P_i(\vec{r}_0, E, \vec{\Omega}, t)\, d\Omega|_{r_0=0,\ E=E_0}, \tag{2.2.57}$$

where P_i, E_0 are the portion (part) of the given type of radiation (i) with the energy E_0 for one decay act.

(6) *Source output* with respect to the radiation i:

$$W_i(t) = 4\pi r^2 \int_{E=0}^{\infty} \int_{\Omega=0}^{4\pi} \varphi_i(\vec{r}, E, \vec{\Omega}, t) dE d\Omega, \tag{2.2.58}$$

etc.

The relationships indicated above express the idealized (irrelative of any particular real conditions) connection between the corresponding quantitics, arising from their definitions.

In each real case it is necessary to take into account the specific conditions of getting information about the input parameters and the quantities of these equations, and to use in a number of cases additional regularities and parameters.

Problems of such a type arise in the process of constructing a particular measuring setup (or method). However, the indicated connections and subordination among various quantities should not be neglected, since they are principal (it follows from the definitions of physical quantities).

It should be noted that in a number of cases the relationships indicated include the parameters (cross-section of reactions, average energy of ion formation, parameters of the decay scheme), which are a product of research experiment, and, therefore, here the importance of reliable reference data in the field of nuclear physics is clearly seen.

Thus, within the framework of the system under consideration in the capacity of initial physical quantities it is possible to adopt the quantities used as the arguments in initial spectral dependence (2.2.52):

$$n, \vec{r}, t, E, \vec{\Omega}. \tag{2.2.59}$$

From the point of view of measurement accuracy, only two problems specific for the given field of the quantity measured are problematic, i.e., n (a number of particles) and E (an energy of particles), since $\vec{r}$, t, and $\vec{\Omega}$, being the space–time characteristics, play the role of classical (rather than quantum-mechanical) quantities in all real cases and, consequently, their measurements relate to the classical field of measurements.

The problem of measurement (a number of units and photons) is reduced, as a matter of fact, to the problem of determining the efficiency of radiation detectors and recording system. Let us note that the second part of this problem is common for all kinds of radiation and mainly depends on n. But at the same time the first part (a detector) strictly depends on both a sort (a kind) of radiation (i) and its energy range.

However the solution of this main problem, i.e., the determination of the detector *efficiency* (the detector sensitivity to various kinds of radiation), may be realized by the way of using radiation sources certified with respect to the activity:

$$\varepsilon_i = \frac{(n_i)_{\text{det}}}{t(A_i)_{\text{source}} \cdot \frac{\Delta\Omega}{4\pi}}, \tag{2.2.60}$$

where ε_i is the detector efficiency with regard to radiation i, $(n_i)_{\text{det}}$ is the speed of counting particles, $(A_i)_{\text{source}}$ is the source activity certified with regard to radiation i, and $\Delta\Omega$ is the collimation solid angle.

At the same time the measurand is only the solid angle and speed of counting in the detector (such as it is, i.e., it does not coincide with the true speed of counting in the given place of the radiation field).

It is quite possible that in the future the activity of sources will be determined on the basis of decay constants and radioactive source mass.

The radionuclide activity is unambiguously connected with its decay constant (N) and amount of radionuclide atoms (N) in a source (preparation) in the form

$$A = \lambda \cdot N, \tag{2.2.61}$$

where $\lambda = \frac{2\ln 2}{T}$ is the decay constant characterizing the radionuclide given and determined by a nuclear structure of an isotope. It can be determined at a rather high accuracy (error $\approx 10^{-4}$) [317] for some isotopes with a convenient half-life period $T/2$ by means of relative measurements of a number of ejecting particles in the certain time intervals.

The determination of the number of radionuclide atoms N in a source is a much more complicated problem. However, taking into account recent successes in preparing pure substances (including those that are applied in radiochemistry and mass-spectrometry) as well as in analyzing their spectral distribution and determining the Avogadro constant, it seems possible that this problem will be solved in the very near future.

The problem is reduced to obtaining a sufficiently pure sample of a monoisotope (with mixtures $\leq 0.01\,\%$), its weighing (even among existing standards in the field of ionizing radiations the radium mass measurement standard has the best accuracy of reproduction and maintenance ($\leq 0.1\,\%$)), recalculating the mass into a number of atoms with the help of the Avogadro constant and atoms of the isotope mass (M):

$$N = \frac{mN_0}{M}. \tag{2.2.62}$$

If one succeeds in preparing a sample (a source) of some isotope with a convenient $T/2$ and in measuring N in it at an error of ≤ 0.1 %, then it will be possible to determine its activity and, consequently, the efficiency of the detectors for some kinds of radiation at the same error.

It should be noted that in both the existing method of determining the activity A and the method suggested for finding the detector efficiency, the necessary element of a priori knowledge is the sufficiently precise knowledge of the radionuclide decay scheme, i.e., knowledge of decay channels, relationships of their intensities, energies of radiation in each channel, and other elements of the decay scheme.

In other words, the determining factor in the development of this field of measurements at any setting is the advanced development of nuclear–spectrometry methods and means of measurements (at least it concerns the nuclear spectroscopy in relative measurements).

The problem of measurement of another quantity, specific for the nuclear physics, the energy E_i, has up to now been mainly solved "inside" nuclear spectroscopy itself predominantly by the method of relative measurements. Only in works, not numerous, which were done abroad, infrequent attempts were made to link measures of the radiation energy (mainly of the photon radiation), used in the nuclear spectroscopy, with measures of length in the field of X-ray spectrum (which until recently has not been connected with the field of optical radiation) or with the annihilation mass (i.e., with such fundamental constants as the rest mass of an electron, $\underset{\sim}{\overline{\overline{m}}}_e$, and light speed $\underset{\sim}{\overline{\overline{C}}}_e$).

Meanwhile, it is known that not long ago a developed optical – X-ray interferometer – allows energy scales of optical and X-ray ranges of the electromagnetic radiation to be linked, and high-precision crystal-diffraction spectrometers existing for a long time allow X-ray and γ-radiation scales to be linked at an accuracy of $\approx 10^6$.

Thus, the possibility of setting a dimension of the energy unit of nuclear monoenergetic radiation through a dimension of one of the base units of mechanical quantities, in particular the meter reproduced on the basis of wavelength of optical radiation, becomes a real one. At the same time, the energy reference points in the field of X-ray (or γ-) radiation become the primary ones with regard to the energy reference points in other kinds of nuclear radiation, since even if they are determined with absolute methods, but the determination accuracy nontheless being much worse than that expected for electromagnetic radiation.

Among charged particles there are methods of measuring energies only for electrons ($\bar{e}$), the accuracy of which can be compared to the accuracy of the methods used for photons.

At the same time three ways are possible for them:

(1) determination of conversion lines energies on E_γ and bonding strength of electrons in an atom (i.e., again on X-ray data):

$$E_e = E_\gamma - E_{\text{bond}}\,;$$

(2) measurement of electron energies on an potential difference:

$$\Delta E = e \cdot \Delta V\,;$$

(3) measurement of conversion lines energies in an uniform magnetic field:

$$E = f(H\rho),$$

where H is the magnetic field strength and ρ is the curvature radius.

Two last methods can be used for controlling the first method (an additional coordination of the measurement units in various fields).

As to other kinds of nuclear radiations (protons, neutrons, and others), the requirements of practice regarding the accuracy of measuring their energies are less strict and the methods are "painted over" by knowledge of the energies of either photon or electron radiation.

The considered system of physical quantities in the field of ionizing radiations (or, more precisely, of the nuclear ones) as well as the analysis of this system from the point of view of measurements (above all of precision measurements) allow the approach to formation of a corresponding system of interconnected measurement standards in the considered field of measurements to be suggested.

In accordance with this system, in the field under consideration the primary standards of the activity and nuclear radiation energy units have to be the base ones. On the basis of these measurement standards it is possible to develop a standard of an initial differential characteristic of fields and sources of nuclear radiations, i.e., the radiation density.

2.3 Summary

The investigations carried out with respect to the *classification of systems of reproducing physical quantity units and transferring their sizes* (see Section 2.1) show the necessity of extending and improving the essence of a whole series of fundamental concepts of metrology connected with the reproduction of measurement units and the transfer of their sizes. Therefore, a rather large part of our work is devoted to these concepts. The main results of these investigations are as follows.

The concept "*metrological system*" was introduced, and with the help of the analysis of the *essence of measurements as the simplest metrological system* the elementary metrological sets are revealed. On the basis of these sets it is possible to construct metrological systems of a more complicated order.

Reasoning from the analysis of the essence of practical metrology, the "*hierarchy of proper metrological systems*" on the RUTS system line is constructed. The characteristic features of proper metrological systems inside this hierarchy are revealed and formalized, in particular the *conditions of correctness* (achievement of the required uaccuracy) as well as the *conditions of* measurement *comparability* which determine the target function of the "system of ensuring measurement quality" and its most important component, the "system of ensuring measurement uniformity".

On the basis of the analysis of the *measurement essence as the process of finding the value* (in particular, basing on the postulates of relationships between a priori and a posteriori information) the content of the concepts "*reproduction*", "*maintenance*," and "*transfer*" of the PQ unit dimensions are revealed. A new formulation of these concepts is also given. The place of the RUTS system is thereby determined in the system of ensuring the measurement uniformity as the system providing the comparison of dimensions of units realized in each particular measuring instrument in the form of a priori measurement information.

The concepts "*total* RUTS system" and "*particular* RUTS system" were introduced and the structure of a particular RUTS system analyzed. The expediency of introducing the generalized concept for informal RUTS system elements, "*metrological measuring instrument*", is shown. The formal part of the RUTS system consists of methods of transferring the unit dimension.

Proceeding from the hierarchical structure of the RUTS system, in metrological measuring instruments three classes are singled out: *reference* metrological measuring instruments, *subordinate* metrological measuring instruments, and *auxiliary* metrological measuring instruments. Each of these classes carries a specific load (a metrological function). From these positions the categories of reference, subordinate, and auxiliary metrological measuring instruments which are legalized at the present time are analysed.

From the point of view of the *interspecific classification* of (particular) RUTS systems (irrespective of a kind of measurements) the means and methods of reproducing the physical quantity units and transferring their dimensions in accordance with a number of classification features (an extent of centralizing the unit reproduction, method of reproducing in time and with respect to the process of transferring), which function today, are analyzed.

The classification *according to the extent of centralization unit reproduction* has made it possible to reveal four types of RUTS systems: the complete (CC), multiple (MC), and local (LC) centralization in a national system of measurements (SSM), as well as with total decentralization (DC). It is shown that only some of the functioning national verification schemes with national standards at the head can be referred to the systems of the type CC, and the remaining ones are various combinations of systems of different types.

Classification *according to the method of reproducing in time* has revealed two types of reference metrological measuring instruments (with the continuous and discrete method of reproduction). At the same time, the necessity for introducing an additional metrological characteristic of reference metrological measuring instruments (in particular, of national measurement standards), i.e., reproducibility, is also shown.

Classification *according to the relation of the unit reproduction* process to the process of transferring its dimension revealed two groups of reference metrological measuring instruments, one of which reproduces the unit only at the moment of transferring (does not contain any PQ source) and, strictly speaking, has to possess a name that differs from the standard name (for example, verification setup of the highest accuracy).

Consideration was given to some aspects concerning the specific classification of RUTS systems (relative to concrete physical quantities) including the general nomenclature of measurable physical quantities, problem of PQ uniformity from the point of view of the RUTS systems, the problem of physical constants for RUTS systems, and the issue related to the RUTS systems for dimensionless quantities (coefficients).

At the same time, in particular,

- it is shown that to establish the complete nomenclature of physical quantities is rather difficult and further analysis and improvement of the concept "physical quantity" are required;
- the hierarchy of the PQ system is demonstrated, including that of physical constants;
- the role of physical constants in reproducing PQ units depending on whether or not a constant is uniform with respect to a given PQ, was defined more precisely. In connection with that the role of a national system of standard reference data is defined more precisely;
- it was revealed that there are about 235 physical quantities, and the analysis of a possible type of RUTS systems for them was made;
- it was proved that the existence of RUTS systems (especially of the type CC) for dimensionless quantities (coefficients) seems poorly grounded.

The problem concerning the "*technical and economic efficiency*" of RUTS systems was considered.

The concept "*metrological effect*" was introduced, with the help of which the analysis of the relative efficiency of RUTS systems of various types ws made. The general conditions *required* and *sufficient* for creating centralized RUTS systems of any type were formulated.

Investigation carried out on the *development of physical-metrological fundamentals of constructing systems for reproducing units of physical quantities and transferring their sizes* (see Section 2.2) reveal the need for using a system approach to the problem being considered, and show the possibility and efficiency of the application of a set-type apparatus for describing metrological systems of various types, and the necessity for considering a wide complex of interrelated issues.

The investigation has allowed the authors to realize mainly the following:

- to analyze the literature on the problems investigated and to determine the main lines of study within the framework of a program of fundamental investigation on metrology;
- to determine the system of fundamental initial concepts of the theory of describing and constructing RUTS systems (the kinds and dimensions of physical quantities, the measurement problem, the measurement method, unit reproduction, unit dimension transfer, metrological measuring instruments, metrological system, and others);
- to consider the *description of measurement* as both the simplest metrological system and the process of solving a measurement problem. As a *system*, any measurement contains a definite set of controllable and uncontrollable (given) elements. As a *process*, any measurement consists of three stages of successive transformation of the formal and informal type. Here the role of a priori information in measurements becomes very apparent;
- to show the general dependence of a measurement result on all its components and to analyze the most considerable links among them, as well as to derive the general *equation of measurement correctness*;
- to show that a formalized description of a separate measurement allows various metrological systems to be successively described in a formalized form; to introduce and make more specific the concepts "*national system of measurements*" and "*system of* ensuring *measurement quality*" with respect to which all proper metrological systems ("*system of metrological assurance*", "*system of ensuring measurement uniformity*", "*system of reproducing physical quantity units and transferring their sizes*") are the subsystems;
- to investigate the general structure of RUTS systems and determine their subsystems and elements, then on the basis of presenting the generalized element of the RUTS system, i.e., *metrological measuring instruments* in the form of measures, measurement instrument, and transducers, to determine the types of possible links between the elements (methods of transferring unit dimensions), and finally to show that the methods of direct comparison (without any comparator) and of indirect measurements cannot serve as methods for transferring unit dimensions;
- to consider the equation of transferring the unit dimension with the help of certification in the language of adopted formalization, as well as to show how the unit dimension of a measuring instrument under certification changes;
- to select the quality of measurements carried out in a national system of measurements as the main criterion of the operation efficiency of all proper metrological systems (including RUTS systems), as well as to consider the main *quality indices*: accuracy, reliability, trueness, precision of measurements, and comparability of measurement results, among which only *accuracy* and *comparability*, characteristic for

estimates of the RUTS system influence on any national system of measurements, are in the opinion of the authors the most general and invariable;

- to formulate for a closed NSM the conditions of providing a necessary level of measurement quality with respect to both indices: the condition of *providing a given (required) accuracy* of measurements within the framework of NSM and the *measurement comparability* condition within the framework of NSM.

These conditions allow the requirements to the system which is the immediate "superstructure" above the NSM, which provides the possibility for choosing the required controllable parameters.

The equivalency of the concepts "*comparability*" and "*measurement uniformity*" was shown. This makes it possible to determine the SEUM as the system making it possible to meet the condition of measurement comparability.

The concept "*problematic situation*" on ensuring the measurement uniformity connected with the system "a manufacturer $\leftrightarrow$ a customer" was introduced, and it was also shown that for a system of ensuring measurement uniformity to be constructed important is the number of problematic situations in it, rather than the number of measurements carried out in an NSM.

It was shown that the condition of measurement comparability can be met not only when the units realized in two different measurement systems are similar, but also when they are close to a "true" dimension of the unit (according to its definition). This has made possible the refinement and formalization of the goal of any SEMU and RUTS system (as a SEMU component).

The general estimates of NSM parameters and dimensions for describing this system (the 10-dimension system with a great number ($\sim 10^{13}$) of realizations), which are necessary for constructing a RUTS system, were given. Since construction of such a system in the general form is not possible, the main *principles* and *methods* of constructing were described, according to which a general approach to the system can be used. Considerations of the *algorithms* that can be used to *construct* the system at various stages are also given.

An analysis of the formalized RUTS system goals has shown the importance of taking into account the relationship of the errors recorded by a customer and manufacturer in the problematic situation on the SEMU, and on the basis of this analysis two practical recommendations on this issue were formulated. On the basis of the goals of the RUTS system and the analysis of costs for realizing the RUTS system as a scientific and technical syste,, the functional of the RUTS system operation was composed. Some optimization tasks of different types and their setting were also considered.

The concept "*extent of centralization*" of reproducing a unit is introduced and the principal possibility of setting and solving the task on optimization according to the extent of centralization taking into account of economic factors were shown.

A separate part of this chapter was devoted to the main properties of the *system of unit* where the input parameters of the system, unit reproduction equation, and comparability condition at a decentralized unit reproduction are analyzed. It was shown that no principal restriction of the metrological character is applied to the extent of the unit reproduction centralization.

Metrological requirements to the means and methods of reproducing base and derived units of physical quantities were analyzed on the basis of the *reproduction axioms* which were also formulated by the authors.

The necessity for introducing into practice the primary reference methods for reproducing some derived units was shown. The minimum required composition of initial metrological instruments was determined.

The problem of measurement result comparability within different parts of a measurand range, which is especially characteristic when a unit is reproduced at "a single point", was studied.

The authors' opinion concerning the interrelationship between a physical quantity and system of physical quantity units from the point of view of choosing base quantities was presented.

It was suggested to choose length, time, angle, mass, electric current, temperature and radionuclide activity as the basic quantities of the system of physical quantities, reflecting the modern physical world view and covering all fields of physics. The importance of the mutually univocal correspondence between quantities and their units has been underlined. At the same time, special attention is drawn to the fact that it is not obligatory for the groups of basic quantities and basic units to coincide with respect to the nomenclature (but they have to be similar in a number of elements).

As an example for the theory of constructing a system of unit reproduction in separate fields of measurements, the field of measuring parameters of ionizing radiation was chosen. The advantages and disadvantages as well as the perspectives of realizing a system of interrelated initial metrological measuring instruments in the field of IRPM based on a choice of the radionuclide activity and radiation energy in the capacity of base units of the system, which, in their turn, rest upon fundamental constants and relative measurements of corresponding quantities, were analyzed in detail.

Thus, the following conclusion can be made. The theory developed and its fundamentals take into account the whole spectrum of problems connected with physical quantity unit reproduction, transfer of unit dimensions and construction of a corresponding RUTS system, as outlined above.

Chapter 3
Potential measurement accuracy

3.1 System approach to describing a measurement

The potential accuracy of measurements is a problem that both experimental physicists realizing high precision physical experiments and metrologists developing measurement standards of units of physical quantities to be measured constantly come across. However, until now, unlike other measurement problems, this problem has not received proper study and analysis in metrology literature. Moreover it was never presented in the form of an separate part (section) in metrology reference books, training aids, and particularly manuals.

This circumstance has affected our text here since it required the development of a particular approach to the problem as well as the introduction of a number of additional concepts.

3.1.1 Concept of a system approach to the formalized description of a measurement task

The main point of the system approach in any study of some object lies, firstly, in an analysis of this object as a *system* consisting of interconnected elements. However, apart from individuating the elements into the object and the determining links between them (i.e., determining the system structure), in this approach it is also very important to determine the boundaries of the system and its interaction with its "environment", to study "inputs and outputs" of the system and the processes of transformation between them. For systems connected with human activities (i.e., organizational and controlled activities) it is also very important to determine the aims and tasks of the system connected to its purpose.

As mentioned in Chapter 2 the most adequate mathematical apparatus used in forming a description and studying various systems is the *set theory apparatus*.

The language of the set theory is a universal mathematical language: any mathematical statement can be formulated as a statement about a certain relationship between sets. Let us remember some basic elements of this "language" required to understand the material below.

The concept of *set* as the original, fundamental concept of set theory is indefinable. Intuitively by a set one means a totality of certain quite distinguishable entities (objects, elements), which is considered as a single whole. Practically it is the equivalent of the concept of "system". Separate entities forming a set are called elements of the set. The general notation of a set is a pair of figure braces $\{\ldots\}$, within which the set elements are located.

Specific sets are denoted with capital letters of the alphabet, whereas set elements are denoted with small letters. Membership of an element "a" with respect to a set "M" is denoted as $a \in M$ ("a belongs to M").

Example 3.1.
$M_1 = \{$Alekseev, ..., Ivanov, ..., Yakovlev$\}$ is the set of students in a group;
$M_2 = \{0, 1, 2, 3, \ldots\}$ is the set of all natural numbers;
$M_3 = \{0, 1, 2, 3, \ldots, N\}$ is the set of all natural numbers not greater than N;
$M_4 = \{x_1, x_2, \ldots, x_n\}$ is the set of n states of any system ;
$M_5 =$ the football team "Zenit" (i.e., the set of its football players);
$M_6 =$ the set of all football teams of the premier league; it is seen that $M_5 \in M_6$.

The last example shows that an element of a set can also be a set.

There are two methods of assigning sets: enumeration and description. Examples of the first method (enumeration of all elements belonging to the set) are the assignments of all sets M_1–M_6 indicated above.

The description method consists in indicating a characteristic property $p(x)$, inherent in all elements of the set $M = \{x | p(x)\}$. Thus, the sets M_2, M_3, M_5 in Example 3.1 can be assigned the following description:

$M_2 = \{x | x$ – integer$\}$;
$M_3 = \{x \in M_2 | x \leq N\}$;
$M_5 = \{x | x$ – football player of the team "Zenit"$\}$.

Two sets are called equal if they consist of the same elements. whereby the order in the set is unessential: $(3, 4, 5\} = \{4, 3, 5\}$. From this it also follows that a set may not have distinguishable (equal) elements: the notation $\{a, b, c, b\}$ is not correct and has to be changed to $\{a, b, c\}$.

The set X is called a *subset* of set Y if any element of X is the element of Y. In the notation:$X \subseteq Y$ the symbol $\subseteq$ is the sign of being contained ("Y contains X"). So, if M_1 is the set of students of a group, and M_1^0 is the set of excellent students of the same group, then $M_1^0 \subseteq M_1$, since an excellent student is simultaneously a student.

Sets can be finite (i.e., they consist of a finite number of elements) and infinite. The number of elements in a finite set M is called a *set power*. The set power is frequently denoted as $|M|$.

Very important is the concept of an *ordered set* (or a *vector*) as the totality of elements in which every element has a definite place, i.e., in which the elements form a sequence. At the same time the elements are called *components* (or *coordinates*) of a vector. Some examples of ordered sets (vectors) are: a set of people standing in a queue; a set of words in a phrase; the numbers expressing the geographical longitude and latitude of a point on a site, etc.

To denote ordered sets (vectors) angle brackets $\langle\ldots\rangle$ or round brackets $(\ldots)$ are used. The power of such sets is called the vector length. Unlike the usual sets, in ordered sets (vectors) there can be similar elements: two similar words in a phrase, etc.

3.1.2 Formalized description of a measurement task

In previous works carried out earlier [37] the elements of the system approach have already been used. In particular some separate components of measurement were considered, a general aim of any measurement was determined, and the stages of successive transformations in the process of solving a measurement task were revealed, etc. (see also Section 2.1.2.3).

Systemizing all previously used elements of the system approach, let us turn to measurement elements (as those of a system) arising at the stage of setting a measurement task.

By *a measurement task* we mean the description (formulation) of a problem measurement situation precedeing the preparations for carrying out a measurement, i.e., formulation of a task: "what do we wish to make?" It is obvious that for this purpose it is necessary to know (to indicate, to determine): what quantity has to be measured, on which objcct should be done, under what environmental conditions, at what accuracy, and within which space–time boundaries all this should be realized.

Using the formalism of set theory, the measurement task can be written in the form of the following ordered set (vector):

$$Z_i \equiv \langle \varphi_i, o_i, \psi_i, \Delta\varphi_i, g_i, t_i, \Delta t_i, \ldots \rangle, \tag{3.1}$$

where
φ_i is the measurand as a quality;
o_i is the object of study (the measurand carrier);
ψ_i is the conditions of measurements (the totality of external influencing factors);
$\Delta\varphi_i$ is the confidence interval within which it is necessary to get a required value of the measurand (at a given confidence probability);
g_i is the form of presenting a measurement result;
t_i is the moment of time when the measurement is carried out;
Δt_i is the time interval within which it is necessary to carry out the measurement.

These are all the most important elements which are significant for setting any measurement task. It is important to note that a complete and correct formulation of the measurement task in itself is complicated, which will be seen from examples given below. At the same time, setting a measurement task in an incorrect and incomplete manner can result in either the impossibility of performing the measurement or its incorrect results.

The setting of a particular measurement task in its turn depends on the practical goal which can be reached for making a decision only on the basis of obtained measurement information. This practical goal has to be connected with the measurement task with a chain (sometimes a very complicated one) of logical arguments of measurement information consumers. Correctness and completeness of the measurement task formulation entirely depend on the experience and skill of a measurement information consumer. However in any case this logical chain should be finished with a set of parameters concretizing the elements of a set (3.1).

Example 3.2. Let the final practical goal be the delivery of a cargo to point A within a given time interval (t_0). To do this, it is necessary to make a decision about when (at what moment of time) (t_n) has to go from a resting position (point B) to a railway station. Making the measurement problem more concrete, let us specify first of all that the measurement task has to be the time, or more precisely, the time interval ΔT_n, which allows the distance B to A to be covered.

In this case an object of study is the process (the process of moving from B to A). It is evident that this process should be detailed according to the method of moving (the kind of transportation and the route). The accuracy at which it is necessary to determine ΔT depends on the deficiency of time. Provided the conditions are such, that it is possible to reach the railway station with five minutes to spare ($T_0 - T_N = \Delta T_n + 5\,\text{min}$), then it is enough to assume a confidence interval equal to $\pm$ 1 min at a confidence probability of 0.997.

Consequently, the root mean square deviation of a measurement result should be $\sigma_{\Delta T} \cong 20\,\text{s}$. A digital form for presenting the result is preferable. The time moment before which the result (T_r) should be obtained has to precede T_n: $T_r \cong T_n$ (hour), n being determined by the daily routine of a goer.

It is obvious that even for one quantity (measurand) there exists an infinite set of different settings of measurement tasks:

$$\begin{aligned} &\{Z_i(\varphi)\} = \langle \{o_i\}, \{\psi_i\}, \{\Delta\varphi_i\}, \{g_i\}, \{t_i\}, \ldots \rangle, \\ &o_i \in \{o_1, o_2, \ldots\}, \ \psi_i \in \{\psi_1, \psi_2, \ldots\}, \ \text{etc.}, \end{aligned} \tag{3.2}$$

determined by a variety of realizations of the elements entering (3.2). In practice, however, only individual groups of measurement tasks are allocated, which are close not only in the parameters of their elements, but also in the methods used for solving these tasks.

3.1.3 Measurement as a process of solving a measurement task

After introducing the measurement problem, a further process as a whole, connected with measurements (more precisely, with the achievement of a final goal of measurement, i.e., finding a value of a measurand) has to be considered as a process of solving this measurement task. In this process three stages are usually distinguished.

In the *first stage*, in accordance with the measurement task, reflecting the question "what should be done?", *a plan of the measuring experiment* is developed which makes the question more exact: "How should it be done?" At this stage one chooses the measurand unit, method and the required measuring instruments, the procedure for using them is defined more exactly, and then the methods and means of experimental data processing are specified. Moreover, at this stage an operator (experimenter) who can perform the experiment is determined.

Information corresponding to the measuring experiment plan is usually provided in the form of *a procedure for performing measurements*. By analogy with (3.1) it can be structurally formalized in the form of an ordered set:

$$U_z \equiv \langle [\varphi], m, s, v, w, \ldots \rangle, \tag{3.3}$$

where
$[\varphi]$ is the unit of the measured physical quantity;
m is the selected method of measurements (or an algorithm of using the selected technical means s *and* w);
s is the measuring instruments used for solving a given measurement task;
v is the operator (experimenter) realizing the plan of a measurement experiment;
w is the means of processing results of the measurement experiment (or some other auxiliary means);
(here everywhere the index "I", meaning a membership of the set U_z and that of each of its elements to the given measurement task z_i, is omitted).

During the same stage, when using the plan which has been set up, a preliminary estimation of the uncertainty of an expected result $\hat{\Delta}\varphi$ (an interval of measurand uncertainty), is made, and then this estimate is compared with the required accuracy of measurement ($\Delta\varphi$).

In the *second stage* of the measurement process the measurement experiment is realized, i.e., the process of real (physical) transformations connected with a physical interaction of the selected measuring instruments with the object and environmental conditions (with participation of the operator) is carried out which results in a response of the measuring instrument to the quantity measured (indication x_s).

Finally, in the *third* and final *stage* of the measurement procedure the obtained response of the measuring instrument is transformed into a value of the measurand, and an a posteriori degree of its uncertainty (in case of multiple measurements) is estimated, thereby achieving the final goal of measurement, i.e., getting a result corresponding to the solution of the posed measurement problem.

All transformations and operations in the second and third stages are realized in accordance with the measurement experiment plan developed at the first stage.

A general scheme of the measurement process (the structure of a measuring chain) is shown in Figure 3.1.

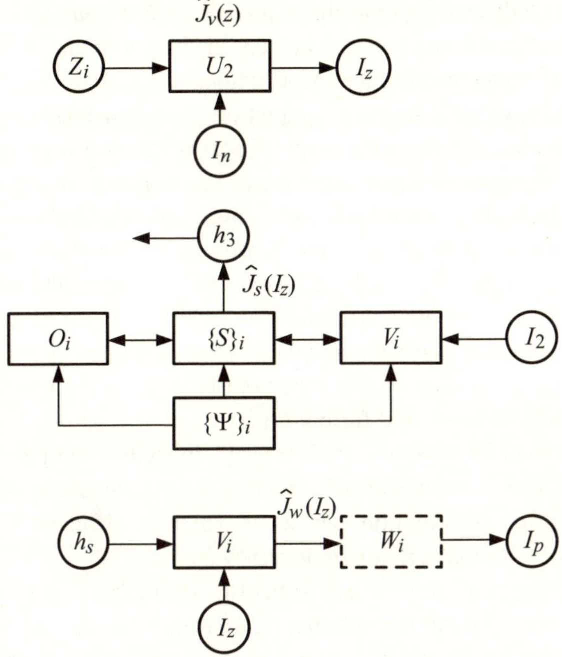

Figure 3.1. Stages of a measurement as those of the process of solving a measurement task. Here: $\hat{J}_v(z)$, $\hat{J}_s(I_z)$, $\hat{J}_w(I_z)$ are the operators of a successive transformation of a priori information I_a into a posteriori information at different stages of the measurement process; the remaining notations are explained in the text.

3.1.4 Formalization of a measurement as a system

From the above description of the measurement chain structure it is seen that it contains both the components of the measurement task (3.1) and the components of the measurement plan (3.3). Combining these sets gives a new set:

$$I_Z \equiv \langle Z_i, U_z \rangle, \tag{3.4}$$

where $Z_i \subseteq I_z$, $U_z \subseteq I_z$, which corresponds to a particular singular measurement, related to the system as a whole.

In equation (3.4) the elements contained in Z_i are *uncontrollable* measurement components (i.e., given in advance in the given measurement task), and elements contained in U_i are *controllable* (selected in planning the measurement experiment).

With regard to formal description (3.1), (3.3), and (3.4) we note the following.

1) None of the indicated sets replaces a detailed substantial analysis of a particular measurement situation which makes necessary the carrying out of a particular mea-

surement. The sets indicated serve as the *guideline for this analysis*, underlining those components of measurement which are present in each of them, and for each of them it is necessary to have quite sufficient information.

2) All components of measurement in equation (3.4) appear (from the point of view of the information about them) at the stage of posing the measurement task and at the stage of planning the measurement experiment, i.e., before carrying out the measurement procedure itself (the experiment). This once again underlines the great – as a matter of fact deciding – role of the a priori information in measurements.

The statement according to which *any particular measurement requires the availability of definite (final) a priori information* about components of measurement (3.4) is so universal that it also plays a part of one other *postulate of metrology*. An adequately convincing illustration of this statement is given in the preceding sections of this volume and will be supported further on.

3) As a consequence of the measurement problem analysis, at the stage of planning the measurement experiment it is understood that the experiment as such is not needed. An amount of a priori information can occur which is sufficient to find a value of measurement information at a required accuracy *by calculation*.

4) In planning the measurement experiment it is possible to reveal more than one version of the solution for the measurement task which satisfies its conditions. In this case, the choice of a final version is based on economic considerations (the version requiring minimal cost is chosen).

3.1.5 Target function of a system

The goal function of a system I_z is determined by the general destination (goal) of measurement, i.e., getting information about a value of the measurand φ with the required accuracy characterized by an interval $\Delta\varphi$. Therefore, the goal of measurement will be achieved when the condition given below will be met:

$$\Delta\varphi_{\text{norm}} \geq \Delta\varphi_{\text{meas}} \cong |\varphi_h^{\text{meas}} - \varphi_l^{\text{meas}}|, \tag{3.5}$$

where φ^{meas} is the value obtained as a result of measurement, and $\Delta\varphi_{\text{meas}}$ is the estimate of the uncertainty of the value obtained for the measurand.

Example 3.3. The measurement task described in Example 3.2 can be solved by a number of methods, depending on the availability of different a priori information.

1) If a clear-cut train schedule for the delivery of transported goods from point B to point A is available (accurate within 1 min), then, using the schedule, a train departure time T_n is calculated from the condition $T_n = T_0 - \Delta T_n - 5\,\text{min}$.

2) If there is no train schedule, then having chosen the kind of transportation (bus, taxi or pedestrian method of moving) based on economic considerations, a method, measuring instrument, and operator are chosen. As a measuring instrument a simple

chronometer can be used (electronic wristwatch) which provides the required accuracy of measurements. At the same time the method of measurement is the method of direct readout performed by the operator, who is the person who wishes to leave from the railway station.

3) If is a question of whether or not the external conditions in moving from point B to point A accidentally influence the time of moving, then multiple measurements are carried out under different conditions , and a dispersion or root-mean-square deviation of the values measured $\Delta T_n(i)$ are found.

4) In the above various extreme situations are not taken into consideration (traffic congestions, unexpected ice-covered ground, absenteeism or nonfunction of transportation, etc.). These anomalies can be neglected if their probability $\ll (1-0.997) = 0.003$. In the opposite case this probability has to be taken into account within a confidence interval, and when that increases it is necessary to correspondingly increase the time reserve for the departure from point B.

3.2 Potential and limit accuracies of measurements

We need to distinguish the potential and limit accuracy of measurements. The *limit accuracy* is understood to be the maximum achievable accuracy at which a measurement of a physical quantity can be carried out at a given developmental stage of science and engineering. The limit accuracy of measurements is realized, as a rule, in national measurement standards reproducing the units of a physical quantity with the maximum accuracy within the framework of the respective country. Frequently it is identified with the ultimate sensitivity of a measuring instrument.

By *potential accuracy* we mean the maximum accuracy at the contemporary developmental stage of scientific knowledge and engineering. The potential accuracy cannot be technically realized at the present-day stage of production engineering (including the measurement standards); in this case the measurements themselves become only potential.

The maximum achievable accuracy appearing in both definitions means just the maximum accuracy of measurements; however this requires an improvement in choosing a particular characteristic (index) of accuracy. This becomes evident in considering the so-called *ultimate measurements*, i.e., the measurements of a certain physical quantity at a level of the sensitivity threshold of an applied measuring instrument.

Since the sensitivity threshold according to its definition is a minimum value of the measurand $\varphi_{\min}(s_i)$ at which a given measuring instrument s_i still gives a noticeable response to the input measurand, then the measurement of small amounts of the quantity, comparable with the sensitivity threshold, has a high accuracy from the point of view of absolute uncertainty, and an extremely low one from the point of view of relative uncertainty (up to 100 %) $\Delta\varphi(s_i)/\varphi_{\min}(s_i) \cong 1$.

In spite of the difference indicated above the potential and limit accuracies of measurements have common features (which is also seen from the definitions). Therefore, considering mainly the *potential* accuracy of measurements we will touch on some aspects of the *limit* accuracy of measurements.

A structural presentation of the measurement as a system is used as the *methodological basis* for systematically studying the potential accuracy (as well as the limit accuracy). This is described in the preceding section.

According to this presentation the measurand value in question, φ, is determined with all components $x_k \in I_z$ which influence it:

$$\varphi^{\text{meas}}(I_z) = f(I_z) = f(\{x_k\}) = f\big(\varphi, o, \psi, g, t, \ldots \,|[\varphi], s, m, v, w, \ldots\big). \quad (3.6)$$

Let the parameters characterizing the properties of a component x_k influencing the measurement result be denoted as $q_{ik} \in x_k, i, k = 1, 2, \ldots, l_k$. If dependence (3.6) for these parameters is presented in an analytical form and the function f itself is differentiable, then by calculating *the influence function* for each of the parameters q_{ik}, as partial derivatives,

$$h_{lk} = \frac{\partial f(I_z)}{\partial q_{ik}}, \quad (3.7)$$

it is possible to make an estimate of the inauthenticity (uncertainty) of the measurement result (the interval within which the value measured at a given or the chosen probability can be given) through the uncertainties of the parameters values q_{ik} of the system I_z:

$$\hat{\Delta}\varphi^{\text{meas}}(I_z) = \sqrt{\sum_k \sum_{i_k=1}^{l_k} \left(\frac{\partial f}{\partial q_{ik}}\right)^2 \cdot \Delta q_{ik}}. \quad (3.8)$$

This scheme for evaluating the measurement accuracy by the degree of the uncertainty of its result is correct for any measurement task.

However, a specific character of the situation when considering the potential accuracy of measurements can allow some simplification which excludes from consideration separate components x_k, connected primarily with the parameters of economical and technical (or technological) character.

Thus, we may very well digress from the possibility of the appearance of appreciable crude errors [323]. These crude errors can be caused by miscalculation in realizing the measurement experiment plan, imperfection of the operator work, the limits imposed on the accuracy connected with the imperfection of methods and means used for processing measurement information, etc. Here we also assume that it is potentially possible to reduce the value of a random error component to a negligible value by repeating the readouts many times, i.e., to leave for consideration only the uncertainties at single readouts.

Thus, in considering the potential accuracy of measurements it is necessary to analyze only the influence of the components φ, o, ψ, $[\varphi]$, s, as well as that of the time–space parameters.

3.3 Accuracy limitations due to the components of a measurement task

This section deals with the influence of the system I_z, the components of which participate in the process of measurement but nominally appear only at the stage of posing the measurement task. Here we have: a measurand (φ), th eobject of measurement (o), external conditions (ψ). Time–space limitations will be considered separately (in Section 3.4.4). All these components are uncontrollable (given) within the framework of the system I_z. However, the accuracy of giving them (or more precisely the accuracy of giving the parameters characterizing them) is not infinite. It is determined by the amount of a priori information about them and the degree of its correspondence to present-day reality.

3.3.1 Measurand and object models

The measurand as a quality is given by its definition in the system of quantities and concepts, i.e., on the basis of agreements. The agreements (if they are sufficiently univocal), in principle, can be implemented with complete precision. However, in practice, as applied to a specific situation and a definite object of study, some assumptions giving a concrete expression to the definition of the measurand arise. Let us consider the above for physical quantities, which is the most important case.

In the system of physical quantities a formalized reflection of the quality of each quantity (its definition) is the quantity dimensionality that reflects the relationship between this quantity and the basic quantities of the system. Such definitions are accurate (univocal) information about the quality (kind) of the quantities measured under the condition of the successful choice of the basic quantities. Thus, the velocity of a body movement is determined by the derivative of a distance with respect to time, the density of substance is determined by the ratio of a mass of this substance to its volume, etc.

However, even for the simplest derived quantities of this kind their definitions are abstractions and are related to idealized objects, since any physical quantity is the property that in a qualitative respect is inherent in many objects.

Thus, all quantities of kinematics (for example, displacement, velocity, acceleration, and others) are introduced for an idealized "material point". The definition of substance density, indicated above, assumes its homogeneity, etc. The matter stands even worse with basic quantities. As a matter of fact, these quantities are indefinable in the system of physical quantities (although there are vague wordings such as "length is a measure of a spatial stretch of bodies", "mass is a measure of the inertial and gravitational properties of a body"). Such is the dialectics of knowledge. The higher the degree of the concept generality, the more abstract this concept is, i.e., it is separated from concreteness.

Therefore, while formulating the measurement problem and then planning the measurement experiment, in addition to understanding the measurand as a quality, it is necessary to detail it as applied to the object under study, having constructed a model of the object analyzed.

It should be added that in constructing such a model, it is necessary to define more exactly the following parameters.

1) To construct a *model of the measurand* of an object under study, i.e., to determine a particular *measurable parameter* of the object uniform with respect to the physical quantity being studied. Usually this parameter is connected with an outside part (surface) of the object. A number of examples of linear dimensions and electrical parameters are given in Section 3.4. Below one more example from the field of mechanics is given.

Example 3.4. When measuring the vibration parameters of linearly oscillating mechanical systems, where the object of study is the oscillation process of a hard body movement for the model of an object serves, as a rule, the harmonic law of material point movement:

$$x(t) = x_m \sin(\omega t + \varphi_0), \tag{3.9}$$

where x is the displacement of a definite point on a body surface with respect to the static (equilibrium) position, x_m is the maximum value (amplitude) of the displacement, ω is the frequency of the oscillation movement, and φ_0 is its starting phase.

Usually as the measurand either a length (linear displacement) or acceleration a, connected with x, are used:

$$a(t) = d^2x/dt^2 = -x_m\omega^2 \cos(\omega t + \varphi_0) = -a_m \cos(\omega t + \varphi_0).$$

In a particular case, depending on the goal of measurement and reasoning from other considerations regarding the object to be measured either an amplitude of displacement x_m, or of acceleration a_m, or an "instantaneous" value of displacement $x(t)$ or acceleration $a(t)$, or their mean values over a definite interval of time are used.

2) As noted in [37], to perform measurements it is necessary to have a priori information not only about the *qualitative* but also the *quantitative characteristics* of the measurand, its dimension. The more precise and consequently more useful the available a priori information about the measurand dimension is, the more forces and means there are for getting more complete a posteriori information for improving the a priori information.

3) A measurement signal entering from the object at an input of a measuring instrument in addition to information about the measurand can contain information about some other *"noninformational" parameters* of the object which influence the readouts of the measuring instrument. This is especially characteristic for measurements in the dynamic mode. Thus, in Example 3.4 one noninformational parameter of the object is the frequency ω at which its oscillations take place.

4) Although the "informative" and "noninformative" parameters of the object considered above reflect its inherent properties, however, being embodied into parameters of the measurement signal, they experience real transformations in the measurement chain.

In addition any object possesses a variety of other "inherent properties" which do not influence the parameters of the measurement signal, but rather the information about it which allows a correct model of the object (for the given measurement problem) to be constructed, which corresponds as much as possible to reality.

Parameters characterizing such properties of the object which influence the model ideas about this object will be called the *model parameters* of the object.

Thus, while carrying out precise measurement of the density or viscosity of a substance it is necessary to know its composition, distribution of admixtures over the whole volume, its temperature, its compressibility, etc.

The fact that the model in Example 3.4 is not ideal usually becomes apparent because of deflection of a real form of oscillation of a point on the surface of a body from the harmonic law (3.9).

To construct a more accurate model it is necessary to know such model parameters of the object as dislocation of the mass center of an oscillating body relative to a point of measurement, elastic properties of the object, etc. A more detailed example illustrating the influence of imperfection of the object model on the measurement accuracy will be considered in Section 3.4.

Provided that the constructed object model is such that it allows the dependence of a measurand on informative (q_{io}), noninformative (q_{no}), and model (q_{mo}) parameters to be expressed analytically, then the uncertainty of a measurement result caused by these parameters can be determined with the help of the corresponding functions of influence (3.7) and the uncertainties of the values of these parameters (q_{io}):

$$\hat{\Delta}_o \cdot \varphi(I_z) = \sqrt{\sum_{i_0=1}^{l_0} \left(\frac{\partial f(q_{i_0})}{\partial q_{i_0} \cdot \Delta q_{i_0}} \right)^2}. \tag{3.10}$$

(here the index "0" accentuates that the system parameters considered belong to the object).

3.3.2 Physical limitations due to the discontinuity of a substance structure

It would seem that improving the values of an object q_{i0} more and more (as knowledge is accumulated, i.e., decreasing their uncertainty Δq_{i_0}) we can gradually, but still continuously, decrease the uncertainty of the measurement result, connected with the parameters $\Delta_{0\varphi} \to 0$ at $\Delta q_{i_0} \to 0$ (see equation (3.10)).

However there is a certain type of physical knowledge about objects of the world surrounding us which leads us to speak of the *principal limits of accuracy* with which we can obtain quantitative information about a majority of physical measurands.

The first circle of such knowledge is connected with molecular-kinetic ideas about a substance structure, the theory of which is described in courses on molecular physics, thermodynamics, and statistical physics.

The *molecular-kinetic theory* proceeds from the assumption that a substance exists in the form of its components, i.e., molecules (microscopic particles), moving according to laws of mechanics. All real physical objects which are met in practice (be it gases, liquids, or hard bodies) have a discrete structure, i.e., they consist of a very large number of particles (molecules, ions, electrons, etc.).

So, for example, one cubic centimeter of metal contains $\sim 10^{22}$ ions and the same numbefr of free electrons; at room temperature and at an atmospheric pressure of 1 cm^3 of air contains $\sim 3 \cdot 10^{19}$ molecules; even in a high vacuum (one million times less than atmospheric pressure) 1 cm^3 of air contains $\sim 10^{12}$ molecules. At the same time the particle dimensions are of the order of $\sim 10^{-10}\,\mathrm{m} = 10^{-8}\,\mathrm{cm}$.

Although the laws of mechanics allow the behavior of each separate particle in the system to be described precisely, nevertheless a huge number of the system particles offer neither the technical nor physical possibility to describe the behavior of the system as a whole. Under these conditions a single possibility is provided only by a *probabilistic description* of a complicated system state (an ensemble of a set of particles).

Statistical physics is based on these ideas and uses the methods and ways of mathematical statistics to establish the *statistical laws* of the behavior of complicated systems consisting of a great number of particles. Using this as a base, it studies the macroscopic properties of these systems.

Owing to the great number of microparticles composing the objects of a physical study, all their properties considered as random quantities, are averaged. This allows these properties to be characterized with some macroscopic (observable) parameters such as density, pressure, temperature, charge, and others, which are related to the *thermodynamic* parameters of the system which do not take into account their link with the substance structure.

Such a link, however, becomes apparent when the indicated thermodynamic parameters are described using *statistical mean values* of the corresponding quantities which are introduced according to usual rules of forming a mean:

$$\bar{\varphi} = \int \varphi \cdot \rho(\varphi) d\varphi, \tag{3.11}$$

where φ is the quantity being averaged and characterizing the system state, and $\rho(\varphi)$ is the probability density function of quantity values φ (of the system states) (integration is performed over all system states).

So the mean kinetic energy of single-atom gas molecules

$$\bar{E}_k = \int_0^\infty E_k \rho(E_k) dE_k = \frac{3}{2} kT,$$

is found from the law of distribution molecules on energies:

$$\rho(E_k) = \frac{2}{\sqrt{\pi}} (kT)^{-3/2} e^{-E_k/kT} \sqrt{E_k},$$

where k is the Boltzmann constant ($k = 1.38 \cdot 10^{23} J/K$), and T is the absolute temperature of gas (K).

Thus, the mean kinetic energy of the linear motion of the molecules of an ideal gas serves as a measure for its absolute temperature (as a thermodynamic parameter of the system).

In any real physical system, even in the case when the macroscopic quantities characterizing its states are considered to be independent in time, i.e., constant, thanks to the continuous motion of microparticles, spontaneous changes of macroscopic (observable) parameters close to their mean equilibrium value $\bar{\varphi}$ occur. These stochastic deviations from the mean characterize the *fluctuations of a physical quantity on an object.* As a measure of fluctuation the root-mean-square deviation from the mean we generally use

$$\Delta\varphi = \sqrt{(\varphi - \bar{\varphi})^2}, \tag{3.12}$$

or a relative fluctuation

$$\delta\varphi = \frac{\Delta\varphi}{\bar{\varphi}} = \sqrt{\frac{(\varphi - \bar{\varphi})^2}{\bar{\varphi}^2}}. \tag{3.13}$$

From the properties of random quantity variable dispersion

$$\overline{(\varphi - \bar{\varphi})^2} = \overline{\varphi^2} - 2\bar{\varphi} \cdot \bar{\varphi} + \bar{\varphi}^2 = \overline{\varphi^2} - \bar{\varphi}^2$$

it follows that to calculate the fluctuation of two quantities are needed: the mean of the square of fluctuations ($\overline{\varphi^2}$) and the square of the mean ($\bar{\varphi}^2$).

Since the fluctuations are connected with a great number of small random deviations from the mean, these deviations are then subordinated to the normal (Gaussian) distribution law. Provided that the value considered as φ is the function of another random quantity x, then they are both subordinated only to the same distribution law, and it is possible, instead of (3.11), to write

$$\overline{\varphi(x)} = \int \varphi(x)\rho(x)\mathrm{d}x, \tag{3.11-1}$$

and a mean square of fluctuation will be determined as

$$\overline{\varphi^2} = \int \varphi(x)\rho(x)\mathrm{d}x. \tag{3.14}$$

In particular, let all $6N$ (where N is the number of particles of the system, and 6 is the number of degrees of freedom for one particle) be microscopic variables of the system. Then, according to the statistical physics for an isothermal (with a constant temperature) system located in a thermostat, the function of distribution called *the Gibbs' canonical distribution* has the following form:

$$\rho(x) = e^{\frac{\psi - H(x,a)}{\theta}}, \tag{3.15}$$

where ψ and θ are certain constants, (x, a) is the total energy of the system (the Hamiltonian function), and a are the external, with regard to the system, parameters.

Any mean quantity at the canonical distribution is determined as a mean over the whole phase space Γ, the element of which is $(\mathrm{d}x)^{6N}$:

$$\bar{\varphi} = \int_{\Gamma} \varphi(x,a) \cdot e^{\frac{\psi - H}{\theta}} (\mathrm{d}x)^{6N}. \tag{3.11-2}$$

Here θ has a sense of the statistical temperature $\theta = kT$, and the parameter ψ has the sense of the "free energy" of the system. An important role is played by the integral:

$$Z = \int_{\Gamma} e^{-\frac{H}{kT}} (\mathrm{d}x)^{6N}, \tag{3.16}$$

called the *integral of states* reflecting the inner state of the system. The free system energy $\psi = -kT \cdot lnZ$, and a number of other thermodynamic parameters and functions of the system are expressed with the help of this integral.

Example 3.5. Let N gas molecules be in volume V under a piston to which an external pressure p is applied. It is required to find out a fluctuation of the volume V. In this case the pressure p is considered to be an external parameter a corresponding to the volume $V(x)$, and the quantity under study V depends only on a coordinate x of the system along which the pressure $p(x)$ acts. Writing the total energy of the system (the Hamiltonian function) as

$$H(p,x) = H(x) + pV(x),$$

and using the properties of the canonical distribution, we obtain

$$\overline{(V - \bar{V})^2} = -kT \frac{\partial V}{\partial p}. \tag{3.17}$$

The derivative $\frac{\partial V}{\partial p}$ is found from the Mendeleyev–Chapeyron equation:

$$pV = NkT, \quad \text{i.e.,} \quad \left(\frac{\partial V}{\partial p}\right)_T = -\frac{NkT}{p^2}.$$

Then, the volume dispersion is equal to

$$\overline{\Delta V^2} = \overline{(V - \bar{V})^2} = \frac{k^2 T^2 N}{p^2} = \frac{V^2}{N},$$

and the relative volume fluctuation is

$$\delta V \equiv \frac{\Delta V}{V} = \sqrt{\frac{\overline{(\Delta V)^2}}{\bar{V}^2}} = \sqrt{\frac{V^2}{N \cdot V^2}} = \frac{1}{\sqrt{N}}, \tag{3.18}$$

i.e., is inversely proportional to the square root of a number of system particles.

Provided that $N \sim 10^{20}$ is taken into account, then the uncertainty of the volume of the object measured (of a gas medium under pressure p) due to its fluctuation will have the order

$$\delta V = \frac{\Delta V}{V} \approx 10^{-10}. \tag{3.18-1}$$

In general, if the system consists of N independent parts, then the relative fluctuation of any function φ of the state of the system is inversely proportional to the square root from the number of its parts:

$$\delta\varphi = \sqrt{\frac{\overline{(\Delta\varphi)^2}}{\bar{\varphi}^2}} \approx \frac{1}{\sqrt{N}}. \tag{3.18-2}$$

This is correct for both the additive (extensive) quantities φ_e, for which $\varphi_e \sim N$, and the nonadditive (intensive) quantitites φ_i, for which $\varphi_i \neq f(N)$. It is easy to make sure that at the same time

$$\overline{(\Delta\varphi_e)^2} \approx N \quad \text{and} \quad \overline{(\Delta\varphi_i)^2} \approx \frac{1}{N}.$$

Some examples of the extensive quantities characterizing the thermodynamic properties of the system are the volume of substance and its energy; the system temperature and pressure are related to the intensive quantities.

Let us consider some examples of the fluctuation dimensions for some other quantities.

Example 3.6 (Length). Hard bodies in reality are not absolutely hard (with an infinite hardness), but possess a definite elasticity. They change their dimensions and form under the influence of external forces. Elastic properties of a rod at small deformations described by the Hooke's law are

$$\frac{\Delta l}{l_0} = \frac{l(F) - l(0)}{l(0)} = \frac{1}{E \cdot s} \cdot F, \tag{3.19}$$

where Δl is the extent of the rod under the influence of the force F, E is the coefficient (modulus) of elasticity of the rod substance, and s is the area of the rod cross section.

The potential energy of the elastic deformation of the rod is

$$\Delta W_n = -F\Delta l = \frac{E \cdot s \cdot \Delta l}{l_0}\Delta l = \frac{E \cdot s}{l_0}(\Delta l)^2.$$

Since only small random changes of the rod length relative to a medium-equilibrium position l_0 are considered, then the mean value of the potential energy on the ensemble

$$\overline{\Delta W_n} = \frac{E \cdot s}{2l_0}\overline{\Delta l^2}$$

is equal to the mean value of the kinetic energy of the elastic oscillations caused by the thermal motion of the microparticles:

$$\overline{\Delta W_k} = \frac{kT}{2}.$$

Equating both expressions, we obtain

$$\overline{(\Delta l)^2} = \frac{kT}{Es}l_0. \tag{3.20}$$

At $T \approx 300\,\mathrm{K}$, $E \approx 10^{11}\,\mathrm{c/m^2}$, $s = 3\,\mathrm{cm} \cdot 3\,\mathrm{cm} \approx 10^{-3}\,\mathrm{m^2}$, $l_0 \approx 1\,\mathrm{m}$ we have $\sqrt{\overline{(\Delta l)^2}} \approx 10^{-14}$, i.e., the mean quadratic fluctuations of the length of the one meter rod, caused by thermal motions at a room temperature and elastic properties of the rod substance (average for noble metals) are $10^{-14}\,\mathrm{m} = 10^{-8}\,\mu\mathrm{m}$, which is significantly less than the atomic dimensions ($\sim 10^{-10}\,\mathrm{m} = 10^{-4}\,\mu\mathrm{m}$) and are comparable to the dimensions of the atomic nuclear.

Let us note that expression (3.20), obtained by a simplified method, coincides with the accurate derivation of rod fluctuation from the laws of statistical physics and thermodynamics.

Let us consider an example of such a derivation using expressions (3.15), (3.11-2), and (3.19).

It should be noted that the integral of distribution (3.15) over the whole phase ensemble Γ of the object being considered is equal to 1:

$$\int_\Gamma e^{\frac{\psi - H}{\theta}} \cdot (\mathrm{d}x)^{6N} = 1. \tag{3.21}$$

Here $\theta = kT$ is the constant (the distribution modulus), the same for all systems of the ensemble; $\psi = \psi(a, \theta)$ depends on external parameters a and modulus θ, i.e., also the same for all systems of the ensemble, namely it does not depend on the coordinates x in the phase space G; $H = H(x, a)$ is the total mechanical energy of one system from the ensemble, which depends on the coordinates x and external parameters a.

Let us take a partial derivative of equation (3.21) with respect to the parameter a:

$$\begin{aligned}\frac{\partial}{\partial}\int_G e^{\frac{\psi-H}{\theta}}(\mathrm{d}x^{6N} &= \int_G \frac{\partial}{\partial a}\left(e^{\frac{\psi(a,\theta)-H(x,a)}{\theta}}\right)(\mathrm{d}x)^{6N} \\ &= \frac{1}{\theta}\int_G \frac{\partial\psi}{\partial a}\cdot e^{\frac{\psi-H}{\theta}}(\mathrm{d}x)^{6N} \\ &= \frac{1}{\theta}\int_G \frac{\partial H}{\partial a}\cdot e^{\frac{\psi-H}{\theta}}(\mathrm{d}x)^{6N} = 0,\end{aligned}$$

hence

$$\int_G \frac{\partial\psi}{\partial a}e^{\frac{\psi-H}{\theta}}(\mathrm{d}x)^{6N} = \int_G \frac{\partial H}{\partial a}e^{\frac{\psi-H}{\theta}}(\mathrm{d}x)^{6N}.$$

Taking into account the determination of the mean with respect to ensemble (3.11-2) as well as the independence of ψ on the coordinate x, we obtain

$$\left(\frac{\partial\psi}{\partial a}\right)_\theta = \overline{\left(\frac{\partial H}{\partial a}\right)_\theta}. \tag{3.22}$$

Let us determine the derivative of mean (3.11-2) of an arbitrary quantity $\varphi(x)$ with respect to a at the canonical distribution (3.15):

$$\begin{aligned}\frac{\partial\bar{\varphi}}{\partial a} &= \frac{\partial}{\partial a}\int_G \varphi(x,a)e^{\frac{\psi-H}{\vartheta}}(\mathrm{d}x)^{6N} \\ &= \int_G\left[\frac{\partial\varphi}{\partial a}e^{\frac{\psi-H}{\vartheta}} + \frac{\varphi(x,\ a)}{\theta}e^{\frac{\psi-H}{\vartheta}}\frac{\partial\psi}{\partial a} - \varphi(x,a)\frac{1}{\theta}e^{\frac{\psi-H}{\vartheta}}\frac{\partial H}{\partial a}\right](\mathrm{d}x)^{6N} \\ &= \frac{\partial\bar{\varphi}}{\partial a} + \frac{1}{\theta}\left(\frac{\partial\psi}{\partial a}\bar{\varphi} - \overline{\varphi\frac{\partial H}{\partial a}}\right)\end{aligned}$$

(here we have once again used the definition of the mean).

Substituting $\frac{\partial\psi}{\partial a}$ from equation (3.22) in the obtained expression we get

$$\left(\frac{\partial\bar{\varphi}}{\partial a}\right)_\theta = \left(\frac{\partial\bar{\varphi}}{\partial a}\right)_\theta + \frac{1}{\theta}\left(\frac{\partial H}{\partial a}\bar{\varphi} - \overline{\varphi\frac{\partial H}{\partial a}}\right),$$

and taking into account the fact that for any random quantities A and B the following expression is correct:

$$\overline{(A-\overline{A})(B-\overline{B})} = \overline{AB} - \overline{A}\cdot\overline{B}, \tag{3.23}$$

finally we have

$$\left(\frac{\partial\bar{\varphi}}{\partial a}\right)_\theta = \overline{\left(\frac{\partial\varphi}{\partial a}\right)_\theta} - \overline{\frac{1}{\theta}(\varphi-\bar{\varphi})\left(\frac{\partial H}{\partial a} - \frac{\partial\bar{H}}{\partial a}\right)}. \tag{3.24}$$

Let us apply the obtained expression to the case being considered, i.e., the case of the fluctuation of the metal (elastic) rod length, when $\varphi(x) = l(x)$, $\bar{\varphi} = l_0$, $a = F$ (force as an external parameter).

The total mechanical energy $H(x, F)$ of one system from the ensemble describing possible elastic states of the rod (the equation of the rod state) will be

$$H(x, F) = F \cdot l(x). \tag{3.25}$$

Here we disregard the thermal expansion of the rod $l(T) = l(0) \cdot [1 + \alpha(T - T_0)]$, since it is insignificant for considering elastic fluctuations, especially because it is possible to assume that $T = T_0$.

Then the mean value of the derivative of $l(x)$ with respect to F in equation (3.24) will be

$$\overline{\left(\frac{\partial l(x)}{\partial F}\right)_\theta} = 0,$$

since $l(x)$ does not depend on F; the derivatives of H with respect to F according to equation (3.25) are equal to

$$\frac{\partial H}{\partial F} = l, \quad \overline{\frac{\partial H}{\partial a}} = l_0,$$

and the derivative of l, as of the macroscopic length of the rod, with respect to F is determined using Hooke's law (3.19):

$$l = l_0 - \frac{l_0}{Es} F, \tag{3.19-1}$$

i.e.,

$$\left(\frac{\partial \bar{l}}{\partial F}\right)_\theta = -\frac{l_0}{Es}.$$

Substituting the new variables and derivatives which have ben determined into equation (3.24), we obtain

$$\frac{l_0}{Es} = \frac{1}{kT}\overline{(l - l_0)(l - l_0)} = \frac{\overline{(\Delta l)^2}}{kT},$$

and hence equation (3.20) is obtained.

Now let us consider the fluctuations in electrical networks caused by the fluctuations of electrical current due to the thermal motion of carriers of an electron charge in a conductor.

Example 2.7 (Fluctuations in electrical networks). Fluctuations of electrical current in a conductor give rise to a *fluctuation electromotive force*, which appears on a resistance

R and is called noise voltage:

$$\varepsilon_n = \sqrt{\overline{\varepsilon_n^2}} = R\sqrt{\overline{(\Delta I)^2}}.$$

Some rather complicated calculations supported by electrodynamics and statistical physics show that

$$\overline{\varepsilon_n^2}(\nu) = 4kT \cdot R(\nu) \cdot \Delta\nu, \tag{3.26}$$

where Δv is the frequency bandwidth of the electromagnetic waves within which the fluctuations of voltage on the resistance is studied (observed).

Expression (3.26) is called *the Nyquist formula* that sometimes is written by way of a fluctuation of power (v), allocated on the resistance R:

$$\sqrt{\overline{P^2(v)}} = \frac{\overline{\varepsilon^2}}{R} = 4kT \cdot \Delta v. \tag{3.26-1}$$

The fluctuation electromotive force is small, although observable on high-value resistors (for example, on photoresistors). Thus, at $R \approx 10^8$ Ohm $=$ 100 MOhm and room temperature in the frequency range of up to several tens of Hz, the value of the electromotive force is of the order of some units $\mu\text{V} = 10^{-6}$ V.

3.3.3 Quantum-mechanical limitations

Apart from the above natural limitations due to the fluctuation phenomena connected with the discrete structure of observable macroscopic objects, there are restrictions applied to measurement accuracy of a more fundamental character. They are connected with the quantum-mechanical (discrete) properties of the behavior of microcosm objects.

The equations given in (3.3.2) are the consequence of the application of the laws of classical physics. The laws of *classical mechanics* and thermodynamics are only generalized and specified with reference to the discrete (molecular-kinetic) nature of the substance structure consisting of a huge number of microparticles which are described as an ensemble of many states (subsystems) of the object under study. Each of the system (object) states corresponds to a point in the phase space of states and *continuously changes* over this space.

Unlike these classical ideas, *quantum physics* takes into account the discrete properties of microobjects, which becomes apparent when some of the quantities characterizing the properties of microparticles can change only discretely by definite portions or quanta.

The system energy E, the vector of angular momentum $\vec{M}$, the vector of the magnetic momentum $\vec{\mu}$, and the projections of these vectors to any selected direction are related to such quantities. In particular a quantum parameter is the so-called *spin* (an intrinsic moment) of an electron. Quantization of these quantities is connected with the availability of a minimum portion (a quantum) of action called the *Planck constant*:

$$h = 6.6^{-34} \mathrm{J} \cdot \mathrm{s}.$$

Thus, in accordance with quantum mechanics the radiation energy of a harmonic oscillator (a particle oscillating under the influence of an elastic force) can receive only definite *discrete* values which differ from one another by an integer of elementary portions, i.e., *quanta of energy*:

$$E_v = hv, \tag{3.27}$$

where v is the radiation frequency connected with the cyclic proper frequency ω of the oscillator oscillation: $\omega = 2\pi v$.

Quantum physics revealed not only discrete (corpuscular) properties of traditionally continuous objects (fields, waves, radiations), but established wave (continuous) properties of discrete formations – particles, having compared every particle with a mass m moving at a speed V with a corresponding wave length:

$$\lambda = \frac{h}{mv} = \frac{h}{p}, \tag{3.28}$$

which is called the *de Broglie wavelength.* Thereby the corpuscular-wave dualism in the properties of microobjects was established as a general demonstration of the interconnection between two forms of substance and field.

In order to describe these microcosm peculiarities with the help of mathematics, it was required to compare each physical quantity with a definite operator Λ, which showed what actions had to be performed with respect to the function ψ, describing a state of the quantum system. This function is called the *wave function* of the system and has a special physical sense, since a square of its modulus $|\psi(q)|^2$, where q is the coordinates of the system, is connected with the probability of finding a quantum system in an element of the space of coordinates $dq = (q_1, q_2, \ldots, q_N)$ by the ratio

$$dw(q_1, q_2, \ldots, q_N) = |\psi|^2 dq_1 dq_2 \ldots dq_N, \tag{3.29}$$

i.e., $|\psi(q)|^2$ is the density of distribution of the probability of the system according to the states $(q_1, q_2, \ldots, q_N)$.

Thus, quantum mechanics already contains *statistical* (*probabilistic*) *conceptions* in its base.

The most important operator in quantum mechanics is the *Hamilton's operator H* representing the total energy of a system:

$$H = E_k + U = \frac{\hbar^2}{2m}\left(\frac{\partial^2}{\partial x^2} + \frac{\partial^2}{\partial y^2} + \frac{\partial^2}{\partial z^2}\right)\psi + U(q), \tag{3.30}$$

where

the first item is the operator of kinetic energy indicating that the wave function should

be twice differentiated with respect to each of the coordinates x, y, z;
$U(q)$ is the potential energy of the system depending on the system coordinates x, y, z, and, generally speaking, on time t;
$\hbar \equiv \frac{h}{2\pi} = 1{,}05 \cdot 10^{-34}\,\mathrm{J \cdot s}$;
m is the mass of a particle.

The main equation of quantum mechanics, which describes the behavior (a motion) of microparticles, is the *Schrödinger equation:*

$$i\hbar\frac{\partial\psi}{\partial t} = H\psi, \tag{3.31}$$

which is an analogue to Newton's equation in the classical description of the motion of a material particle (in the form of the canonical Hamilton equations).

For stationary (i.e., unchangeable with time) processes equation (3.31) is written in the form

$$H\psi_n = E_n\psi_n, \tag{3.31-1}$$

where E_n is the proper values of the operator H,and ψ_n is the proper wave functions of the particle (or of a system of particles) which are found by solving equation (3.31-1).

While describing the systems of many particles (i.e., macrosystems) the Schrödinger equation allows in the best case only a spectrum of the possible states of the system, $\{\psi_n\}$, to be obtained. However, it is impossible to answer the question: In what state is the quantum system at a given moment?

Just as when describing the system of microparticles in classical mechanics, this requires a knowledge of the distribution law of probabilities of separate states, with the help of which it is possible to determine the mean values of various parameters (physical quantities) of the system. The methods of mathematical statistics (as applied to quantum systems) can be of assistance in this case.

The wave function describing the state of a quantum system of N particles can depend only on either the coordinates $\psi(q_1, q_2, \ldots, q_N)$ or pulses $\psi(p_1, p_2, \ldots, p_N)$.

A particular feature of quantum statistics is the correspondence of one state of the quantum system in the phase space of states G to a certain minimum volume of this space $G_{\min}$, rather than a point. This is caused by a corpuscular-wave dualism in the behavior of microparticles, since the concept "coordinate" for the wave is devoid of a definite physical sense. The concept "trajectory" of a particle is also lacking the usual sense. Any microparticle turns out to be as if it were "smeared" over the space of states.

The principal invisibility following from here or the *identity of particles* in quantum mechanics leads to a relation reflecting a *principle uncertainty* in defining the values of a number of physical quantities being measured. In fact, the minimum phase state volume, corresponding to one state of the system, is connected with the Planck constant by a relation:

$$G_{\min} = h^{3N},$$

where $3N$ is the number of generalized coordinates or pulses of the system from N particles.

And since the volume of any element of the phase space is the product of the dimensions of the coordinates and pulses (i.e., the product of all parameters of the state) of this element,

$$G = (\Delta p_1 \Delta q_1)(\Delta p_2 \Delta q_2) \dots (\Delta p_N \Delta q_N), \quad \text{then}$$
$$(\Delta p_1 \Delta q_1)(\Delta p_2 \Delta q_2) \dots (\Delta p_N \Delta q_N) \geq h^{3N}. \tag{3.32}$$

Applying relation (3.32) for the case of one particle ($N = 1$), moving in the direction of one coordinate x, we get

$$\Delta x \Delta p_x \geq h. \tag{3.32-1}$$

Inequalities (3.32) and (3.32-1), linking the uncertainties of a simultaneous determination of the coordinate Δx and the projection of the particle pulse Δp_x are called a *Heisenberg's uncertainty relation*. This relation indicates that the more accurate the particle coordinates having been determined (i.e., the less Δx or $\Delta y, \Delta z$ are), the less accurate are the values of projections of its pulse (i.e., the more Δp_x, or $\Delta p_y, \Delta p_z$).

Relation (3.32-1) gives rise to a number of similar relations of uncertainties between other quantities called the canonically conjugate quantities. In particular, the following relation has a place:

$$\Delta E \cdot \Delta t \geq h, \tag{3.33}$$

which indicates that a decrease in the time (Δt) (or its uncertainty) during which the system is under observation results in an increase of the uncertainty of the energy value of this system; and vice versa, the system under study can have accurate values of energy only for indefinably large time intervals (i.e., only in the stationary cases).

Relation (3.33) imposes principal limitations on the accuracy of measurements in which the *methods of spectral analysis* of substance radiation are used. This is connected with the fact that any substance radiation is a consequence of transformations of state in the quantum-mechanical system, i.e., it is caused by a transition of the system from one discrete state with an energy E_1 into another discrete state with an energy E_2 (Figure 3.2):

$$E_v = hv = H_1 - H_2 = \Delta E_{1-2}.$$

Provided the state E_2 is the main state of the system, i.e., the state with a minimum possible energy, then $\Delta E_2 = 0$, since the system can remain in this state infinitely long and $\Delta t = \infty$. However any other energy state E_i of the quantum-mechanical system is its excited state in which it can remain during a limited (finite) time $\tau_i = \Delta t_i$ (the *life time* of an exited state) and in accordance with relation (3.33) has a finite energy uncertainty ΔE_i.

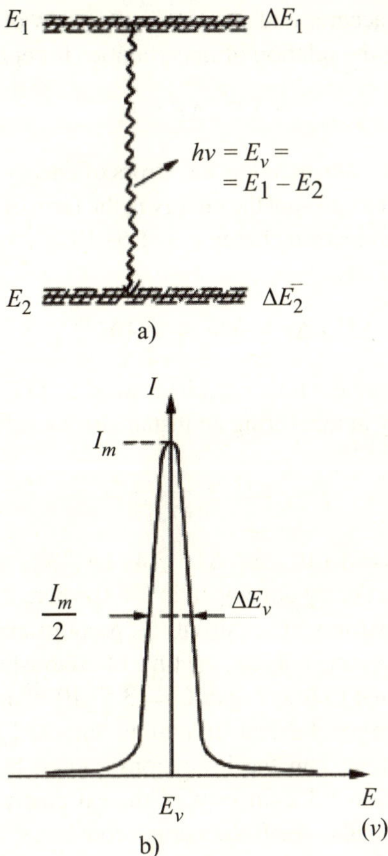

Figure 3.2. Energy jumps between two possible states of the quantum-mechanical systems (a) and a width of the corresponding spectral line of radiation (of a quantum hv), caused by this jump (b).

So, in atomic systems the life time of exited states is $\Delta t_i \sim 10^{-8}$ s, which corresponds to an interval of uncertainty $\Delta E_i \sim 10^{-7}$ eV $\sim 1.6 \cdot 10^{-26}$ J (1 eV $= 1.6 \cdot 10^{-19}$ J). Spectral lines of electromagnetic radiation of an atom (in an optical range of the spectrum where $v \sim 10^{14}$ Hz, and $hv \sim$ eV) have the same uncertainty. In nuclear systems the same electromagnetic interactions result in γ-radiation (hard electromagnetic radiation) with the energy of $hv \sim$ MeV and $v \sim 10^{20}$ Hz; the life time of nuclear levels is $\tau_l \sim 10^{-8}$–10^{-15} s, which corresponds to the uncertainty $\Delta E = \Delta(hv) \sim 10^{-7}$–$10^{14}$ eV. These quantities are very small.

Example 2.8 (Potential accuracy of joint measurements of the time dependence of electrical voltage). Using the relation of uncertainties for energy and time (3.33)

$$\Delta E \cdot \Delta t \geq h,$$

where ΔE and Δt can be considered as the errors of energy and time measurements, correspondingly [452], let us present the energy in the form of a product $q \cdot U$. Choosing an elementary charge (an electron charge $e \sim 1.6 \cdot 10^{-19}$) as q, it is possible to write this formula in the form

$$\Delta U \cdot \Delta t \geq h/e \sim 4 \cdot 10^{-15}\,\mathrm{V \cdot s}.$$

Substituting the measurement time τ_{meas} instead of Δt [37] we obtain that the potentially achievable accuracy in measuring an instantaneous value of electrical voltage is

$$\Delta U \geq \frac{h}{e\tau_{\text{meas}}},$$

i.e., for example, at $\tau_{\text{meas}} = 1 \cdot 10^{-9}\,\mathrm{s}$, $\Delta U \geq 4 \cdot 10^{-6}\,\mathrm{V}$.

This correlates with the theory developed in [368] where it is shown that the energy level of the measuring instrument sensitivity is proportional to its crror, the power consumed from the measurement object, the time of establishment of the measurement result, and cannot decreased to lower than $C = 3.5 \cdot 10^{-20}\,\mathrm{J}$.

In general, all quantum-mechanical limitations imposed on the measurement accuracy are significantly less than the limitations caused by molecular-kinetic substance structure (see Section 3.3.2) in view of the extremely small Planck's constant $h = 6.6 \cdot 10^{-34}\,\mathrm{J}$ as compared to the Boltzmann's constant $k = 1.37 \cdot 10^{-23}\,\mathrm{J/K}$ (if this comparison is performed for real macroscopic systems where $t \sim \mathrm{s}$, and $T \sim$ degree).

However the fluctuations of measurands which are caused by the classical (thermal) motion of particles of the thermodynamic system are decreased with decreasing its temperature, whereas the quantum fluctuations can reveal quite different (nonthermal) nature and in a real system there can be conditions under which the quantum fluctuations become predominant.

The above is clearly seen from the *generalized Nyquist formula* for the fluctuation (noise) power in the thermodynamically equilibrium system with discrete levels of energy $E_i = h v_i$:

$$\sqrt{\overline{p_n^2}} = \left[\frac{2h v_i}{e^{h v_i/kT} - 1} + h v_i\right] \Delta v. \tag{3.34}$$

Here the first item determines a proper thermal noise, and the second one is caused by "zero oscillations of vacuum". Under the condition $h v_i > kT$ (which, in particular, is true for the range of optical frequencies and higher) formula (3.34) is transformed and obtains the form

$$\sqrt{\overline{p_n^2}} = (hv + 2kT)\Delta v \approx hv\Delta v, \tag{3.34-1}$$

i.e., practically it does not contain any dependence on the temperature. In the general case the quantum fluctuations begin to become apparent when nearly $h\nu \geq T$.

3.4 Influence of external measurement conditions

The role of the influence of external factors including measurement conditions on a measurement result habe been previously shown in some examples. Separate methods of taking into account the external factors were considered. Here it is important to emphasize once again that there is no object or no system which can be completely isolated from its environment. This relates to all material components of measurement: an object under study, applied measuring instruments realizing a measurement experiment, subject (operator) or auxiliary technical means adjusting the process of the experiment.

Let us consider this question methodically in the general form, first as applied to the object under study.

The influence of some parameter ψ_i, characterizing the environment (of the object) on the measurand φ is determined in an explicit form by the dependence between them:

$$\varphi = f(\psi_i)$$

and can be taken into account through generalized functions of influence:

$$h_i = \frac{\partial f(\psi_i)}{\partial \psi_i}.$$

If during the measurement time Δt the parameter ψ_i remains unchanged (constant), i.e., $\Delta\psi_i = 0$, then the measurement result φ^{meas} can be considered to be accurate with respect to this parameter for the posed measurement problem. However in practice during the time Δt some final uncontrolled changes of the parameter ψ_i take place, and the measurement result has a certain uncertainty due to the influence of this parameter:

$$\Delta_i \varphi^{\text{meas}} = \frac{\partial f(\psi_i)}{\partial \psi_i} \Delta\psi_i. \tag{3.35}$$

Let the components of the uncertainty of the obtained measurement result be arranged according to different parameters of the external conditions in the course of their diminution:

$$\frac{\partial f(\psi_1)}{\partial \psi_1} \Delta\psi_1 > \frac{\partial f(\psi_2)}{\partial \psi_2} \Delta\psi_2 > \cdots > \frac{\partial f(\psi_i)}{\partial \psi_i} \Delta\psi_i > \cdots \tag{3.36}$$

For any particular measurement problem in which the level of uncertainty for the result $(\Delta\varphi)$, series (3.36) obtains a final character, since applying the rule of the negligible component [323] leads to the fact that a real contribution into a resulting error is provided only by a number of the first members of the series (usually no more than 3–5).

However in analyzing the potential measurement accuracy we need to keep in mind that series (3.36) is not principally limited and that as accuracy increases, i.e., as the successive registration, exclusion or compensation of the most significant factors influencing the measurement result (i.e., the first members of the series (3.36)) progressively new and new components begin coming to the forefront. This is a visual evidence of the philosophy principle of our inexhaustible knowledge of the world surrounding us.

From the above an important methodological rule is derived according to which, while performing accurate measurements, the value of a measurand has to be accompanied with indication of values of essential parameters of external influencing conditions to which the result obtained is reduced (or at which the measurements are performed). Otherwise, the measurements carried out with similar objects will have incommensurable results.

3.5 Space–time limitations

It is known that the space and time are the most general, universal characteristics of material systems and are considered in philosophy as integral properties (forms of existence) of substance. From the point of view of studying the potential accuracy of measurements it is important to emphasize (as an axiom) that all physical systems being really studied have *a finite extent both in space and time*.

The finite extent does not mean the infinite and zero extents, even if we are trying to localize as accurately as possible in the space-temporal continuum both the object property being studied and measuring instrument being applied.

Consequently, all measurands and parameters of the systems under study are the *averaged* ones with respect to a definite space volume (Δp) and time interval (Δt). This was mentioned earlier when the measurements were considered as applied to random processes and fields. In Section 3.3.3 Heisenberg's relations (3.3.32-1) and (3.33) are given. These relations link the uncertainty of the space or time localization of a micro-object, correspondingly, with the uncertainty of its pulse or energy, which is connected with a random nature of the behavior of quantum-mechanical objects.

However the similar relations take place for the case of determinated processes in macroscopic systems too. Let two typical situations be considered in studying harmonic oscillations.

Example 3.9. Let the oscillation of some object take place during a limited time interval Δt, but within this interval it obeys the harmonic law (Figure 3.3a):

$$x(t) = \begin{cases} x_0 e^{i2\pi v_0 t} & \text{on } -\dfrac{\Delta t}{2} \le t \le \dfrac{\Delta t}{2}; \\ 0 & \text{on other } t. \end{cases} \qquad (3.37)$$

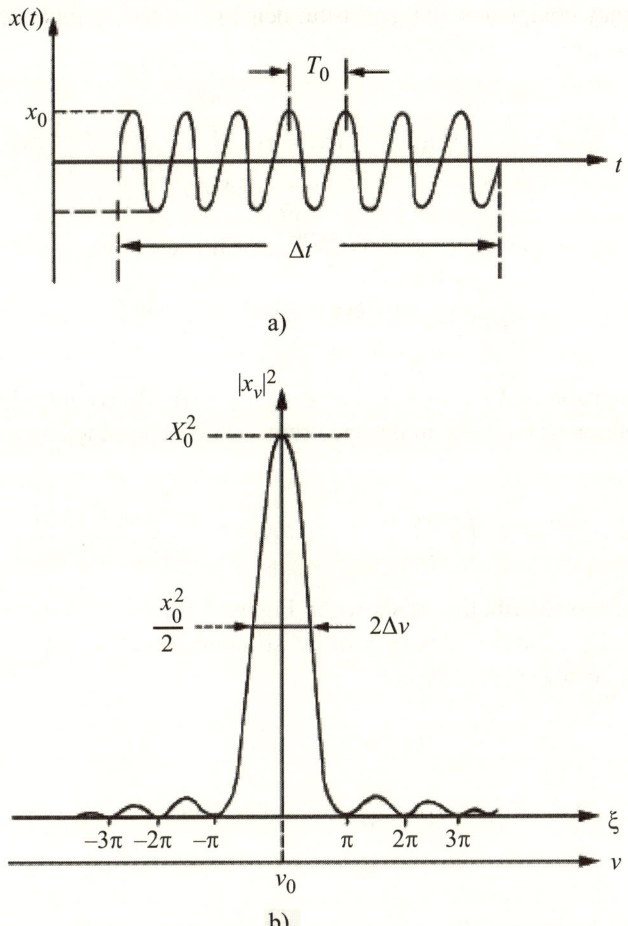

Figure 3.3. Time-limited oscillatory quasi-monochromatic process (a) and its spectrum (b).

Such an oscillation process, limited with time, happens at an act of radiation of a "team" of waves and practically in any real oscillatory process which, as we usually consider, is harmonic (or monochromatic, i.e., with a single frequency v_0). As a matter of fact, it is already not only nonmonochromatic, but noncyclic as a whole.

For such (noncyclic) processes spectral analysis (expansion in a frequency spectrum) is performed by means of the *Fourier integral*:

$$x(t) = \int_{-\infty}^{\infty} x_v e^{i2\pi vt} dv, \tag{3.38}$$

$$x_v = \int_{-\infty}^{\infty} x(t) \cdot e^{-i2\pi vt} dt. \tag{3.39}$$

The frequency component (the spectrum density) x_v taking into account equation (3.37) has the form

$$x_v = \int\limits_{-\Delta t/2}^{\Delta t/2} x(t)e^{-i2\pi vt}dt = x_0 \int\limits_{-\Delta t/2}^{\Delta t/2} e^{-i2\pi(v_0-v)t}dt.$$

Since $\int e^{ay}dy = \frac{1}{a}e^{ay}$ and $a = \frac{e^{iz}-e^{-iz}}{2i} = \sin z$, we have

$$x_v = x_0 \frac{\sin \pi(v_0 - v)\Delta t}{\pi(v_0 - v)} = X_0 \frac{\sin \xi}{\xi}, \tag{3.40}$$

where the designations $X_0 = x_0 \cdot \Delta t$ and $\xi = (v_0 - v) \cdot \Delta t$ are introduced.

The distribution of intensity in the spectrum $x(t)$ is determined by the square of its density, i.e.,

$$I(v) = |x_v|^2 = X_0^2 \left(\frac{\sin \xi}{\xi}\right)^2 = (x_0 \cdot \Delta t)^2 \left[\frac{\sin \pi(v_0 - v)\Delta t}{\pi(v_0 - v)\Delta t}\right]^2. \tag{3.41}$$

The form of this distribution is shown in Figure 3.3b; $I(v)$ has its maximum equal to X_0^2 at $v = v_0$, i.e., at the frequency of quasi-monochromatic oscillations and turns into the first zero at $\xi = \pm\pi$. Hence,

$$\Delta v_1 \equiv v_0 - v_1 = \frac{1}{\Delta t}.$$

At $\xi = \pm 2\pi$ $\Delta v_1 \equiv v_0 - v_1 = \frac{1}{\Delta t}$, and so on. Thus,

$$\Delta v \cdot \Delta t \geq 1. \tag{3.42}$$

Relation (3.42) obtained for a completely delineated process of persistant quasi-harmonic oscillations (3.37) means that the degree of monochromaticity of such a process is the higher, the larger the time interval Δt of its realization or observation is. From this it follows that the accuracy of determining the frequency of the quasi-harmonic process is inversely proportional to the time of its existence or observation.

Example 3.10. Let us consider another case where the "structure" of a periodical process is studied, i.e., the dependence itself $x = x(t)$ is studied by means of successive measurements of the x_i quantity values in the moment of time t_i.

Due to the finiteness of measurement time Δt the values measured x_i, which refer to the moment t_i, are the averaged values $\overline{x(t_i)}$ within the interval from $t_i - \frac{\Delta t}{2}$ to $t_i + \frac{\Delta t}{2}$ (see Figure 3.4), i.e.,

$$\overline{x(t_i)} = \frac{1}{\Delta t} \int_{t_i - \frac{\Delta t}{2}}^{t_i + \frac{\Delta t}{2}} x(t)dt, \tag{3.43}$$

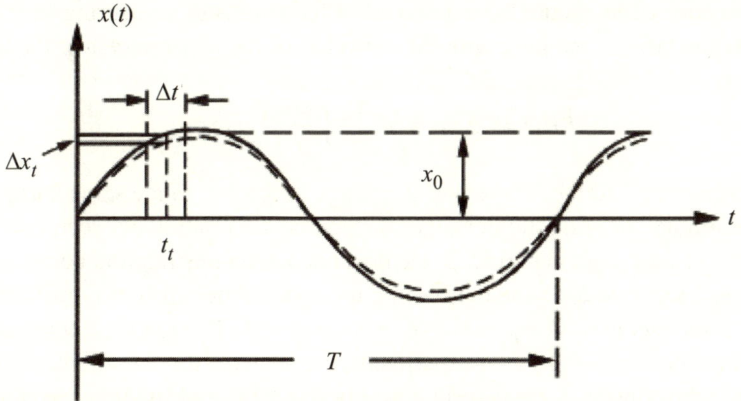

Figure 3.4. Model of measuring "instantaneous" values of a quantity changing according to the harmonic law.

at that $X(t_i) = x_0 \cdot \sin \omega t_i$ are real values of the measurand X in the moments of time t_i in the case of a purely harmonic process:

$$\overline{x(t_i)} = \frac{x_0}{\Delta t} \int_{t_i - \frac{\Delta t}{2}}^{t_i + \frac{\Delta t}{2}} \sin \omega t dt = \frac{x_0}{\omega \Delta t} \left[\cos \omega \left(t - \frac{\Delta t}{2} \right) - \cos \left(t + \frac{\Delta t}{2} \right) \right].$$

Since

$$\cos(\alpha - \beta) - \cos(\alpha + \beta) = -2 \sin \frac{\alpha - \beta + \alpha + \beta}{2} \cdot \sin \frac{\alpha - \beta - \alpha - \beta}{2} = 2 \sin \alpha \cdot \sin \beta,$$

we obtain

$$\overline{x(t_i)} = \frac{2x_0}{\omega \Delta t} \sin \omega t_i \cdot \sin \omega \frac{\Delta t}{2}. \tag{3.43-1}$$

Let the difference between the real and measured (averaged) values x_i be found:

$$\Delta x_i \equiv x(t_i) - \overline{x(t_i)} = x_0 \sin \omega t_i - \frac{2x_0}{\omega \Delta t} \sin \omega t_i \cdot \sin \omega \frac{\Delta t}{2} = x_0 \sin \omega t_i \left(1 - \frac{\sin \omega \frac{\Delta t}{2}}{\omega \frac{\Delta t}{2}} \right). \tag{3.44}$$

Since $\sin_{\alpha \to 0} \alpha = \alpha - \frac{\alpha^3}{3!} + \frac{\alpha^5}{5!} - \cdots$, i.e., $\sin \alpha \cong \alpha(1 - \frac{\alpha^2}{3!}) = a(1 - \frac{\alpha^2}{6!})$, we obtain

$$\Delta x = \frac{x_0}{6} \sin \omega t \left(\frac{\omega \Delta t}{2} \right)^2 = x_0 \frac{\pi^2}{6} \left(\frac{\Delta t}{T} \right)^2 \sin \omega t. \tag{3.45}$$

Thus, a correction should be introduced in the results of measuring the "instantaneous" values of x_i, changing with the period T of the x parameter of the harmonic process:

$$\theta(t_i) = \Delta x_i = x_0 \frac{\pi^2}{6} \left(\frac{\Delta t}{T}\right)^2 \sin \omega t_i. \tag{3.45-1}$$

This correction and the measurand X change according to the same harmonic law, but the amplitude of correction changes depend on the relation between the measurement time Δt and process period T, the dependence on this relation being quadratic.

There is a separate question concerning the space-time uniformity of fundamental physical constants used in equations of measurements. Though their constancy is established on the basis of physical experiments connected with measurements, however postulating the constancy henceforth need not result in an oblivion (especially for the case of potential measurements) of the fact that the degree of this constancy of fundamental physical constants is established experimentally by means of measurements and has a finite uncertainty.

3.6 Summary

The difference between the maximum attainable accuracy and potential accuracy that has not been hitherto attained at present-day development of science, technique and technology was marked.

Under *the maximum attainable accuracy* we mean the maximum accuracy with which a measurement of a physical quantity at a given stage of science and technique development can be carried out. The maximum accuracy of measurements is realized, as a rule, in national measurement standards which attain the reproduction of a physical quantity unit with the highest accuracy in the country. It is frequently identified with the maximum sensitivity of a measuring instrument.

Under the *potential accuracy of measurements* we mean the maximum not attainable accuracy.

A specific character of the situation in considering the potential accuracy of measurements allows some simplification which excludes from the consideration separate components of the measurement problem, e.g.: to ignore the fact that crude errors can arise; imperfection of the operator, of methods and means of measurement information processing, etc.

Thus, on the basis of the system approach with application of the set-theoretical mathematical apparatus for considering the potential accuracy of measurements it is necessary to analyze only the influence of components: φ (a measurand as the quality), o (object of the study as the carrier of a measurand), ψ (measurement conditions or totality of external influence factors), $[\varphi]$ (the unit of the physical measurand), s (measuring instruments used for solving a given measurement problem), as well as space–time parameters.

It was emphasized that *any concrete measurement needs the availability of definite a priori information* about the listed measurement components.

In order to carry out measurements it is necessary to have a priori information not only *about qualitative* but also *about quantitative characteristics* of the measurand, its dimension. The more accurate and, consequently, the more extensive a priori information about dimension of the measurand available, the more forces and means remain for getting more complete a posteriori information for improving the a priori information.

Improving more and more the values of parameters of the object under study (i.e., decreasing more and more their uncertainty) it is appears to be possible to gradually but continuously decrease the measurement result uncertainty caused by these parameters.

However, there is a certain type of physical knowledge about objects of the world surrounding us which obliges us to speak about *principle limits of accuracy* with which we can obtain quantitative information about a majority of measurable physical quantities.

First of all, such knowledge is connected with molecular-kinetic views concerning the substance structure.

The estimates of physical quantity fluctuations were given by the examples of a volume of gas under a piston, the length of a metal rod and electrical current voltage due to thermal motion of charge carriers, i.e., electrons, in a conductor.

Besides the natural limitations due to fluctuation phenomena connected with a discrete structure of observed macroscopic objects, there are limitations of the measurement accuracy of a more fundamental character. They are connected with quantum-mechanical (discrete) properties of the behavior of microworld objects.

The invisibility or *identity of particles* in quantum mechanics leads to the relation that reflects the *principle uncertainty* of determination of values of a number of physical measurands.

It is necessary to note that all the quantum-mechanical limitations for accuracy are significantly less than the limitations caused by the molecular-kinetic substance structure in view of the extremely small Planck's constant as compared to the Boltzmann's constant.

However, the measurand fluctuations caused by the classical (thermal) motion of particles of the thermodynamic system decrease with the decrease in temperature, while the quantum fluctuations prove to be of a quite different (nonthermal) nature. In a real system there can be conditions under which the quantum fluctuations become predominant. The above was illustrated by examples.

Considering the role of external factors influence, including measurement conditions on a measurement result, it is important to emphasize that neither the objects nor the systems can be completely isolated from the environment.

This relates to all material components of a measurement: an object under study, applied measuring instruments realizing a measurement experiment, subject (operator) or auxiliary technical means controlling the process of the experiment.

However, in analyzing the potential accuracy of measurements one should keep in mind that the series constructed is not principally limited and as the accuracy increases, i.e., as the successive registration is accomplished, exclusion or compensation of the most important factors influencing the result of measurement is performed, and progressively new components begin to advance to the forefront.

From the point of view of studying the potential accuracy of measurements it is important to emphasize that all physical systems studied have a *finite extent both in space and time*, and all measurands or parameters are *averaged* with respect to these space volume and time interval and obey the Heisenberg relations.

The presented Heisenberg relations link the uncertainty of the space or time localization of the microobject, respeectively, with the uncertainty of its pulse or energy, which accentuates the random (principally probabilistic) character of the behavior of quantum-mechanical objects.

However, similar relations take place in the case of determinate processes in macroscopic systems. It is also shown that a correction should be inserted into the results of measuring "instantaneous" values of x_i changing with the period T of the parameter x of a harmonic process. This correction changes according to the same harmonic law as in the case of the measurand X, but with an amplitude that depends on the relation of the measurement time Δt and process period T, the dependence on this relation being the quadratic one.

Chapter 4

Algorithms for evaluating the result of two or three measurements

4.1 General ideas

One important part of metrology is processing the results of repeated measurements of an unknown scalar quantity. In particular, it can be a question of output signals of a number of sensors measuring one and the same quantity, for example, temperature, pressure, or navigation parameters. To obtain an estimate the measurement data from sensor outputs are subjected to processing in accordance with a selected algorithm and are then transmitted for further use.

The choice of algorithm for processing measurement data depends significantly on the portability of a sample characterizing a general totality of observations or obtained knowledge of a measurand or its parameter. From the point of view of the portability of a sample and the accuracy of the results, the measurements can be divided into three categories: metrological, laboratorial, and technical.

Metrological measurements [323] are the measurements of the highest accuracy. They are characterized, as a rule, by large sample portability, careful performance of a measurement procedure, high qualification of a personnel, long-term study of stability and reliability of measurement results, scrupulous registration of all factors influencing the results, assurance of fixed normal conditions of equipment operation as well as by traceability of measurement results to national measurement standards.

Technical measurements [338, 555, 558] are characterized by requirements established in advance for the desired accuracy of a measurement result, small volume of a sample, reasonable requirements to personnel qualification, quickness of obtaining a result, working conditions of measurements. The traceability to measurement standards is certainly present, however, no particular attention is paid to it.

Laboratorial measurements hold an intermediate position between metrological and technical measurements.

Obviously, the main part of resources spent in any country for measurements is spent on carrying out technical measurements. The simplest example of technical measurements with a minimum sample is the situation where a purchaser who wishes to buy something goes to scale in a shop and weighs his purchase. He immediately has a final result which does not require any additional processing with an error that does not exceed an accuracy class of the scale if it has been verified and established correctly.

Similar situations arise in measuring parameters of unique processes and phenomena, for example, the characteristics of a nuclear explosion. In such cases it is impossible to rely on a large volume of sampling, and a "small" sample [176] has to be applied which consists of one, two, three, or a few more measured values. At the same time a powerful apparatus of mathematical statistics appears to be useless.

The key comparisons of national measurement standards aimed at determining the degree of their equivalence have proven to be a rather unexpected example of this kind [96]. In these comparisons a relatively small number of laboratories take part. For some types of measurements the number of competent and independent national laboratories participating in the comparisons is limited to 2 or 3. Such an example is given by V. Bremser [326] for substantiating the necessity to find an estimate of the "self-sufficient" nonstatistic reference value of a measurand in the key comparisons.

In this chapter a situation of exactly this type is considered. It can be called "the problem of three measurements", even though many problems and their results apply to larger samples. To obtain an estimate in similar cases it is necessary to use certain algorithms of measurement data processing. One of the main criteria of evaluating measurement results with the help of a selected algorithm, in addition to its complexity and the reliability of obtained estimates, is the error or uncertainty of the result.

Let it be noted that not long ago the status of international document was given to the "Guide to the Expression of Uncertainty in Measurement" (known as the GUM)" [243]. At present two approaches for describing the accuracy of measurements are used: the traditional concept of errors and the concept of measurement uncertainty introduced in accordance with [243]. A certain reduction in the contradictions between the mentioned approaches has taken place with the advent of the third edition of the International Vocabulary on Metrology (VIM) [246], in which these two concepts are both used.

In spite of the methodical and organizational importance of these concepts they do not significantly influence the choice of the evaluation algorithm for small samples (two or three measurements). Here the choice is made on the basis of an optimizable criterion that unambiguously leads to a certain evaluation algorithm, for example to an arithmetical weighed mean, harmonic or median mean, or some other one. This chapter is devoted to the description, systematization, and analysis of such evaluation algorithms.

One of the classical problems of metrology is known as the problem of three measurements. It is distinguished by the simplicity of its formulation in combination with its depth of content and the originality of its results. In the simplest setting this problem is reduced to the following. We have the results of three measurements x_1, x_2, x_3 of an unknown quantity x. A more accurate evaluation $\hat{x}$ of the quantity x needs to be made, if possible.

This problem is extremely widespread and is met with in various fields of human activities: science, technology, industry, economics, medicine, sports, everyday life, etc.

The task might be, e.g., (a) determining the weight of an object on the basis of the results of three repeated measurements, (b) calculating an average monthly salary for

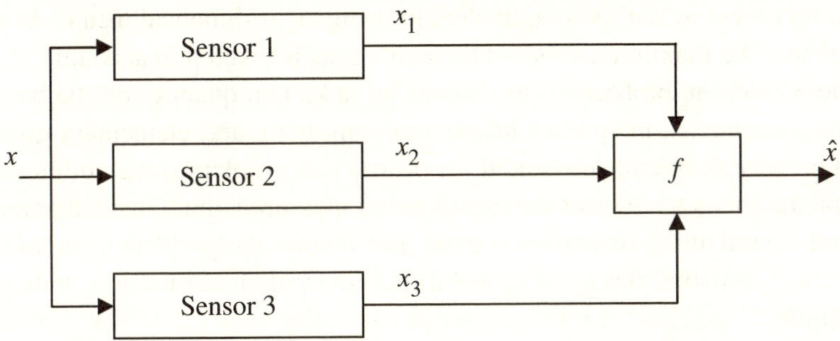

Figure 4.1. Structural interpretation of the problem of three measurements (x is the measurand or parameter, $\hat{x}$ is the estimate obtained).

a quarter, (c) determining the average price of potatoes on the market, (d) estimating a sportsman's performance in a competition based on the number of his attempts, (e) comparison of national measurement standards, and others.

In technology, the task might be e.g., in order to increase the reliability of equipment a parallel switching-on of three single-type blocks the output signals of which are averaged. In other words, for this purpose a structural or hardware redundancy is introduced. A similar approach is widely used for improving the reliability of flight equipment: actuators, automatic pilots, measurement sensors, digital flight computers, etc. [440].

In Figure 4.1 a structure is shown which contains three sensors for measuring one and the same parameter x.

Or, it could be the task of jointly using three airplane velocimeters or of evaluating the temperature by the readouts of three thermometers. The sensor output signals x_1, x_2, x_3 are processed according to the algorithm selected in block f. As a result of this an estimate $\hat{x}$ of the measurand or parameter is formed.

In all of these cases, the central task is the choice of a kind of functional dependence $\hat{x} = f(x_1, x_2, x_3)$. Consider the example where time is measured with the help of three chronometers. One of the natural methods is the calculation of the arithmetical mean $\hat{x} = \frac{1}{3}(x_1 + x_2 + x_3)$. This method is quite satisfactory when all three readouts of the chronometers are close. If one of the chronometers fails, then its arithmetical mean estimate will give an unequal result. A more reliable estimate in this sense is the estimate of the type of a sample median, according to which extreme measurements are disregarded and the readout of a "middle" chronometer is taken as the estimate.

It should be noted that a similar approach, when the extreme values are discarded and the remaining ones are averaged, is frequently used in evaluating the quality of sports performance, e.g., in figure skating or other physical performance where a result is evaluated by a panel of judges. In other kinds of sports some other evaluation algorithms are applied. For example, for target shooting the results of all the attempts are summed up (in essence, this is equivalent to using an arithmetical mean). In jumping and throwing, the maximum value of three attempts is taken into account.

On the whole, the problem of evaluating an unknown quantity on the basis of repeated measurements, in spite of its external simplicity and elementary quality, appears to be very deep and substantial. In posing this problem some additional information on the characteristics of the signals being measured, the statistical properties of noise, and the reliability of sensors is used, and to solve the problem a combination of mathematical statistics, theory of probabilities, interpolation, filtration, and optimization is applied.

Many outstanding scientists have been engaged in solving the problem of processing results of repeated measurements since as long ago as ancient times. It is known that the teachings of Pythagoras and his disciples on harmonics and gold proportions were connected with three kinds of mean values: arithmetic, geometrical, and harmonic. Eratosthenes marked out even more, i.e., seven kinds of means, besides the Pythagorean ones.

A great contribution to this study was made in the 19th century by German mathematician K. F. Gauss [177, 178] and Belgian statistician L. A. J. Quetelet, and in the 20th century by Russian mathematician Yu. V. Linnik [306] and Italian mathematical statistician K. Gini [183, 184]. We also should mention the valuable contributions by French mathematician O. L. Cauchy and Russian mathematician A. N. Kolmogorov [270], who developed the classical theory of average estimates (arithmetical, geometric, and power mean, and others) on which many evaluation algorithms are based.

The importance of this subject for technology is confirmed by the wide range of complexation methods, structural reservation and technical diagnostics [339] which require that output signals be formed on the basis of direct or indirect measurement of various parameters. Moreover, the real possibility of using the computer rather than complicated algorithms of evaluation appeared, where earlier this was merely of theoretical interest.

The expediency of studying the "problem of three measurements" is confirmed by the Russian standard "National System for Ensuring the Uniformity of Measurements. Direct Measurements with Multiple Observations. Methods of Processing the Results of Observations. Basic Principles", where it is noted that "by direct measurements with multiple observations are implied those cases where no less than four measurements are realized". From this it follows that the case $n = 3$ and $n = 2$ is specific and requires a special consideration.

It is known that at present there are many estimation algorithms with various levels of accuracy, reliability, etc. Descriptions of a majority of them are dissipated in journals, monographs, patents and other sources (see, in particular, [12, 16, 18, 22–24, 29, 51, 64–68, 74, 79, 80, 83, 87-89, 93, 94, 124, 125, 133, 137, 160, 167, 169–171, 183, 184, 206, 208, 214, 215, 219, 221, 226, 241, 296, 300, 305, 312, 314–316, 318, 319, 329, 350, 351, 354, 359–361, 363, 364, 370, 374, 378, 386, 387, 391–393, 397, 399,

408, 420–422, 424–426, 433, 434, 444, 486, 492, 494, 497, 512–520, 524, 539, 540, 549, 562], and many others).

In this chapter more than a hundred various estimation algorithms are collected, described, systematized, and analyzed. In order to be able to compare them clearly, a geometrical interpretation of estimation functions in the form of two-dimensional diagrams and three-dimensional surfaces is used.

The main attention is given to the algorithms of processing three and two measurements, although a majority of results allow a natural generalization in the multidimensional case.

4.2 Evaluation problem and classical means

4.2.1 Classification of measurement errors

Any measurement, no matter how careful it has been carried out, is characterized by errors (uncertainties) which define the accuracy of parameter estimation which are of interest [195, 323, 403, and others].

All errors, based on their origin, are distinguished in the following way: personal, instrumental, external, and methodical errors, as well as errors due to the inadequacy of a model and errors caused by classification mistakes.

The source of a resulting error is presented in the diagram shown in Figure 4.2.

Personal, subjective, or crude errors refer to errors depending on the physical or physiological features of an operator (observer): his/her qualification, fatigue capacity,

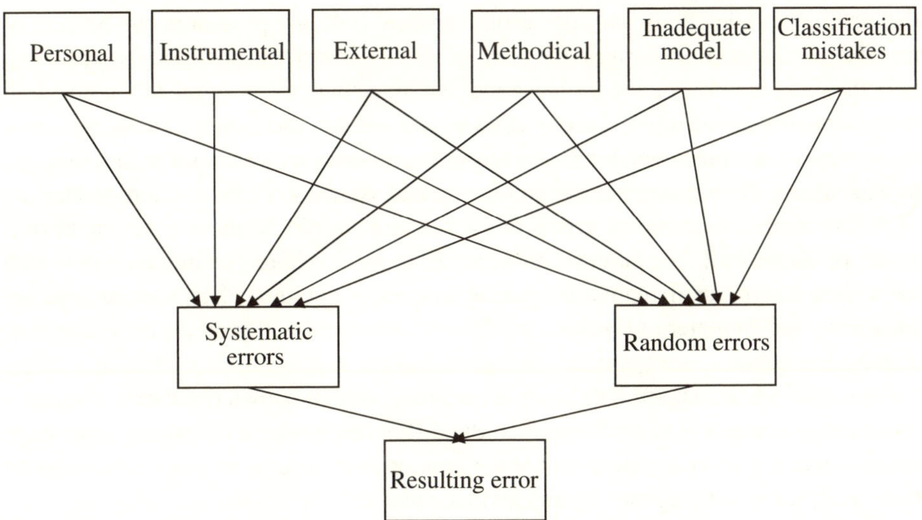

Figure 4.2. Character of the formation of a resulting error.

etc., which determine the values and characteristic features of the erros. By crude or gross error is meant [323] the error of an individual measurement entering a series of measurements. For given conditions this error sharply differs from the remaining results of this series.

Instrumental errors means the errors (breakdowns, failures, faults) arising due to imperfection of measuring instruments. Usually the reasons for their appearance are thoroughly analyzed, and if it is not possible to eliminate them, they are taken into account with the aid of various corrections. Instrumental errors include those arising due to the variation and drift of readouts of a measuring instrument, a zero shift or insensibility of a measuring instrument, and others.

External errors are connected to the influence that physical quantities which are characteristics of an environment but do not represent an object to be measured, exert on a measuring instrument. Pushes, shocks, vibrations, strong wind, humidity, variation in temperature, and other deviations of environmental parameters from normal conditions under which a calibration of a measuring instrument was made also lead to the generation of errors in its readouts. There are normal, normalized, working, and extreme conditions of measurements or operation of measuring instruments, as well as normal and working fields of influencing quantity values.

Methodical errors, i.e., errors caused by the imperfection of a chosen measurement method, are of great importance. Most often these errors arise due to frequently used simplifications in equations for measurements. They are also generated by various approximations, roundings, truncating members of the highest order in a series expansion, and not taking into account other factors which influence the results of measurement data processing.

Errors caused by the *inadequacy of the model used* are connected with the fact that an object under study and its various physical links are present in the process of measurement data processing in the form of certain abstract ideas (models) reflecting only the main features of a real object and real links, but never completely coinciding with them. The adequacy of the object model means that the model reflects just those properties of the object which are of interest to a researcher, which allow him to judge to what extent these properties are significant and which of them can be neglected.

A very simple example of errors of such a kind would be the measurement of a radius of a steel ball. We perceive this ball as an ideal sphere, the diameter of which we wish to determine. In reality the real ball does not coincide with its idealized image. Measuring the diameter of this ball in different directions will produce noncoincident results.

If a certain set of experimental data subordinate to an unknown regularity is approximated by a curve of a given form, then the difference between a formula expression of this curve and the true law of variable variation is the source of error which should be considered an error caused by model inadequacy.

The errors caused by *classification mistakes* appear when it is possible to refer the measurement of an extraneous object to the object under study. Such mistakes sometimes arise, for example, while observing artificial satellites, when a launch vehicle or some other satellite is mistaken for the satellite under study. In order to exclude classification mistakes, usually a preliminary check of whether or not all the conditions for solving a posed problem are satisfied is made. Checks of this kind are part of a particular segment of probability theory, i.e., the theory of statistical solutions.

According to theor origin, each of the errors listed above can be referred to as either systematic or random.

An error is called *systematic* when it expresses significant links which arise during measurement or measurement data processing. Such an error inevitably appears every time certain conditions are established. By definition [403] a systematic error is the component of measurement error which remains constant or varies in a predictable manner in replicate measurements of the same physical quantity.

An error is called *random* when it has a stochastic nature and reflects less significant links. Such an error cannot be accurately reproduced by creating the same or some other conditions. This component of a measurement error varies in an unpredictable manner (with regard to sign and value) in repeated measurements of one and the same constant physical quantity carried out with similar thoroughness.

An error is called *crude* (failure, transient error, high level error, anomalous value of measurement) when it is a particular kind of a random error and by far exceeds nominal (passport) characteristics of the measuring instrument. Usually the crude error is connected with an abrupt distortion of experiment conditions, breakage or malfunction of the measuring instrument, mistakes in algorithms, as well as with personnel miscounts. In particular, classification errors always belong to the category of crude errors.

By a *resulting error* they imply the error of a measurement result which consists of a sum of random and systematic errors which are not eliminated and taken for random ones.

It should be noted that the line between the systematic and random errors is rather fuzzy. On and the same error can be considered either systematic or random, depending on the way it is spread (over all or a part of measurements).

The number of measurements, as a rule, depends on the extent of the time interval within which the measurements are carried out. If some factor acts over the time of the whole interval, then an error caused by the action of this influence factor should be considered systematic. If the action of this factor becomes apparent acting over a short time interval significantly less than the whole interval, then it should be considered random.

Sometimes it is possible to study the action of such a factor and take into account the corresponding error by introducing a correction. In this case the error of this type is practically excluded from the resulting sum of errors. In some situations the action of a factor is neglected, and then the error is considered to be random (as in the case of the remainder of the systematic errors) and just in this capacity it is present at measurement data processing. Thus, one and the same error can be considered either random or systematic, depending on a concrete content of an experiment performed.

The classification described can be represented in terms of the concept of uncertainty of measurement. In 1999 at the D. I. Mendeleyev Institute for Metrology Recommendation MI 2552-99 "State system of ensuring the uniformity of measurements. Application of the Guide to the expression of uncertainty in measurement" was developed, which became a basis for creating document [403].

This document contains the statement of the main regulations of the Guide [243] and recommendations on their practical use, a comparable analysis of two approaches to describe the accuracy characteristics of measurements, as well as an illustration of the correspondence between the forms of presenting measurement results used in normative documents based on the concept of errors, and the form used in the Guide mentioned above.

A universal method to decrease the random and resulting errors consists of repeated the measurement of one and the same physical quantity. In practice repeated measurements or a number of measurement sensors of different types are widely used. At the same time the problem of averaging the values of measurement results for obtaining a resulting estimate arises.

Later on, an analysis of classical averaging algorithms will bc madc. Thc main attention will be given to the case of three measurements.

4.2.2 Problem definition and classification of evaluation methods

Let x_1, x_2, x_3 be the results of three measurements of an unknown parameter x, and e_1, e_2, e_3 are the unknown measurement error values. Then a link between them is described by the following set of equations:

$$\begin{aligned} x + e_1 &= x_1, \\ x + e_2 &= x_2, \\ x + e_3 &= x_3. \end{aligned} \tag{4.2.1}$$

In the case of indirect measurements the set of equations (4.2.1) takes a more general form:

$$\begin{aligned} \xi_1(x) + e_1 &= x_1, \\ \xi_2(x) + e_2 &= x_2, \\ \xi_3(x) + e_3 &= x_3, \end{aligned} \tag{4.2.2}$$

where ξ_1, ξ_2, ξ_3 are the functions of the apparatus sensors.

The systems of equations (4.2.1) and (4.2.2) are not quite usual, since the number of unknowns x, e_1, e_2, e_3, exceeds the number of equations. Therefore, the problem of estimating the parameter x, for all its seeming simplicity, has attracted the attention of researchers for a long time. Researchers have suggested quite a number of methods for estimating this parameter.

The basis of one of the first approaches to solving this problem is the assumption that all values of the measurement error are equal to zero, i.e., $e_1 = e_2 = e_3 = 0$. Then instead of the underdetermined system of equations (4.2.1) a redundant system of equations (since the number of equations in this case exceeds the number of the unknowns) which is generally contradictory is obtained:

$$x = x_1, \quad x = x_2, \quad x = x_3. \tag{4.2.3}$$

This contradiction is of an artificial nature, since it is connected with a deliberately false assumption that all values of the error are equal to zero.

On the basis of such an approach there have been developed methods of obtaining the estimate $\hat{x}$ of the parameter x from the system of equations (4.2.1) or (4.2.2), which are now classical. These methods are intended for obtaining an arithmetic mean, minimax (Chebyshev's) and median estimates.

The method of obtaining a *mean arithmetical estimate* at three measurements has been known since ancient times and is so natural that it could be hardly attributed to some concrete author:

$$\hat{x} = \frac{x_1 + x_2 + x_3}{3}. \tag{4.2.4}$$

In a more general case of posing the problem for an arbitrary number of measurements and parameters a similar estimate was obtained by K. F. Gauss [177, 178] at the beginning of the 19th century. Under more detailed consideration it appears that the estimate obtained minimizes a quadratic criterion:

$$J_1 = (x_1 - \hat{x})^2 + (x_2 - \hat{x})^2 + (x_3 - \hat{x})^2 \to \min_{\hat{x}}. \tag{4.2.5}$$

In fact, by differentiating J_1 with respect to $\hat{x}$and making the result equal to zero, we obtain $x_1 + x_2 + x_3 - 3\hat{x} = 0$, and from this formula (4.2.4) immediately follows.

Thus, the mean arithmetical estimate is optimal in the sense of the least-squares method.

The *minimax* or *Chebyshev's estimate* of the parameter x has the form

$$\hat{x} = \frac{\max(x_1, x_2, x_3) + \min(x_1, x_2, x_3)}{2}. \tag{4.2.6}$$

This estimate is called minimax since it minimizes the criterion:

$$J_2 = \min_{\hat{x}} \max \left(|x_1 - \hat{x}|, |x_2 - \hat{x}|, |x_3 - \hat{x}|\right). \tag{4.2.7}$$

A defect of the estimates cited is the deficiency of operability in the presence of crude errors (failures, transient errors, high level errors) of measurements.

It is possible to protect from failures with the help of a *median estimate* that minimizes *the modulus criterion*:

$$J_3 = |x_1 - \hat{x}| + |x_2 - \hat{x}| + |x_3 - \hat{x}| \to \min_{\hat{x}}. \tag{4.2.8}$$

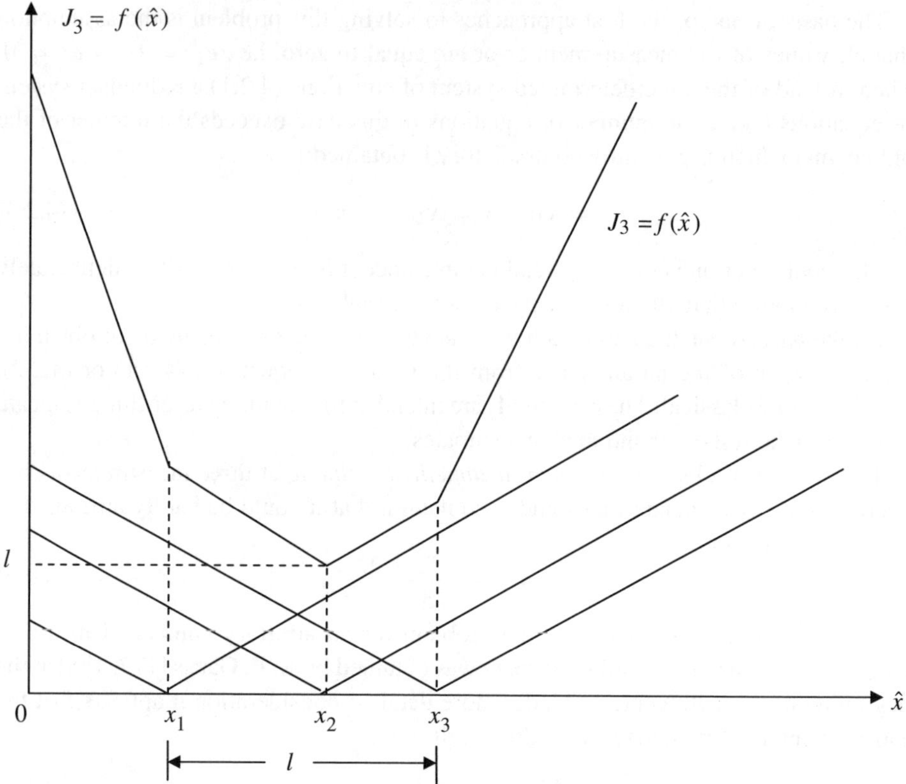

Figure 4.3. Graphical-analytical method for determining a median estimate.

It is possible to obtain an optimal estimate $\hat{x}$ minimizing this criterion with the help of the graphical-analytical method, i.e., by constructing a diagram of $J_3(\hat{x})$ (Figure 4.3).

From the diagram it is seen that the optimal estimate coincides with the mean of three measured values $\hat{x} = x_2$. At the same time a change of the x_1 value in the range $[-\infty, x_2]$ and a change of the x_3 value in the range $[x_2, \infty]$ do not lead to a change of the $\hat{x}$ estimate. This is evidence of the good noise immunity of such an estimate, which is sometimes is the *sample median*.

At present a great number of methods for obtaining mean estimates with respect to both the direct and indirect measurements are known. According to the way of grounding the obtained results, these methods can be divided in the following groups:

(1) methods for obtaining estimates optimizing probabilistic criteria;

(2) methods for obtaining estimates optimizing deterministic criteria;

(3) diagnostic methods;

(4) heuristic methods.

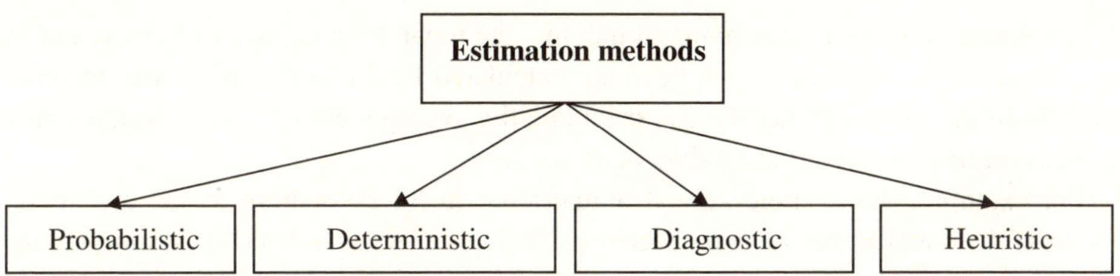

Figure 4.4. Classification of estimation methods.

The classification of estimation methods is illustrated schematically in Figure 4.4.

It should be noted that when obtaining estimates heuristics is present in all four groups. It becomes apparent when a criterion is chosen, a density function of measurement value probability distribution is set, as well as when a priori information about a measurement parameter value is taken into account.

Before considering each of the groups, let us turn our attention to the details of the classical estimates such as arithmetical mean, geometrical mean, quadratic mean, and some of their generalizations.

4.2.3 Classical means and their properties

The number of mean estimates, known as Pythagorean, includes the arithmetic mean, the geometric mean, and the harmonic mean. Later on Eudoxus, Nicomah, and other mathematicians in Ancient Greece, relying on the theory of proportions, described a series of other means, particularly, contraharmonic and contrageometric means, which are united today known under the general name "Greek means".

4.2.3.1 Arithmetical mean

An arithmetical mean of the numbers $x_1, \dots, x_n$ is the quantity:

$$\hat{x} = A(x_1, \dots, x_n) = \frac{x_1 + x_2 + \cdots + x_n}{n}. \tag{4.2.9}$$

For the case of two and three measurements the arithmetical mean is determined by the formulae

$$\hat{x} = \frac{1}{2}(x_1 + x_2), \quad \hat{x} = \frac{1}{3}(x_1 + x_2 + x_3).$$

Its name is connected with the fact that each term of an arithmetical progression beginning from the second one is equal to an arithmetical mean of its adjacent terms.

The arithmetical mean is the most known and widespread of all means. It is widely applied in scientific research, technology, industry, and private life.

The mean quantity of any national income, the mean crop capacity of any country, the mean food consumption per head are calculated by the arithmetic mean formula (4.2.9). In the same manner we can calculate the average annual air temperature in a city, averaging the observation data of many years.

For example, the average annual temperature in St. Petersburg as the arithmetic mean of the data for the 30 years between 1963 and 1992 is +5.05 °C. It summarizes the average annual temperature values varying from +2.90 °C in 1976 to +7.44 °C in 1989.

Example 4.2.1. Let us calculate the average age of students in a group of 20 men on the basis of the data indicated in Table 4.1.

Table 4.1. The age of students in a group.

N°	1	2	3	4	5	6	7	8	9	10
Age (years)	18	19	19	20	19	20	19	19	19	20

N°	11	12	13	14	15	16	17	18	19	20
Age (years)	22	19	19	20	20	21	19	19	19	19

The average age is found by formula (4.2.9):

$$\hat{x} = \frac{1}{20}\bigl(18 + 19 + 19 + \cdots + 21 + 19 + 19 + 19 + 19\bigr) = \frac{388}{20} = 19.4 \text{ years.}$$

4.2.3.2 Geometrical mean

The geometrical mean of some positive numbers $x_1, \ldots, x_n$ is the quantity:

$$\hat{x} = G(x_1, \ldots, x_n) = \sqrt[n]{x_1 x_2 \cdots x_n}. \tag{4.2.10}$$

The name of this mean is connected to the fact that each term of a geometrical progression with positive terms beginning with the second one is equal to the geometrical mean of its adjacent terms.

When the values of n are large, the geometrical mean is usually calculated with the help of the logarithm finding the first arithmetical mean of logarithms:

$$\ln G(x_1, \ldots, x_n) = \frac{\ln x_1 + \cdots + \ln x_n}{n},$$

and then performing the exponentiation. This is precisely the reason why the geometrical mean is sometimes called the *logarithmical mean.*

In accordance with the psychophysical Weber–Fechner law the estimate of a certain quantity which we directly perceive with the help of our organs of sense (vision, hearing, and others) is proportional to the logarithm of this quantity. Therefore, if the estimate is based on our perceptions, it would be correct to apply a geometric mean rather than an arithmetic mean.

For the case of two and three measurements, the geometrical mean is determined by the formulae

$$\hat{x} = \sqrt{x_1 x_2}, \quad \hat{x} = \sqrt[3]{x_1 x_2 x_3}.$$

In mathematics the expression $b^2 = ac$ is known as the geometrical proportion. It expresses the geometrical mean of two numbers and the geometrical progression with the denominator (coefficient) $q = a/b = b/c$. The geometric mean of two numbers is, because of its mean, also called proportional.

Note that the altitude of a triangle, pulled down on the hypotenuse of this triangle, is equal to the geometrical mean of the segments of the triangle base (Figure 4.5a). This gives a geometrical method for constructing the geometrical mean of the length values of two segments: it is necessary to construct a circumference on the basis of a sum of these two segments as on the diameter; then the altitude $x =$ OA, reestablished from a point of their connection up to its intersection with the circle, will give a geometrical mean estimate $\frac{x}{a} = \frac{b}{x}$ or $x = \sqrt{a\,b}$.

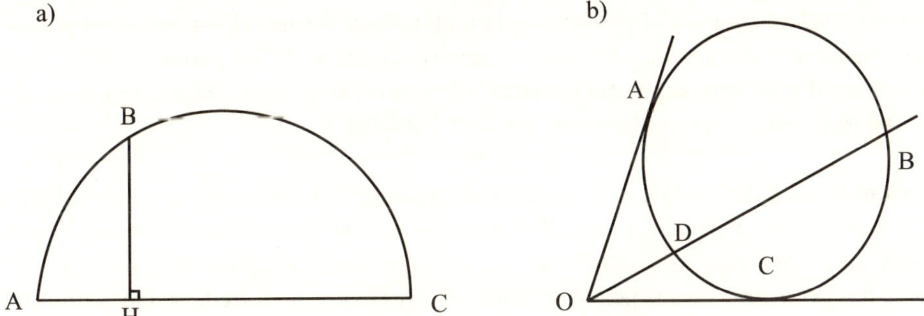

Figure 4.5. Construction of the geometrical mean of the lengths of two segments.

Figure 4.5a allows the arithmetical and geometrical mean of two numbers to be compared by sight. From this figure it is seen that if a is not equal to b, then the arithmetical mean is always greater than the geometrical one.

Another method for constructing the geometrical mean of the length values of two segments is based on the theorem, according to which the tangent squared to a circumference is equal to the product of any secant line by its any external part: $\mathrm{OA}^2 = \mathrm{OB} \cdot \mathrm{OD}$ (Figure 4.5b). When constructing this, it is handy to take a secant line passing through the centre of the circumference and to assume that $\mathrm{OB} = a$, $\mathrm{OD} = b$, then $x = \mathrm{OA} = \sqrt{\mathrm{OB} \cdot \mathrm{OD}}$.

The geometrical mean appears in a natural manner under solution of many mathematical problems.

Example 4.2.2. Let the probem of determining a cube edge d be considered, the cube volume being equal to that of a standard brick with the sides $a = 25$ cm, $b = 12$ cm, $c = 6.5$ cm.

The arithmetic mean quantity $d = (25 + 12 + 6.5) : 3 = 14.5$ cm will give a cube of an overshot volume $V = 3048.625\,\text{cm}^3$. The correct result $V = 1950\,\text{cm}^3$ is obtained, using the formula of a geometrical mean:

$$d = \sqrt[3]{abc} = \sqrt[3]{25 \cdot 12 \cdot 6.5} \approx 12.5\,\text{cm}.$$

When the values n are large, the geometrical mean is usually calculated with the help of the logarithm, finding first the arithmetical mean of the logarithms:

$$\ln G(x_1, \ldots, x_n) = \frac{\ln x_1 + \cdots + \ln x_n}{n},$$

and then performing the exponentiation. This is the reason why the geometrical mean is sometimes referred to as the *logarithmical mean.*

In applied statistics the geometrical mean is useful when a measurement scale is nonlinear. The geometrical mean is most often applied when determining the average growth rates (the means of growth coefficients) where the individual values of a measurements are presented in the form of relative quantities. The geometrical mean is also used if it is necessary to find a mean between the minimum and maximum values of the measurements (e.g., between 100 and 1 000 000).

Example 4.2.3. In [183] K. Gini gives as an example of the issue discussed in Galilei's correspondence with Noccolino. If a horse is valued by two appraisers at 10 and 1000 units, then what is the preferred mean of the quantity representing its real cost? Is it the geometric mean, i.e., 100 units, which Galilei preferred, or the arithmetic mean, i.e., 505 units, as it was proposed by Noccolino? Most scientists are of Galilei's opinion.

Example 4.2.4. As an example of *applying the geometrical mean* in practice, let us consider a curious extract from the memoirs of a known Russian astrophysicist, I. S. Shklovsky, which he published under the title *The State Secret.*

Once (in the Soviet period) he spoke with the journalist O. G. Chaikovskaya, well known from her papers on criminal and judicial subjects.

"Olga Georgiyevna, how many people condemned for various criminal acts are in prison and camps?"

"Alas, I don't know. What I can propose to you is my personal observations in the city of Rostov where for some years I was a chief of a press office of the "Izvestiya"(one of the central newspapers of the USSR). So, it appears that the law courts of that city pass approximately 10 000 sentences per year".

"Fine!" I exclaimed. "We shall rather arbitrarily say that the law courts in Rostov condemn on average each of the accused persons to five years. Thus, we can assert that about 50 000 people are constantly in Soviet prisons and camps.

"So, we have no choice but to estimate Rostov's contribution to the total criminality in the Soviet Union. The simplest thing is to consider that it is equal to Rostov's share of the total population in our country. This is about 1/300. If this estimate were accepted, we would obtain a highly improbable number of prisoners in our country.

"One should not make the estimation in this way. Rostov is a classical bandit (gangster) city about which many well-known songs about thieves have been written. But on the other hand, compared to the absolute quantity of sentences passed Rostov is inferior to our giant cities Moscow and Leningrad.

"However, it is clear that to ascribe 10 % of the criminality of the entire Soviet Union Rostov seems very high. On the other hand, consider it to be equal to 1 % is obviously low. A mistake in estimating will be minimized if a logarithmical mean is taken between these extreme values. This mean is equal to the root of ten, i.e., approximately 3 %.

"From this it follows that about 1.5 million people are kept in camps and prisons of the USSR at any one time. I think that the probable error of this estimate is equal to some tenths of a percent, which is not so bad".

Mr. E. Bogat who was present at this discussion, exclaimed, "Where do you known this from? Everyone knows that it is the state secret!"

Example 4.2.5 (Operating speed of an automatic control system). In an analysis of control systems the concept of a mean geometric root w is used. Consider the characteristic equation of a dynamic system of the form

$$p^n + a_{n-1}p^{n-1} + \cdots + a_1 p + a_0 = 0,$$

where $p_1, \ldots, p_n$ are its roots.

Let us denote the geometrical mean of these roots as ω, then

$$\omega = \sqrt[n]{a_0} = \sqrt[n]{p_1 p_2 \ldots p_n}.$$

Let the characteristic equation be reduced to the normalized form, making the substitution $p = q\sqrt[n]{a_0} = \omega q$:

$$q^n + \frac{a_{n-1}}{a_0}\omega^{n-1}q^{n-1} + \cdots + \frac{a_1}{a_0}\omega q + a_0 = 0.$$

The increase of ω results in a proportional radial bias of the roots on the plane of complex numbers. At the same time, the form of a transient process remains unchanged, and only its time scale will change. Therefore, the mean geometrical root ω can serve as the measure of the response speed of the automatic control system.

Let us notice that the formula of a geometrical mean of two numbers $G = \sqrt{x_1 x_2}$ can be generalized for the case of n numbers in another manner, namely by extracting the square root from the arithmetic mean of n pair products of these numbers:

$$G_1 = \sqrt{\frac{x_1 x_n + x_2 x_{n-1} + x_3 x_{n-2} + \cdots + x_n x_1}{n}}.$$

Considering all possible pair products, we get the formula of the *combinatoric mean*:

$$G_2 = \sqrt{\frac{x_1 x_2 + x_1 x_3 + \cdots + x_1 x_n + x_2 x_3 + \cdots + x_2 x_n + \cdots + x_{n-1} x_n}{C(n, 2)}},$$

where $C(n,2)$ is the binomial coefficient (n choose 2).

At $n = 3$ both formulae coincide:

$$G_1 = G_2 = \sqrt{\frac{x_1 x_2 + x_1 x_3 + x_2 x_3}{3}}. \tag{4.2.11}$$

Example 4.2.6 (Surface of the parallelepiped and cube). Figure 4.6 presents a brick with the edges $l_1 = 250\,\text{mm}$, $l_2 = 160\,\text{mm}$, $l_3 = 65\,\text{mm}$. What must the edge l of a cubic brick with an cqual surface be?

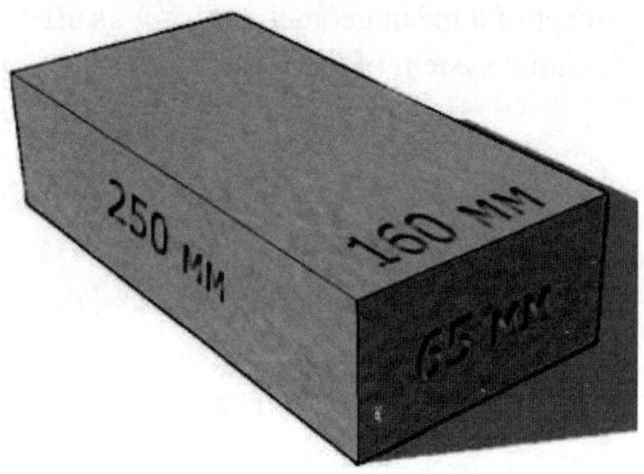

Figure 4.6. Surfaces of the parallelepiped and cube.

To get the answer, let us calculate the total area of all facets:

$$S = 2(l_1 l_2 + l_1 l_3 + l_2 l_3) = 133\,300,$$

and equate its surfaces of the cube sought as $6l^2$.

As a result we get the formula of the combinatoric mean (4.2.11):

$$l = \sqrt{\frac{l_1 l_2 + l_1 l_3 + l_2 l_3}{3}} = \sqrt{\frac{66\,650}{3}} \approx 149\,\text{mm}.$$

4.2.3.3 Harmonic mean

By the harmonic mean of n numbers $x_1, \cdots, x_n$ is meant the quantity

$$\hat{x} = H(x_1, \ldots, x_n) = \frac{n}{\frac{1}{x_1} + \frac{1}{x_2} + \cdots + \frac{1}{x_n}}. \tag{4.2.12}$$

Thus, the harmonic mean is the number the reciprocal value of which is the arithmetical mean of the reciprocal values of the numbers given. Because of this it is also called the *reciprocal arithmetic mean.*

For the case of two and three measurements the harmonic mean is determined by the formulae

$$\hat{x} = \frac{2}{\frac{1}{x_1} + \frac{1}{x_2}} = \frac{2x_1x_2}{x_1 + x_2}, \quad \hat{x} = \frac{3}{\frac{1}{x_1} + \frac{1}{x_2} + \frac{1}{x_3}} = \frac{3x_1x_2x_3}{x_1x_2 + x_1x_3 + x_2x_3}.$$

The name "harmonic mean" is connected with the harmonic series, which is well known in mathematics:

$$1 + \frac{1}{2} + \frac{1}{3} + \cdots + \frac{1}{n} + \cdots .$$

Every term of this series beginning from the second one is equal to the harmonic mean of its adjacent terms, e.g., $1/n$ is the harmonic mean of fractions $1/(n-1)$ and $1/(n+1)$. However, there is also an opposing point of view, according to which the name of this series comes from from the harmonic mean.

The harmonic mean is used for calculating a mean velocity at the time of a relay race if the velocity values at its separate stages and the length values of all stages are known. The harmonic mean is also used in calculating an average lifetime, average price of products when sale volumes at a number of trade shops are known. This mean is also applied in the case of determining labor inputs or material costs per unit of production for two or more enterprises.

Let us now look at two problems of calculating a mean speed.

Problem 1. A car moves at a speed $v_1 = 20\,\text{km/h}$ during the first hour. What is the velocity during the second hour so that the average velocity is 40 km/h?

Problem 2. A car moves from point A to point B at a speed $v_1 = 20\,\text{km/h}$. What is the speed v_2 of the car on its return from point B to point A so that the average speed is 40 km/h?

The above problems differ from each other in the following way: in the first case the time of movement is divided into two equal parts, and in the second case the two parts of the journey are equal.

Simple calculation shows that in the first case the average speed is the arithmetical mean of the speeds on the separate legs of the journey. Thus, we obtain the result: $v_2 = 60\,\text{km/h}$.

In the second case the average speed is the harmonic mean of the speeds on the separate legs of the journey:

$$v_{\text{mean}} = \frac{2}{\frac{1}{v_1} + \frac{1}{v_2}} = 2\frac{v_1 v_2}{v_1 + v_2}.$$

Substitution of the problem data in this formula results in an equality $20 + v_2 = v_2$, implemented only at $v_2 \to \infty$.

In this way we obtain two rules for calculating the average speed:

(1) If the speeds $v_1, \dots, v_n$ belong to equal time intervals, then the formula of the arithmetical mean $v_m = \frac{1}{n}(v_1 + \dots + v_n)$ should be used.

(2) If the speeds $v_1, \dots, v_n$ belong to equal parts of the journey, then the formula of the harmonic mean $v_m = \frac{n}{\frac{1}{v_1} + \dots + \frac{1}{v_n}}$ should be used.

Example 4.2.7. We want to calculate the average speed of two cars which made the same journey, but at different speeds. The first car moved at the speed of 100 km/h and the second at the speed of 90 km/h.

To calculate the average speed the formula of the harmonic mean is used:

$$\hat{x} = \frac{2}{\frac{1}{100} + \frac{1}{90}} = \frac{1\,800}{19} = 94.7\,\text{km/h}.$$

Example 4.2.8. Two typists are given the task of reproducing a report. The more experienced of them could do the entire job in two hours, the less experienced in 3 hours. How much time would be necessary to reproduce the report if the typists divide the work between them to finish it in the shortest possible time?

The following formula gives the correct answer:

$$T = \frac{1}{\frac{1}{T_1} + \frac{1}{T_2}} = \frac{1}{\frac{1}{2} + \frac{1}{3}} = \frac{6}{5} = 1\,\text{h}\;12\,\text{min}.$$

This is the half of the harmonic mean.

Example 4.2.9 (Waiting for a bus). At a bus station buses from three routes make a stop. The intervals of their movement are 5, 9, and 12 min, respectively. What is the average time of waiting for a bus?

It is intuitively clear that this time is twice less than 5 min. But an accurate answer can be calculated by the formula

$$T = \frac{1}{\left(\frac{1}{T_1} + \frac{1}{T_2} + \frac{1}{T_3}\right)} = 2.535\,\text{min},$$

i.e., the average time is equal to a third of the harmonic mean.

The formula of a harmonic mean can be applied when the arguments have different signs; however, then the result can exceed the maximum of the arguments.

Example 4.2.10. To fill a bath tub with cold water takes 8 minutes, and with hot water 10 minutes. The water from the full tub drains in 5 minutes. How much time is necessary to fill the tub when the drain is open?

Let the bathtub volume be taken as the unit, then the speed of filling the tub with cold water will be $v_1 = \frac{1}{8}$, and with hot water $v_2 = \frac{1}{10}$, and the draining speed $v_3 = -\frac{1}{5}$. Multiplying the total speed by the time sought, we obtain

$$(v_1 + v_2 + v_3)T = 1,$$

Hence

$$T = \frac{1}{v_1 + v_2 + v_3} = \frac{1}{\frac{1}{8} + \frac{1}{10} - \frac{1}{5}} = \frac{1}{0.025} = 40\,\text{min.}$$

This is one-third of the harmonic mean of the quantities 8, 10, and –5.

Example 4.2.11. Let us consider a cylindrical container with height h and radius r. We need to find the ratio of its volume V to its full surface S:

$$V = \pi r^2 h, \quad S = 2\pi r^2 + 2\pi r h, \quad \frac{V}{S} = \frac{rh}{2r + 2h} = \frac{1}{4} H(r, h).$$

It is one-fourth of the harmonic mean of the height and radius of the cylinder.

Example 4.2.12 (Harmonic quadruple of points). Consider four points A, M, B, N on a straight line (Figure 4.7).

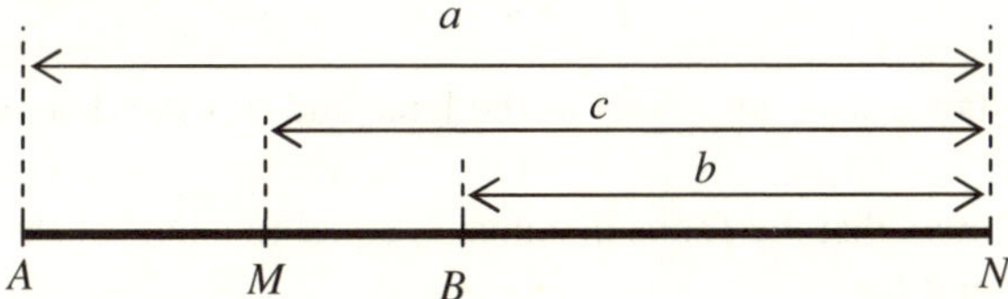

Figure 4.7. Harmonic quadruple of points.

It is said that point M divides the segment AB in the ratio of $\frac{AM}{MB} = k_1$ in the internal way, and the point N divides the same segment in the ratio of $\frac{AN}{NB} = k_2$ in the external way. If the condition $k_1 = k_2$, is met, then the quadruple of the points is called *harmonic*. For example, if the segments AB and BN are of equal length, then to fulfill the condition of harmony point M has to divide the segment AB in the ratio of $2:1$. The concept of a harmonic quadruple of points plays an important role in projective geometry, being one of its invariants.

In Figure 4.7 it is possible to see that the segment MN is shorter than AN, but longer than BN. Let us show that the length values of these segments for the harmonic quadruple are linked by the formula of harmonic mean.

Let the lengths of the segments AN, BN, MN be denoted by the letters a, b, c, and c be expressed through a and b.

From the relation $\frac{AM}{MB} = \frac{AN}{BN}$ we have

$$\frac{AM}{MB} = \frac{a}{b}, \quad \frac{AM + MB}{MB} = \frac{a+b}{b}, \quad \frac{a-b}{MB} = \frac{a+b}{b},$$

hence

$$MB = b\frac{a-b}{a+b}, \quad MB + b = \frac{2ab}{a+b}.$$

Taking into account that $MB + b = c$ we get

$$c = \frac{2ab}{a+b},$$

which coincides with the formula of harmonic mean.

Thus, the segment MN represents the harmonic mean of the segments AN and BN.

Let us note that the formulae of the focal distance of a lens, the connection of parallel resistors, and the hardness of springs connected in a series, respectively, have a form close to the harmonic mean.

Example 4.2.13 (Focal distance of a lens). To calculate the focus distance of the lens f in geometrical optics, the formula given below is used:

$$\frac{1}{f} = \frac{1}{a} + \frac{1}{b},$$

where a is the distance from an object to the lens, and b is the distance from the lens to the image.

From this it follows that the focus distance is equal to a half of the harmonic mean of the numbers a and b.

There is an elegant method which allows any value to be found from the three values f, a, and b (where two are unknown) without calculations using a nomogram with three simple scales located at an angle of 60° relative to each other (Figure 4.8).

If a thread placed against the scales of such a type passes through two given points (as shown in Figure 4.8) then by the given numbers a and b (or f) it is immediately possible to determine f (or, correspondingly, b). Halving the scale along the vertical axis is a simple method for determining the geometrical mean of two numbers.

Example 4.2.14 (Parallel connection of resistors). Let us consider the two circuits shown in Figure 4.9. We want to determine the quantity R at which the resulting resistances of both circuits are equal.

The R quantity in question is the harmonic mean of resistances R_1, R_2, R_3:

$$R = \frac{3}{\frac{1}{R_1} + \frac{1}{R_2} + \frac{1}{R_3}}.$$

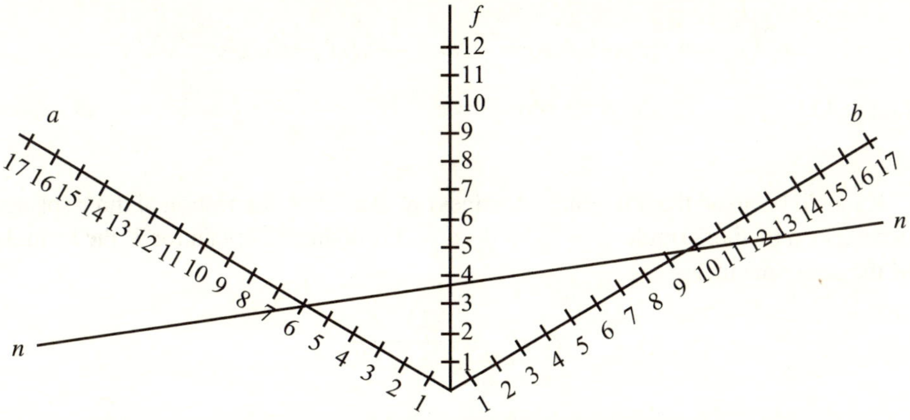

Figure 4.8. Finding the harmonic mean of two numbers.

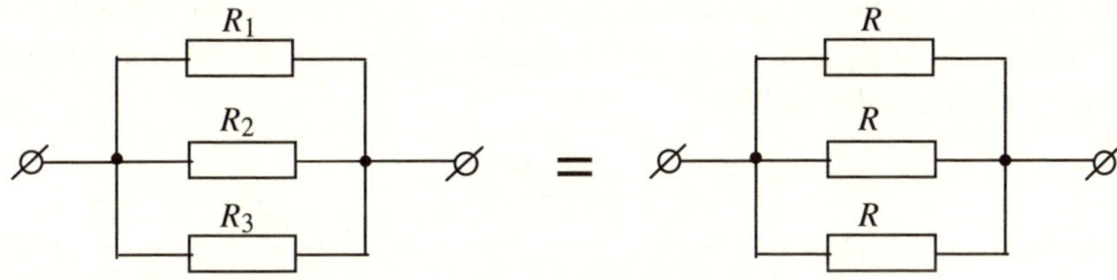

Figure 4.9. Harmonic mean of resistance.

Example 4.2.15 (Connection of springs in series). In Figure 4.10 we depict a connection in a series of two elastic springs with the values of hardness k_1 and k_2.

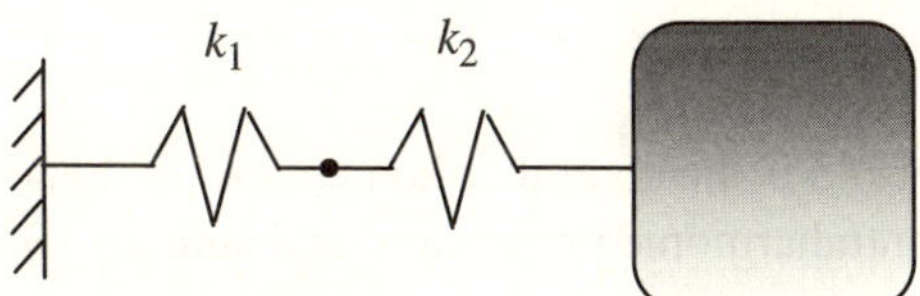

Figure 4.10. Connection of springs in series.

The hardness K of an equivalent spring is determined by the formula $\frac{1}{K} = \frac{1}{k_1} + \frac{1}{k_2}$, which is similar to the formula for the parallel connection of resistors.

Let us consider the two systems of springs shown in Figure 4.11. We want to determine the hardness of the spring K at which the general hardness values of both spring systems are equal.

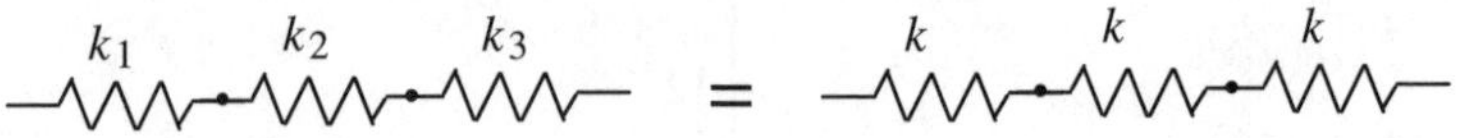

k_1 k_2 k_3 = k k k

Figure 4.11. Harmonic mean of the spring force K.

It should be noted that the general hardness K of a series connection of three springs is described by the formula $\frac{1}{K} = \frac{1}{k_1} + \frac{1}{k_2} + \frac{1}{k_3}$. From this we obtain for K the formula of the harmonic mean:

$$K = \frac{3}{\frac{1}{k_1} + \frac{1}{k_2} + \frac{1}{k_3}}.$$

4.2.3.4 Contraharmonic mean

Contraharmonic mean of n positive numbers $x_1, \cdots, x_n$ is the term for a quantity equal to the ratio of the arithmetic mean of the squares of these numbers to an arithmetic mean of the numbers themselves:

$$\hat{x} = C(x_1, x_2, \ldots, x_n) = \frac{\left(\frac{x_1^2 + x_2^2 + \cdots + x_n^2}{n}\right)}{\left(\frac{x_1 + x_2 + \cdots + x_n}{n}\right)} = \frac{x_1^2 + x_2^2 + \cdots + x_n^2}{x_1 + x_2 + \cdots + x_n} \quad (4.2.13)$$

For two and three measurements the contraharmonic mean is determined by

$$\hat{x} = \frac{x_1^2 + x_2^2}{x_1 + x_2}, \quad \hat{x} = \frac{x_1^2 + x_2^2 + x_3^2}{x_1 + x_2 + x_3}.$$

The graph of the two variables is the quadratic cone surface $x^2 + y^2 - xz - yz = 0$, as one may infer its level curves $(x - c/2)^2 + (y - c/2)^2 = c^2/2$.

Example 4.2.16. Integer 5 is the contraharmonic mean of 2 and 6, as well as of 3 and 6, i.e., 2, 5, 6 are in contraharmonic proportion, and similarly 3, 5, 6:

$$(2^2 + 6^2)/(2 + 6) = 40/8 = 5; \quad (3^2 + 6^2)/(3 + 6) = 45/9 = 5.$$

The word “contraharmonic” is explained by the fact that for two measurements the contraharmonic is greater than the arithmetic mean as much as the arithmetic mean is greater than the harmonic mean: $C - A = A - H$.

Hence it follows that the arithmetic mean of two numbers a, b is equal to the arithmetic mean of their harmonic and contraharmonic means:

$$A(a, b) = A(H(a, b), C(a, b)) \quad \text{or} \quad A = (H + C)/2.$$

Thus, at $n = 2$ the contraharmonic mean is a function complementary to the harmonic mean.

As a gets closer to 0, then $H(a,b)$ also gets closer to 0. The harmonic mean is very sensitive to low values. On the contrary, the contraharmonic mean is sensitive to larger values, so as a approaches 0 then $C(a,b)$ approaches b (so their average remains $A(a,b)$).

The contraharmonic mean can be applied in algorithms of image processing (the contraharmonic mean filter is well suited for reducing or virtually eliminating the effects of salt-and-pepper noise).

The generalization of the contraharmonic mean is the so called *Lehmer mean* that at $n = 3$ is determined by the formula

$$\hat{x} = \frac{x_1^p + x_2^p + x_3^p}{x_1^{p-1} + x_2^{p-1} + x_3^{p-1}}.$$

At $p = 1$ there is an arithmetic mean, at $p = 2$ a contraharmonic mean, and at $p \to \infty$ the maximum of three measurements is obtained.

4.2.3.5 Quadratic mean

A quadratic mean of numbers $x_1, \cdots, x_n$ is called the quantity

$$\hat{x} = Q(x_1, \ldots, x_n) = \sqrt{\frac{x_1^2 + x_2^2 + \cdots + x_n^2}{n}}. \tag{4.2.14}$$

For the case of two and three measurements the quadratic mean is determined by the formulae

$$\hat{x} = \sqrt{\frac{x_1^2 + x_2^2}{2}}, \quad \hat{x} = \sqrt{\frac{x_1^2 + x_2^2 + x_3^2}{3}}.$$

The quadratic mean is used in many fields. In particular, concepts from probability theory and mathematical statistics such as dispersion and mean square deviation are determined with the help of the quadratic mean. The quadratic mean is applied if according to the problem conditions it is necessary to keep the sum of squares unchanged at their replacement by a mean. The measurement of the diameter of a tree, which is conventionally done in forestries, is a quadratic mean diameter, rather than an arithmetic mean diameter [133].

Example 4.2.17. Given three square lots with sides $x_1 = 100\,\text{m}$; $x_2 = 200\,\text{m}$; $x_3 = 300\,\text{m}$. We want to replace the different values of the side lengths by a mean reasoning in order to have the same general area for all the lots.

The arithmetic mean of the side length $(100 + 200 + 300)\colon 3 = 200\,\text{m}$ does not satisfy this condition, since the joint area of the three lots with the side of 200 m would be equal to

$$3(200\,\text{m})^2 = 120\,000\,\text{m}^2.$$

At the same time, the area of the given three lots is equal to

$$(100\,\mathrm{m})^2 + (200\,\mathrm{m})^2 + (300\,\mathrm{m})^2 = 140\,000\,\mathrm{m}^2.$$

The quadratic mean will be the correct answer:

$$\hat{x} = \sqrt{\frac{(100)^2 + (200)^2 + (300)^2}{3}} = 216\,\mathrm{m}.$$

For two measurements $x_1 = a, x_2 = b$ the quadratic mean can be expressed through an arithmetic, geometric, and contraharmonic means with the help of the formula

$$Q(a,b) = G(A(a,b), C(a,b)),$$

or for short, $Q = G(A, C)$.

Its correctness follows from the chain of equalities:

$$G(A(a,b), C(a,b)) = G\left(\frac{a+b}{2}, \frac{a^2+b^2}{a+b}\right) = \sqrt{\frac{a+b}{2} \cdot \frac{a^2+b^2}{a+b}}$$
$$= \sqrt{\frac{a^2+b^2}{2}} = Q(a,b).$$

4.2.4 Geometrical interpretation of the means

The arithmetic, geometric, and harmonic means are known as the Pythagorean means. One of the geometric presentations which allow the Pythagorean means of two numbers to be visually compared is shown in Figure 4.12. They are shown in the form of the segments a and b, forming a diameter of a semicircumference. Its radius is equal to the arithmetic mean of the numbers a and b, and the length of the perpendicular G is equal to their geometric mean. The length of the leg H of a right-angle triangle with the hypotenuse G is equal to the harmonic mean. It should be noted that, since $H + C = 2A$, the contraharmonic C is equal to the remaining part of the diameter on which the segment $H$is located.

From Figure 4.12 it is seen that if a is not equal to b, the geometric mean will always be greater than the harmonic mean, but less than the arithmetic mean $C > A > G > H$.

There are many interesting problems connected with the Pythagorean means. One of them consists of finding such pairs of the positive integers a, b, for which all three A, G, H are the integers. There are a great many of such pairs. Some of them are indicated in Table 4.2; others can be found at http://i-eron.livejournal.com/55678.html.

There are also other fine geometric representations of classical means [444]. Let A be the arithmetic mean of two positive numbers a and b, G is their geometric mean, H is the harmonic mean, Q is the quadratic mean, m is the minimum number and M is the maximum number of the numbers a and b.

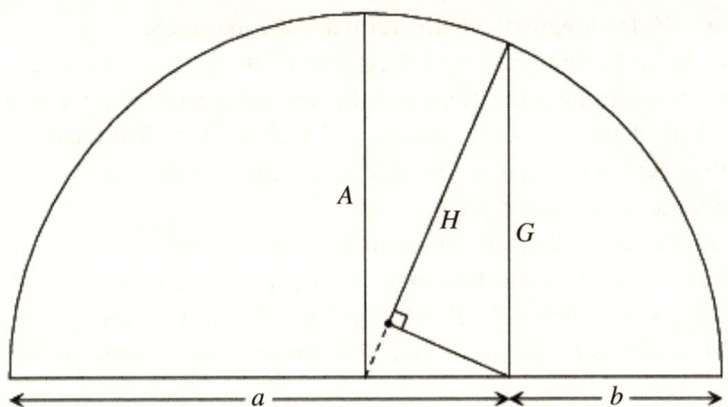

Figure 4.12. Pythagorean means.

Table 4.2. Pairs of positive integers a, b, for which the means A, G, H are the integers.

a	b	A	G	H
5	45	25	15	9
10	40	25	20	16
30	120	75	60	48
13	325	169	65	25

The geometric interpretation of these means, e.g., some segments in a trapezium with the bases a and b, is illustrated in Figure 4.13. All segments H, G, A, Q are parallel to the trapezoid bases.

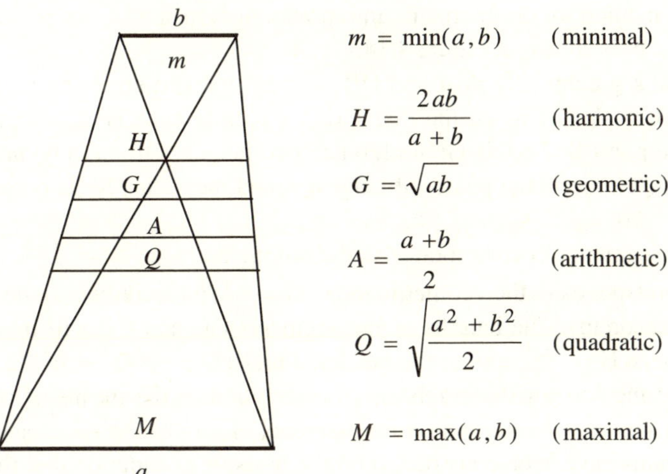

Figure 4.13. Trapezoid of the means.

The segment H, the length of which is equal to the harmonic mean, passes through a cross point of the trapezium diagonals. Segment G, the geometrical mean, divides the trapezium into two similar parts. Segment A is the median of the trapezium. Its length is equal to a half of the sum of the bases, i.e., their arithmetical mean. The segment Q, the length of which is equal to the quadratic mean, divides the trapezium into two equal (with respect to the area) parts.

Segment C, corresponding to the contraharmonic mean, is at the same distance from the median line A as the harmonic segment H (the center of the dotted line circumference lies on segment A). Finally the bases themselves, equal to the maximum and minimum numbers of all numbers a, b, represent the extreme cases of the mean values.

From Figure 4.13 it can be seen that the means satisfy the chain of inequalities:

$$m \le H \le G \le A \le Q \le C \le M. \tag{4.2.15}$$

Thus, the contraharmonic mean occupies the closest place with respect to the large base of the trapezoid, after it the quadratic mean and arithmetic mean follow, further on the geometric mean goes and, at last, the harmonic mean.

It is clear that the lengths of these segments do not depend on the form of the trapezoid. For all trapezoids with given lengths of bases a and b the means indicated are equal correspondingly. When $a = b$, then the trapezoid becomes a parallelogram, and all seven means coincide and become equal to a.

Details of the geometric construction of all indicated segments can be found in [444] and at the site http://jwilson.coe.uga.edu/EMT668/EMAT6680.2000/Umberger/EMAT6690smu/.

The third method for the geometric interpretation of means of two positive numbers a and b, $a < b$, is shown in Figure 4.14.

The initial segments OA $= a$ and OB $= b$ are marked on a straight line. On the segment AB $= b - a$ as on the diameter a circumference is constructed with the center at the point S. To this circumference two tangent OP and OQ are lined. The segment PCQ connects the points of tangency, and the radius SE is perpendicular to the diameter AB.

Then the segment OS corresponds to the arithmetic mean A $= \frac{a+b}{2}$, and the segment OP corresponds to the geometric mean G $= \sqrt{ab}$ (according to the tangent and secant line theorem). The lengths of the segments OC and OE will be equal to the harmonic mean H $= \frac{2ab}{a+b}$ and to the quadratic mean Q $= \sqrt{(a^2 + b^2)/2}$.

One more method for the graphical representation of the means of two numbers is shown in Figure 4.15. In this figure diagrams of curves for six means are given. These diagrams have been constructed on the basis of assumption that the number b is constant and the number a possesses various values. At $a = b$ all the curves are intersected. Hatch is applied to “forbidden” areas within which no means of two num-

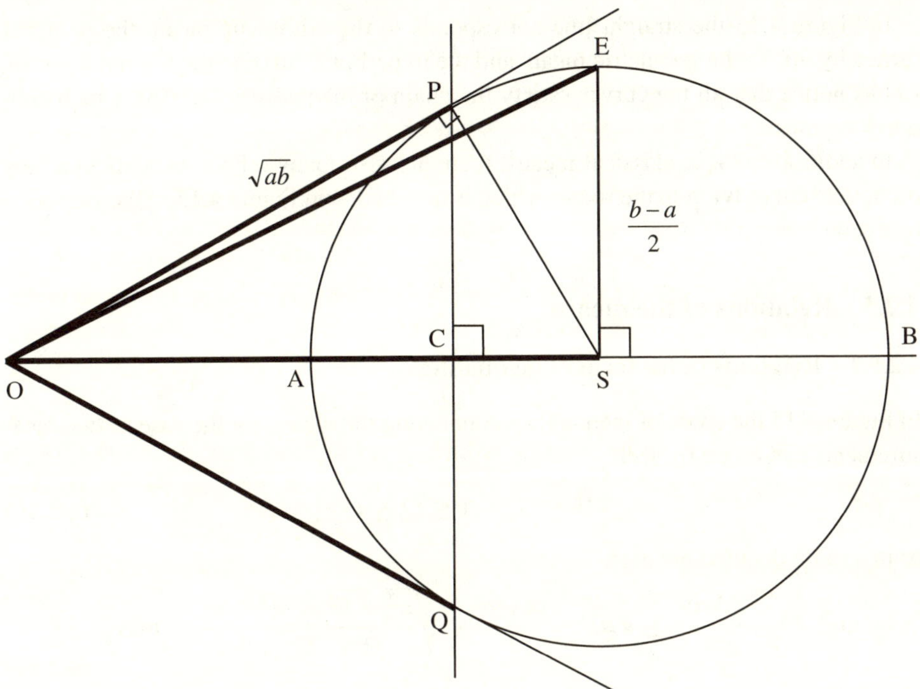

Figure 4.14. Circumference of the means.

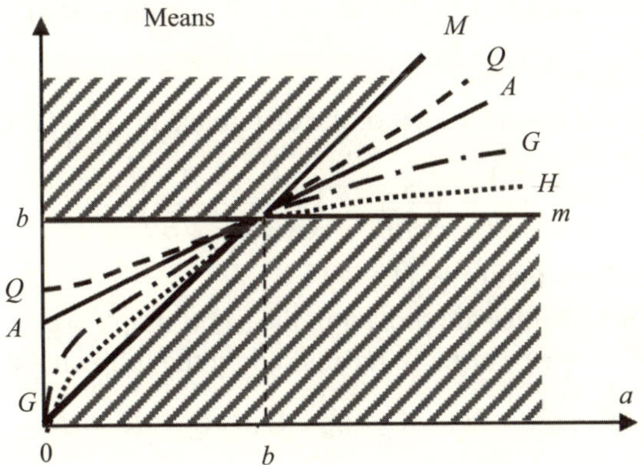

Figure 4.15. Diagrams of the classical means.

bers can be located (according to the definition the mean has to be located between the numbers a and b).

In Figure 4.15 the straight line corresponds to the arithmetic mean, the parabola turned by 90° to the geometric mean, and the hyperbolic curve to the harmonic mean. Let us notice that all the curves satisfy the chain of inequalities (4.2.15), which connects the means.

In addition to these classical means, there are also many others. In particular, any monotone curve lying in the sector which is not shaded in Figure 4.15 will correspond to a certain mean.

4.2.5 Relations of the means

4.2.5.1 Relations in the form of inequalities

In Figure 4.15 the chain of inequalities connecting the means for the case of two measurements can be easily seen:

$$m \le H \le G \le A \le Q \le C \le M, \tag{4.2.16}$$

or in a more detailed notation

$$\min(a,b) \le \frac{2ab}{a+b} \le \sqrt{ab} \le \frac{a+b}{2} \le \sqrt{\frac{a^2+b^2}{2}} \le \frac{a^2+b^2}{a+b} \le \max(a,b). \tag{4.2.17}$$

The algebraic proof of these inequalities can be done without difficulty.

For example, considering the difference A–G, we get

$$\frac{a+b}{2} - \sqrt{ab} = \left(a - 2\sqrt{ab} + b\right)/2 = \left(\sqrt{a} - \sqrt{b}\right)^2/2 \ge 0,$$

and hence it follows that $A \ge G$.

In the same way the consideration of the difference $(1/H)$–$(1/G)$ gives

$$\frac{1}{a} + \frac{1}{b} - \frac{2}{\sqrt{ab}} = \left(\frac{1}{\sqrt{a}} - \frac{1}{\sqrt{b}}\right)^2 \ge 0,$$

and from here it follows that $G \ge H$.

The remaining inequalities are proved in the same simple way.

The pair-wise differences of these means satisfy the following three inequalities:

$$G - A \le H - Q, \quad H - A \le Q - G, \quad Q - H \le H - G. \tag{4.2.18}$$

Inequalities (4.2.15) are true with any number of arguments. They, as well as their generalizations (the Cauchy, Minkowski, and Jensen inequalities) are often used in mathematics to solve various problems.

Example 4.2.18. The task is to determine the minimum value of the sum $a^2 + b^2$ if $a + b = 2$, the numbers a and b being positive.

This problem can be solved with the help of the inequality between the quadratic mean and the arithmetic mean $Q^2 \ge A^2$. In accordance with the problem condition

$A = \frac{a+b}{2} = 1$, therefore, $Q^2 = \frac{a^2+b^2}{2} \geq 1$. Consequently, the minimum sum $a^2 + b^2$ is 2.

Example 4.2.19. The task is to prove that for any numbers a, b, and c with similar signs the relation $\frac{a}{b} + \frac{b}{c} + \frac{c}{a} \geq 3$ is true. Let the inequality between the arithmetic and geometric means of three numbers $\frac{a}{b}, \frac{b}{c}, \frac{c}{a}$ be used:

$$\frac{\frac{a}{b} + \frac{b}{c} + \frac{c}{a}}{3} \geq \sqrt[3]{\frac{a}{b} \cdot \frac{b}{c} \cdot \frac{c}{a}} = 1.$$

The second member of the inequality is equal to 1; hence the required relation is obtained.

4.2.5.2 Relations in the form of equalities

Let us examine the question of how to find algebraic combinations of the means invariant with respect to measurements; e.g., the arithmetic mean, quadratic mean and contraharmonic mean:

$$A = \frac{1}{n} \sum_{i=1}^{n} x_i, \qquad Q = \sqrt{\frac{1}{n} \sum_{i=1}^{n} x_i^2}, \qquad C = \frac{\sum_{i=1}^{n} x_i^2}{\sum_{i=1}^{n} x_i},$$

at any x_i satisfy the invariant relation

$$AC - Q^2 = 0. \tag{4.2.19}$$

It is obvious that the number of such relations depend on both the set of the means being considered and the number of measurements x_i. The less n is, the greater is the number of invariant relations which can be expected.

The simplest method for finding the invariant relations is as follows. Let us take the totality $m = n + 1$ means from n arguments x_i:

$$y_i = F_i(x_1, \ldots, x_n), \quad i = \overline{1, m}. \tag{4.2.20}$$

Then, excluding from the system of $(n+1)$ equations (4.2.20) n variables $x_1, \ldots, x_n$, we obtain the invariant relation

$$f(y_1, \ldots, y_m) = 0,$$

which does not depend on the measurement values $x_1, \ldots, x_n$.

For example, excluding the pair a, b from three Pythagorean means

$$A = \frac{a+b}{2}, \quad G = \sqrt{ab}, \quad H = \frac{2ab}{a+b},$$

causes us to get the invariant relation $AH - G^2 = 0$.

If $m = n + 2$, then removing alternately one of the equations from system (4.2.20) and acting in the manner described, it is possible to obtain the totality of $(n + 2)$ invariant relations:

$$f_i(y_1, \ldots, y_m) = 0, \ i = \overline{1, m}.$$

Similarly, (m choose 2) invariant relations are obtained for $m = n + 3$. Note that among the relations found, there can be relations, functionally dependent, which are the combinations of other relations. One of the approaches for revealing such combinations is based on the construction of the Jacobi matrix of the function system under study and analysis of its rank.

Let us look more closely at the invariant relations for two and three measurements.

4.2.5.3 Invariant relations for two measurements

Let five means of two numbers a and b be considered: arithmetic, geometric, harmonic, contraharmonic, and quadratic. At $n = 2$ the formulae for them have the form

$$A = \frac{a+b}{2}, \quad G = \sqrt{ab}, \quad H = \frac{2ab}{a+b},$$

$$C = \frac{a^2+b^2}{a+b}, \quad Q = \sqrt{\frac{a^2+b^2}{2}}. \tag{4.2.21}$$

In order to find the algebraic relations connecting them to each other, use the method of excluding the parameters a, b from different triplets of means.

Altogether, we have 10 versions of such triplets, obtained by removing the following pairs of the means from formulae (4.2.21):

$$(A, Q), (G, Q), (H, Q), (C, Q), (G, C), (H, C), (A, C), (A, H), (G, H), (A, G). \tag{4.2.22}$$

Performing such calculation operations, we obtain (5 choose 2) = 10 algebraic relations connecting different triplets of means (see Table 4.3).

In each line of the table we give invariant combinations of the given triplet of means, which do not depend on the arguments (the numbers a, b), formula connecting these means, and the corresponding functional equation.

All invariant relations are uniform, whereby one of them is linear, seven are quadratic, and two (the seventh and the eighth ones) are of the fourth order. For the purpose of uniformity the linear equation can be transformed into the quadratic one, multiplying it by A:

$$2A^2 - AC - AH = 0.$$

Since the number of equations is twice the number of arguments, then the relations have to be functionally dependent. Let us write the Jacobi matrix for the left parts of

Table 4.3. Invariant combinations of the means at $n = 2$.

Nº	Means	Invariant relations of different triplets of means	Formulae of connecting triplets of means	Functional equation
1.	(G,H,C)	$H^2+CH-2G^2=0$	$H(C+H)=2G^2$	$G(A(C,H)+H=G$
2.	(A,H,C)	$2A-C-H=0$	$C+H=2A$	$A(C,H)=A$
3.	(A,G,C)	$2A^2-AC-G^2=0$	$AC+G^2=2A^2$	$A(AC,G^2)=A^2$
4.	(A,G,H)	$AH-G^2=0$	$AH=G^2$	$G(A,H)=G$
5.	(A,H,Q)	$2A^2-AH-Q^2=0$	$AH+Q^2=2A^2$	$A(AH,Q^2)=A^2$
6.	(A,G,Q)	$G^2+Q^2-2A^2=0$	$G^2+Q^2=2A^2$	$Q(G,Q)=A$
7.	(G,H,Q)	$-G^2H^2-H^2Q^2+2G^4=0$	$H^2(G^2+Q^2)=2G^4$	$G(H,Q(G,Q))=G$
8.	(G,C,Q)	$C^2G^2+C^2Q^2-2Q^4=0$	$C^2(G^2+Q^2)=2Q^4$	$G(C,Q(G,Q))=Q$
9.	(A,C,Q)	$AC-Q^2=0$ for any n	$AC=Q^2$	$G(A,C)=Q$
10.	(H,C,Q)	$C^2+CH-2Q^2=0$	$C(C+H)=2Q^2$	$G(A(C,H),C)=Q$

the invariant relations:

$$J=\begin{bmatrix} 0 & -4G & 2H+C & H & 0 \\ -C+4A-H & 0 & -A & -A & 0 \\ -C+4A & -2G & 0 & -A & 0 \\ H & -2G & A & 0 & 0 \\ 4A-H & 0 & -A & 0 & -2Q \\ 4A & -2G & 0 & 0 & -2Q \\ 0 & -2GH^2+8G^3 & -2G^2H-2HQ^2 & 0 & -2H^2Q \\ 0 & 2C^2G & 0 & 2G^2C+2Q^2C & 2QC^2-8Q^3 \\ -C & 0 & 0 & -A & 2Q \\ 0 & 0 & C & H+2C & -4Q \end{bmatrix}.$$

Its analysis allows the following linear dependencies between the lines S_i of this matrix be revealed:

$$S_2+S_4=S_3, \quad S_4+S_5=S_6, \quad S_5+S_9=S_2.$$

Such dependencies exist between the corresponding invariant relations shown in Table 4.3; e.g., summing up the fourth and fifth relations one gets the sixth one, and so on.

4.2.5.4 Invariant relations in case of three measurements

Consider the same five means: arithmetic, geometric, harmonic, contraharmonic, and quadratic. At $n = 3$ the formulae for them have the form:

$$A = \frac{a+b+c}{3}, \quad G = \sqrt[3]{abc}, \quad H = \frac{3}{\frac{1}{a}+\frac{1}{b}+\frac{1}{c}},$$
$$C = \frac{a^2+b^2+c^2}{a+b+c}, \quad Q = \sqrt{\frac{a^2+b^2+c^2}{3}}. \tag{4.2.23}$$

The algebraic relations, connecting these means to each other, will be found by the way of excluding the parameters a, b, c from the equations for the following five groups of the means indicated:

$$(A, G, H, Q),\ (A, G, H, C),\ (G, H, C, Q),\ (A, G, C, Q),\ (A, (H, C, Q)).$$

Performing the corresponding computation, one gets four invariant relations connecting different quadruples of means (see Table 4.4).

Table 4.4. Invariant combinations of the means at $n = 3$.

N°	Means	Invariant relations of different quadruples of means	Formulae connecting four means
1.	A, G, H, Q	$3A^2H - HQ^2 - 2G^2 = 0$	$2G^3 + HQ^2 = 3A^2H$
2.	A, H, C	$3A^2H - AHC - 2G^3 = 0$	$2G^3 + ACH = 3A^2H$
3.	A, G, C	$C^2HQ^2 + 2G^3C^2 - 3HQ^4 = 0$	$C^2HQ^2 + 2G^3C^2 = 3HQ^4$
4.	A, H, G	$AC - Q^2 = 0$ for any n	$AC = Q^2$

The invariant relations indicated in Table 4.4 are not independent. For example, the substitution of $Q^2 = AC$ into the third relation gives the second one.

To make an analysis of the functional dependence of invariant relations, let us construct Jacobi matrix for the four functions

$$f_1 = 3A^2H - HQ^2 - 2G^2,$$
$$f_2 = 3A^2H - AHC - 2G^3,$$
$$f_3 = C^2HQ^2 + 2G^3C^2 - 3HQ^4,$$
$$f_4 = H(AC - Q^2),$$

of five arguments (A, G, H, C, Q):

$$J = \begin{bmatrix} 6AH & -6G^2 & -Q^2 + 3A^2 & 0 & -2QH \\ -HC + 6AH & -6G^2 & -CA + 3A^2 & -AH & 0 \\ 0 & 6G^2C^2 & G^2C^2 - 3Q^4 & 2CHQ^2 + 4G^3C & 2C^2HQ - 12HQ \\ HC & 0 & CA - Q^2 & AH & -2QH \end{bmatrix}.$$

The matrix rank J is equal to 3, which reveals the linear dependence of its lines: the first line is equal to the sum of the second and fourth one. The similar dependence of functions $f_1 = f_2 + f_4$ is in agreement with this.

Thus, the five means considered appear to be connected by ten invariant relations at $n = 2$, and by four relations at $n = 3$. At $n = 4$ and higher they have only one relation (4.2.14) $AC = Q^2$.

Let us note that knowledge of invariant relations appears to be useful for saving calculation operations in the process of finding the totality of means, for computing operations control, and in the treatment of indirect measurements.

The number of relations connecting means can be increased if they include the quantities being averaged. The following relations can serve as an example:

$$Q^2 = A^2 + q^2, \quad C = A + \frac{1}{A}q^2,$$

where q means the quadratic mean of centered measurements

$$q = \sqrt{\frac{1}{n}\sum_{i=1}^{n}(x_i - A)^2}.$$

In addition to that, sometimes approximate relations between the means are considered. So, the approximate equality $N \approx A - 3(A - M)$ between the mode N and the arithmetic mean A and maximum M.

4.2.6 Inverse problems in the theory of means

The direct problem of the theory of mean quantities consists of determining a mean (or a number of means) of a given set of quantities measured. For small samples the inverse problem consisting of determining (calculating) unknown quantities by the known values of their means, is of interest.

In metrology this corresponds to the case of indirect measurements, when indications of instruments measuring various means of physical quantities (for example, an arithmetic mean or quadratic mean) are known and we wish to determine the physical quantities themselves.

Another example is the problem of determining the geometric dimensions of a body, if its volume, surface area, or some other means characteristics are known.

It is clear that to solve the inverse problem it is necessary that the number of known means be not less than a number of unknown quantities. In the general case the inverse problem can be formulated in the following manner: there are given m different known means of a certain totality $n \leq m$ of quantities; it is necessary to find these quantities.

So long as the cases of two and three unknown quantities are of the greatest interest for practice, let us consider these cases in more detail, limiting ourselves to only five means: arithmetic, geometric, harmonic, contraharmonic, and quadratic.

4.2.6.1 Inverse problems for $n = 2$

For two unknown quantities a, b the algebraic equations for five mentioned means have the form

$$A = \frac{a+b}{2}, \quad G^2 = ab, \quad H = \frac{2ab}{a+b}, \quad C = \frac{a^2+b^2}{a+b}, \quad Q^2 = \frac{a^2+b^2}{2}. \tag{4.2.24}$$

To determine two unknown quantities a, b, it is necessary to know the values of two means. Looking through all the pairs from five means (4.2.24), one gets 10 inverse problems for $n = 2$, where 10 is the binomial coefficient (5 choose 2).

Below, for each of them the algebraic equation is given, the roots of which give the values of the quantities sought a, b.

Problem 1 for $n = 2$ (Recovery of the quantities a, b on the basis of the arithmetic and geometric means). The quantities sought are equal to the roots of the quadratic equation

$$x^2 - 2Ax + G^2 = 0, \quad a = A - \sqrt{A^2 - G^2}, \quad b = A + \sqrt{A^2 - G^2}.$$

Problem 2 for $n = 2$ (Recovery of the quantities a, b on the basis of the arithmetic and harmonic means). The quantities sought are equal to the roots of the quadratic equation

$$x^2 - 2Ax + AH = 0, \quad a = A + \sqrt{A(A-H)}, \quad b = A - \sqrt{A(A-H)}.$$

Problem 3 for $n = 2$ (Recovery of the quantities a b on the basis of the arithmetic and contraharmonic means). The quantities sought are equal to the roots of the quadratic equation

$$x^2 - 2Ax + A(2A - C) = 0, \quad a = A + \sqrt{A(C-A)}, \quad b = A - \sqrt{A(C-A)}.$$

Problem 4 for $n = 2$ (Recovery of the quantities a, b on the basis of the arithmetic and quadratic means). The quantities sought are equal to the roots of the quadratic equation

$$x^2 - 2Ax + 2A^2 - Q^2 = 0, \quad a = A - \sqrt{Q^2 - A^2}, \quad b = A + \sqrt{Q^2 - A^2}.$$

Problem 5 for $n = 2$ (Recovery of the quantities a, b on the basis of the geometric and harmonic means). The quantities sought are equal to the roots of the quadratic equation

$$x^2 - 2\frac{G^2}{H}x + G^2 = 0, \quad a = \frac{G}{H}(G - \sqrt{G^2 - H^2}), \quad b = \frac{G}{H}(G + \sqrt{G^2 - H^2}).$$

Problem 6 for $n = 2$ (Recovery of the quantities a, b on the basis of the geometric and contraharmonic means). The quantities sought are equal to the positive roots of the equation of the fourth order

$$x^4 - Cx^3 - CG^2x + G^4 = 0.$$

Introducing the designation $r^2 = C^2 + 8G^2$, the latter can be represented in the form $P_1(x) \cdot P_2(x) = 0$, where $P_1(x) = 2x^2 - (C + r)x + 2G^2$, $P_2(x) = 2x^2 - (C - r)x + 2G^2$.

The roots of the polynomial $P_1(x)$ are positive; they are equal to the quantities sought a, b:

$$a = \frac{1}{4}\left(C + r - \sqrt{2(r^2 + Cr - 12G^2)}\right),$$

$$b = \frac{1}{4}\left(C + r + \sqrt{2(r^2 + Cr - 12G^2)}\right).$$

Problem 7 for $n = 2$ (Recovery of the quantities a, b on the basis of the geometric and quadratic means). The quantities sought are equal to the positive roots of the biquadratic equation

$$x^4 - 2Q^2x^2 + G^4 = 0,$$

which can be represented in the form

$$\left(x^2 - x\sqrt{2G^2 + 2Q^2} + G^2\right)\left(x^2 + x\sqrt{2G^2 + 2Q^2} + G^2\right) = 0.$$

The roots of the first multiplier are positive and equal to the quantities sought a, b:

$$a = \frac{\sqrt{2}}{2}(\sqrt{Q^2 + G^2} - \sqrt{Q^2 - G^2}), \quad b = \frac{\sqrt{2}}{2}(\sqrt{Q^2 + G^2} + \sqrt{Q^2 - G^2}).$$

Another record of these roots has the form

$$a = \sqrt{Q^2 - \sqrt{Q^4 - G^4}}, \quad b = \sqrt{Q^2 + \sqrt{Q^4 - G^4}}.$$

Problem 8 for $n = 2$ (Recovery of the quantities a, b on the basis of harmonic and contraharmonic means). The quantities sought are equal to the roots of the quadratic equation

$$x^2 - (C + H)x + \frac{H}{2}(C + H) = 0,$$
$$a = \frac{1}{2}(C + H - \sqrt{C^2 + H^2}), \quad b = \frac{1}{2}(C + H + \sqrt{C^2 + H^2}).$$

Problem 9 for $n = 2$ (Recovery of the quantities a, b on the basis of harmonic and quadratic means). The quantities sought are equal to the positive roots of the fourth order

$$2x^4 - 2Hx^3 - (4Q^2 - H^2)x^2 + 4Q^2Hx - H^2Q^2 = 0.$$

Introducing the designation $r^2 = H^2 + 8Q^2$, the latter can be represented in thc form $P_1(x) \cdot P_2(x) = 0$, where

$$P_1(x) = 4x^2 - 2(H + r)x + H(H + r), \quad P_2(x) = 4x^2 + 2(H + r)x + H(H + r).$$

The quantities sought are equal to the roots of the polynomial $P_1(x)$:

$$a = \frac{1}{4}(H + r - \sqrt{2(4Q^2 - H^2 - rH)}), \quad b = \frac{1}{4}(H + r + \sqrt{2(4Q^2 - H^2 - rH)}).$$

Problem 10 for $n = 2$ (Recovery of the quantities a, b on the basis of contraharmonic and quadratic means). The quantities sought are equal to the roots of the quadratic equation

$$C^2x^2 - 2CQ^2x + Q^2(2Q^2 - C^2) = 0,$$
$$a = \frac{Q}{C}(Q - \sqrt{C^2 - Q^2}), \quad b = \frac{Q}{C}(Q + \sqrt{C^2 - Q^2}).$$

Thus, the solution of all ten inverse problems at $n = 2$ is reduced to finding the roots of quadratic equations, the coefficients of which are expressed algebraically in terms of the mean quantities or with the help of radicals.

The results obtained are given in Table 4.5, where the equations and formulae for determining the unknown quantities a, b by the given pairs of means are also indicated.

Example 4.2.20. Let the geometric and contraharmonic means of two numbers $G = 3$, $C = 8.2$ are known. We wish to find these numbers.

Table 4.5. Inverse problems for $n = 2$.

N°	Means	Quadratic equations for determining the quantities a, b	Formulae for the quantities a, b
1.	A, G	$x^2 - 2Ax + G^2 = 0$	$A \pm \sqrt{A^2 - G^2}$
2.	A, H	$x^2 - 2Ax + AH = 0$	$A \pm \sqrt{A(A - H)}$
3.	A, C	$x^2 - 2Ax + A(2A - C) = 0$	$A \pm \sqrt{A(C - A)}$
4.	A, Q	$x^2 - 2Ax + 2A^2 - Q^2 = 0$	$A \pm \sqrt{Q^2 - A^2}$
5.	G, H	$x^2 - 2\frac{G^2}{H}x + G^2 = 0$	$\frac{G}{H}(G \pm \sqrt{G^2 - H^2})$
6.	G, C	$2x^2 - (C + r)x + 2G^2 = 0$, $r = \sqrt{C^2 + 8G^2}$	$\frac{C + r \pm \sqrt{2(r^2 + Cr - 12G^2)}}{4}$
7.	G, Q	$x^2 - x\sqrt{2G^2 + 2Q^2} + G^2 = 0$	$\frac{\sqrt{Q^2 + G^2} \pm \sqrt{Q^2 - G^2}}{\sqrt{2}}$
8.	H, C	$x^2 - (C + H)x + \frac{H}{2}(C + H) = 0$	$\frac{1}{2}(C + H \pm \sqrt{C^2 - H^2})$
9.	H, Q	$4x^2 - 2(H + r)x + H(H + r) = 0$, $r = \sqrt{H^2 + 8Q^2}$	$\frac{H + r \pm \sqrt{2(4Q^2 - H^2 - rH)}}{4}$
10.	C, Q	$C^2x^2 - 2CQ^2x + Q^2(2Q^2 - C^2) = 0$	$\frac{Q}{C}(Q \pm \sqrt{C^2 - Q^2})$

In accordance with inverse problem 4.6 let us write the equation of the fourth order as

$$x^4 - Cx^3 - CG^2x + G^4 = 0.$$

After substitution of the numerical data and factorization this equation accepts the form

$$x^4 - 8.2x^3 - 73.8x + 81 = (x - 1)(x - 9)(x^2 + 1.8x + 9) = 0.$$

Dropping the third multiplier (its roots are complex), one obtains the values of the numbers sought: $a = 1, b = 9$.

It should be noted that Table 4.5 can be used not only for calculating the quantities a, b on the basis their means, but also for deriving the invariant relations indicated in Table 4.3. The fact is that the roots of all quadratic equations of Table 4.5 are equal to the same numbers a, b. Consequently, the coefficients of these equations at similar powers x must also be equal to each other. Equating them gives a totality of invariant relations between the means at $n = 2$.

In particular, the conditions of the equality of the free members of the equations give, for example, 6 invariant relations.

The remaining invariant relations can be obtained by equating the coefficients at x.

4.2.6.2 Inverse problems for $n = 3$

Let us proceed to the problem of recovering three quantities a, b, c with the help of the known means of these quantities. Let the same five means given above be considered. At $n = 3$ the algebraic equations have the form

$$A = \frac{a+b+c}{3}, \qquad G^3 = abc, \qquad H = \frac{3}{\frac{1}{a} + \frac{1}{b} + \frac{1}{c}},$$
$$C = \frac{a^2+b^2+c^2}{a+b+c}, \qquad Q^2 = \frac{a^2+b^2+c^2}{3}. \tag{4.2.25}$$

To "recover" the three quantities a, b, c it is necessary to know at least three means.

Looking through all five triplets form means (4.19), ten inverse problems are obtained for $n = 3$ (the number of combinations from 5 by 3: 5 choose 3).

Below for each of ten problems the algebraic equation is given, the roots of which give the values of the quantities sought a, b, c.

Problem 1 for $n = 3$ (Recovery of the quantities a, b, c by the arithmetic, geometric, and harmonic means). The quantities sought are equal to the roots of the cubic equation

$$x^3 - 3Ax^2 + 3\frac{G^3}{H}x - G^3 = 0.$$

Problem 2 for $n = 3$ (Recovery of the quantities a, b, c by the arithmetic, geometric, and contraharmonic means). The quantities sought are equal to the roots of the cubic equation

$$x^3 - 3Ax^2 + \frac{3}{2}A(3A - C)x - G^3 = 0.$$

Problem 3 for $n = 3$ (Recovery of the quantities a, b ,c by the arithmetic, geometric, and quadratic means). The quantities sought are equal to the roots of the cubic equation

$$x^3 - 3Ax^2 + \frac{3}{2}A(3A^2 - Q^2)x - G^3 = 0.$$

Problem 4 for $n = 3$ (Recovery of the quantities a, b, c by the arithmetic, harmonic, and contraharmonic means). The quantities sought are equal to the roots of the cubic

equation
$$x^3 - 3Ax^2 + \frac{3}{2}A(3A^2 - C)x - \frac{1}{2}AH(3A - C) = 0.$$

Problem 5 for $n = 3$ (Recovery of the quantities a, b, c by the arithmetic, harmonic, and quadratic means). The quantities sought are equal to the roots of the cubic equation
$$x^3 - 3Ax^2 + \frac{3}{2}(3A^2 - Q)x - \frac{H}{2}(3A^2 - Q^2) = 0.$$

Problem 6 for $n = 3$ (Recovery of the quantities a, b, c by the arithmetic, contraharmonic, and quadratic means). The quantities indicated are connected by the relation $Q^2 = AC$, therefore the recovery of measurements is impossible. The problem has no solution (the degenerated case).

Problem 7 for $n = 3$ (Recovery of the quantities a, b, c by the geometric, harmonic, and contraharmonic means). The quantities sought are equal to the roots of the sixth order equation
$$x^6 - Cx^5 - \frac{G^3}{H}(3C + 2H)x^3 + \frac{G^3}{H^2}(9G^3 + CH^2)x^2 - 6\frac{G^6}{H}x + G^6 = 0.$$

By performing the factorization of the first member of $P = P_1P_2$ and designating $R^2 = C^2H^2 + 24G^3H$ we get
$$P_1 = x^3 - \frac{R + CH}{2H}x^2 + \frac{3xG^3}{H} - G^3,$$
$$P_2 = x^3 + \frac{R + CH}{2H}x^2 + \frac{3xG^3}{H} - G^3.$$

The quantities sought a, b, c are the roots of the polynomial P_1:
$$P_1 = x^3 - \frac{HC - \sqrt{H(c^2H + 24G^3)}}{2H}x^2 + \frac{3G^3}{H}x - G^3.$$

Problem 8 for $n = 3$ (Recovery of the quantities a, b, c by the geometric, harmonic and quadratic means). The quantities sought are equal to the roots of the equation of the sixth order
$$x^6 - 3Q^2x^4 - 2G^3x^3 + 9\frac{G^6}{H^2}x^2 - 6\frac{G^6}{H}x + G^6 = 0.$$

The factorization of the first member of $P = P_1P_2$ is performed:
$$P_1 = x^3 + Rx^2 + \frac{3G^3}{H} - G^3, \quad P_2 = x^3 - Rx^2 + \frac{3G^3}{H} - G^3.$$

where $R^2 = 6G^3/H + 3Q^2H$.

The quantities sought a, b, c are the roots of the cubic equation

$$x^3 - Rx^2 + \frac{3G^3}{H}x - G^3 = 0.$$

Problem 9 for $n = 3$ (Recovery of the quantities a, b, c by the geometic, contraharmonic and quadratic means). The quantities sought are equal to the roots of the cubic equation

$$x^3 - \frac{3Q^2}{C}x^2 + \frac{3Q^2}{2C^2}(3Q^2 - C^2)x - G^3 = 0.$$

Problem 10 for $n = 3$ (Recovery of the quantities a, b, c by the harmonic, contraharmonic and quadratic means). The quantities sought are equal to the roots of the cubic equation

$$x^3 - \frac{3Q^2}{C}x^2 + \frac{3Q^2}{2C^2}(3Q^2 - C^2)x - \frac{HQ^2}{2C^2}(3Q^2 - C^2) = 0.$$

Thus, the solution of all inverse problems at n =3 is reduced to finding the roots of the cubic equations, the coefficients of which are expressed algebraically through the mean quantities or with the help of radicals.

The obtained resulats are given in Table 4.6, with the cubic equations for determining the unknown quantities a, b, c by the given triplets of means.

Example 4.2.21. The parallelepiped has the following known dimensions: volume V = 8, area of its surface S = 28, total length of its edges L = 28.

What are the lengths of its edges a, b, c?

In this case the arithmetic mean is $S = 7/8$, and their geometric mean is $G = 2$. The harmonic mean of the edge lengths are found by the formula

$$H = \frac{3abc}{ab + ac + bc},$$

dividing the trebled volume of the parallelepiped by a half of its surface area gives $H = 12/7$.

In this way the problem is reduced to the first problem of the inverse ones. In accordance with the first line of Table 4.6 the edge lengths being sought are equal to the roots of the cubic equation

$$x^3 - 3Ax^2 + 3\frac{G^3}{H}x - G^3 = 0.$$

After substitution of the numerical data and factorization the equation has the form

$$x^3 - 7x^2 + 14x - 8 = (x - 1)(x - 2)(x - 4) = 0.$$

Consequently, the lengths of edges are $a = 1, b = 2, c = 4$.

Table 4.6. Inverse problems at $n = 3$.

N°	Means	Cubic equations for determining the quantities a, b, c
1.	A, G, H	$x^3 - 3Ax^2 + 3\frac{G^3}{H}x - G^3 = 0$
2.	A, G, C	$x^3 - 3Ax^2 + \frac{3}{2}A(3A - C)x - G^3 = 0$
3.	A, G, Q	$x^3 - 3Ax^2 + \frac{3}{2}A(3A^2 - Q^2)x - G^3 = 0$
4.	A, H, C	$x^3 - 3Ax^2 + \frac{3}{2}A(3A - C)x - \frac{1}{2}AH(3A - C) = 0$
5.	A, H, Q	$x^3 - 3Ax^2 + \frac{3}{2}(3A - Q^2)x - \frac{H}{2}(3A^2 - Q^2) = 0$
6.	A, C, Q	Determination of the quantities a, b, c is impossible.
7.	G, H, C	$x^3 - \frac{HC - \sqrt{H(C^2H + 24G^3)}}{2H}x^2 + \frac{3G^3}{H}x - G^3 = 0$
8.	G, H, Q	$x^3 - \sqrt{\frac{6G^3 + 3Q^2H}{H}}x^2 + \frac{3G^3}{H}x - G^3 = 0$
9.	G, C, Q	$x^3 - \frac{3Q^2}{C}x^2 + \frac{3Q^2}{2C^2}(3Q^2 - C^2)x - G^3 = 0$
10.	H, C, Q	$x^3 - \frac{3Q^2}{C}x^2 + \frac{3Q^2}{2C^2}(3Q^2 - C^2)x - \frac{HQ^2}{2C^2}(3Q^2 - C^2) = 0$

In the same way as at $n = 2$, Table 4.6 can be used to derive the invariant relations given in Table 4.4. This is achieved by equating the coefficients of the cubic equations from Table 4.6 at the similar powers x. For example, writing the equality condition of free terms of the first and fourth equations, we get the invariant relation from the second line of Table 4.4.

4.2.7 Weighted means

All the mean estimates considered above have the property of symmetry with respect to measurement values. The permutation of the arguments $x_1, \dots, x_n$ in these means does not change the resulting estimate.

In practice such means can be applied where we have equally accurate measurements. If the measurements are not equally accurate, then they should be included into the formulae using their multiplication by the corresponding coefficients, which takes into account error values of separate measurements. The estimates hereby obtained are referred to as weighted and have no symmetry property.

4.2.7.1 The weighted arithmetic mean

The weighted arithmetic mean of a set of numbers $x_1, \dots, x_n$ with the positive weights $a_1, \dots, a_n$, the sum of which is equal to 1, is determined as

$$\hat{x} = \sum_{i=1}^{n} a_i x_i, \qquad \sum_{i=1}^{n} a_i = 1.$$

If the sum of weights is not equal to 1, then in the formula given the normalizing factor is added:

$$\hat{x} = \frac{1}{\sum_{i=1}^{n} a_i} \sum_{i=1}^{n} a_i x_i. \tag{4.2.26}$$

Usually the weights a_i are taken inversely proportional to the squares of the corresponding standard (quadratic mean) deviations.

For two and three measurements the weighted arithmetic mean is determined by the formulae

$$\hat{x} = a_1 x_1 + a_2 x_2; \quad a_1 + a_2 = 1;$$

$$\hat{x} = a_1 x_1 + a_2 x_2 + a_3 x_3; \quad \sum_{i=1}^{3} a_i = 1.$$

Example 4.2.22 (Average speed). A man was was walking at the speed $v_1 = 4.6$ km/h. for 2 hours and at the speed $v_2 = 5.1$ km/h. for 3 hours. At what speed must he walk to cover the same distance within the same time?

The average speed is determined by finding the weighted arithmetic mean:

$$v = (2v_1 + 3v_2)/(2 + 3) = 24.5/5 = 4.9\,(\text{km/h}).$$

Example 4.2.23. The average blood artery pressure P of an adult person is determined by the weighted arithmetic mean formula

$$P = \frac{1}{3}(P_1 - P_2) + P_2 = \frac{1}{3}P_1 + \frac{2}{3}P_2,$$

where P_1 and P_2 are the systolic and diastolic arterial blood pressure.

For example, if the pressure is 126/75 mm Hg, then the mean pressure will be

$$P = \frac{(126 - 75)}{3} + 75 = 92\,\text{mm Hg}.$$

The standard of the World Health Organization is the pressure of 140/90 mm Hg, this corresponds to the average pressure of 107 mm Hg.

Example 4.2.24. Table 4.7 contains the data related to the monthly mean prices and the sales of potatoes in a supermarket for a yearly quarter.

Table 4.7. Monthly mean prices and amount of sale of potatoes in a supermarket.

	January	February	March
Monthly mean price per kg (roubles)	25	30	35
Realized (kg)	9000	10000	8500

What is the mean monthly price P for potatoes?

To calculate the mean monthly price, let the formula of the weighted arithmetic mean be used:

$$P = \frac{25 \cdot 9000 + 30 \cdot 10000 + 35 \cdot 8500}{9000 + 10000 + 8500} = 29.9 \text{ roubles.}$$

Example 4.2.25. Let the initial data of Table 4.1 be divided into groups, separating the groups of the same age. The data was given in Table 4.8.

Table 4.8. Number of students of the same age per group.

Age X, years	18	19	20	21	22	Sum total
Number of students	2	11	5	1	1	20

Now the mean age of the students of a group can be calculated by formula (4.2.20):

$$\hat{X} = \frac{18 \cdot 2 + 19 \cdot 11 + 20 \cdot 5 + 21 \cdot 1 + 22 \cdot 1}{2 + 11 + 5 + 1 + 1} = \frac{36 + 209 + 100 + 21 + 22}{20}$$
$$= \frac{388}{20} = 19.4.$$

The answer has not changed, but now the algorithm for calculation is more compact.

As a result of grouping, a new index has been obtained, i.e., the relative frequency, which indicates the number of students of X years old.

The maximum frequency corresponds to the age of 19 years. This is the so-called mode of the given set of numbers. Considering the middle of the ordered totality of ages, one obtains the median which is also equal to 19 years.

Example 4.2.26 (Coordinates of the center of gravity). Let the system of n material points $P_1(x_1, y_1)$; $P_2(x_2, y_2)$; $\ldots$, $P_n(x_n, y_n)$ with the masses $m_1, m_2, \ldots, m_n$ be given on the plane Oxy. The products $m_i x_i$ and $m_i y_i$ are called the static mass moments m_i with respect to the axes Oy and Ox.

The coordinates of the centre of gravity of this system of points are determined by the formula

$$X_c = \frac{m_1 x_1 + m_2 x_2 + \cdots + m_n x_n}{m_1 + m_2 + \cdots + m_n} = \frac{\sum_{i=1}^n m_i x_i}{\sum_{i=1}^n m_i},$$

$$Y_c = \frac{m_1 y_1 + m_2 y_2 + \cdots + m_n y_n}{m_1 + m_2 + \cdots + m_n} = \frac{\sum_{i=1}^n m_i y_i}{\sum_{i=1}^n m_i}.$$

Both formulae have the form of the weighted arithmetic mean; they are used in searching for the centers of gravity of various figures and bodies. For example, the center of gravity of a triangle is the point of intersection of its medians. It is the arithmetic mean of apexes of the triangle.

Example 4.2.27 (Simpson's rule). This rule allows the volume of seven geometric bodies to be calculated, namely the volume of a prism, pyramid, frustum pyramid, cylinder, cone, frustum cone, and sphere (Figure 4.16).

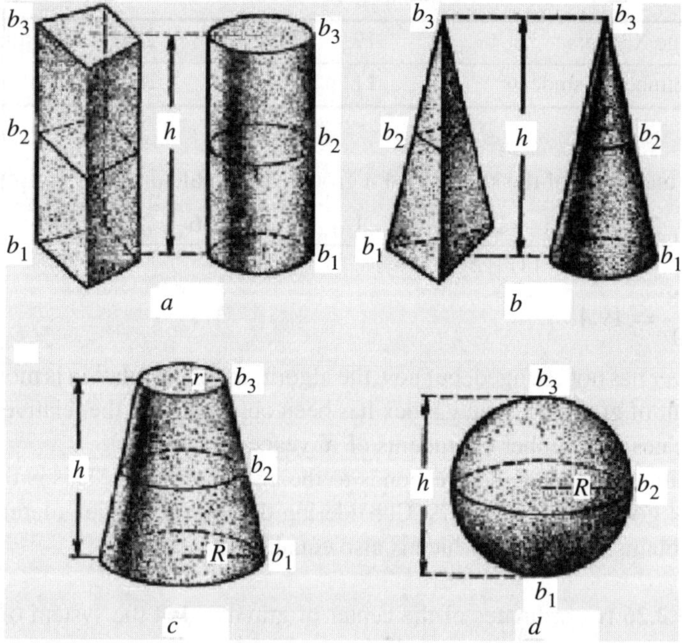

Figure 4.16. View of geometric bodies: prism, pyramid, frustum pyramid, cylinder, cone, frustum cone, and sphere.

Simpson's formula has the form

$$v = \frac{h}{6}(b_1 + 4b_2 + b_3),$$

where h is the height of the body, b_1 is the area of the bottom base, b_2 is the section area of the body at the middle of its height, and b_3 is the area of the upper base.

Simpson's formula can be applied for the approximate calculation of integrals:

$$\int_{x_{i-1}}^{x_i} f(x)dx = \frac{h}{6}\sum_{i=1}^{n}\left[f(x_{i-1}) + 4f\left(\frac{x_{i-1}+x_i}{2}\right) + f(x_i)\right].$$

This gives a more accurate result as compared to the formula of trapezoids in which the simple arithmetic mean is used:

$$\int_a^b f(x)dx = h\left(\frac{f(x_0)+f(x_n)}{2} + \sum_{i=1}^{n-1} f(x_i)\right).$$

A peculiar feature of applying Simpson's formula is the fact that the number of dissections of the integration segments is even. If the number of the dissection segments is odd, then for the first three segments the formula, which uses the parabola, raised to the third power and passing through the first four points, should be applied for the approximation of the integrand:

$$I = \frac{3h}{8}(y_0 + 3y_1 + 3y_2 + y_3). \tag{4.2.27}$$

This is Simpson's rule of "three-eighths".

As another example for applying the weighted mean in higher mathematics we cite the Runge–Kutta method, which is applied for the numerical solution of ordinary differential equations. If an initial value of the function y_0 is given, then the approximated value at the subsequent points is calculated by the iteration formula

$$y_{n+1} = y_n + \frac{1}{6}(k_1 + 2k_2 + 2k_3 + k_4).$$

4.2.7.2 The weighted geometric mean

The weighted geometric mean of a set of the positive numbers $x_1, \dots, x_n$ with the positive weights $a_1, \dots, a_n$ is determined as

$$\hat{x} = \left(\prod_{i=1}^{n} x_i^{a_i}\right)^{1/\sum_{i=1}^{n} a_i} = \exp\left(\frac{1}{\sum_{i=1}^{n} a_i}\sum_{i=1}^{n} a_i \ln x_i\right).$$

For two and three measurements we have that

$$\hat{x} = (x_1^{a_1}x_2^{a_2})^{\frac{1}{a_1+a_2}}, \quad \hat{x} = (x_1^{a_1}x_2^{a_2}x_3^{a_3})^{\frac{1}{a_1+a_2+a_3}}.$$

In practice it is easier to calculate this with the help of logarithms:

$$\ln \hat{x} = \frac{1}{\sum_{i=1}^{n} a_i} \sum_{i=1}^{n} a_i \ln x_i.$$

4.2.7.3 The weighted harmonic mean

The weighted harmonic mean of a set of the numbers $x_1, \ldots, x_n$ with the positive weights $a_1, \ldots, a_n$ is determined by the formula

$$\hat{x} = \frac{\sum_{i=1}^{n} a_i}{\sum_{i=1}^{n} \frac{a_i}{x_i}}. \tag{4.2.28}$$

For two and three measurements we have that

$$\hat{x} = \frac{(a_1 + a_2)x_1x_2}{a_2x_1 + a_1x_2}, \quad \hat{x} = \frac{(a_1 + a_2 + a_3)x_1x_2x_3}{a_3x_1x_2 + a_2x_1x_3 + a_1x_2x_3}.$$

Example 4.2.28. Table 4.9 contains the data for the potatoes sales at three supermarkets in St. Petersburg for the period of one month.

Table 4.9. Data for potatoes sales at three supermarkets in St. Petersburg for one month.

Hypermarket	Cost of the goods sold, thousands of roubles	Price for 1 kg, in roubles
Sennoy	140	25
Kuznechniy	200	35
Vasileostrovskiy	175	28

The task is to determine the mean price for 1 kg of potatoes.

Let us use formula (4.2.28) of the weighted harmonic mean:

$$P = \frac{140 + 200 + 175}{\frac{140}{25} + \frac{200}{35} + \frac{175}{28}} = \frac{51.5}{1.755} = 29.3 \text{ roubles}.$$

Here the cost of the goods sold is applied as the weighted coefficients.

The *weighted contraharmonic mean* of a set of the numbers $x_1, \ldots, x_n$ with the positive weights $a_1, \ldots, a_n$ is determined by the formula

$$\hat{x} = \frac{a_1x_1^2 + a_2x_2^2 + \cdots + a_nx_n^2}{a_1x_1 + a_2x_2 + \cdots + a_nx_n}.$$

For two and three measurements we have that

$$\hat{x} = \frac{a_1 x_1^2 + a_2 x_2^2}{a_1 x_1 + a_2 x_2}, \quad \hat{x} = \frac{a_1 x_1^2 + a_2 x_2^2 + a_3 x_3^2}{a_1 x_1 + a_2 x_2 + a_3 x_3}.$$

The *weighted Lehmer mean* with the positive weights a_i is defined as

$$\hat{x} = \frac{a_1 x_1^p + a_2 x_2^p + \cdots + a_n x_n^p}{a_1 x_1^{p-1} + a_2 x_2^{p-1} + \cdots + a_n x_n^{p-1}}.$$

4.2.7.4 The weighted quadratic mean

The weighted quadratic mean of a set of the numbers $x_1, \dots, x_n$ with the positive weights $a_1 \dots, a_n$ is calculated by the formula

$$\hat{x} = \sqrt{\frac{\sum_{i=1}^n a_i x_i^2}{\sum_{i=1}^n a_i}}.$$

For two and three measurements we have that:

$$\hat{x} = \sqrt{\frac{a_1 x_1^2 + a_2 x_2^2}{a_1 + a_2}}, \quad \hat{x} = \sqrt{\frac{a_1 x_1^2 + a_2 x_2^2 + a_3 x_3^2}{a_1 + a_2 + a_3}}.$$

Example 4.2.29. From 1998–2002 Prof. I. Kon did research in Russia on one anatomical dimension of 8267 men over the age of 18 years. The data he obtained are given in percentage in Table 4.10.

Table 4.10. Results of research by Prof. I. Kon.

Dimension (cm)	6	8	10	12	13	14	15	16	17	18	19	22	24	26	28
Men altogether (%)	1	1	1	6	12	15	20	14	13	9	4	1	1	1	1

Notice that the elements of the second line give the sum of 100 %. When calculating the weighted means they are applied as the weight coefficients. Let us find four weighted mean estimates: arithmetic, geometric, harmonic, and quadratic.

The weighted arithmetic mean is determined by the formula

$$\hat{x} = 0.01(6 + 8 + 10 + 6 \cdot 12 + \cdots + 4 \cdot 19 + 24 + 26 + 28) \approx 15.45.$$

The weighted geometric mean is found by the formula

$$\hat{x} = (6 \cdot 8 \cdot 10 \cdot 12^6 \cdot \ \cdots \ \cdot 19^4 \cdot 22 \cdot 24 \cdot 26 \cdot 28)^{\frac{1}{100}} \approx 15.17.$$

The weighted harmonic mean is found by the formula

$$\hat{x} = \frac{100}{\frac{1}{6} + \frac{1}{8} + \frac{1}{10} + \frac{6}{12} + \cdots + \frac{4}{19} + \frac{1}{24} + \frac{1}{26} + \frac{1}{28}} \approx 14.6.$$

The weighted quadratic mean is found by the formula

$$\hat{x} = 0.1\sqrt{6^2 + 8^2 + 10^2 + 6 \cdot 12^2 + \cdots + 4 \cdot 19^2 + 22^2 + 24^2 + 26^2 + 28^2}$$
$$\approx 15.73.$$

The means obtained satisfy the chain of inequality (4.2.16).

Notice that in the given case the so-called *structural means*, i.e., the *median* (the mean sample value) and *mode* (the most frequently obtained value) coincide and are equal to 15.

4.2.7.5 The weighted means of a number of means

One of the ways to get a new means is to average the means already known. The characteristic example of this is the Heronian mean of two numbers a and b. It is determined by the formula

$$H_1 = \frac{1}{3}(a + \sqrt{ab} + b), \tag{4.2.29}$$

and used in finding the volume of a frustum of a pyramid or cone. Their volume is equal to the product of the height of the frustum and the Heronian mean of the areas of the opposing parallel faces.

Here, in contrast to the Simpson formula (see Example 4.2.27) the area of the average section of the body does not need to be known.

The Heronian mean can be considered to be

- the weighted arithmetic mean of two quantities, arithmetic mean and geometric mean:
$$H_1 = \frac{2}{3} \cdot \frac{a+b}{2} + \frac{1}{3}\sqrt{ab} = \frac{2}{3}A + \frac{1}{3}G;$$
- the arithmetic mean of three quantities, the minimum, maximum, and geometric means:
$$H_1 = \frac{1}{3}(m + M + G).$$

This leads to two versions of its generalization for three and more arguments.

The simple way for getting a new means is to add to the measured quantities one or more means and then to average the expanded set.

For example, the Heronian formula (4.2.29) can be interpreted as the mean of three quantities: numbers a, b, and their geometric mean:

$$H_1(a,b) = A(a,b,G(a,b)).$$

Replacing the quadratic mean by the arithmetic mean in this formula, we arrive at a new mean:

$$H_2 = Q(a,b,G(a,b)) = \sqrt{\frac{a^2+ab+b^2}{3}}.$$

Use of the power mean instead of the quadratic mean gives the formula

$$H_p = \left(\frac{a^p + (ab)^{p/2} + b^p}{3}\right)^{1/p},$$

which at $p = 1,2,3,\ldots$ describes the whole family of means.

Another generalization of the Heronian mean has the form

$$H_\omega(a,b) = \frac{a+b+\omega\sqrt{ab}}{2+\omega},$$

where $\omega \geq 0$ is the arbitrary parameter [206, 241].

In particular, at $\omega = 1$ one gets the usual Heronian mean (4.2.29).

From the materials of this section we can see that there are many evaluation algorithms which can be applied for averaging data or measurement results in different situations. They are given above without any scientific explanation. In the next sections we considered approaches for proving and obtaining these and other evaluation methods in accordance with the classification scheme in Figure 4.4.

4.3 Algorithms of optimal evaluation

4.3.1 Probability approach

The optimal, in the sense of their probability criteria, estimates are the most grounded from a theoretical point of view. However, to use them it is necessary to know the probability properties of the signals being measured and of noise, which in turn usually requires a long and hard effort to collect experimental statistical material.

The *maximum likelihood* estimates, as well as the *Bayesian* and *Markovian* estimates, have received the greatest publicity within the framework of the stochastic approach.

4.3.1.1 Method of maximum likelihood

This method was suggested by K. F. Gauss [177, 178], generalized by R. A. Fisher [166], and has become very widely used. The idea of the method is as follows. Let x_1, x_2, x_3 be the measured values of a certain quantity x and having the distribution density of probabilities $g((x_1, x_2, x_3)|x)$. For independent measurements the joint density can be written as the product of three densities:

$$g\left((x_1, x_2, x_3)\,|x\right) = f\left(x_1\,|x\right) f\left(x_2\,|x\right) f\left(x_3\,|x\right). \tag{4.3.1}$$

The function $g((x_1, x_2, x_3)|x)$ is called the *function of likelihood*. The value x, for which the likelihood function achieves the maximum, is called *the maximum likelihood estimate* $\hat{x}$. Usually to simplify the computation scheme the logarithm rather than the likelihood function itself is maximized:

$$L\left(x_1, x_2, x_3\,|x\right) = \ln g\left(x_1, x_2, x_3\,|x\right). \tag{4.3.2}$$

When the likelihood function is rather "smooth", the determination of the estimate is reduced to the solution of the nonlinear equation of the form

$$\left.\frac{\partial L\left(x_1, x_2, x_3\,|x\right)}{\partial x}\right|_{x=\hat{x}} = 0 \tag{4.3.3}$$

relative to an unknown parameter x.

Example 4.3.1. Let $f(x_i, x) = \frac{1}{\sqrt{2\pi}} \exp[-\frac{(x_i - x)^2}{2}]$ be the normal density of probability distribution with the mean value x and unit dispersion. We wish to determine the estimate for x by a sample consisting of independent observations x_1, x_2, x_3.

In this case the likelihood function will have the form

$$L\left(x_1, x_2, x_3\,|x\right) = 3\ln\left(\frac{1}{\sqrt{2\pi}}\right) - 0.5\sum_{i=1}^{3}\left(x_i - x\right)^2,$$

since for independent observations $g((x_1, x_2, x_3)|x) = f(x_1|x) f(x_2|x) f(x_3|x)$. To get the estimate of the parameter x it is necessary to solve equation (4.3.3):

$$\left.\frac{\partial L\left(x_1, x_2, x_3\,|x\right)}{\partial x}\right|_{x=\hat{x}} = (x_1 - \hat{x}) + (x_2 - \hat{x}) + (x_3 - \hat{x}) = 0,$$

hence

$$\hat{x} = \frac{1}{3}(x_1 + x_2 + x_3).$$

Thus, the evaluation according to the maximum likelihood method at the heuristic assumption with respect to the normal law of measurement error distribution and independence of the measurement values x_1, x_2, x_3 coincides with estimate (4.2.4), obtained according to the least-squares method.

Provided that the probability density function of independent measurement values is described by the Laplace distribution law

$$f\,(x_i - x) = c_1 \exp[c_0\,|x_i - x|],$$

then the maximization of the likelihood function

$$L\,(x_1, x_2, x_3\,|x\,) = 3 \ln c_1 - c_0\,(|x_1 - x| + |x_2 - x| + |x_3 - x|) \tag{4.3.4}$$

is reduced to the minimization of criterion (4.2.8):

$$J_3 = |x_1 - \hat{x}| + |x_2 - \hat{x}| + |x_3 - \hat{x}| \rightarrow \min_{\hat{x}},$$

i.e., to the search for a *mean module* estimate.

If the values of measurement are correlated between themselves and described by the normal law of distribution with the known covariation matrix N, then the maximization of the likelihood function leads to the problem of minimizing the criterion

$$J = [(x_1 - \hat{x})\,(x_2 - \hat{x})\,(x_3 - \hat{x})]\,N^{-1}\,[(x_1 - \hat{x})\,(x_2 - \hat{x})\,(x_3 - \hat{x})]^T \rightarrow \min_{\hat{x}}. \tag{4.3.5}$$

At the same time, the optimal estimate, called the *Markovian* estimate, is calculated by the formula

$$\hat{x} = \left\{[1\;1\;1]\,N^{-1}\begin{bmatrix}1\\1\\1\end{bmatrix}\right\}^{-1}\,[1\;1\;1]\,N^{-1}\begin{bmatrix}x_1\\x_2\\x_3\end{bmatrix} \tag{4.3.6}$$

and coincides, as will be shown further on, with the estimate obtained using the generalized least-squares method.

4.3.1.2 Bayesian estimate

This estimate is obtained as a result of minimization of the criterion of a mean risk:

$$J = \int c\,(x, \hat{x})\,g\,(x\,|x_1, x_2, x_3)\mathrm{d}x \rightarrow \min_{\hat{x}}. \tag{4.3.7}$$

The required condition of the minimum has the form

$$\frac{\partial}{\partial x}\int c\,(x, \hat{x})g\,(x\,|x_1, x_2, x_3)\,\mathrm{d}x\,|_{x=\hat{x}} = 0. \tag{4.3.8}$$

In the formulae applied, $c(x, \hat{x})$ is the function of losses, and $g(x|x_1, x_2, x_3)$ is the a posteriori probability density of the parameter x under the given measurement results x_1, x_2, x_3. The function $g(x|x_1, x_2, x_3)$ is connected with the probability density in the maximum likelyhood method $g(x_1, x_2, x_3|x)$ by the Bayesian formula

$$g(x|x_1, x_2, x_3) = \frac{g(x_1, x_2, x_3|x)\, g(x)}{g(x_1, x_2, x_3)}, \tag{4.3.9}$$

where $g(x)$ is the a priori probability density of the parameter x.

By assigning the loss function heuristically and presenting a priori information about the parameter in the form of the function $g(x)$, equation (4.3.8) is obtained.

For example, having assigned $c(\hat{x}, x) = (\hat{x} - x)^2$, it is possible to show that the optimal estimate $\hat{x} = \int x g(x|x_1, x_2, x_3) \mathrm{d}x$ represents a conventional mathematical expectation of the quantity x. If a priori information about the parameter x is absent and $c(x, \hat{x}) = \text{const}$, then one usually chooses the value $\hat{x}$, coinciding with the value x maximizing $g(x|x_1, x_2, x_3)$, i.e., the estimate x being the maximum likelihood estimate.

Thus, assigning the loss function, it is possible to get one or another of the probability criteria, the minimization of which by the parameter will give the estimate of the unknown quantity.

4.3.2 Deterministic approach

4.3.2.1 Idea of the deterministic approach

Let us consider the second approach, which does not use information about the probability laws of error distribution in an explicit form. Accordance to this approach the problem of evaluation is interpreted as a deterministic problem of approximation. It consists of the fact that point $\hat{x}$ (an estimate) has to be found, which is the nearest to the three given points x_1, x_2, x_3 of the measurement values (Figure 4.17). For a strict formulation of the problem it is necessary to introduce a remoteness measure of the estimate from the values measured $J = F(\hat{x}, x_1, x_2, x_3)$, which will serve as the criterion to be minimized. It is natural to require that this measure not be negative and could be transformed into zero only when all points coincide.

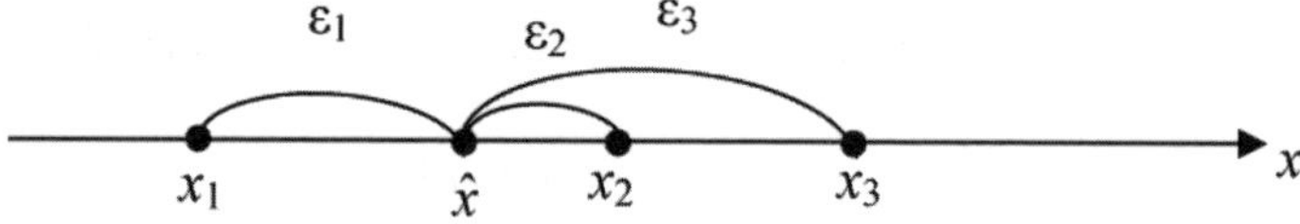

Figure 4.17. Finding an estimate nearest to three measured values.

In the simplest case it is possible to minimize the total distance $|\varepsilon_1| + |\varepsilon_2| + |\varepsilon_3|$ from the values x_1, x_2, x_3 to point $\hat{x}$ (Figure 4.17).

This results in the following problem statement with respect to optimization. For the obtained measurement values x_1, x_2, x_3 it is necessary to find the number $\hat{x}$ (the estimate), at which the function $J = F(\hat{x}, x_1, x_2, x_3)$ is minimal.

If the function F can be differentiated, the solution of the problem can be obtained from the condition

$$\frac{\partial}{\partial \hat{x}} F(\hat{x}, x_1, x_2, x_3) = 0. \tag{4.3.10}$$

The central moment in this approach is the choice of the kind of criterion J, realized for heuristic reasons by taking into account the character of the physical or technical problem to be solved. The mostly widely-used criteria include the quadratic, module, Chebyshev, and others. It should be noted that situations are possible where different criteria lead to the same estimate. This can be explained by the fact that different functions F can have the extremum at a common point.

Below, some concrete criteria will be treated.

4.3.2.2 Application of classical criteria

The quadratic criterion

According to this criterion the sum of the squares of the distances from the point $\hat{x}$ to the points x_1, x_2, x_3, i.e., criterion (4.2.5), is minimized:

$$J = (x_1 - \hat{x})^2 + (x_2 - \hat{x})^2 + (x_3 - \hat{x})^2.$$

By differentiating and equating the derivative to zero, the following equation is obtained:

$$2(x_1 - \hat{x}) + 2(x_2 - \hat{x}) + 2(x_3 - \hat{x}) = 0,$$

from which

$$\hat{x} = \frac{1}{3}(x_1 + x_2 + x_3).$$

Thus, the mean arithmetic estimate is optimal by the quadratic criterion. As shown above, this estimate is the best one which can be obtained with the method of maximum likelihood under equally accurate measurements with Gauss errors. These facts in combination with the computational simplicity of evaluation is the reason of its wide use in practice.

The weighted quadratic criterion

If the measurements are not equally accurate (for example, when performed with different sensors) and their dispersions $\sigma_1^2, \sigma_2^2, \sigma_3^2$ are known, then it is possible to use the weighted quadratic criterion:

$$J = \frac{1}{\sigma_1^2}(x_1 - \hat{x})^2 + \frac{1}{\sigma_2^2}(x_2 - \hat{x})^2 + \frac{1}{\sigma_3^2}(x_3 - \hat{x})^2. \tag{4.3.11}$$

Differentiating the derivatie and and equating it to zero leads to the weighted mean arithmetic estimate:

$$\hat{x} = \left(\frac{1}{\sigma_1^2} + \frac{1}{\sigma_2^2} + \frac{1}{\sigma_3^2}\right)^{-1} \left(\frac{x_1}{\sigma_1^2} + \frac{x_2}{\sigma_2^2} + \frac{x_3}{\sigma_3^2}\right), \tag{4.3.12}$$

which coincides with the previous one at $\sigma_1 = \sigma_2 = \sigma_3$.

The generalizing quadratic criterion

If the measurements are not equally accurate and dependent, and the covariation matrix N of the measurement errors,

$$N = \begin{bmatrix} \sigma_1^2 & k_{12} & k_{13} \\ k_{21} & \sigma_2^2 & k_{23} \\ k_{31} & k_{32} & \sigma_3^2 \end{bmatrix},$$

the elements of which characterize their mutual correlation, is known, then it is possible to apply the criterion of Markovian evaluation (4.3.5), which corresponds to the generalized method of least squares:

$$J = [x_1 - \hat{x}, x_2 - \hat{x}, x_3 - \hat{x}]\, N^{-1} \begin{bmatrix} x_1 - \hat{x} \\ x_2 - \hat{x} \\ x_3 - \hat{x} \end{bmatrix}.$$

This criterion leads to estimate (4.3.6), generalizing both previous estimates:

$$\hat{x} = \left([1\ 1\ 1]\, N^{-1} \begin{bmatrix} 1 \\ 1 \\ 1 \end{bmatrix} \right)^{-1} [1\ 1\ 1]\, N^{-1} \begin{bmatrix} x_1 \\ x_2 \\ x_3 \end{bmatrix} = \frac{1}{d}\left(a_1 x_1 + a_2 x_2 + a_3 x_3\right).$$

The normalizing coefficient d in the last formula is equal to the sum of the elements of the inverse matrix N. The second multiplier represents a linear combination of the measurements, and its coefficients are formed by summing up the lines of this matrix.

Example 4.3.2. Let us consider the case when the first value of measurements is independent with respect to $k_{12} = k_{13} = 0$ and other two have the same error:

$$d = 2\sigma_1^2 + \sigma_2^2 + k_{23}, \quad a_1 = \sigma_2^2 + k_{23}, \quad a_2 = a_3 = \sigma_1^2,$$

i.e., the estimate assumes the form

$$\hat{x} = \frac{\left(\sigma_2^2 + k_{23}\right) x_1 + \sigma_1^2 \left(x_2 + x_3\right)}{2\sigma_1^2 + \sigma_2^2 + k_{23}}.$$

If all three measurements are assumed to be equally accurate, $\sigma_1 = \sigma_2 = \sigma_3 = \sigma$, then the formula will become more simple:

$$\hat{x} = \frac{k_{23} x_1 + \sigma^2 \left(x_1 + x_2 + x_3\right)}{3\sigma^2 + k_{23}}.$$

At $k_{23} = 0$ this estimate becomes the usual arithmetic mean.

The module criterion

This criterion has the form of equation (4.2.8):

$$J = |x_1 - \hat{x}| + |x_2 - \hat{x}| + |x_3 - \hat{x}|,$$

i.e., it is equal to the sum of the distances $\varepsilon_1, \varepsilon_2, \varepsilon_3$ from the measurement values of the estimate (Figure 4.17). In this case the function F cannot be differentiated; therefore its minimum has to be found by some other method, rather than the one used before.

This can be achieved most quickly by reasoning as follows. The function J is not negative; moreover, it is clear that $J \geq l$, where l is the length of the minimal segment, containing points, i.e., values of measurements (Figure 4.3). The diagram in Figure 4.3 shows the dependence of the criterion J on $\hat{x}$. The quantity $J = l$ is achieved if $\hat{x}$ and the middle of three measurement values coincide, i.e., if $\hat{x} = \text{med}(x_1, x_2, x_3)$ is taken. Such an estimate is called the sample median (terms such as "majority function" or "function of voting" are also in use) and has many interesting properties.

In particular, it is weakly sensitive to possible variations of the error distribution laws, i.e., it is robust.

As mentioned above, this is optimal for the errors distributed according to Laplace's law. Robust means that such an estimate will give values close to the optimal ones and under other distribution laws.

It is important that this estimate can be used at errors (crude errors, errors of a high level, for example, at a single failure of sensors), where one of the values measured strongly differs from two others. This is called the reliability or steadiness of the estimate. It should be noted that the estimates indicated above have no such property.

Weighted module criterion

This criterion is constructed by analogy with the weighted quadratic mean and has the form

$$J = \frac{1}{\sigma_1}|x_1 - \hat{x}| + \frac{1}{\sigma_2}|x_2 - \hat{x}| + \frac{1}{\sigma_3}|x_3 - \hat{x}|. \tag{4.3.13}$$

Its minimization leads to the estimate coinciding with one of the measurement values (either with the median value or with the most accurate one).

Minimax (Chebyshev) criterion

When using Chebyshev criterion (4.2.7)

$$J = \max\left(|x_1 - \hat{x}|, |x_2 - \hat{x}|, |x_3 - \hat{x}|\right) \to \min_{\hat{x}},$$

the maximum deviation from $\varepsilon_1, \varepsilon_2, \varepsilon_3$ is minimized. The minimum of this criterion is represented by the estimate located just in the middle of the segment (Figure 4.18).

Here, the middle point is ignored ("disregarded") like in the previous evaluation, where the extreme points were disregarded.

The square mean is described by the formula

$$\hat{x} = \sqrt{\frac{x_1^2 + x_2^2 + x_3^2}{3}}. \tag{4.3.14}$$

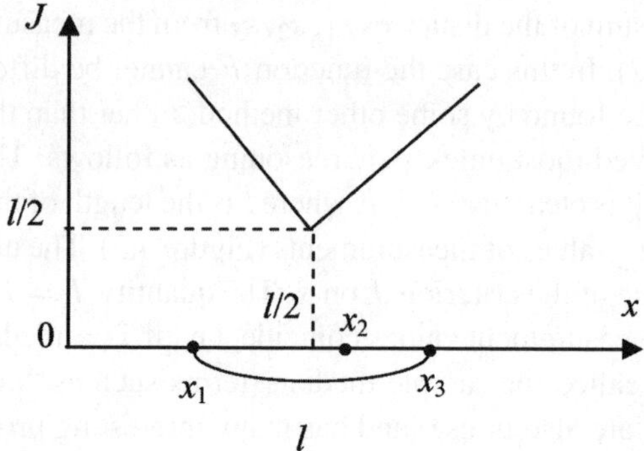

Figure 4.18. Minimization of deviations according to the Chebyshev criterion.

It minimizes the criterion

$$J = \left(x_1^2 - \hat{x}^2\right)^2 + \left(x_2^2 - \hat{x}^2\right)^2 + \left(x_3^2 - \hat{x}^2\right)^2. \tag{4.3.15}$$

It is possible to make sure it will perform the differentiation and equating the derivative to zero.

The *geometric mean* of three positive measurement values has the form

$$\hat{x} = \sqrt[3]{x_1 x_2 x_3}. \tag{4.3.16}$$

It minimizes the sum of logarithmic deviations squares:

$$J = (\ln x_1 - \ln \hat{x})^2 + (\ln x_2 - \ln \hat{x})^2 + (\ln x_3 - \ln \hat{x})^2. \tag{4.3.17}$$

The *harmonic mean* is obtained from the arithmetic mean for inverse quantities:

$$\frac{1}{\hat{x}} = \frac{1}{3}\left(\frac{1}{x_1} + \frac{1}{x_2} + \frac{1}{x_3}\right).$$

This leads to the formula

$$\hat{x} = \frac{3}{\dfrac{1}{x_1} + \dfrac{1}{x_2} + \dfrac{1}{x_3}}, \tag{4.3.18}$$

which minimizes the sum of the squares of the inverse quantities deviations:

$$J = (x_1^{-1} - \hat{x}^{-1})^2 + (x_2^{-1} - \hat{x}^{-1})^2 + (x_3^{-1} - \hat{x}^{-1})^2. \tag{4.3.19}$$

This criterion can be rewritten in the equivalent form

$$J = \left(1 - \frac{\hat{x}}{x_1}\right)^2 + \left(1 - \frac{\hat{x}}{x_2}\right)^2 + \left(1 - \frac{\hat{x}}{x_3}\right)^2. \tag{4.3.20}$$

Power means

A separate class of estimates is assigned by the formula

$$\hat{x} = \sqrt[k]{\frac{1}{3}\left(x_1^k + x_2^k + x_3^k\right)}, \qquad (4.3.21)$$

where k is any number, $x_i \geq 0$.

They can be obtained when minimizing the criterion:

$$J = (x_1^k - \hat{x}^k)^2 + (x_2^k - \hat{x}^k)^2 + (x_3^k - \hat{x}^k)^2. \qquad (4.3.22)$$

Some of the estimates given above also belong to this class; in particular, the quadratic mean (4.3.14) is obtained at $k = 2$, arithmetic mean (4.1.4) is obtained at $k = 1$, and at $k = -1$ the harmonic mean (4.3.18) is obtained. At $k = -\infty$ and $k = \infty$ two new formulae are obtained:

$$\hat{x} = \min(x_1, x_2, x_3) \qquad (4.3.23)$$

and

$$\hat{x} = \max(x_1, x_2, x_3), \qquad (4.3.24)$$

at which the smallest or largest value is taken from three measurement values.

Assuming that $k = 3$, we come to the cubic mean of the form

$$\bar{x} = \sqrt[3]{\frac{\sum x_i^3}{n}}. \qquad (4.3.25)$$

The cubic mean is applied when because of the problem conditions it is necessary to keep the sum of the cube quantities unchanged when replacing them with a mean quantity. This estimate is applied, for example, to determine the mean diameters of tree trunks, stocks of timber in warehouses, and on woodlots, for calculating the mean height of one-type vessels while preserving the general volume.

Weighted power estimates

Another still wider class of estimates can be obtained moving to the weighted criterion of the form

$$J = \frac{1}{\sigma_1^2}\left(x_1^k - \hat{x}^k\right)^2 + \frac{1}{\sigma_2^2}\left(x_2^k - \hat{x}^k\right)^2 + \frac{1}{\sigma_3^2}\left(x_3^k - \hat{x}^k\right)^2, \qquad (4.3.26)$$

which is constructed similarly to criterion (4.3.22).

It leads to the weighted power mean

$$\hat{x} = \sqrt[k]{\left(\frac{1}{\sigma_1^2} + \frac{1}{\sigma_2^2} + \frac{1}{\sigma_3^2}\right)^{-1}\left(\frac{x_1^k}{\sigma_1^2} + \frac{x_2^k}{\sigma_2^2} + \frac{x_3^k}{\sigma_3^2}\right)}, \qquad (4.3.27)$$

which generalizes the estimates obtained earlier (4.2.4), (4.3.12), (4.3.14), (4.3.21), (4.3.23), (4.3.24).

For example, at $k = 3$ the weighted cubic mean is obtained:

$$\bar{X} = \sqrt[3]{\frac{\sum x_i^3 a_i}{\sum a_i}}.$$

Further generalizations of criterion (4.3.26) are possible.

4.3.2.3 Use of compound criteria

A large group of estimates can be obtained using so-called compound criteria, which represent definite combinations of the simple criteria considered earlier. This allows the estimates with assigned properties with respect to accuracy and reliability to be obtained.

For example, we can consider the combinations of the quadratic $J_Q = J_1$ (4.2.5) and module $J_M = J_3$ (4.2.8) criteria. They relate to the class of criteria which symmetrically depend on the differences $x_1 - \hat{x}, x_2 - \hat{x}, x_2 - \hat{x}$:

$$J = \rho(x_1 - \hat{x}) + \rho(x_2 - \hat{x}) + \rho(x_3 - \hat{x}), \tag{4.3.28}$$

where ρ is the so-called loss function (at the same time being the function of contrast, weight, or penalty).

The form of the loss function for the quadratic criterion is shown in Figure 4.19a, and for the module criterion it is shown in Figure 4.19b.

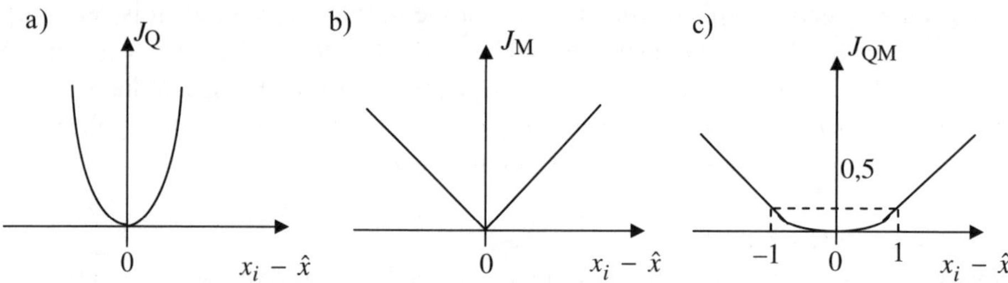

Figure 4.19. Examples of compound criteria functions.

Figure 4.19c illustrates the function of the compound (quadratic-module) criterion J_{QM}. The lower part of the diagram shown in the figure (at $|x| < 1$) is formed by a parabola, and the upper part is formed by the straight line segments:

$$\rho(x_i - \hat{x}) = \begin{cases} 0.5(x_i - \hat{x})^2 & \text{if } |x_i - \hat{x}| \le a \\ a\,|x_i - \hat{x}| - 0,5a^2 & \text{if } |x_i - \hat{x}| > a. \end{cases} \tag{4.3.29}$$

At a loss function of this kind, one assigns to the small deviations, which do not exceed the quantity a, the same weight as in the case of the quadratic criterion, and deviations larger in value are taken into account with the less weight. This leads to an estimate equal to the arithmetic mean of the measurement values with a small deviation and to the sample median when the deviation is large.

In Figure 4.20 the diagrams of the sensitivity function of the three criteria under consideration which are obtained by differentiating the loss function $u = \frac{\partial \rho}{\partial (x_i - \hat{x})}$ are shown.

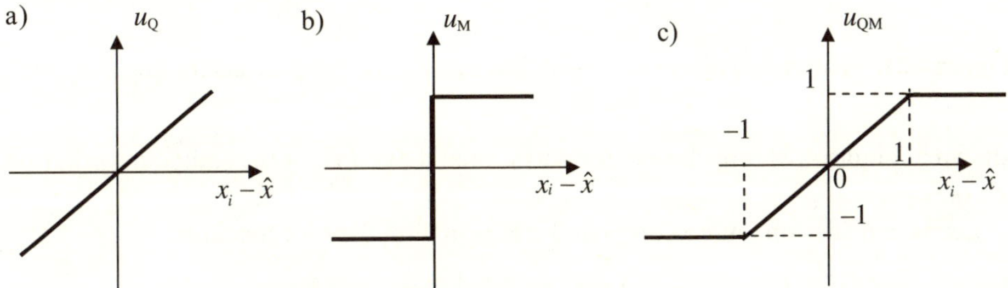

Figure 4.20. Sensitivity functions for the compound criterion.

Note that the analytic criterion is known which by its character is close to the compound criterion described above. It is used in a hybrid computer of the "Extrema" type and has form (4.3.27) with the loss function:

$$\rho\,(x_i - \hat{x}) = \sqrt{(x_i - \hat{x})^2 + a^2} - a, \tag{4.3.30}$$

where a is a certain constant.

At small (as compared to a) values of the difference $|x_i - \hat{x}|$ the function ρ (4.3.30), in the same way as function (4.3.29), is near to a parabola:

$$\rho = a\left(\sqrt{1 + \left(\frac{x_i - \hat{x}}{a}\right)^2} - a\right) \approx \frac{1}{2a}(x_i - \hat{x})^2. \tag{4.3.31}$$

At small (as compared to a) values of the difference $|x_i - \hat{x}|$ the function ρ (4.3.30), in the same way as function (4.3.29), has the linear character

$$\rho \approx |x_i - \hat{x}| - a.$$

Correspondingly, the derivative of the loss function is almost linear at $|x_i - \hat{x}| \ll a$ and modulo close to 1 at $|x_i - \hat{x}| \gg a$.

The diagrams of function (4.3.31) and sensitivity function

$$u = \frac{\partial \rho}{\partial \hat{x}} = \frac{x_i - \hat{x}}{\sqrt{(x_i - \hat{x})^2 + a^2}},$$

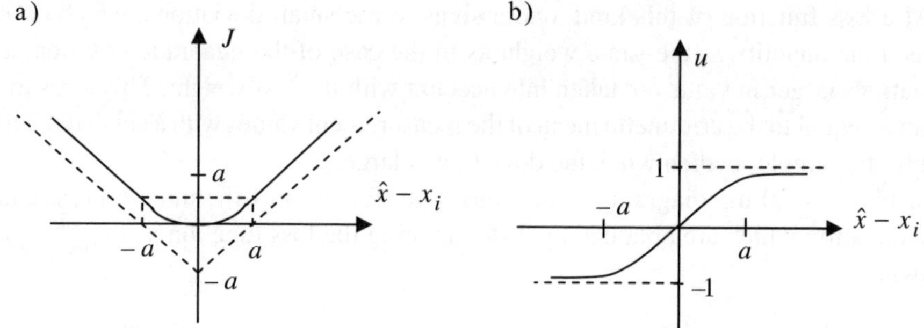

Figure 4.21. View of criterion (4.3.27) with loss function (4.3.29) (a) and its sensitivity (b).

shown in Figure 4.21, are close in form to similar diagrams for the compound criterion J_{QM}.

Let us consider one more compound criterion with the loss function:

$$\rho(x_i - \hat{x}) = \begin{cases} 0.5\,(x_i - \hat{x})^2 & \text{at } |x_i - \hat{x}| \le a \\ 0.5a^2 & \text{at } |x_i - \hat{x}| > a. \end{cases} \tag{4.3.32}$$

The answer to this is found by averaging with similar weights of measurement results satisfying the condition $|x_i - \hat{x}| \le a$. Measurements exceeding this threshold are replaced by the constant $0.5a^2$. Diagrams of the loss function and its derivative for this case are shown in Figure 4.22.

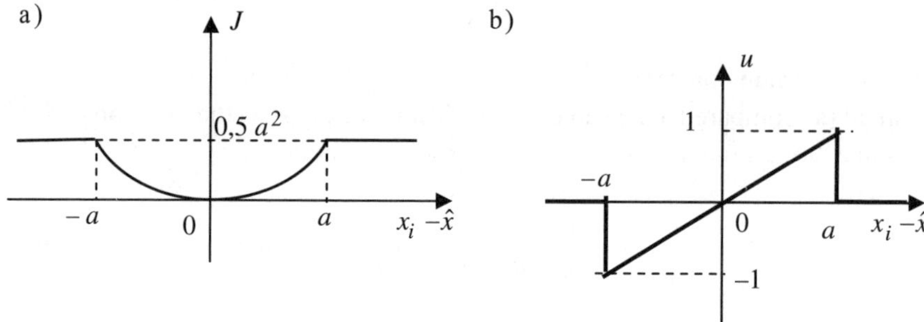

Figure 4.22. Loss function (a) and its derivative (b) for criterion (4.3.32).

Above some examples of constructing compound criteria are given. It is clear that the number of possible examples is unlimited, because the number of possible versions of the compound criteria construction is infinite.

4.3.2.4 Use of combined and other criteria

Combined criteria differ from compound criteria by the use of different loss functions with respect to different variables. The following three criteria are examples of this:

$$J = (x_1 - \hat{x})^2 + (x_2 - \hat{x})^2 + |x_3 - \hat{x}|, \tag{4.3.33}$$

$$J = |x_1 - \hat{x}| + \left(x_2^2 - \hat{x}^2\right)^2 + (x_3 - \hat{x})^2, \tag{4.3.34}$$

$$J = |x_1 - \hat{x}| + |x_2 - \hat{x}| + |x_3 - \hat{x}| + (x_1 - \hat{x})^2 + (x_2 - \hat{x})^2 + (x_3 - \hat{x})^2, \tag{4.3.35}$$

as well as many others.

Every criterion of this type uniquely defines an optimal estimate, no explicit analytical expressions existing for it. In such cases in order to obtain an estimate, the numerical methods of finding the extrimum of the criterion minimized are applied.

Other criteria
At the beginning of this section the general criterion formula is given as

$$J = F(\hat{x}, x_1, x_2, x_3), \tag{4.3.36}$$

and then its special cases are considered, using several examples. When classifying them it is possible to indicate a number of typical situations (for simplicity's sake let us restrict ourselves to symmetrical criteria).

First of all, let the criteria of type (4.3.27) be noted, which depend on the differences $(x_i - \hat{x})$:

$$J = \rho(x_1 - \hat{x}) + \rho(x_2 - \hat{x}) + \rho(x_3 - \hat{x}).$$

Examples of the loss functions ρ are given above (Figures 4.19–4.22). A more general case will be obtained using the expression of the form $g(x_i) - g(\hat{x})$ as an argument of the loss function, rather than the difference $(x_i - \hat{x})$:

$$J = \rho\left[g(x_1) - g(\hat{x})\right] + \rho\left[g(x_2) - g(\hat{x})\right] + \rho\left[g(x_3) - g(\hat{x})\right], \tag{4.3.37}$$

where ρ is a monotone function of any kind.

For example, choosing $g(x) = x^k$ and a quadratic loss function, one obtains criterion (4.3.22).

Assuming that under the same conditions $g(x) = e^x$ one obtains

$$J = (e^{x_1} - e^{\hat{x}})^2 + (e^{x_2} - e^{\hat{x}})^2 + (e^{x_3} - e^{\hat{x}})^2. \tag{4.3.38}$$

Minimization of this criterion leads to the estimate

$$\hat{x} = \ln \frac{e^{x_1} + e^{x_2} + e^{x_3}}{3}. \tag{4.3.39}$$

A still more general kind of the criterion is

$$J = \rho(x_1, \hat{x}) + \rho(x_2, \hat{x}) + \rho(x_3, \hat{x}). \tag{4.3.40}$$

Here the nonnegative function ρ characterizes the measure of closeness of two points, and its minimum has to be achieved under using the same arguments.

An example of this criterion is

$$J = \ln^2 \frac{\hat{x}+a}{x_1+a} + \ln^2 \frac{\hat{x}+a}{x_2+a} + \ln^2 \frac{\hat{x}+a}{x_3+a}, \tag{4.3.41}$$

to which the following estimate corresponds:

$$\hat{x} = \sqrt[3]{(x_1+a)+(x_2+a)+(x_3+a)} - a. \tag{4.3.42}$$

It would be possible to add to the list of algorithms obtained by the means of the optimization of the deterministic criteria. It is clear that the general number of such algorithms is infinitely large, since the amount of possible criteria is practically infinite, and each one of them generates its own estimate.

4.4 Heuristic methods for obtaining estimates

4.4.1 Principles of heuristic evaluation

The majority of the methods described above, intended for obtaining optimal estimates, are complicated in their realization. Many of them require a priori information of a statistical character.

At the same time, their ability to be optimal is very relative, since the choice of one or another criterion (probabilistic or deterministic) is to a great extent arbitrary. The estimates optimal with respect to one criterion can be far from the optimum in the sense another one. Therefore, in practice the heuristic approach is often used, whereby the evaluation algorithm is first constructed for one or another reason, and then its properties are studied, and a check of its operability is fulfilled.

In spite of the well known pragmatism of such an approach, one cannot deny its definite logic. The point is that elements of heuristics are inevitably present in any method of evaluation. In particular, in the deterministic approach they appeark when choosing an optimization criterion, and in the probabilistic approach they manifest themselves when choosing a concrete law of measurement error distribution and principle of optimization (likelihood ratio, mean risk, and others). From this point of view the heuristic choice of a concrete evaluation algorithm simply means the transfer of the heuristic from the upper levels to a lower level.

In substantiating such an approach it is possible to add that for practical purposes it is unlikely that the reasoning and acceptability, e.g., of the sample median algorithm (extreme measurements are ignored and the mean is accepted as the estimate), significantly increase due to the additional information that at the same time the mean module criterion is minimized, and the estimate obtained is optimal if the measurements are equally accurate, values measured are independent and their density of probabilities distribution is subordinated to Laplace's law. The decisive arguments are, most probably, the simplicity of the algorithm and the fact that it keeps its operability in the presence of at least single mistakes (failures, malfunctions).

Moreover, the present stage of development in mathematics allows us to indicate the deterministic or even probabilistic criterion for any heuristic algorithm which is optimized by. In other words, the heuristics are raised by one or two "floors".

Before moving on to a statement of particular heuristic algorithms for getting estimates, let us consider the general restrictions which they must meet.

4.4.1.1 Restrictions on heuristic estimates

Let, as before, x_1, x_2, x_3 be the experimental measurement values, according to which it is necessary to construct the estimate $\hat{x}$ of an unknown scalar quantity x. Let us assume that a priori information is absent. Then designating the evaluation function by f, it is possible to write

$$\hat{x} = f(x_1, x_2, x_3).$$

Then the evaluation function has to meet some general restrictions. For example, it is logically correct to require the affiliation of the estimate $\hat{x}$ with the same segment l of the axis x as the measurement values x_1, x_2, x_3:

$$\min(x_1, x_2, x_3) \leq \hat{x} \leq \max(x_1, x_2, x_3). \tag{4.4.1}$$

Hence, as a consequence, the estimate has to coincide with all three measurement values if they are equal:

$$f(x_1, x_1, x_1) = x_1. \tag{4.4.2}$$

Inequality (4.4.1), from the geometric point of view, can be illustrated with the help of Figure 4.23, in which the position of the measurement value points is indicated along the x-axis , and along the y-axis that of the estimate $\hat{x} = f(x_1, x_2, x_3)$.

In this and the next figures it is shown that when constructing diagrams it is assumed that the measurements x_1, x_2 marked on the x-axis by small vertical lines are fixed, and at the same time the measurement of x_3 passes through all the possible values in the range $0 \leq x_3 < \infty$. In the figures the domain not satisfying inequality (4.4.1) is shaded with lines. The domain boundaries are determined by the functions

$$\begin{aligned} f_1 &: \quad \hat{x} = \max(x_1, x_2, x_3), \\ f_2 &: \quad \hat{x} = \min(x_1, x_2, x_3), \end{aligned}$$

representing two extreme cases of estimates.

In th diagrams all the possible evaluation functions have to be located in the non-shaded domain. As an example, in Figure 4.23 the diagram is shown which corresponds to the mean arithmetic estimate:

$$f_3 : \hat{x} = \frac{1}{3}(x_1 + x_2 + x_3).$$

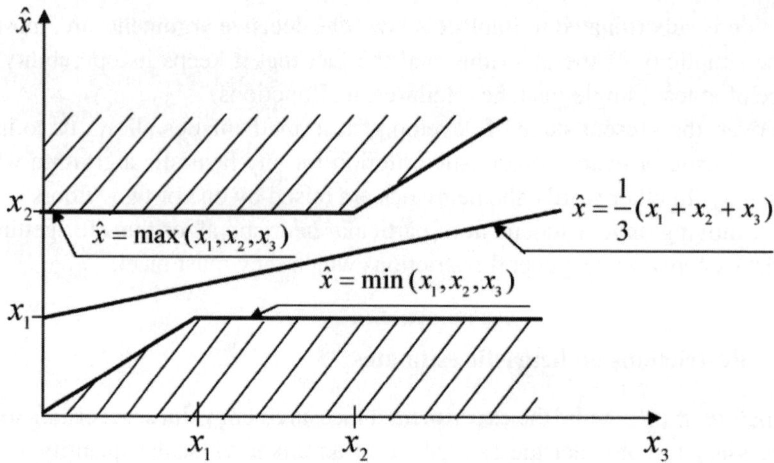

Figure 4.23. Diagram of the arithmetic mean of three measurements.

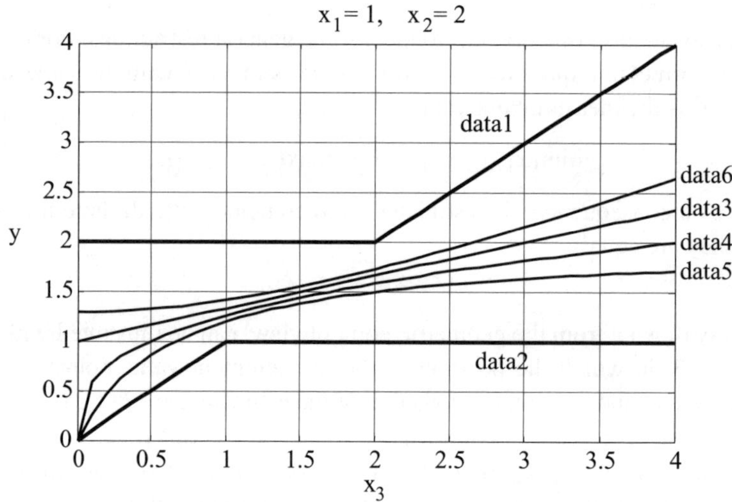

Figure 4.24. Six classical means. Here: greater than three data1; less than three data2; arithmetic mean data3; geometric mean data4; harmonic mean data5; quadratic mean data6.

The diagrams of functions f of the six classical means for three values of measurements are shown in Figure 4.24.

Similar diagrams for two measurements are shown above (see Figure 4.15).

4.4.1.2 Mean quantities according to Cauchy and Kolmogorov

Condition (4.4.1) was first introduced the first half of the 19th century by the French mathematician A. L. Cauchy. He gave the following definition for mean quantity, also known as a "weak" definition.

Definition 1. The mean quantity of real numbers $x_1, x_2, \dots, x_n$ is any function $f(x_1, x_2, \dots, x_n)$ such that at any possible values of arguments the value of this function is not less than the minimal number of the numbers $x_1, x_2, \dots, x_n$ and not greater than the maximal number of these numbers:

$$\min(x_1, \dots, x_n) \le f(x_1, \dots, x_n) \le \max(x_1, \dots, x_n). \tag{4.4.3}$$

A function of this form is called the Cauchy mean. Let us notice that the mean of similar numbers is equal to their common value.

All the kinds of means considered above are Cauchy means. Much stronger requirements are made with respect to the function $f(x_1, x_2, \dots, x_n)$ in a "strong" definition of means, which was done by the Russian mathematician A. N. Kolmogorov.

Definition 2. The continuous real function $f(x_1, \dots, x_n)$ of n nonnegative variables is called a mean if for any $x_1, \dots, x_n, \lambda \ge 0$ the conditions given below are met:

(1) $\min\{x_1, \dots, x_n\} \le f(x_1, \dots, x_n) \le \max\{x_1, \dots, x_n$, i.e., the function f "averages" any set of n nonnegative numbers *(Cauchy's averaging property)*;

(2) $x_1 \le y_1, \dots, x_n \le y_n \Rightarrow f(x_1, \dots, x_n) \le f(y_1, \dots, y_n)$, i.e., a greater value of the function f corresponds to a "greater" set of arguments *(the property of increasing)*;

(3) At any permutation of the numbers $x_1, \dots, x_n$ the value of the function f does not change *(the property of symmetry)*;

(4) $f(\lambda x_1, \dots, \lambda x_n) = \lambda f(x_1, \dots, x_n)$ *(the property of uniformity)*.

In 1930 A. N. Kolmogorov proved [270] that the function $f(x_1, x_2, \dots, x_n)$, meeting these conditions, has the form

$$f(x_1, \dots, x_n) = \varphi^{-1}\left(\frac{\varphi(x_1) + \dots + \varphi(x_n)}{n}\right), \tag{4.4.4}$$

where φ is the continuous strongly monotone function, and φ^{-1} is the inverse function with respect to φ.

The function of such a form is called the Kolmogorov's mean. This function is continuous and monotone for each argument x_i.

Two properties of means according to Kolmogorov are:

- as previously, the mean from similar numbers is equal to their general value;
- a certain group of values can be replaced by their mean without changing the general mean.

Let us also note some important particular cases of the function φ:

- at $\varphi(x) = x$ the arithmetic mean is obtained;
- at $\varphi(x) = \ln x$ the geometric mean is obtained;
- at $\varphi(x) = x^{-1}$ the harmonic mean is obtained;
- ïðè $\varphi(x) = x^2$ the quadratic mean is obtained;
- ïðè $\varphi(x) = x^\alpha, \quad \alpha \neq 0$ the power mean is obtained.

It is clear that the mean according to Kolmogorov is a particular case of the Cauchy's mean from which it was required that it possess only the property of averaging. In particular, any weighted means cannot be represented in the form according to Kolmogorov, since they have no properties of symmetry.

Definition 3 [359]. The quasi-mean of nonnegative numbers $x_1, \dots, x_n$ is the quantity of the form

$$M(x_1, \dots, x_n) = f^{-1}\left[\sum_{i=1}^{n} p_i f(x_i)\right], \quad \text{where} \quad p_i > 0,\ i = 1, \dots, n,\ \sum_{i=1}^{n} p_i = 1,$$

under the condition that the function f is continuous and monotone in the interval containing x_i.

In particular, at $f(x) = x$ the weighted arithmetic mean is obtained, at $f(x) = \ln x$ the geometric mean is obtained and at $f = x^r$ the weighted power mean is obtained.

It is evident that the quasi-means also include the usual means (nonweighted), if $p_i = 1/n$ is taken for all numbers of i, as well as the same functions: $f = \ln x$, $f = x^r$; these particular cases of the quasi-means meet all the conditions for a "*strong*" definition of the mean quantity.

Let the class of functions f_i, satisfying the definition according to Cauchy, be denoted by F. This class is extremely broad and contains as special cases the Kolmogorov means and quasi-means. The considerations leading to the choice of one or the other evaluation algorithm is frequently formulated in the form of heuristic principles such as the principle of voting, the principle of an excluded mean, the hypothesis of compactness, the principle of confidence in the majority, principles of diagnostics and correction, the method of redundant variables, and others.

Depending on the kind of the function f_i one can distinguish linear, quasi-linear, and nonlinear estimates. Below several tens of functions $f_i \in F$ and the evaluation algorithms corresponding to them are given, some of which coincides with the mean estimates considered earlier.

4.4.2 Linear and quasi-linear estimates

4.4.2.1 Linear estimates

Linear evaluation algorithms are obtained when using the functions f of the form

$$\hat{x} = a_1 x_1 + a_2 x_2 + a_3 x_3. \tag{4.4.5}$$

From conditions (4.4.1) and (4.4.2) it follows that the constants a_1, a_2, a_3 must be positive and meet the relation

$$a_1 + a_2 + a_3 = 1. \tag{4.4.6}$$

Typical representatives of estimates of this class are the arithmetic mean (see the function f_3) and the weighted arithmetic mean:

$$f_4: \quad \hat{x} = \left(\frac{1}{\sigma_1^2} + \frac{1}{\sigma_2^2} + \frac{1}{\sigma_3^2}\right)^{-1} \left(\frac{x_1}{\sigma_1^2} + \frac{x_2}{\sigma_2^2} + \frac{x_3}{\sigma_3^2}\right). \tag{4.4.7}$$

Assuming that one or two of the coefficients a_1, a_2, a_3 in formula (4.4.5) are equal to zero (or directing the corresponding error dispersions in formula (4.4.7) to infinity), we obtain linear estimates which do not take into account separate measurements, for example

$$f_5: \quad \hat{x} = \frac{1}{2}(x_1 + x_2),$$

$$f_6: \quad \hat{x} = x_1,$$

$$f_7: \quad \hat{x} = \left(\frac{1}{\sigma_1^2} + \frac{1}{\sigma_3^2}\right)^{-1} \left(\frac{x_1}{\sigma_1^2} + \frac{x_3}{\sigma_3^2}\right),$$

and others.

4.4.2.2 Quasi-linear estimates

Provided the coefficients a_i in formula (4.4.5) are not constant and depend on the values of measurements, then the estimates obtained (and the function f_i) are called quasi-liner. They have the form

$$\hat{x} = a_1(x_1, x_2, x_3) \cdot x_1 + a_2(x_1, x_2, x_3) \cdot x_2 + a_3(x_1, x_2, x_3) \cdot x_3, \tag{4.4.8}$$

at the same time, as before, at any x_1, x_2, x_3 condition of normalizing (4.4.6) has to be met.

In particular, assuming that $a_1 = \gamma x_1, a_2 = \gamma x_2, a_3 = \gamma x_3$, where γ is the general normalizing multiplier $\gamma = (x_1 + x_2 + x_3)^{-1}$, we get the quasi-linear estimate

$$f_8: \quad \hat{x} = \frac{x_1^2 + x_2^2 + x_3^2}{x_1 + x_2 + x_3},$$

coinciding with the contraharmonic one.

The estimate f_8 is biased with respect to the arithmetic mean to the side of the maximum measurement. To prove this let us subtract from it the arithmetic mean and convince ourselves that the difference obtained is nonnegative:

$$f_8 - f_3 = \frac{3(x_1^2 + x_2^2 + x_3^2) - (x_1 + x_2 + x_3)^2}{3(x_1 + x_2 + x_3)}$$
$$= \frac{2x_1^2 + 2x_2^2 + 2x_3^2 - 2x_1x_2 - 2x_1x_3 - 2x_2x_3}{3(x_1 + x_2 + x_3)}$$
$$= \frac{(x_1 - x_2)^2 + (x_1 - x_3)^2 + (x_2 - x_3)^2}{3(x_1 + x_2 + x_3)} \geq 0.$$

As another example the evaluation function f_9, obtained at $a_1 = \gamma x_2$, $a_2 = \gamma x_3$, $a_3 = \gamma x_1$ we have:

$$f_9 : \quad \hat{x} = \frac{x_1x_2 + x_2x_3 + x_1x_3}{x_1 + x_2 + x_3}.$$

At an unlimited increase of x_3 this estimate, unlike the previous one, remains finite and does not exceed the level $x_1 + x_2$. In order to show that it is never less than the arithmetic mean, let us consider the difference (f_9–f_3):

$$f_9 - f_3 = \frac{3(x_1x_2 + x_2x_3 + x_1x_3) - (x_1 + x_2 + x_3)^2}{3(x_1 + x_2 + x_3)}$$
$$= -\frac{x_1^2 + x_2^2 + x_3^2 - x_1x_2 - x_1x_3 - x_2x_3}{3(x_1 + x_2 + x_3)}$$
$$= -\frac{1}{2}(f_8 - f_3) \leq 0.$$

Diagrams of the evaluation functions f_8 and f_9 are shown in Figure 4.25. They lie on different sides of the dotted line corresponding to the arithmetic mean estimate.

From these diagrams we can see that the estimate f_8 is more preferable at crude errors (failures, misses) decreasing one of the measurement values (of the type of signal missing), and the estimate f_9 is preferable when one of the measurement values (the error of a high level) increases.

The next important example of a quasi-linear estimate is obtained assuming that

$$a_1 = \gamma\frac{1}{x_1}, a_2 = \gamma\frac{1}{x_2}, a_3 = \gamma\frac{1}{x_3},$$

where as before the multiplier γ is chosen from the condition of normalizing (4.4.6):

$$f_{10} : \quad \hat{x} = \frac{3}{\frac{1}{x_1} + \frac{1}{x_2} + \frac{1}{x_3}} = \frac{3x_1x_2x_3}{x_1x_2 + x_1x_3 + x_2x_3}.$$

This is the harmonic mean of three measurement values. It is not very sensitive to high-level errors, since the greatest measurement value is contained in the estimate with the smallest weight and the weight of the smallest measurement is maximum. If one of the values of measurements is equal to zero, then the estimate is also equal to zero.

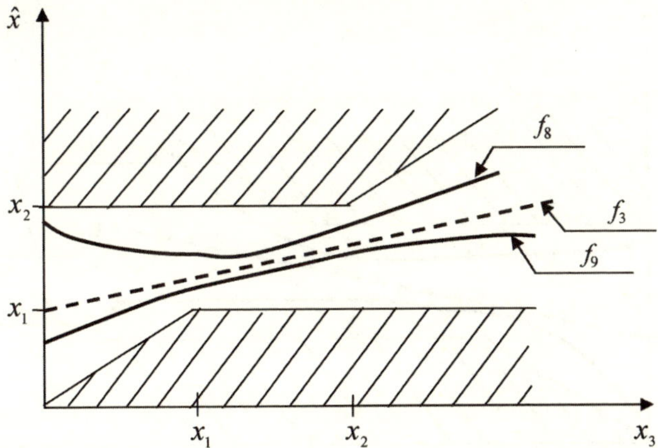

Figure 4.25. Diagrams of the evaluation functions f_8 and f_9 with respect to the arithmetic mean f_3.

A diagram of this estimate is given in Figure 4.26. It lies below the dotted line, which corresponds to the arithmetic mean, since at positive measurement values the difference

$$f_3 - f_{10} = \frac{(x_1 + x_2 + x_3)(x_1x_2 + x_1x_3 + x_2x_3) - 9x_1x_2x_3}{3(x_1x_2 + x_1x_3 + x_2x_3)}$$

$$= -\frac{x_1^2x_2 + x_1^2x_3 + x_1x_2^2 + x_2^2x_3 + x_1x_3^2 - x_2x_3^2 - 6x_1x_2x_3}{3(x_1x_2 + x_1x_3 + x_2x_3)}$$

$$= \frac{x_1(x_2 - x_3)^2 + x_2(x_1 - x_3)^2 + x_3(x_1 - x_2)^2}{x_1x_2 + x_1x_3 + x_2x_3}$$

is nonnegative.

The estimates f_8 and f_{10} relate to the family of Lehmer means that for three measurements are determined by the formula

$$f_{11}: \quad \hat{x} = \frac{x_1^{k+1} + x_2^{k+1} + x_3^{k+1}}{x_1^k + x_2^k + x_3^k},$$

where k is any real number.

If $k = -1$, to which the harmonic mean corresponds, at k =0 the arithmetic mean is obtained, and at k =1 the estimate f_8 is quasi-linear quadratic. Figure 4.26 also shows the diagram of the evaluation function obtainable at $k = 3$:

$$f_{12}: \quad \hat{x} = \frac{x_1^4 + x_2^4 + x_3^4}{x_1^3 + x_2^3 + x_3^3}.$$

It lies above the arithmetic mean and is closed between the function f_8 and the upper boundary of the permissible domain.

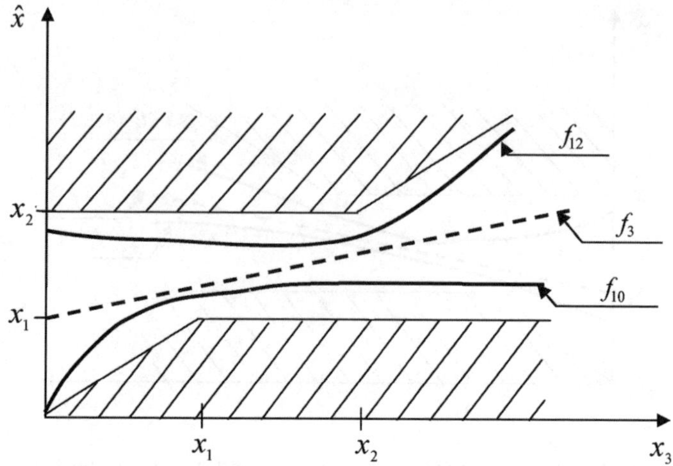

Figure 4.26. Diagrams of the evaluation functions f_{10} and f_{12} with respect to the arithmetic mean f_3.

As k increases to infinity the diagram of the function f_{11} approaches without any restriction the upper boundary, and the estimate itself f_{11} becomes the estimate $f_1 = \max(x_1, x_2, x_3)$. At $k \to -\infty$ the plot aspires to the bottom boundary of the permissible domain, and the estimate passes into $f_2 = \min(x_1, x_2, x_3)$. Thus, the estimates f_1 and f_2 can be considered to be the extreme cases of quasi-linear estimates.

Above the quasi-linear estimates the coefficients are considered, of which a_i are taken as proportional to the different powers of the measurement values.

Estimates of the following form are more general in character:

$$\hat{x} = \frac{x_1\psi(x_1) + x_2\psi(x_2) + x_3\psi(x_3)}{\psi(x_1) + \psi(x_2) + \psi(x_3)},$$

where $\psi(x)$ is the monotone function of a certain kind.

For example, assuming that $\psi(x) = e^{ax}$, we obtain the estimate

$$f: \quad \hat{x} = \frac{x_1 \cdot e^{ax_1} + x_2 \cdot e^{ax_2} + x_3 \cdot e^{ax_3}}{e^{ax_1} + e^{ax_2} + e^{ax_3}},$$

the diagrams for which for $a = 1$ and $a = -1$ are given in Figure 4.27.

4.4.3 Difference quasi-linear estimates

The estimates of the coefficients a_i, which depend on the differences $(x_i - x_j)$, form a separate group. As was noted previously, the algorithms with such a formation of the weight coefficients have, in general, a heuristic character. Among these we note estimates which anticipe scrapping (rejecting as defective) doubtful measurement re-

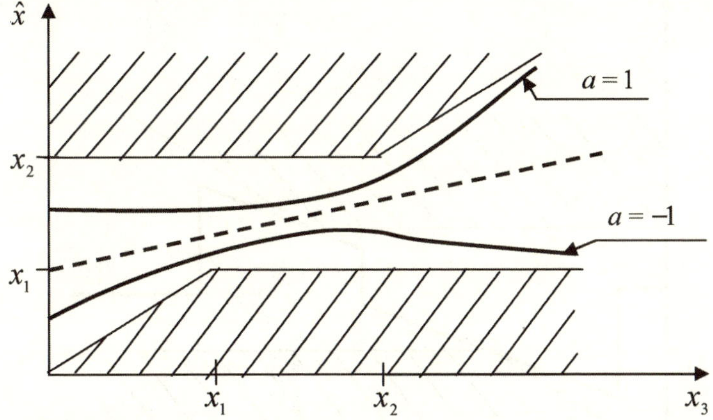

Figure 4.27. Diagrams of the estimates for the functions $\psi(x) = e^{ax}$ $a = 1$ and $a = -1$.

sults (or taking them into account with small weights), as well as estimates based on averaging the measurement values, mostly close to each other. Let us consider two algorithms of this kind.

The algorithm of averaging the two nearest measurement values consists of the following: one of three differences $x_1–x_2$, $x_1–x_3$, $x_2–x_3$, the smallest one with respect to the module, is selected, and the arithmetic mean of the measurements which it includes is accepted as an estimate:

$$f_{13}: \quad \hat{x} = \frac{x_i + x_j}{2}, \quad \text{if} \quad |x_i - x_j| = \min\big(|x_1 - x_2|, |x_1 - x_3|, |x_2 - x_3|\big).$$

Such an estimate is quasi-linear and meets condition (4.4.6), two of the coefficients a_1, a_2, a_3 being equal to $\frac{1}{2}$ and the third one equal to zero (its index is not known in advance and depends on the differences in measurement values). The diagram of this estimate shown in Figure 4.28a, has a “relay” character (contains ruptures of the first kind).

The horizontal sections in Figure 4.28a, in its initial and end part, demonstrate the insensitivity of the algorithm to single high-level errors. This is seen more clearly in the diagram of the sensitivity function (the derivative of the estimate) in Figure 4.28b. Note that similar diagrams for sensitivity functions can be constructed for each of the estimates mentioned.

The congenial principle of confidence in the two nearest values of measurements is realized with the quasi-linear estimate. Its coefficients depend on the difference of measurement values in the following way:

$$a_1 = \gamma(x_2 - x_3)^2; \quad a_2 = \gamma(x_1 - x_3)^2; \quad a_3 = \gamma(x_1 - x_2)^2.$$

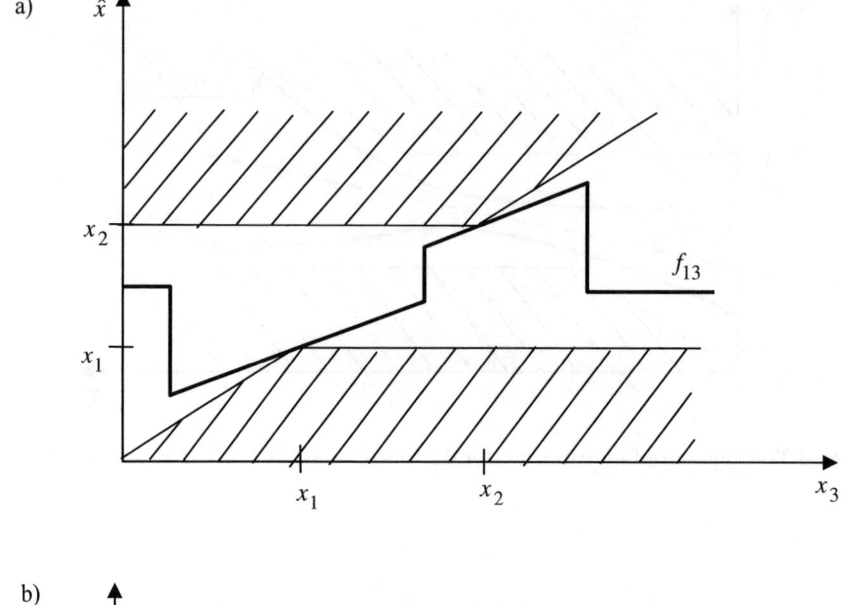

Figure 4.28. Diagrams of the estimate f_{13}.

Such a choice of weight coefficients strengthens the influence of the average value of measurements and weakens the remaining ones proportionally from moving off from it. Having found the normalizing multiplier and substituting the coefficients a_1, a_2, a_3 in expression (4.4.8) it is possible to get

$$f_{14}: \quad \hat{x} = \frac{x_1(x_2 - x_3)^2 + x_2(x_1 - x_3)^2 + x_3(x_1 - x_2)^2}{(x_1 - x_2)^2 + (x_1 - x_3)^2 + (x_2 - x_3)^2}.$$

The distinctive feature of this estimate is its insensitivity to single high-level errors (fails) when the evaluation function is of an analytical character. A diagram of this estimates is given in Figure 4.29a. It resembles the diagram of the preceding estimate, but has no breaks.

Note that both diagrams touch the dotted zone at angular points and have the horizontal asymptote $\hat{x} = \frac{1}{2}(x_1 + x_2)$. The first of these facts means that at the equality of two values of measurements the estimate coincides with them (since the weight coefficient a_i for the third measurement appears to be equal to zero). The second fact means

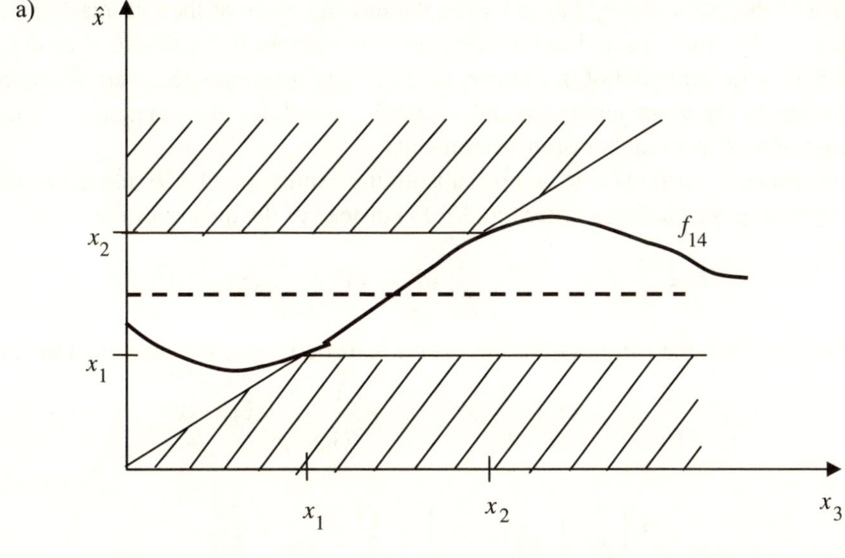

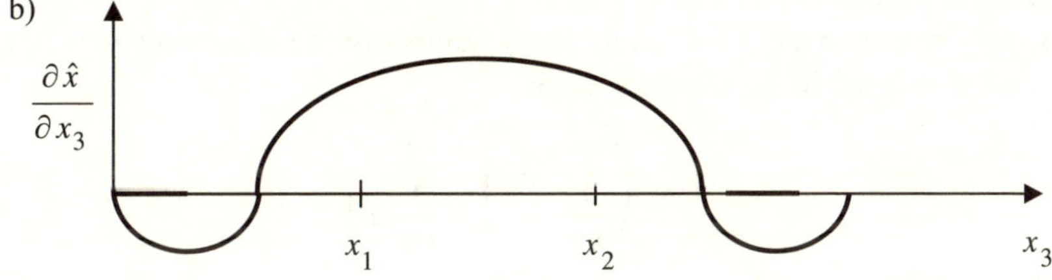

Figure 4.29. Diagrams of the estimate f_{14}.

that at an unlimited increase of one of the measurement values the estimate becomes equal to the arithmetic mean of the remaining two values.

The diagram of the sensitivity function (Figure 4.29b) has a definite similarity with the diagram in Figure 4.28b. However, unlike Figure 4.28b, it does not contain any breaks and smoothly tends to zero when one of the measurement values increases.

Comparison of the quasi-linear algorithm f_{14} with the linear estimates shows that there is a definite similarity with the estimates of the weighted arithmetic mean f_4. This becomes more evident if the expression for f_{14} is rewritten in the form

$$\hat{x} = \left(\frac{1}{\hat{\sigma}_1^2} + \frac{1}{\hat{\sigma}_2^2} + \frac{1}{\hat{\sigma}_3^2}\right)^{-1} \left(\frac{x_1}{\hat{\sigma}_1^2} + \frac{x_2}{\hat{\sigma}_2^2} + \frac{x_3}{\hat{\sigma}_3^2}\right), \tag{4.4.9}$$

where

$$\hat{\sigma}_1 = |x_1 - x_2||x_1 - x_3|; \hat{\sigma}_2 = |x_2 - x_1||x_2 - x_3|; \quad \hat{\sigma}_3 = |x_3 - x_1||x_3 - x_2|. \tag{4.4.10}$$

Each of the quantities $\hat{\sigma}_i^2$ characterizes the moving away of the i-th measurement value from the other two and can be considered as an empirical estimate of its dispersion. Such representation of f_{14} allows us, firstly, to determine the kind of criterion minimized by the given algorithm and, secondly, to indicate how to generalize it for the case of a greater number of measurements.

The desired criterion is obtained by substituting expression (4.4.9) into the formula of weighted quadratic mean criterion (4.3.11) instead of the quantities $\hat{\sigma}_i$:

$$J = \frac{1}{\hat{\sigma}_1^2}(x_1 - \hat{x})^2 + \frac{1}{\hat{\sigma}_2^2}(x_2 - \hat{x})^2 + \frac{1}{\hat{\sigma}_3^2}(x_3 - \hat{x})^2 .$$

In order to find the extremum of this criterion its derivative is made equal to zero:

$$\frac{2}{\hat{\sigma}_1^2}(x_1 - \hat{x}) + \frac{2}{\hat{\sigma}_2^2}(x_2 - \hat{x}) + \frac{2}{\hat{\sigma}_3^2}(x_3 - \hat{x}) = 0,$$

hence

$$\hat{x}\left(\frac{1}{\hat{\sigma}_1^2} + \frac{1}{\hat{\sigma}_2^2} + \frac{1}{\hat{\sigma}_3^2}\right) = \frac{x_1}{\hat{\sigma}_1^2} + \frac{x_2}{\hat{\sigma}_2^2} + \frac{x_3}{\hat{\sigma}_3^2},$$

which directly leads to formula (4.4.9).

To apply this estimate f_{14} to an arbitrary number of measurements $n > 3$, let formula (4.4.9) be used in the rewritten form

$$\hat{x} = \left(\sum_{i=1}^{n} \frac{1}{\hat{\sigma}_i^2}\right)^{-1} \sum_{i=1}^{n} \frac{x_i}{\hat{\sigma}_i^2},$$

where the expressions for the empirical estimates of dispersions $\hat{\sigma}_i^2$ are written by analogy with (4.4.10):

$$\hat{\sigma}_i = \prod_{\substack{j=1 \\ j \neq i}}^{n} |x_i - x_j|, i = \overline{1, n}.$$

Generalizing the functions f_{14} for arbitrary powers of differences leads to the estimate

$$f_{15}: \ \hat{x} = \frac{x_1|x_2 - x_3|^k + x_2|x_1 - x_3|^k + x_3|x_1 - x_2|^k}{|x_1 - x_2|^k + |x_1 - x_3|^k + |x_2 - x_3|^k},$$

where k is any real number.

It is interesting that the estimate obtained at $k = 1$ (proposed by M. I. Bim)

$$f_{16}: \ \hat{x} = \frac{x_1|x_2 - x_3| + x_2|x_1 - x_3| + x_3|x_1 - x_2|}{|x_1 - x_2| + |x_1 - x_3| + |x_2 - x_3|},$$

coincides with the function of the sample median. Really, at any measurement values the denominator of the expression given is equal to $2l$, i.e., to the doubled length of the segment of the axis x, which contains the measurement values x_1, x_2, x_3. In order to evaluate the nominator, let the minimum, average and maximum values x_1, x_2, x_3 of all measurement ones be denoted as x'_1, x'_2, x'_3.

Opening the brackets, one obtains

$$x'_1(x'_3 - x'_2) + x'_2(x'_3 - x'_1) + x'_3(x'_2 - x'_1) = 2x'_2(x'_3 - x'_1) = 2x'_2 l.$$

Hence a somewhat different expression is obtained for the same estimate:

$$f'_{16}: \quad \hat{x} = \frac{2x'_2 l}{2l} = x'_2 = \text{med}(x_1,\ x_2,\ x_3).$$

Diagrams of this estimate and its sensitivity function are given in Figure 4.30. They have an obvious resemblance with similar curves in Figure 4.29, essentially in its average part.

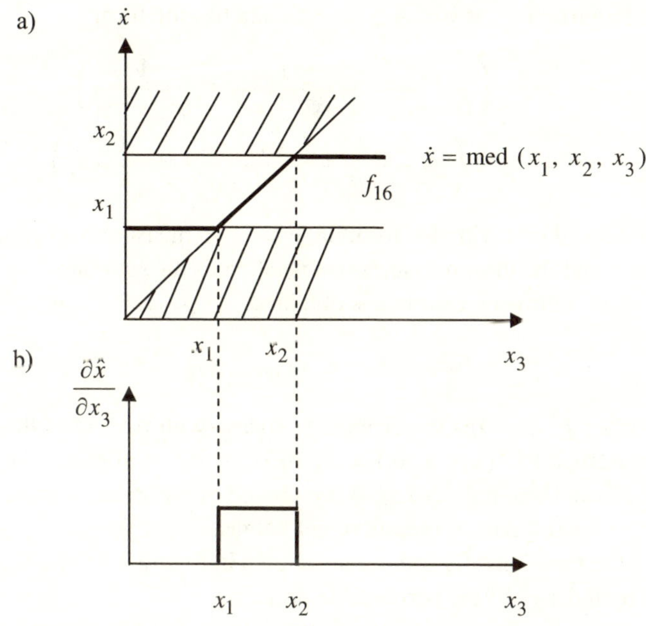

Figure 4.30. Diagrams of the estimate f_{16}.

A comparison of Figures 4.28, 4.29, and 4.30 allows us to conclude that the estimate f_{14} (the principle of confidence in the majority) occupies the intermediate position between the estimates f_{13} (the principle of averaging the two nearest estimates) and f_{16} (the principle of dropping the extreme estimates).

Let us note that the possibility of writing the function of the sample median f'_{16} in the form f_{16} is interesting, at least from two points of view. Firstly, due to this the affiliation of the sample median with the class of quasi-linear estimates with the coefficients depending on the difference of measurement values is stated. Secondly, writing the equation for f_{16} gives a new algorithm for median calculation which does not use the comparison operations.

Further analysis of the estimate f_{15} shows that at an increase of the parameter k ($k =, 4, \ldots$) the x-coordinates of the extreme points of the characteristic (Figure 4.28) become more distant from the points x_1, x_2, and the ordinates of these points approach the levels x_1 and x_2, correspondingly. At $k \to \infty$ the characteristic takes a form given in Figure 4.29, i.e., again turning into the sample median estimate:

$$\lim_{k\to\infty} f_{15}(k) = f_{15}(\infty) = \operatorname{med}(x_1, x_2, x_3).$$

When the values of the parameter k are negative the evaluation function f_{15} begins to realize the principle of "distrust of the majority" (principle of "nonconformism"), which at $k \to -\infty$ transfers to the principle of dropping off the two nearest values of measurement. In particular, at $k = -1$ the estimate has the form

$$f_{17}: \quad \hat{x} = \left(\frac{1}{|x_3 - x_1|} + \frac{1}{|x_3 - x_2|} + \frac{1}{|x_2 - x_1|}\right)^{-1} \times \left(\frac{x_1}{|x_3 - x_1|} + \frac{x_2}{|x_3 - x_2|} + \frac{x_3}{|x_2 - x_1|}\right).$$

Note that, in accordance with this formula, when two measurement values coincide, they are dropped and the third measurement is taken as the estimate.

At $k \to -\infty$ the following estimate is obtained:

$$f_{18}: \quad \hat{x} = L''(x_1, x_2, x_3),$$

where the operator L'' realizes the dropping process with respect to the two nearest values of measurement: $L''(x_1, x_2, x_3) = x_k$, if $|x_i - x_j| = \min(|x_1 - x_2|, |x_1 - x_3|, |x_2 - x_3|)$, i.e., from three indices i, j, k the choice of the index is made which does not belong to the nearest pair of measurement values.

Diagrams of the functions f_{17} and f_{18} are shown in Figure 4.31, the second of them entirely at the boundaries of the permissible domain.

In the given quasi-linear estimates the coefficients a_i are taken proportional to different powers with respect to the differences $(x_i - x_j)$.

A broader class consists of the estimates having the form

$$\hat{x} = \frac{x_1\psi(x_2 - x_3) + x_2\psi(x_1 - x_3) + x_3\psi(x_1 - x_2)}{\psi(x_2 - x_3) + \psi(x_1 - x_3) + \psi(x_1 - x_2)},$$

where $\psi(x)$ is a certain function.

Assuming, for example, that $\psi(x) = e^{|x|}$, we obtain the difference quasi-linear estimate

$$f: \quad \hat{x} = \frac{x_1 e^{|x_2 - x_3|} + x_2 e^{|x_1 - x_3|} + x_3 e^{|x_1 - x_2|}}{e^{|x_2 - x_3|} + e^{|x_1 - x_3|} + e^{|x_1 - x_2|}},$$

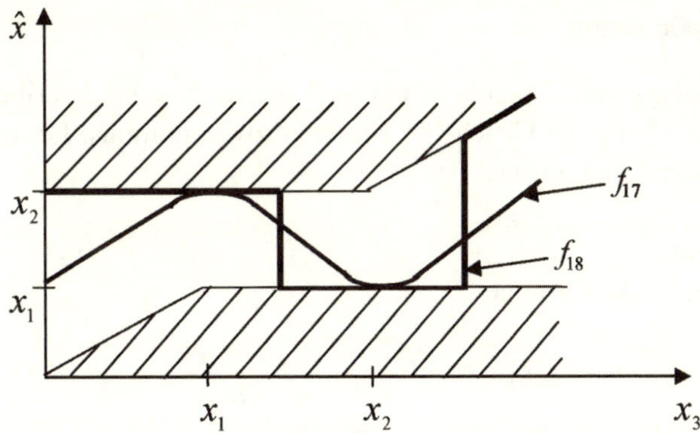

Figure 4.31. Diagrams of the evaluation functions f_{17} and f_{18}.

which realizes one of the versions of the principle of confidence in the majority. From the point of view of calculation, this estimate is slightly more complicated than the estimate f_{14}; however in it there is not the problem of division by zero when the measurement values coincide.

It is possible to choose some other function $\psi(x)$, e.g.,

$$\psi(x) = e^{\frac{1}{|x|}}, \quad \psi(x) = e^{ax}, \quad \text{and so on.}$$

Thus, the heuristic approach permits construction of an unlimited number of various estimates satisfying the additional requirements and possessing a moderate calculation complexity. Theoretical analysis of the obtained evaluation algorithms makes it possible to study their distinctive features and to reveal the conditions of their applicability.

4.4.4 Heuristic means for $n = 2$

The case of two measurements occupies a special place in the theory of evaluation as well as in metrology. On the one hand it is the simplest one and leads to the coincidence of some means. For example, the minimum (Chebyshev) estimate coincides with the arithmetic mean, and the combined mean estimate coincides with the geometric mean.

On the other hand, at $n = 2$ there exists a series of means which have no evident analogs in the case of three and more measurements. Due to this fact the case with $n = 2$ is of special interest for mathematicians, who have managed to obtain for such a case a series of fine and deep results.

In the known methods of calculation the estimates at $n = 2$ can be divided into two groups: analytical and iteration. Let us consider each of these groups.

4.4.4.1 Analytic means

Let a and b be the positive numbers and their mean value be described by the function $f(a,b)$, satisfying the Cauchy or Kolmogorov conditions. Let us also give the description of some such means.

Centroidal mean

This mean is set by the formula

$$T(a,b) = \frac{2}{3} \cdot \frac{a^2 + ab + b^2}{a+b}.$$

In a trapezoid with bases of lengths "a" and "b", (see Figure 4.11) the centroidal mean is the length of the segment that is parallel to the bases and that also passes through the centroid of area of the trapezoid.

Heron means

The Heron means form the single-parametric family of means, described by the formula

$$H_p(a,b) = \left(\frac{a^p + (ab)^{p/2} + b^p}{3}\right)^{1/p} \quad \text{for } p \neq 0 \text{ and } H_0 = \sqrt{ab}.$$

In particular, at $p = 1$ the classical Heron mean is obtained:

$$H_1 = \frac{1}{3}(a + \sqrt{ab} + b),$$

used for calculating the volume of a frustum of a pyramid or cone.

Let the inequalities connecting different means of the real numbers a, b, assuming that $a \neq b$:

$$H < G < H_1 < A < T < Q < C,$$

where H is the harmonic mean, G is the geometric mean, H_1 is the Heron mean, A is the arithmetic mean, T is the centroidal mean, Q is the quadratic mean, and C is the contraharmonic mean.

Seiffert's mean

Seiffert's mean, proposed by H. J. Seiffert [51, 426], has the form

$$P(a,b) = \frac{a-b}{4\arctan\left(\sqrt{a/b}\right) - \pi} = \frac{a-b}{2\arcsin\frac{a-b}{a+b}}.$$

Heinz' mean

The Heinz mean belongs to the family of means also known as the symmetric means. It is described by the formula

$$SH_p(a,b) = \frac{a^p b^{1-p} + a^{1-p} b^p}{2} \quad \text{with} \quad 0 \le p \le 1/2.$$

This mean satisfies the inequality

$$\sqrt{ab} = SH_{1/2}(a,b) < SH_p(a,b) < SH_0(a,b) = \frac{a+b}{2}.$$

Identric mean

The identric mean is set by the formula

$$I(a,b) = \frac{1}{e}\left(\frac{b^b}{a^a}\right)^{1/(b-a)}.$$

Along with it the mean J is used, which is defined by

$$J(a,b) = (a^a b^b)^{1/(a+b)} \quad (a,b > 0).$$

Logarithmic mean

The logarithmic mean of two positive quantities a, b is determined by the formula

$$L(a,b) = \frac{a-b}{\ln a - \ln b}.$$

The means $I(a,b)$ and $L(a,b)$ satisfy the inequality

$$\min \le H \le G \le L \le I \le A \le \max.$$

There is a connection of the logarithmic mean to the arithmetic mean:

$$\frac{L(a^2,b^2)}{L(a,b)} = \frac{a+b}{2} = A(a,b).$$

The logarithmic mean may be generalized to 3 variables:

$$L_{MV}(a,b,c) = \sqrt{\frac{(a-b)(b-c)(c-a)/2}{(b-c)\ln a + (c-a)\ln b + (a-b)\ln c}}.$$

Power logarithmic mean

The power logarithmic mean, or p-logarithmic mean, is defined by

$$L_p(a,b) = \left(\frac{a^{p+1} - b^{p+1}}{(p+1)(a-b)}\right)^{1/p}.$$

The formula sets a single-parametric family of means. Its particular cases at $p = 1$, $p = 0$ and $p = -1$ are the logarithmic mean, the identric mean, and the logarithmic mean:

$$L_1(a,b) = A, L_0(a,b) = \frac{1}{e}\left(\frac{a^a}{b^b}\right)^{1/(a-b)} = I(a,b),$$

$$L_{-1}(a,b) = \frac{a-b}{\ln a - \ln b} = L(a,b), L_{-2}(a,b) = G(a,b).$$

For a particular application of logarithmic and identric means in deciding the issue concerning the distribution of places in the US Congress, see the paper [5].

Power difference mean

This mean defined as

$$D_p(a,b) = \frac{p}{p+1} \cdot \frac{a^{p+1} - b^{p+1}}{a^p - b^p}.$$

The particular special values of are understood as $D_1 = A$, $D_{-1/2} = G$, $D_{-1} = G/L$.

Stolarsky means

The Stolarsky means of order (p,q) are defined by the formula

$$S_{p,q}(a,b) = \left(\frac{p}{q} \cdot \frac{a^q - b^q}{a^p - b^p}\right)^{1/(q-p)}.$$

This includes some of the most familiar cases in the sense

$$S_{p,0}(a,b) = \left(\frac{1}{p} \cdot \frac{a^p - b^p}{\ln a - \ln b}\right)^{1/p},$$

$$S_{1,q+1}(a,b) = L_q(a,b), \qquad S_{p,p+1}(a,b) = D_p(a,b).$$

Dual means [399]

For the given mean M, which is symmetric and homogeneous, the following relation is set:

$$M^*(a,b) = \left(M(a^{-1},b^{-1})\right)^{-1}.$$

It is easy to see that M^* is also a mean, called the *dual mean* of M.

This mean is described by

$$M^*(a,b) = \frac{ab}{M(a,b)}$$

which we briefly write as $M^* = G^2/M$.

Every mean M satisfies $M^{**} = M$ and, if M_1 and M_2 are two means such that $M_1 \le M_2$, then $M_1^* \le M_2^*$.

The mean M is called self-dual if $M^* = M$. It is clear that the arithmetic and harmonic means are mutually dual, and that the geometric mean is the unique self-dual mean.

The dual of the logarithmic mean is given by

$$L^* = L^*(a,b) = ab\frac{\ln b - \ln a}{b-a}, \quad L^*(a,a) = a,$$

while that of the identric mean is

$$I^* = I^*(a,b) = e\left(\frac{a^b}{b^a}\right)^{1/b-a}, \quad I^*(a,a) = a.$$

The following inequalities are immediate from the above:

$$\min \le H \le I^* \le L^* \le G \le L \le I \le A \le \max.$$

4.4.4.2 Iterative means

There is a special kind of means which we can introduce with the help of some iteration procedures. The best known of them are the arithmetic-geometric, arithmetic-harmonic, and geometric-harmonic means of two quantities.

Arithmetic-geometric mean

The procedure of constructing the arithmetic-geometric mean (the so-called *AGM-procedure*) is as follows.

Let two positive numbers a and b, $a < b$, be given. First the arithmetic and geometric means of the numbers a, b are calculated:

$$a_1 = \frac{a+b}{2}, \quad b_1 = \sqrt{ab}.$$

Then for the numbers a_1 and b_1 the same means are calculated, and a_2 and b_2 are obtained:

$$a_2 = \frac{a_1+b_1}{2}, \quad b_2 = \sqrt{a_1 b_1}.$$

After that the procedure is repeated. As a result two sequences of numbers (a_n) and (b_n) are formed.

It is not difficult to show that they arrive a common limit located between the initial numbers a and b. This is called the *arithmetic-geometric mean* of these numbers and is designated as $AGM(a, b)$.

Example 4.4.1. Let $a = 1$, $b = 3$. Applying the AGM-procedure, we get

$$\begin{aligned} a_1 &= 2; & b_1 &= \sqrt{3} \approx 1.732050808; \\ a_2 &\approx 1.866025404; & b_2 &\approx 1.861209718; \\ a_3 &\approx 1.863617561; & b_3 &\approx 1.863616006; \\ a_4 &\approx 1.863616784; & b_4 &\approx 1.863616784. \end{aligned}$$

We see that the sequences (a_n) and (b_n) quickly draw together.

Calculating the arithmetic-geometric mean for $2 \leq b \leq 5$, we obtain

$$\begin{aligned} AGM(1,2) &\approx 1.45679103104690686\ldots \\ AGM(1,3) &\approx 1.86361678324489654\ldots \\ AGM(1,4) &\approx 2.24302858028760257\ldots \\ AGM(1,5) &\approx 2.60400819053094028\ldots \end{aligned}$$

A diagram of the arithmetic-geometric mean for $0 \leq b \leq 20$ is shown in Figure 4.32.

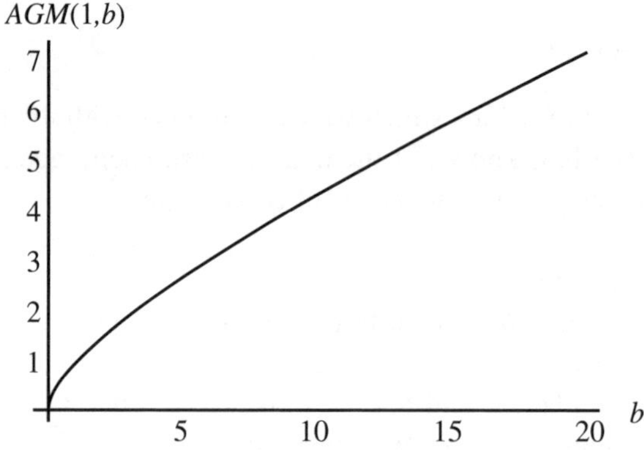

Figure 4.32. Diagram of the function $AGM(1, b)$.

To find an explicit expression for the $AGM(a, b)$ by terms of a and b is not simple. This was first done by C. F. Gauss. He showed that the arithmetic-geometric mean is expressed in terms of an elliptic integral with the help of the formula

$$AGM(a,b) = \frac{\pi}{2\displaystyle\int_0^{\pi/2} \frac{dx}{\sqrt{a^2 \sin^2 x + b^2 \cos^2 x}}}.$$

Thanks to the quick coincidence of the iteration *AGM*-procedure in modern calculation practice, this formula is applied for calculating elliptical integrals, transcendental functions, and even for calculating the number π (see, for example, [66]).

From Gauss' formula the following properties of the arithmetic-geometric mean are readily apparent:

- *homogeneity*: $\lambda AGM(a,b) = AGM(\lambda a, \lambda b)$;
- *invariance*: $AGM(a,b) = AGM(\frac{1}{2}(a+b), \sqrt{ab})$,
  $AGM(1, \sqrt{1-x^2}) = AGM(1+x, 1-x)$.

In mathematics Gauss' constant is applied, which is defined by the formula

$$\frac{2}{\pi}\int_0^1 \frac{dt}{\sqrt{1-t^4}} = \frac{1}{AGM\left(1, \sqrt{2}\right)} = 0.8346268416740731862 8\ldots .$$

Arithmetic-harmonic mean

The process of constructing the arithmetic-harmonic mean of two numbers a and b is similar to the *AGM*-procedure, whereby the geometric mean is replace by the harmonic mean.

Thus, the terms of the sequences (a_n) and (b_n) are determined by formulae

$$a_{n+1} = \frac{a_n + b_n}{2}, \quad b_{n+1} = \frac{2a_n b_n}{a_n + b_n}, \quad (n = 0, 1, 2, \ldots;\ a_0 = a, b_0 = b). \tag{4.4.11}$$

These sequences (a_n) and (b_n) are monotone, restricted, and have a common limit. This limit is called the *arithmetic-harmonic mean* of the numbers a and b and is designated as $AHM(a,b)$. From (4.4.11) it follows that

$$a_{n+1} \cdot b_{n+1} = a_n \cdot b_n = \ldots = a_1 \cdot b_1 = ab,$$

therefore $AHM(a,b)^2 = ab$.

Thus, $AHM(a,b) = \sqrt{ab}$, i.e., the arithmetic-harmonic mean coincides with the geometric mean.

Geometric-harmonic mean

This mean is obtained if the sequences (a_n) and (b_n) are constructed with the help of the harmonic and geometric means:

$$a_{n+1} = \frac{2a_n b_n}{a_n + b_n}, \quad b_{n+1} = \sqrt{a_n b_n}, \quad (n = 0, 1, 2, \ldots;\ a_0 = a,\ b_0 = b). \tag{4.4.12}$$

The common limit of these sequences is called the *geometric-harmonic mean* of the numbers a and b and is designated as $GHM(a,b)$.

It is not difficult to show that

$$\frac{1}{GHM(a,b)} = AGM\left(\frac{1}{b}, \frac{1}{a}\right),$$

or, owing to Gauss' formula, that

$$GHM(a,b) = \frac{2ab}{\pi} \int_0^{\pi/2} \frac{dx}{\sqrt{a^2 \cos^2 x + b^2 \sin^2 x}}.$$

Schwab–Borchardt mean

Of the three means arithmetic-geometric, arithmetic-harmonic, and geometric-harmonic only the second is elementary, expressed in terms of the initial numbers a and b. It is interesting that a small change of the AGM-procedure leads the sequences whose common limit is elementary, expressed in terms of a and b.

Let us have

$$a_{n+1} = \frac{a_n + b_n}{2}, b_{n+1} = \sqrt{a_{n+1} b_n}, \quad (n = 0, 1, 2, \dots; \ a_0 = a, \ b_0 = b). \tag{4.4.13}$$

The common limit α of these sequences is called the Schwab–Borchardt mean (SBM). It is described by the formula

$$SBM = \frac{\sqrt{b^2 - a^2}}{\arccos \dfrac{a}{b}}.$$

In particular, at $a = \frac{1}{2}$, $b = \frac{1}{\sqrt{2}}$ we get $SBM = \frac{2}{\pi}$.

Considering various iterative procedures it is possible to obtain a number of other means. Let us describe three of them.

Lemniscatic mean [363]

Let the iterative procedure be considered in which the terms of the sequences (a_n) and (b_n) are determined by formulae

$$a_{n+1} = \frac{1}{2}(a_n + b_n), \quad b_{n+1} = \sqrt{a_n a_{n+1}}.$$

These sequences converge on a common limit, called the *lemniscatic mean* and designated as $LM(a,b)$. It is expressed in terms of the inverse Gauss' arc lemniscate sine function:

$$\text{arcsl}x = \int_0^x \frac{dx}{\sqrt{1 - x^4}} dx, |x| \le 1$$

and in terms of the inverse hyperbolic arc lemniscatic sine function

$$\text{arcslh}x = \int_0^\infty \frac{dx}{\sqrt{1 + x^4}}.$$

The lemniscatic mean is determined by the formulae

$$LM(a,b) = \begin{cases} \sqrt{\dfrac{a^2 - b^2}{\operatorname{arcsl}\left(1 - \frac{b^2}{a^2}\right)}}, & a > b \\ \sqrt{\dfrac{b^2 - a^2}{\operatorname{arcslh}\left(\frac{b^2}{a^2} - 1\right)}}, & a < b. \end{cases}$$

If $a = b$, then $LM(a,a) = a$.

Some properties of the lemniscatic are

homogeneity: $LM(\lambda a, \lambda b) = \lambda\, LM(a,b)$,

nonsymmetry: $LM(a,b) \neq LM(b,a)$,

invariance: $LM(a,b) = LM\left(\frac{1}{2}(a+b), \sqrt{\frac{a}{2}(a+b)}\,\right)$.

Logarithmic mean

Let two positive numbers a and b, $a > b$, be given. Below the iterative sequences will be considered which are determined by the formulae

$$a_{n+1} = \frac{1}{2}(a_n + \sqrt{a_n b_n}), b_{n+1} = \frac{1}{2}(b_n + \sqrt{a_n b_n}).$$

They converge on the common limit equal to the logarithmic mean of the initial numbers $a_0 = a$, $b_0 = b$:

$$L(a,b) = \frac{a-b}{\ln a - \ln b}.$$

The logarithmic mean possesses the uniformity property $L(\lambda a, \lambda b) = \lambda L(a,b)$ and is closed between the arithmetic and geometric means:

$$G(a,b) < L(a,b) < A(a,b).$$

Family of iterative means

Let the iterative procedure be considered in which the sequences (a_n) and (b_n) are determined by the formulae

$$a_{n+1} = f_i(a_n, b_n), \quad b_{n+1} = f_j(a_n, b_n), \quad i \neq j,$$

where

$$f_1(a,b) = \frac{1}{2}(a+b), \qquad f_2(a,b) = (ab)^{1/2},$$

$$f_3(a,b) = \left(a\frac{a+b}{2}\right)^{1/2}, \qquad f_4(a,b) = \left(b\frac{a+b}{2}\right)^{1/2}.$$

For each of the twelve choices of i and j, $i \neq j$, the common limit of a_n and b_n as $n \to \infty$ is $L_{ij}(x, y)$. Thus, the family of 12 iterative means $L_{ij}(a, b)$ is obtained. For example, L_{14} is the Schwab–Borchardt mean *SBM* [89].

The natural question arises about the possibility of applying the iterative procedures in the case of three and more arguments. Let us mention two possible approaches.

Arithmetic-Geometric-Harmonic mean

One obvious extension of the iterative means of more than two arguments is to consider the iterated Arithmetic-Geometric-Harmonic mean (*AGH*) of three numbers $a_0 = a$, $b_0 = b$, $c_0 = c$, defined by the recurrence

$$a_{n+1} = A(a_n, b_n, c_n), \quad b_{n+1} = G(a_n, b_n, c_n), \quad c_{n+1} = H(a_n, b_n, c_n).$$

Then the *AGH*-mean will be equal to the common limit of these sequences.

Similarly, making use of other means we can have iterations of many elements.

Symmetric and "super-symmetric" means

Another approach to the generalization of the means to an arbitrary number of arguments is based on the use of elementary symmetric polynomials [78].

Recall that the elementary symmetric polynomials of $x_1, x_2, \ldots, x_n$ are

$$\begin{aligned} S_1 &= x_1 + x_2 + \cdots + x_n, \\ S_2 &= x_1x_2 + x_1x_3 + \cdots + x_{n+1}x_n, \\ &\ldots\ldots\ldots\ldots\ldots\ldots\ldots\ldots \\ S_n &= x_1x_2 \ldots x_n. \end{aligned}$$

We can define the k-th "symmetric mean" of n objects as

$$M_k = \left(\frac{S_k}{C(n,k)}\right)^{1/k},$$

where $C(n, k)$ is the binomial coefficient (n choose k).

Of course, M_1 is the arithmetic mean, M_2 is the combinatoric mean, and M_n is the geometric mean.

The generalization of the *AGM* to n elements is the "super-symmetric mean" of the n values $x_1(0), x_2(0), \ldots, x_n(0)$, defined as the convergent value given by the succession of iterations:

$$x_j(k+1) = M_j(x_1(k), x_2(k), \ldots, x_n(k)), \quad j = 1, 2, \ldots, n.$$

Obviously, the original values of $x_j(0)$ are the roots of the polynomial

$$f(x) = x^n - S_1x^{n-1} + S_2x^{n-2} - \ldots \pm S_n,$$

and the next set of values $x_j(1)$ are the roots of another polynomial, the coefficients of which are the symmetric functions used on the next iteration, and so on.

Thus, we have the sequence of polynomials converging on one of the form $F(x) = (x - R)^n$, where R is the supersymmetric mean of the original x's.

4.5 Structural and diagnostic methods for obtaining estimates

4.5.1 Structural means

Almost all the means considered above represent the smooth differentiable functions for each of the arguments, with one exception, which is represented by the so-called *structural means*; examples of this are various kinds of the order statistics: mode, median, minimum and maximum values, as well as midpoint (a half-sum of a minimum and maximum values).

The characteristics of the structural means include

- dropping a part of measurements when forming the mean;
- preliminary putting in order or some other structuring of measurements;
- use of logical operations in measurement data processing.

Due to the above, the structural means, although they are means in the Cauchy's sense, do not satisfy the conditions for means required by Kolmogorov.

Among the merits of the structural means, thanks to which they have found wide application in practice, we should mention increased noise immunity, i.e., robustness.

To evaluate the degree of robustness the quantitative index or so-called breakdown point is used. It indicates what part of the measurements per sample can be badly distorted, for example, to increase it infinitely without changing the mean quantity.

So, throwing away the maximum and minimum measurement results from the sample containing n terms and averaging the remaining terms, we will obtain a structural mean insensitive to a crude distortion of any of measurements. The breakdown point of such an estimator is $1/n$. The maximum breakdown point is 0.5 and there are estimators which achieve such a breakdown point. The means with high breakdown points are sometimes called *resistant* means.

Below the best known structural means are listed [497, 545].

The *maximum and minimum* estimates of the measurement set $x_1, \dots, x_n$ are described by formulae

$$\hat{x} = \max(x_1, \dots, x_n), \hat{x} = \min(x_1, \dots, x_n).$$

The first of these is protected from errors decreasing the quantity of a part of the measurements; the second one is protected from errors increasing a part of the measurements.

The *median estimate* or *sample mean* $\hat{x} = \text{med}(x_1, \ldots, x_n)$.

At odd n this estimate is formed as a central term of a variational series (of the increase-ordered totality of measurements). If the number of measurements is even, for the median an arithmetic mean with numbers $n/2 - 1$ and $n/2 + 1$ are taken. The median has a breakdown point of 0.5.

In all three cases, as the estimate of one of measurements is taken, all remaining measurements are dropped.

A *mode* is the quantity which is most often met or reiterated. For example, the age student, which is more often seen in a given group, is modal. When deciding practical problems the mode expressed in the form of an interval is frequently used.

Truncated mean

To calculate this mean the data of an ordered sample are averaged after removing from both sides a definite portion of terms (usually within the limits of 5–25 %). The truncated means are widely applied in summing up the results of some forms of sport competitions, when the maximum and minimum estimates of the judges are dropped. The extreme case of the truncated means is the median. The X % truncated mean has the breakdown point of X %.

Winsorized mean

To calculate the value of this mean the initial sample is ordered in the course of increasing, then on each side a part of the data is cut off (usually 10 % or 25 %) and the cut-off data is replaced by the values of extreme numbers from those which remain.

The Winsorized mean is much like the mean and median, and even more similar to the truncated mean.

Example 4.5.1. For a sample of 10 numbers (from x_1, the smallest, to x_{10}, the largest) the 10% Winsorized mean is $(2x_2 + x_3 + \ldots + x_8 + 2x_9)/10$. The key is the repetition of x_2 and x_9: the extra substitution for the original values x_1 and x_{10}, which have been discarded and replaced.

Trimean

Trimean is defined as an average value of the median and midhinge. The trimean quantity is calculated by the following formula:

$$\hat{x} = Q_2 + \frac{Q_1 + Q_3}{2} = \frac{Q_1 + 2Q_2 + Q_3}{4},$$

where Q_2 is the median of the data and Q_1 and Q_3 are, respectively, the lower and upper quartiles.

Thus, the trimean is the weighted sum of quartiles. Like the median, it is a resistant estimator with a breakdown point of 25 %.

Example 4.5.2. For a sample of 15 numbers (from x_1, the smallest, to x_{15}, the largest) the trimean is $(x_4+2x_8 + x_{12})/4$.

Notice that not all structural means are robust. The maximum, minimum, and half-sum of the maximum and minimum are not robust, and their breakdown point is equal to zero. This characteristic of the robust estimates is equal to 50 % for the median, and in other cases it is less and depends on the percent used for dropping data.

4.5.2 Use of redundant variables to increase evaluation accuracy

A number of evaluation algorithms can be obtained using the principles and methods of technical diagnostics and, above all, the ideas of functional diagnosis in systems with algebraic invariants, as well as the method of redundant variables (MRV).

Let us recall that in accordance with [339] the relate those systems to the systems with algebraic invariants, the output signals $x_1, \dots, x_n$ of which satisfy at least one algebraic relation of the form

$$\Delta = M(x_1, \dots, x_n) = 0,$$

Hereby, when there are no errors, this relation has to be fulfilled for any input signals and at any moment of time.

For the case considered the system under study is described by equations (4.2.1):

$$\begin{aligned} x_1 &= x + e_1, \\ x_2 &= x + e_2, \\ x_3 &= x + e_3, \end{aligned} \tag{4.5.1}$$

where a part of the input signal is played by the unknown quantity x, and a part of the output signals is played by the values of measurements x_1, x_2, x_3. In the absence of errors (misses, mistakes, fails) e_i the output signals of this system satisfy two independent linear algebraic relations (algebraic invariants):

$$\begin{aligned} \Delta_1 &= x_1 - x_2 = 0, \\ \Delta_2 &= x_1 - x_3 = 0. \end{aligned} \tag{4.5.2}$$

The latter equations can be written in a more convenient form,

$$\begin{bmatrix} \Delta_1 \\ \Delta_2 \end{bmatrix} = \begin{bmatrix} 1 & -1 & 0 \\ 1 & 0 & -1 \end{bmatrix} \begin{bmatrix} x_1 \\ x_2 \\ x_3 \end{bmatrix} = 0 \tag{4.5.3}$$

or, even shorter,

$$\Delta = MX = 0.$$

Under real conditions the errors $e_i \neq 0$, therefore the discrepancy vector $\Delta = Me$, where $e = [e_1, e_2, e_3]^T$ will also be nonzero. In order to increase the accuracy and reliability it is be natural to attempt to use information about the unknown errors e_1, e_2, e_3 which the vector Δ contains. Within the framework of MRV two approaches to use of such information have been studied:

- correction of measurement values containing small errors;
- detection, localization, and screening (exclusion, "dropping") of unreliable measurement values containing errors of a high level.

In [339, 343, 344, 347, and others] are the corresponding results obtained for dynamic systems with arbitrary invariants. Some of their instantiation is given below for the system described by equations (4.5.1) and (4.5.2).

Let us assume that the errors e_i have small values (are within "tolerance" limits) and it is possible to neglect the probability of any appearance of unreliable measurement results (failures of sensors). Then, to increase the accuracy of the evaluation one can use the principle of error correction which is applied in MRV. In the given case it is reduced to the following.

Substituting in relation (4.5.2) the real signal values for x_i from (4.5.1), one obtains

$$\Delta_1 = e_1 - e_2, \Delta_2 = e_1 - e_3$$

or

$$\Delta = Me. \tag{4.5.4}$$

Thus, the signal Δ carries information about the vector of error e contained in the measurement values. The correction idea consists of extracting the estimate $\hat{e}$ (refining the correction) from the vector of measurement values X. The simplest estimate $\hat{e}$ is obtained by pseudo-inversion of the system (4.5.4):

$$\hat{e} = \mathrm{M}^{+}\Delta = \mathrm{M}^{T}(\mathrm{M}\mathrm{M}^{T})^{-1}\Delta. \tag{4.5.5}$$

After calculations for $M = [\begin{smallmatrix} 1 & -1 & 0 \\ 1 & 0 & -1 \end{smallmatrix}]$, one obtains the pseudo-inversion matrix

$$\begin{aligned}
\mathrm{M}^{+} &= \begin{bmatrix} 1 & 1 \\ -1 & 0 \\ 0 & -1 \end{bmatrix} \left(\begin{bmatrix} 1 & -1 & 0 \\ 1 & 0 & -1 \end{bmatrix} \begin{bmatrix} 1 & 1 \\ -1 & 0 \\ 0 & -1 \end{bmatrix} \right)^{-1} \\
&= \begin{bmatrix} 1 & 1 \\ -1 & 0 \\ 0 & -1 \end{bmatrix} \begin{bmatrix} 2 & 1 \\ 1 & 2 \end{bmatrix}^{-1} \\
&= \begin{bmatrix} 1 & 1 \\ -1 & 0 \\ 0 & -1 \end{bmatrix} \frac{1}{3} \begin{bmatrix} 2 & -1 \\ -1 & 2 \end{bmatrix} \\
&= \frac{1}{3} \begin{bmatrix} 1 & 1 \\ -2 & 1 \\ 1 & -2 \end{bmatrix}.
\end{aligned}$$

Consequently, the corrected vector of the measurement values is determined by the expression

$$\hat{X} = X - \hat{e} = \begin{bmatrix} x_1 \\ x_2 \\ x_3 \end{bmatrix} - \frac{1}{3} \begin{bmatrix} 1 & 1 \\ -2 & 1 \\ 1 & -2 \end{bmatrix} \begin{bmatrix} \Delta_1 \\ \Delta_2 \end{bmatrix},$$

or in the scalar form it will be

$$\begin{aligned} \hat{x}_1 &= x_1 - \frac{1}{3}(\Delta_1 + \Delta_2), \\ \hat{x}_2 &= x_2 + \frac{1}{3}(2\Delta_1 - \Delta_2), \\ \hat{x}_3 &= x_3 - \frac{1}{3}(\Delta_1 - 2\Delta_2). \end{aligned} \tag{4.5.6}$$

Substituting here x_i from (4.5.1) and Δ_i from (4.5.2), one obtains

$$\begin{aligned} \hat{x}_1 &= x + \frac{1}{3}(e_1 + e_2 + e_3), \\ \hat{x}_2 &= x + \frac{1}{3}(e_1 + e_2 + e_3), \\ \hat{x}_3 &= x + \frac{1}{3}(e_1 + e_2 + e_3). \end{aligned}$$

Thus, as a consequence of correction due to the redistribution of errors the values $\hat{x}_1$, $\hat{x}_2$, $\hat{x}_3$ have become equal. Therefore, any of them can be taken as the required estimate, i.e., the resultant evaluation algorithm can be written in the form

$$f_{19}: \quad \hat{x} = x_1 - \frac{1}{3}(\Delta_1 + \Delta_2), \quad \Delta_1 = x_1 - x_2, \quad \Delta_2 = x_1 - x_3.$$

The estimate obtained with this algorithm coincides with the arithmetic mean, which fact can be checked with the help of the substitution of the latter two equations in the evaluation function. This is explained by the fact that the correction on the basis of pseudo-inversion is equivalent to the application of the least squares method that, as we know, leads to the arithmetic mean estimate.

Provided the errors e_i are independent and have similar dispersions, then such correction is optimal and provides a decrease in errors by 3 times. If the measurements are not equally accurate and the correlation matrix of errors $R = M\{e \cdot e^T\}$ is known, then the minimum dispersion is peculiar to the estimate obtained in accordance with the algorithm

$$f_{20}: \quad \hat{x} = x_1 - RM^T(MRM^T)^{-1}\Delta,$$

where the vector Δ is defined by relation (4.5.3).

It can be shown that this estimate coincides with the estimate described earlier by the *Markovian* estimate. There can also be other versions of linear as well as nonlinear algorithms of correction.

The linear algorithm of correction has the form

$$\hat{X} = X - K\Delta, \quad \Delta = MX, \tag{4.5.7}$$

where K is the rectangular matrix with the dimension of 3×2. It is selected in such a manner that the corrected vector of changing $\hat{X}$ turns algebraic invariants (4.5.3) into zero:

$$M\hat{X} = M(X - K\Delta) = (M - MKM)X = 0.$$

Hence, the matrix K has to satisfy the matrix equation

$$MK = E \quad \text{or} \quad \begin{bmatrix} 1 & -1 & 0 \\ 1 & 0 & -1 \end{bmatrix} \begin{bmatrix} k_{11} & k_{12} \\ k_{21} & k_{22} \\ k_{31} & k_{32} \end{bmatrix} = \begin{bmatrix} 1 & 0 \\ 0 & 1 \end{bmatrix}. \tag{4.5.8}$$

This means that 2 from 6 elements k_{ij} can be selected arbitrarily. Denoting $k_{11} = a$, $k_{12} = b$, it is possible to present the matrix K in the form

$$K = \begin{bmatrix} a & b \\ a-1 & b \\ a & b-1 \end{bmatrix},$$

where the elements a, b are any constant or variable coefficients. For example, assuming that $a = b = 1/3$, one can obtain equation (4.1.4), for which the estimate corresponds by the criterion with respect to f_{19}. The substitution of this matrix into formula (4.5.7) gives

$$\hat{X} = (E - KM)\,X = \begin{bmatrix} 1-a-b & a & b \\ 1-a-b & a & b \\ 1-a-b & a & b \end{bmatrix} \begin{bmatrix} x_1 \\ x_2 \\ x_3 \end{bmatrix}.$$

Consequently, at any matrix K that satisfies condition (4.5.8), correction algorithm (4.5.7) is used to obtain the vector $\hat{X}$ with equal components $\hat{x}_1 = \hat{x}_2 = \hat{x}_3$. Hence, the evaluation function on the basis of correction algorithm (4.5.7) has the form

$$f_{21}: \quad \hat{x} = (1-a-b)\,x_1 + ax_2 + bx_3.$$

This evaluation function sets a totality of estimates which are obtained at various values a and b. Assuming, for example, that $a = b = 1$, the following estimate is obtained:

$$f_{22}: \quad \hat{x} = -x_1 + x_2 + x_3.$$

An estimate of this kind, and other estimates similar to it, can only be used at close values of measurements, and this corresponds to the assumption accepted at the beginning of this subsection. They satisfy Cauchy's condition (4.4.3), but at great dispersions of measurements they cannot satisfy Kolmogorov's condition (4.4.4).

4.5.3 Application of diagnostic algorithms for screening a part of measurements

Let us look at the situation, where just as with small errors some single errors of a high level, caused by, for example, failure of sensors, can take place. From the point of view of technical diagnostics in such a situation it is useful to determine the number of unreliable measurement values and to "eliminate" this, forming the estimate on the basis of the two remaining values.

This approach allows a whole group of estimates to be constructed which differ in the diagnostic algorithm used for determining the index of an unreliable measurement value, as well as in the method of forming an estimate on the basis of the remaining values. Below we will give a description of some estimates developed within the framework of the method of redundant variables and systems with algebraic invariants.

4.5.3.1 Rejecting one measurement by minimum discrepancy

In the theory of systems with algebraic invariants it is proved that to diagnose single errors it is necessary to have two independent algebraic invariants, which in the linear case can be written as

$$\Delta = MX = 0,$$

where M is the rectangular matrix having two lines.

Let us pass from the initial algebraic invariants by means of linear combination of them to the system of dependent invariants

$$\bar{\Delta} = \bar{M} X = 0, \tag{4.5.9}$$

where $\bar{M}$ is the square matrix with a zero diagonal.

For systems (4.5.1) and algebraic invariants (4.5.3) equation (4.5.9) will have the form

$$\begin{bmatrix} \Delta_1 \\ \Delta_2 \\ \Delta_3 \end{bmatrix} = \begin{bmatrix} 0 & 1 & -1 \\ 1 & 0 & -1 \\ 1 & -1 & 0 \end{bmatrix} \begin{bmatrix} x_1 \\ x_2 \\ x_3 \end{bmatrix}. \tag{4.5.10}$$

Substituting for $x_i = x + e_i, i = \overline{1,3}$, one obtains

$$\begin{bmatrix} \Delta_1 \\ \Delta_2 \\ \Delta_3 \end{bmatrix} = \begin{bmatrix} 0 & 1 & -1 \\ 1 & 0 & -1 \\ 1 & -1 & 0 \end{bmatrix} \begin{bmatrix} e_1 \\ e_2 \\ e_3 \end{bmatrix}.$$

If one of the errors e_i is significantly greater than the others, this will lead to a significant deviation of all the components of the discrepancy vector, except the one into which it enters with the zero coefficient. This circumstance allows the index of the unreliable measurement value to be determined and rejected.

Thus, the algorithm of diagnostics consists of the fact that from three measurement values x_1, x_2, x_3 only one is dropped, the index of which coincides with the index of the minimum quantity from quantities $|\Delta_1|$, $|\Delta_2|$, $|\Delta_3|$. The estimate is formed on the basis of the two remaining values of measurements (let them be called x_1', x_2'), for example, by means of calculating their arithmetic, geometric, and other means.

The transfer from the measurement values x_1, x_2, x_3 to the values x_1', x_2' can be represented in a formal manner:

$$\begin{bmatrix} x_1' \\ x_2' \end{bmatrix} = L_m' \begin{bmatrix} x_1 \\ x_2 \\ x_3 \end{bmatrix},$$

where a rejection operator on a minimum discrepancy L_m' is set by the matrix (2×3), in which each line one element is equal to 1 and the remaining ones are equal to zero. For example, the rejection matrix of the first measurement has the form

$$L_m' = \begin{bmatrix} 0 & 1 & 0 \\ 0 & 0 & 1 \end{bmatrix}.$$

Using different methods of evaluation of the remaining measurement values it is possible to obtain the following estimates:

$$\left.\begin{aligned} f_{23}: &\quad \hat{x} = \frac{1}{2}\left(x_1' + x_2'\right) \\ f_{24}: &\quad \hat{x} = \sqrt{x_1' \cdot x_2'} \\ f_{25}: &\quad \hat{x} = 2\left(\frac{1}{x_1'} + \frac{1}{x_2'}\right)^{-1} \\ f_{26}: &\quad \hat{x} = \sqrt{\frac{1}{2}(x_1'^2 + x_2'^2)} \\ f_{27}: &\quad \hat{x} = \frac{x_1'^2 + x2'^2}{x_1' + x_2'} \end{aligned}\right\}, \quad \begin{bmatrix} x_1' \\ x_2' \end{bmatrix} = L_m' \begin{bmatrix} x_1 \\ x_2 \\ x_3 \end{bmatrix}.$$

4.5.3.2 Rejection of one measurement on the basis of a maximum discrepancy

The idea of this algorithm consists of getting an estimate of the vector of errors on the basis of an analysis of discrepancies of algebraic invariants as well as of rejection of a measurement value possessing the maximum error estimate.

The optimal, in the sense of least squares method, estimate of the error vector is given by formula (4.5.5):

$$\hat{e} = M^+\Delta = M^T(MM^T)^{-1}MX,$$

hence, for the matrix M^+ one obtains

$$\begin{bmatrix} \hat{e}_1 \\ \hat{e}_2 \\ \hat{e}_3 \end{bmatrix} = \frac{1}{3}\begin{bmatrix} 2 & -1 & -1 \\ -1 & 2 & -1 \\ -1 & -1 & 2 \end{bmatrix}\begin{bmatrix} x_1 \\ x_2 \\ x_3 \end{bmatrix}. \tag{4.5.11}$$

From three measurement values x_1, x_2, x_3 only that value will be eliminated whose index coincides with the index of the maximum quantity from the quantities $|\hat{e}_1|$, $|\hat{e}_2|$, $|\hat{e}_3|$. Denoting as before the remaining values through x'_1, x'_2, it is possible to write

$$\begin{bmatrix} x'_1 \\ x'_2 \end{bmatrix} = L'_M \begin{bmatrix} x_1 \\ x_2 \\ x_3 \end{bmatrix},$$

where L'_M is the operator of rejection according to the minimum estimate of error, similar to the operator L'_m.

In the same way as before, depending on the method of evaluation of the remaining measurement values, it is possible to obtain different estimates:

$$\left.\begin{aligned} f_{28}: \quad \hat{x} &= \frac{1}{2}(x'_1 + x'_2) \\ f_{29}: \quad \hat{x} &= \sqrt{x'_1 \cdot x'_2} \\ f_{30}: \quad \hat{x} &= 2\left(\frac{1}{x'_1} + \frac{1}{x'_2}\right)^{-1} \\ f_{31}: \quad \hat{x} &= \sqrt{\frac{1}{2}(x'^2_1 + x'^2_2)} \\ f_{32}: \quad \hat{x} &= (x'_1 + x'_2)^{-1}(x'^2_1 + x'^2_2) \end{aligned}\right\}, \quad \begin{bmatrix} x'_1 \\ x'_2 \end{bmatrix} = L'_M \begin{bmatrix} x_1 \\ x_2 \\ x_3 \end{bmatrix}.$$

In addition to handling measurement data with the help of the operators L'_M and L'_m, many other rejection methods are known. For example, it is possible to consider a measurement value as unreliable and to reject it, owing to the fact that this value is the most distant from the arithmetic mean of three values, from their geometric mean, or from any other estimate from those indicated earlier.

4.5.3.3 Estimates with rejection of two measurement values

A separate class is formed by algorithms of evaluation which use the rejection of two of three measurement values. These evaluation algorithms allow not only single but also some double failures to be "parried". In accordance with these algorithms one of the measurement values is taken as the estimate, and the two remaining values are not present in an explicit form. This certainly, however, does not mean that they do not influence the formation of the estimate, since its choice depends on the relationship of all three measurement values.

Classical examples of algorithms with rejection of two measurement values are the following algorithms:

$$\hat{x} = \max(x_1, x_2, x_3),$$
$$\hat{x} = \min(x_1, x_2, x_3),$$
$$\hat{x} = \operatorname{med}(x_1, x_2, x_3),$$

which use as the estimate the maximum, minimum, or average value measurement. Other versions are also possible, when, for example, as the estimate a value of measurement nearest to the arithmetic means, geometric means, or some other mean estimate, is taken.

Let us now describe two evaluation algorithms of this kind, which are supported by the diagnostics procedures indicated above.

The first consists of calculating discrepancy vector (4.5.10) and throwing out two of three measurement values x_1, x_2, x_3, the indices of which coincide with the indices of the discrepancies having a less absolute quantity. Thereby a measurement, to which the greatest discrepancy from $|\Delta_1|$, $|\Delta_2|$, $|\Delta_3|$ corresponds, is taken as the estimate.

Let the rejection operator determined in such a manner be denoted as L''_m (the number of dashes indicates the number of measurements rejected); then it is possible to write

$$f_{33}: \quad \hat{x} = L''_m(x_1, x_2, x_3).$$

According to the second algorithm, two measurement values are rejected, for which the estimate of the errors obtained by formula (4.5.11) has the maximum absolute quantity. Hereby a measurement with a minimum error estimate is taken as the estimate. Denoting the corresponding operator as L''_M, it is possible to write

$$f_{34}: \quad \hat{x} = L''_M(x_1, x_2, x_3).$$

4.5.4 Systematization and analysis of evaluation algorithms

4.5.4.1 Systematization of evaluation algorithms

A great number of algorithms for evaluating the scalar quantity x on the basis of its values of two or three measurements x_1, x_2, x_3 have been described above. They were obtained on the basis of four approaches, i.e., probabilistic, deterministic, heuristic, and diagnostic; most of them have been collected in Tables 4.11 and 4.12. On the whole, both tables contain more then 100 algorithms, divided into separate groups.

In drawing up the tables preference was given to nonthreshold algorithms that do not require a priori information of the probabilistic or statistical character.

Tables 4.11 and 4.12 can be used as a source for the analysis of algorithms and their comparison by various feature signs. To the number of algorithms it is possible to refer the algorithm character and method of setting it, convenience of technical realization, accuracy and reliability of the estimate obtained, and others. Let us briefly consider each listed sign.

4.5.4.2 Analysis of evaluation algorithms

Algorithm character

Judging from estimates the appearance the algorithms included in the tables are divided into 9 groups: classical means, linear and quasi-linear estimates, difference and nonlinear estimates, algorithms set by criteria, estimates with rejection of one or two measurement values, and superpositions or combinations of estimates. In the last group only one superposition is indicated from an infinitely great number of possible superpositions. The arithmetic means of all estimates included in Table 4.11, as well as their geometric means, sample medians, are additional examples.

When classifying algorithms, some other characteristics of them are also used. In particular, they distinguish threshold and nonthreshold algorithms, inertial and uninertial, symmetrical and asymmetrical algorithms, and so on.

Methods of setting an algorithm

In order to set an algorithm, it is possible to indicate the estimate function f, its derivative f', minimized criterion J or its derivative J', which, in turn, can be set in analytical, graphical forms as well as in the form of tables or algorithms.

Thus, nearly fifteen methods of setting estimates exist. From the point of view of technical realization, as well as in analyzing the properties of estimates, each of them appears to be useful.

When the analytical description of an evaluation function is too complicated or even does not exist at all, we are obliged to set the algorithm with the help of a criterion. However, in the cases where the analytical description is known, sometimes it is useful to ascertain the kind of criterion for analyzing such properties of the estimate obtained as sensitivity, accuracy, and reliability.

For the geometric interpretation of criteria, four kinds of diagrams are applied. Two of them are the diagrams of criterion and its derivative dependencies on a variated variable x. An estimate looked for is the point $x = \hat{x}$, at which the diagram $J(x)$ has the minimum and the diagram $J'(x)$ intersects the x-axis. An example of the diagram $J(x)$ for the mean module criterion is illustrated in Figure 4.19.

Table 4.11. Evaluation of the scalar quantity by three measurements x_1, x_2, x_3.

N°	Formula or criterion	Notes
Classical means and their generalizations		
1	$\hat{x} = \frac{x_1 + x_2 + x_3}{3}$	Arithmetic mean (A)
2	$\hat{x} = \sqrt[3]{x_1 x_2 x_3}$	Geometric mean (G)
3	$\hat{x} = \frac{3}{\frac{1}{x_1} + \frac{1}{x_2} + \frac{1}{x_3}} = \frac{3x_1 x_2 x_3}{x_1 x_2 + x_1 x_3 + x_2 x_3}$	Harmonic mean (H)
4	$\hat{x} = \frac{x_1^2 + x_2^2 + x_3^2}{x_1 + x_2 + x_3}$	Contraharmonic mean (C)
5	$\hat{x} = \sqrt{\frac{1}{3}\left(x_1^2 + x_2^2 + x_3^2\right)}$	Quadratic mean, or root mean square (Q)
6	$\hat{x} = \frac{a_1 x_1 + a_2 x_2 + a_3 x_3}{a_1 + a_2 + a_3}$	Weighted arithmetic mean
7	$\hat{x} = (x_1^{a_1} x_2^{a_2} x_3^{a_3})^{\frac{1}{a_1 + a_2 + a_3}}$	Weighted geometric mean
8	$\hat{x} = \frac{a_1 + a_2 + a_3}{\frac{a_1}{x_1} + \frac{a_2}{x_2} + \frac{a_3}{x_3}}$	Weighted harmonic mean
9	$\hat{x} = \frac{a_1 x_1^2 + a_2 x_2^2 + a_3 x_3^2}{a_1 x_1 + a_2 x_2 + a_3 x_3}$	Weighted contraharmonic mean
10	$\hat{x} = \sqrt{\frac{a_1 x_1^2 + a_2 x_2^2 + a_3 x_3^2}{a_1 + a_2 + a_3}}$	Weighted quadratic mean
Linear and quasi-linear estimates		
11	$\hat{x} = -x_1 + x_2 + x_3$	Versions of correction on RVM
12	$\hat{x} = (1 - a - b)x_1 + a\, x_2 + b\, x_3$	$\Delta = MX,\ M = \begin{bmatrix} 1 & -1 & 0 \\ 1 & 0 & 1 \end{bmatrix}$
13	$\hat{x} = x_1 - RM^T(MRM^T)^{-1}M\Delta$	R is the covariance matrix of error
14	$\hat{x} = \frac{x_1 x_2 + x_2 x_3 + x_1 x_3}{x_1 + x_2 + x_3}$	Symmetric mean
15	$\hat{x} = \frac{x_1^{k+1} + x_2^{k+1} + x_3^{k+1}}{x_1^k + x_2^k + x_3^k}$	Lehmer means L_k (k-family); L_{-1}=H, L_0=A, L_1=C
16	$\hat{x} = \frac{a_1 x_1^{k+1} + a_2 x_2^{k+1} + a_3 x_3^{k+1}}{a_1 x_1^k + a_2 x_2^k + a_3 x_3^k}$	Weighted Lehmer means (k-family)
17	$\hat{x} = \frac{x_1 e^{x_1} + x_2 e^{x_2} + x_3 e^{x_3}}{e^{x_1} + e^{x_2} + e^{x_3}}$	Exponential weighting
18	$\hat{x} = \frac{x_1 \cdot e^{ax_1} + x_2 \cdot e^{ax_2} + x_3 \cdot e^{ax_3}}{e^{ax_1} + e^{ax_2} + e^{ax_3}}$	a-family

Table 4.11. (cont.)

Nº	Formula or criterion	Notes
19	$\hat{x} = \dfrac{\frac{x_1}{\lvert x_3-x_2\rvert} + \frac{x_2}{\lvert x_3-x_1\rvert} + \frac{x_3}{\lvert x_2-x_1\rvert}}{\frac{1}{\lvert x_3-x_2\rvert} + \frac{1}{\lvert x_3-x_1\rvert} + \frac{1}{\lvert x_2-x_1\rvert}}$	Principle of no confidence in the nearest values of measurements
20	$\hat{x} = \dfrac{x_1 \lvert x_2 - x_3\rvert + x_2 \lvert x_1 - x_3\rvert + x_3 \lvert x_1 - x_2\rvert}{\lvert x_2 - x_3\rvert + \lvert x_1 - x_3\rvert + \lvert x_1 - x_2\rvert}$	Coincides with the sample median
21	$\hat{x} = \dfrac{x_1 (x_2 - x_3)^2 + x_2 (x_1 - x_3)^2 + x_3 (x_1 - x_2)^2}{(x_2 - x_3)^2 + (x_1 - x_3)^2 + (x_1 - x_2)^2}$	Principle of confidence in two nearest measurement values
22	$\hat{x} = \dfrac{x_1 e^{\lvert x_2-x_3\rvert} + x_2 e^{\lvert x_1-x_3\rvert} + x_3 e^{\lvert x_1-x_2\rvert}}{e^{\lvert x_2-x_3\rvert} + e^{\lvert x_1-x_3\rvert} + e^{\lvert x_1-x_2\rvert}}$	
23	$\hat{x} = \dfrac{x_1 \lvert x_2 - x_3\rvert^k + x_2 \lvert x_1 - x_3\rvert^k + x_3 \lvert x_1 - x_2\rvert^k}{\lvert x_2 - x_3\rvert^k + \lvert x_1 - x_3\rvert^k + \lvert x_1 - x_2\rvert^k}$	k-family
Nonlinear estimates		
24	$\hat{x} = \sqrt[3]{(x_1 + a)(x_2 + a)(x_3 + a)} - a$	Biased geometric mean
25	$\hat{x} = \sqrt{\dfrac{3x_1x_2x_3}{x_1 + x_2 + x_3}}$	
26	$\hat{x} = \sqrt[3]{\dfrac{x_1^3 + x_2^3 + x_3^3}{3}}$	Root-mean-cube
27	$\hat{x} = \sqrt[3]{\dfrac{a_1x_1^3 + a_2x_2^3 + a_3x_3^3}{a_1 + a_2 + a_3}}$	Weighted root-mean-cube
28	$\hat{x} = \sqrt[k]{\dfrac{1}{3}\left(x_1^k + x_2^k + x_3^k\right)}$	Power mean (k-family)
29	$\hat{x} = \sqrt{\dfrac{a_1x_1^k + a_2x_2^k + a_3x_3^k}{a_1 + a_2 + a_3}}$	Weighted power mean
30	$\hat{x} = \sqrt{\dfrac{x_1x_2 + x_1x_3 + x_2x_3}{3}}$	Combinatoric mean
31	$\hat{x} = \sqrt{\dfrac{(a-b)(b-c)(c-a)}{2((b-c)\ln a + (c-a)\ln b + (a-b)\ln c}}$	Logarithmic mean
32	$\hat{x} = \ln \dfrac{e^{x_1} + e^{x_2} + e^{x_3}}{3}$	Quasi-arithmetic mean
33	$\hat{x} = f^{-1}\left(\dfrac{f(x_1) + f(x_2) + f(x_3)}{3}\right)$	Kolmogorov's mean (a family)

Table 4.11. (cont.)

Nº	Formula or criterion	Notes
34	$\hat{x} = f^{-1}\left(\dfrac{a_1 f(x_1) + a_2 f(x_2) + a_3 f(x_3)}{a_1 + a_2 + a_3}\right)$	Weighted Kolmogorov's mean (a family)
Estimates, minimizing the criteria of the form $J = \rho(x_1 - \hat{x}) + \rho(x_2 - \hat{x}) + \rho(x_3 - \hat{x})$		
35	$\rho(x_i - \hat{x}) = \sqrt{(x_i - \hat{x})^2 + a^2} - a$	
36	$\rho(x_i - \hat{x}) = \lvert x_i - \hat{x}\rvert + (x_i - \hat{x})^2$	Combined criterion
37	$\rho(\hat{x} - x_i) = \begin{cases} 0.5\,(\hat{x} - x_i)^2 & \text{at } \lvert\hat{x} - x_i\rvert < a \\ a\,\lvert\hat{x} - x_i\rvert - 0,5a^2 & \text{at } \lvert\hat{x} - x_i\rvert \geq a \end{cases}$	Compound criterion
38	$\rho(\hat{x} - x_i) = \begin{cases} 0.5\,(\hat{x} - x_i)^2 & \text{at } \lvert\hat{x} - x_i\rvert < a \\ 0.5a & \text{at } \lvert\hat{x} - x_i\rvert \geq a \end{cases}$	Compound criterion
Iterative means ($a_0 = x_1$, $b_0 = x_2$, $c_0 = x_3$, $n = 0, 1, 2\ldots$)		
39	$a_{n+1} = A(a_n, b_n, c_n)$, $b_{n+1} = G(a_n, b_n, c_n)$, $c_{n+1} = H(a_n, b_n, c_n)$	Arithmetic-Geometric-Harmonic mean (Iterative *AGH*-mean)
40	$a_{n+1} = M_1(a_n,\ b_n,\ c_n)$, $b_{n+1} = M_2(a_n,\ b_n,\ c_n)$, $c_{n+1} = M_3(a_n,\ b_n,\ c_n)$	A family of iterative means; M_1, M_2, M_3 – different means
Estimates with rejection of two measurement values		
41	$\hat{x} = \max(x_1, x_2, x_3)$	Choice of the maximum
42	$\hat{x} = \min(x_1, x_2, x_3)$	Choice of the minimum
43	$\hat{x} = \mathrm{med}(x_1, x_2, x_3)$	Choice of the median
44	$\hat{x} = L''(\mathrm{x}_1, \mathrm{x}_2, \mathrm{x}_3)$	Dropping off the nearest two measurement values
45	$\hat{x} = L''_m(\mathrm{x}_1, \mathrm{x}_2, \mathrm{x}_3)$	Rejection on the basis of diagnostics algorithms RVM
46	$\hat{x} = L''_M(\mathrm{x}_1, \mathrm{x}_2, \mathrm{x}_3)$	
Estimates with rejection of one measurement value and averaging of the remaining ones		
47–51	Dropping off $\max(x_1, x_2, x_3)$	The remaining values of measurements x'_1 and x'_2 are evaluated by one of the formulae 1) $0.5(x'_1 + x'_2)$ 2) $\sqrt{x'_1 x'_2}$ 3) $2\left(\frac{1}{x'_1} + \frac{1}{x'_2}\right)^{-1}$ 4) $\sqrt{\frac{1}{2}(x_1'^2 + x_2'^2)}$ 5) $\left(x'_1 + x'_2\right)^{-1}\left(x_1'^2 + x_2'^2\right)$
52–56	Dropping off $\min(x_1, x_2, x_3)$	
57–61	Dropping off $\mathrm{med}(x_1, x_2, x_3)$	
62–66	Dropping off $L''\ (x_1, x_2, x_3)$	
67–71	Dropping off $L''_m\ (x_1, x_2, x_3)$	
72–76	Dropping off $L''_M\ (x_1, x_2, x_3)$	
77–81	$\hat{x} = L'_m(\mathrm{x}_1, \mathrm{x}_2, \mathrm{x}_3)$	
82–86	$\hat{x} = L'_M(\mathrm{x}_1, \mathrm{x}_2, \mathrm{x}_3)$	

Table 4.11. (cont.)

N°	Formula or criterion	Notes
	Superposition or combination of estimates	
87	$\hat{x} = \frac{1}{4}\left[\frac{1}{3}(x_1+x_2+x_3) + \sqrt[3]{x_1x_2x_3} + 3\left(\frac{1}{x_1}+\frac{1}{x_2}+\frac{1}{x_3}\right)^{-1} + \sqrt{\frac{1}{3}\left(x_1^2+x_2^2+x_3^2\right)}\right]$	Arithmetic mean of four classical means

Table 4.12. Evaluation of the scalar quantity by two measurements a, b.

N°	Formula or criterion	Notes
1	$T(a,\ b) = \frac{2}{3}\cdot\frac{a^2+ab+b^2}{a+b}$	Centroidal mean
2	$H_1(a,\ b) = \frac{1}{3}(a+\sqrt{ab}+b)$	Heron mean
3	$H_\omega(a,\ b) = \frac{a+b+\omega\sqrt{ab}}{2+\omega}$	Heron means (ω-family)
4	$H_p(a,\ b) = \left(\frac{a^p+(ab)^{p/2}+b^p}{3}\right)^{1/p}$	Heron means (p-family); $H_0 = \sqrt{ab}$
5	$Q_p(a,b) = \frac{a^pb^{1-p}+a^{1-p}b^p}{2}$	Heinz or symmetric means (p-family), $0 \le p \le 1/2$. $Q_0 = A$, $Q_{1/2} = G$
6	$L(a,b) = \frac{a-b}{\ln a - \ln b}$	Logarithmic mean
7	$I(a,b) = \frac{1}{e}\left(\frac{b^b}{a^a}\right)^{1/(b-a)}$	Identric or exponential mean
8	$J(a,b) = (a^ab^b)^{1/(a+b)}$	$a, b > 0$
9	$P(a,b) = \frac{a-b}{4\arctan\left(\sqrt{a/b}\right)-\pi}$	Seiffert's mean
10	$L_p(a,b) = \left(\frac{a^{p+1}-b^{p+1}}{(p+1)(a-b)}\right)^{1/p}$	Power logarithmic means (p-family) $L_1 = A$, $L_0 = I$, $L_{-1} = L$, $L_{-2} = G$

Table 4.12. (cont.)

N°	Formula or criterion	Notes
11	$D_p(a,\ b) = \dfrac{p}{p+1} \cdot \dfrac{a^{p+1} - b^{p+1}}{a^p - b^p}$	Power difference means (p-family) $D_1 = A,\ D_{-1/2} = G,\ D_{-1} = G/L$
12	$S_{p,q}(a,\ b) = \left(\dfrac{p}{q} \cdot \dfrac{a^q - b^q}{a^p - b^p}\right)^{1/(q-p)}$	Stolarsky means of order (p,q) (p,q-family) $S_{1,q+1} = L_q, S_{p,p+1} = D_p$
Iterative means ($a_0 = a, b_0 = b, n = 0, 1, 2\ldots$)		
13	$a_{n+1} = \dfrac{a_n + b_n}{2}, b_{n+1} = \sqrt{a_n b_n}$	Arithmetic-Geometric mean (AGM) $AGM(a,b) = \dfrac{\pi}{2\displaystyle\int_0^{\pi/2} \dfrac{dx}{\sqrt{a^2 \sin^2 x + b^2 \cos^2 x}}}$
14	$a_{n+1} = \dfrac{a_n + b_n}{2},\quad b_{n+1} = \dfrac{2a_n b_n}{a_n + b_n}$	Arithmetic-Harmonic mean (AHM) $AHM(a,b) = \sqrt{ab}$
15	$a_{n+1} = \dfrac{2a_n b_n}{a_n + b_n},\quad b_{n+1} = \sqrt{a_n b_n}$	Geometric-Harmonic mean (GHM) $GHM(a,b) = \dfrac{2ab}{\pi} \displaystyle\int_0^{\pi/2} \dfrac{dx}{\sqrt{a^2 \cos^2 x + b^2 \sin^2 x}}$
16	$a_{n+1} = \dfrac{a_n + b_n}{2},\quad b_{n+1} = \sqrt{a_{n+1} b_n}$	Schwab–Borchardt mean (SBM) $SBM = \dfrac{\sqrt{b^2 - a^2}}{\arccos \frac{a}{b}}$
17	$a_{n+1} = \dfrac{1}{2}(a_n + b_n),$ $b_{n+1} = \sqrt{a_n a_{n+1}}.$	Lemniscatic mean (LM) $LM(a,b) = \begin{cases} \sqrt{\dfrac{a^2 - b^2}{\mathrm{arcsl}\left(1 - \frac{b^2}{a^2}\right)}} & a > b \\ \sqrt{\dfrac{b^2 - a^2}{\mathrm{arcslh}\left(\frac{b^2}{a^2} - 1\right)}} & a < b \end{cases}$
18	$a_{n+1} = \dfrac{1}{2}(a_n + \sqrt{a_n b_n}),$ $b_{n+1} = \dfrac{1}{2}(b_n + \sqrt{a_n b_n})$	Logarithmic mean (L) $L\,(a,b) = \dfrac{a-b}{\ln a - \ln b}$

Table 4.12. (cont.)

N°	Formula or criterion	Notes
19	$a_{n+1} = f_i(a_n, b_n)$, $b_{n+1} = f_j(a_n, b_n), \quad i \neq j$ $f_1(a,b) = \frac{1}{2}(a+b)$, $f_2(a,b) = (ab)^{1/2}$, $f_3(a,b) = \left(a\frac{a+b}{2}\right)^{1/2}$, $f_4(a,b) = \left(b\frac{a+b}{2}\right)^{1/2}$	A family of twelve iterative means $L_{ij}(a,b)$
20	$M^*(a,b) = (M(a^{-1}, b^{-1}))^{-1}$	Dual mean of M-means
21	$L^*(a,b) = ab\frac{\ln b - \ln a}{b-a}$	Dual mean of logarithmic mean $L(a,b)$
22	$I^*(a,\ b) = e\left(\frac{a^b}{b^a}\right)^{1/b-a}$	Dual mean of identric mean $I(a,b)$

Two other diagrams are the diagrams of the loss function ρ and its derivative. They are used for the criteria of the form $J = \rho(x_1 - x) + \rho(x_2 - x) + \rho(x_3 - x)$. Examples of the dependences $\rho(x_i - x)$ and $\rho'(x_i - x)$ for the quadratic mean and mean module criteria are given in Figures 4.26 and 4.27. At the point $x = x_i$ the loss function is minimal and the sensitivity function is $u = \rho'(x_i - x) = 0$.

Setting an estimate with the help of the evaluation function whcih for three measurements represents the function of three arguments is more used:

$$\hat{x} = f(x_1, x_2, x_3). \tag{4.5.12}$$

Equation (4.5.12) describes the surface in a four-dimensional space, the character of which depends on the function f form. A definite idea about this surface can be obtained if two arguments are fixed and the dependence of the estimate $\hat{x}$ on the third argument is depicted in the form of a diagram. Just in this way the diagrams, constructed under the condition $x_1 = \text{const}$, $x_2 = \text{const}$ and shown in Figures 4.25–4.31, were obtained. Note that a similar approach is used in electronics and radio engineering, when setting a function of several arguments (for example of transistor characteristics) is performed with the help of a family of flat curves.

A more complete idea about the evaluation surface can be obtained if only one argument (for example, x_3) is considered and the function $\hat{x}$ is regarded as a function of the two remaining arguments. In a geometric manner this will be in line with surfaces

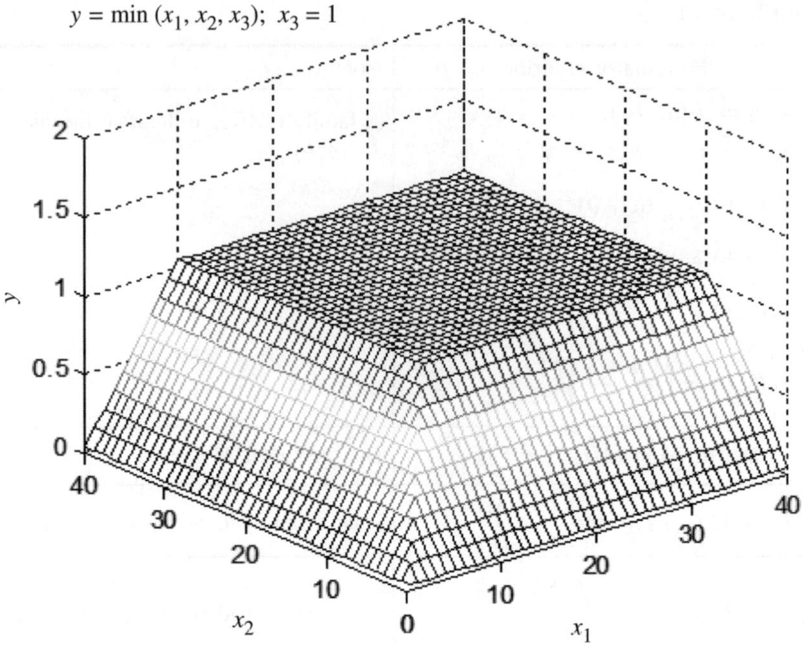

Figure 4.33. Diagram of the evaluation surface for $\hat{x} = y = \min(x_1, x_2, x_3)$ at $x_3 = 1$.

in a three-dimensional space with the coordinate axes $\hat{x}$, x_1, x_2. Figures 4.33 and 4.34 illustrate the surfaces for the functions constructed under condition that $x_3 = 1$:

$$\hat{x} = \min(x_1, x_2, x_3), \hat{x} = \max(x_1, x_2, x_3), \tag{4.5.13}$$

Superimposing surfaces (4.5.13) on one diagram, we obtain Figure 4.35. A single common point a of both surfaces lies on a bisectrix of the first octant and has coordinates $(1, 1, 1)$.

In accordance with condition (4.4.1) Figure 4.35 in the explicit form sets a domain to which the evaluation functions can belong. The surface corresponding to any function of such a kind has to pass through point a and to lie between surfaces (4.5.13), depicted in Figures 4.33 and 4.34. Domains lower than the level $\hat{x} = \min(x_1, x_2, x_3)$ and higher than $\hat{x} = \max(x_1, x_2, x_3)$ are forbidden.

In Figure 4.36 the surface is shown for a sample median constructed for $x_3 = 1$:

$$\hat{x} = \operatorname{med}(x_1, x_2, x_3). \tag{4.5.14}$$

The two-dimensional diagrams for functions (4.5.13) and (4.5.14), which are shown in Figures 4.25 and 4.31, can be obtained by crossing the surfaces in Figures 4.33–4.36 with a vertical plane set by the equation $x_2 = 2$. For example, the diagram for the

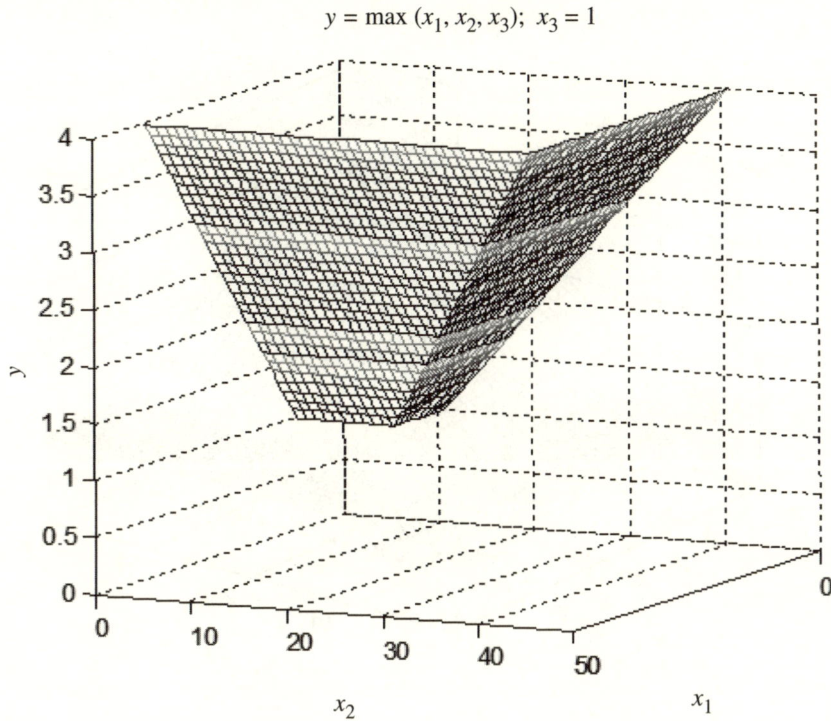

Figure 4.34. Diagram of the evaluation surface for $\hat{x} = y = \max(x_1, x_2, x_3)$ at $x_3 = 1$.

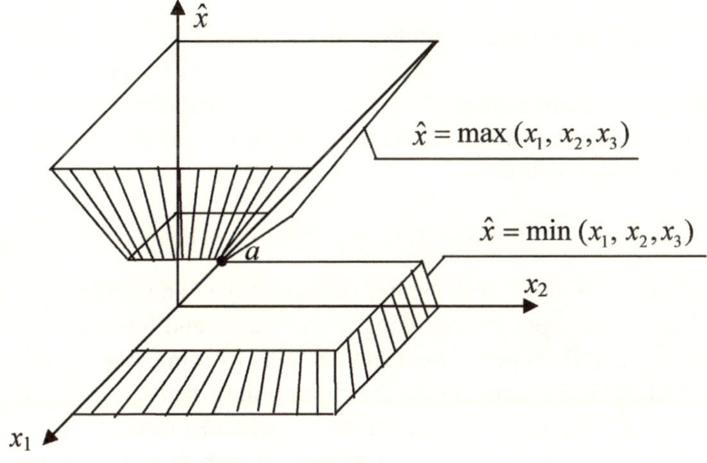

Figure 4.35. Domain of the evaluation surfaces existence.

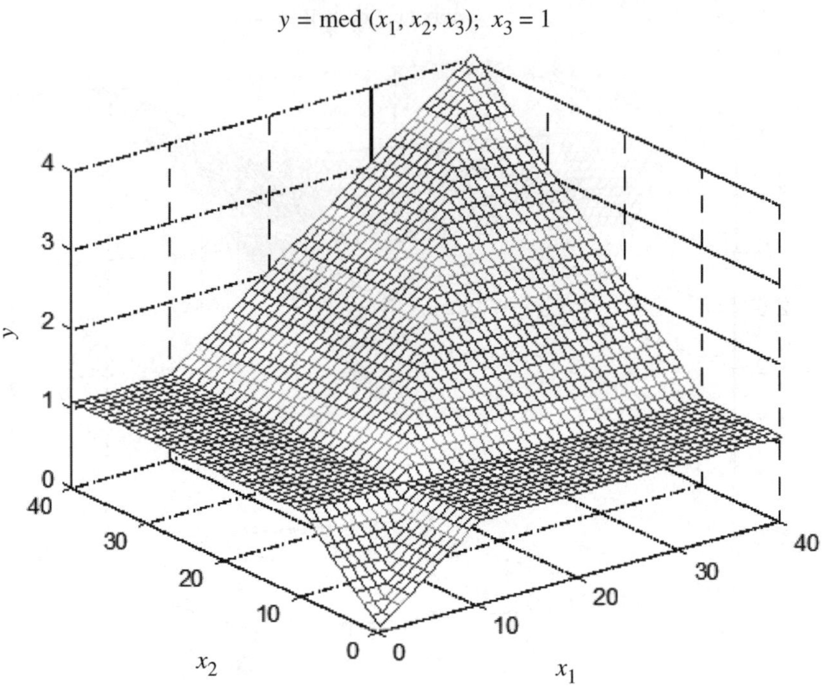

Figure 4.36. Diagram of the surface of the sample mean $\hat{x} = y = \mathrm{med}(x_1, x_2, x_3)$ at $x_3 = 1$.

sample median in Figure 4.31a coincides with the intersection of the surface depicted in Figure 4.36 and the plane which is perpendicular to the axis x_2.

Note that for two measurements $\hat{x} = y = f(x_1, x_2)$ the three-dimensional picture gives a useful idea of the evaluation function. The domain of permissible surfaces for this case is set by the condition

$$\min(x_1, x_2) \le \hat{x} \le \max(x_1, x_2),$$

the geometric sense of which is illustrated in Figures 4.37 and 4.38.

In these figures surfaces corresponding to the higher and lower boundaries of inequality are depicted. By superposing them onto one working drawing we can see the permissible and forbidden domains for the evaluation surfaces.

In particular, any evaluation surface for two measurements must contain a bisectrix of the first octant, since it is the line of contiguity of both limiting surfaces.

In Figures 4.39a and 4.39b the surfaces intended to evaluate the geometric mean $\hat{x} = \sqrt{x_1 x_2}$ and harmonic mean $\hat{x} = \frac{2}{\frac{1}{x_1} + \frac{1}{x_2}} = \frac{2x_1 x_2}{x_1 + x_2}$ are shown.

For the case of an arbitrary number of measurements, preference should be given to two-dimensional diagrams, since on the one hand they are not too complicated with

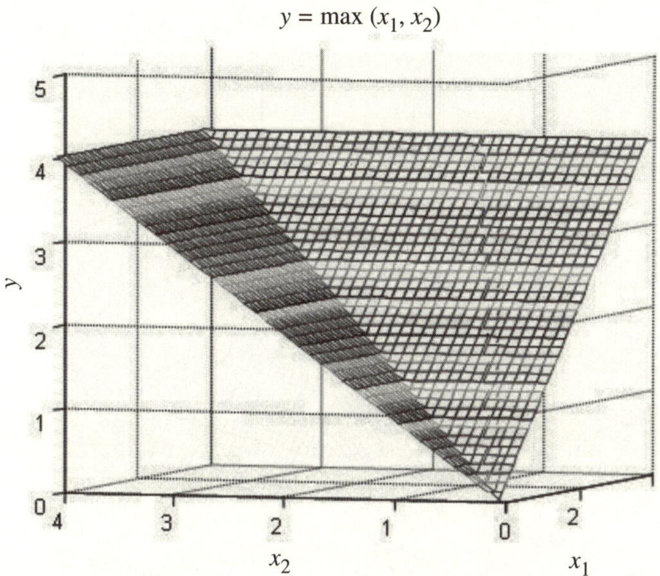

Figure 4.37. Diagram of the evaluation surface for $\hat{x} = y = \max(x_1, x_2)$.

regards to construction, and on the other hand they give a sufficient idea about the character of the evaluation function. If necessary, they can be easily supplemented by sensitivity diagrams with additional information about properties of estimates.

To illustrate the above, in Figure 4.40a and Figure 4.40b the two-dimensional diagrams of the evaluation functions for a sample median and for averaging of three central measurement values at $n = 5$, and in Figure 4.41a and Figure 4.41b the diagrams of their sensitivity functions $\frac{\partial \hat{x}}{\partial x_5}$ are given.

Ease of technical realization

Technically, any estimate is realized either on the basis of software, or with a certain device processing the input signals x_1, x_2, x_3 and forming the output signal $\hat{x}$. This device has to meet the usual technical requirements, such as simplicity, reliability, quick response, operability within a wide range of input signals, and others. Taking them into account, the choice of one of the possible versions for realizing the accepted estimate is performed.

As a starting point for getting an estimate either the evaluation function itself we can use f, or the criterion function F, or its derivative. At the same time, in all cases a hardware or software realization is possible.

Block diagrams of getting the estimate by every method from those listed above are shown in Figure 4.42a,b,c.

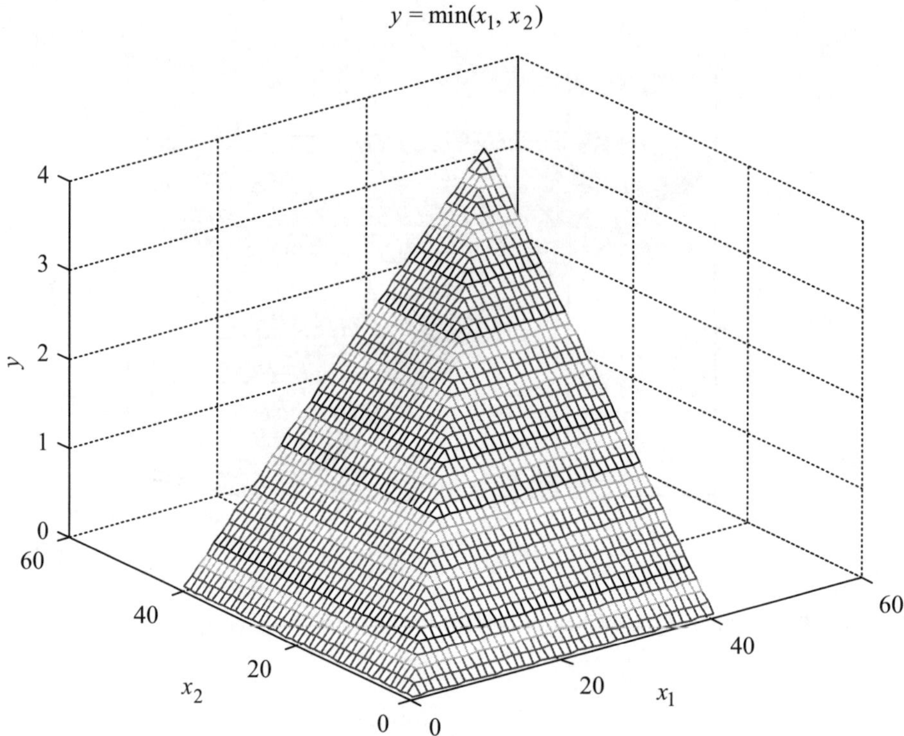

Figure 4.38. Diagram of the evaluation surface for $\hat{x} = y = \min(x_1, x_2)$.

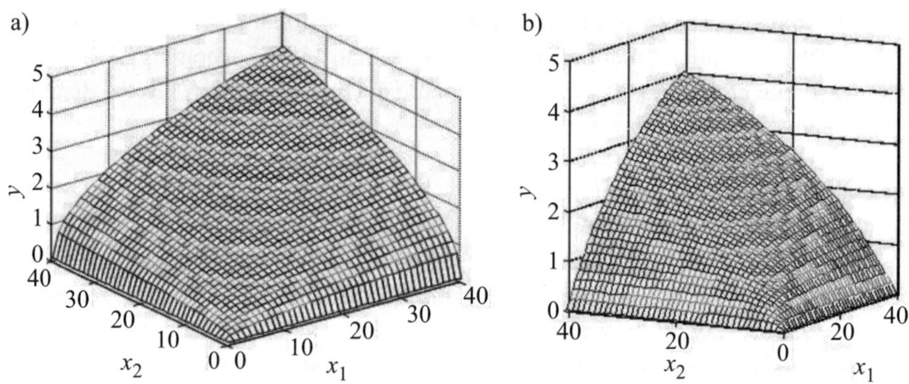

Figure 4.39. (a): surface of the geometric mean $\hat{x} = \sqrt{x_1 x_2}$; (b): surface of harmonic mean $y = \frac{2x_1x_2}{x_1+x_2}$.

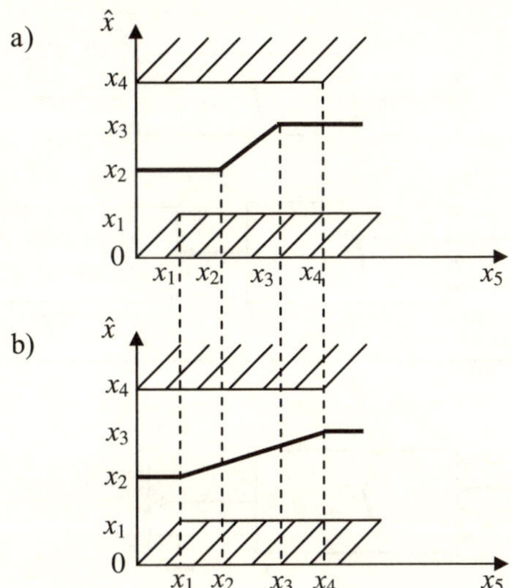

Figure 4.40. Diagrams of the evaluation functions at $n = 5$.

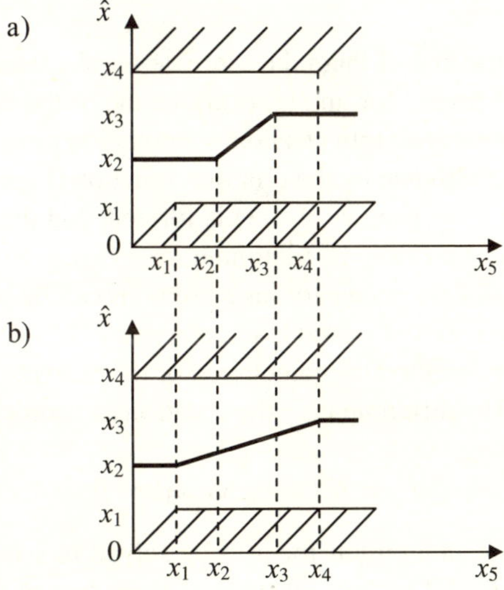

Figure 4.41. Diagrams of the sensitivity functions at $n = 5$.

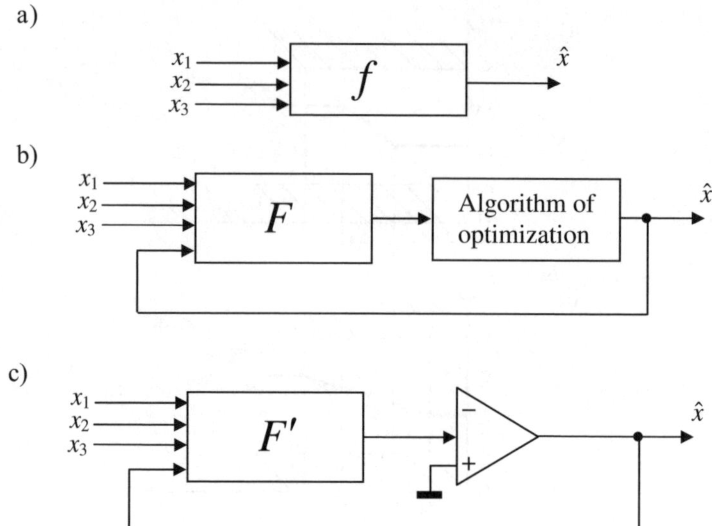

Figure 4.42. Block-schemes for getting an estimate by the evaluation function (a), criterion function (b), and its derivative (c).

In accordance with the first of them the estimate is calculated directly by the evaluation function. For example, for an algorithm of the arithmetic mean the block f (Figure 4.42a) will represent simply an adder (summation unit).

When calculating the estimate by the criterion function (Figure 4.42b), first a value of minimized criterion $J = F(x, x_1, x_2, x_3)$ is formed, and then an automatic search for the quantity $x = \hat{x}$, providing its minimum, is realized. At the same time any of the numerical methods of searching for an extremum can be used, for example, the gradient methods, Gauss–Zeidel method, and others.

According to the third method the calculation of the estimate is performed by solving an equation formed by differentiating the criterion and equating the result obtained to zero:

$$J' = F'(x, x_1, x_2, x_3) = 0. \tag{4.5.15}$$

The structure of an analog circuit for it to be solved (a scheme of a quorum element) is shown in Figure 4.42c. It is constructed according to the method of implicit functions, widely used in analog computing technology for converting functions and equations set in the implicit form.

In accordance with this method the block of forming derivative $J' = F'$ (4.5.15) is included into the feedback of an operational amplifier. Thanks to a significantly large amplification coefficient of the latter, the voltage value at its input is automatically kept close to zero. Consequently, the output signal of the amplifier will approximately coincide with the solution of equation (4.5.15).

Let us explain the versions described by the example of realizing the algorithm of the sample median, starting from the estimate obtained directly by the evaluation function.

Using the operations for selecting the maximum and minimum signals of a number of signals, the function $\hat{x} = \operatorname{med}(x_1, x_2, x_3)$ can be written in the form

$$\hat{x} = \max\left[\min(x_1, x_2), \min(x_1, x_3), \min(x_2, x_3)\right] \tag{4.5.16}$$

or in the tantamount form

$$\hat{x} = \min\left[\max(x_1, x_2),\ \max(x_1, x_3),\ \max(x_2, x_3)\right]. \tag{4.5.17}$$

The block diagram shown in Figure 4.43 is in accordance with these formulas [72, 73].

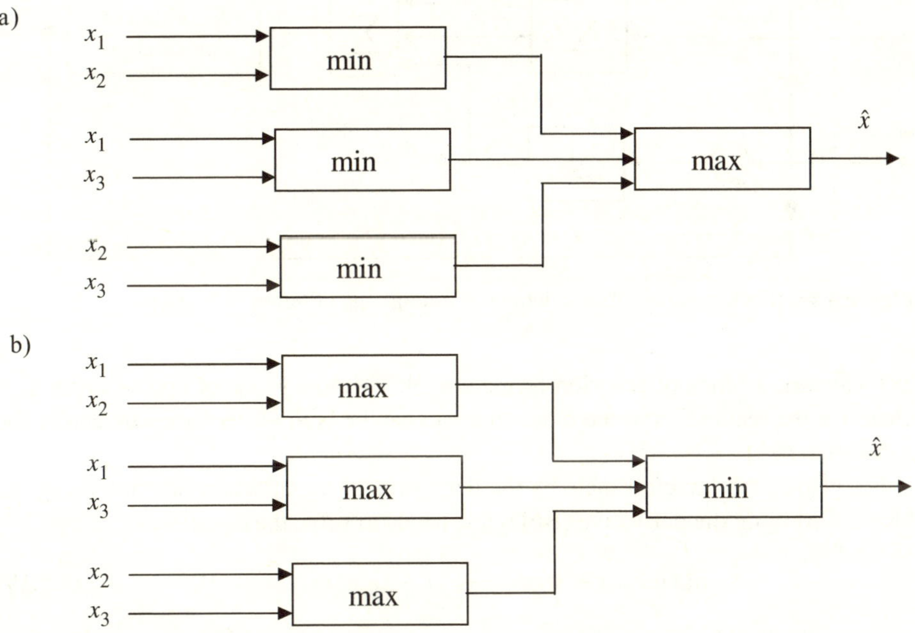

Figure 4.43. Block-diagrams of getting estimates by formulae (4.5.16) and (4.5.17).

Each of the operations on choosing the maximum and minimum can be performed with the help of comparatively simple diode circuits. This allows an analog device to be constructed on the basis of an operational amplifier, diodes and resistors.

The software realization of formulae (4.5.16) and (4.5.17) with the help of an computer is also rather simple and requires only five operations of comparison.

Getting the median estimate by the criterion function is connected with minimization of the criterion value J:

$$J = |x_1 - x| + |x_2 - x| + |x_3 - x| . \tag{4.5.18}$$

For the estimate the value $x = \hat{x}$, providing this criterion with the minimum, is taken. As a matter of fact this method of calculating the sample median is based on the use of numerical methods of optimization. The corresponding block diagram of calculations is shown in Figure 4.44 [72].

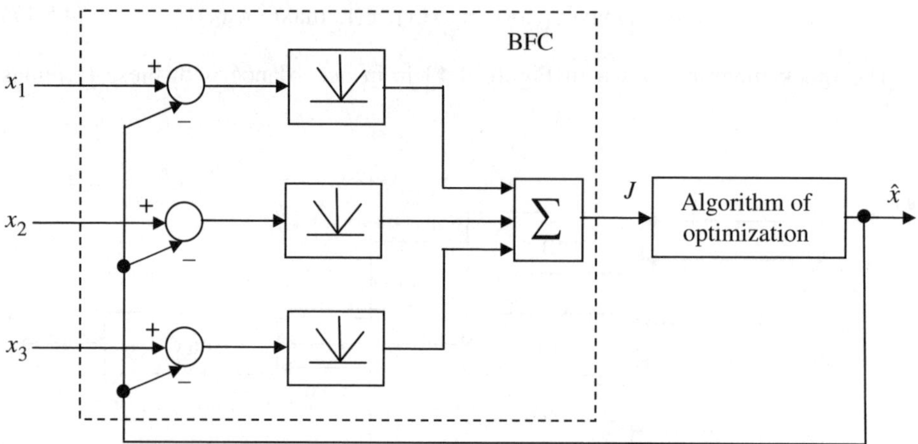

Figure 4.44. Block-diagram of calculations at minimizing criterion (4.5.18).

It contains a block of criterion formation (BCF) and a block of optimization. It is clear that the realization of the block diagram can be both of the hardware and of the software type.

For getting the same estimate by the derivative of the criterion function using the block of forming the derivative (BFD) one needs to solve the equation

$$J' = \text{sign}(x_1 - x) + \text{sign}(x_2 - x) + \text{sign}(x_3 - x) = 0, \tag{4.5.19}$$

which is obtained by differentiating criterion (4.5.18) and equating the result obtained to zero. Here the sign z is used to denote the signature function equal to 1 at $z > 0$; zero at $z = 0$, and -1 at the negative z.

The analog circuit for solving equation (4.5.19) with respect to x, constructed according to the method of implicit functions is shown in Figure 4.45.

This scheme of obtaining estimates and other schemes similar to it are named quorum elements [72]. Their common feature is the application of the method of implicit functions for solving equations $J' = 0$. Let us note that if in the scheme shown in Figure 4.7 the nonlinear blocks of the relay type are replaced by the linear ones, then at

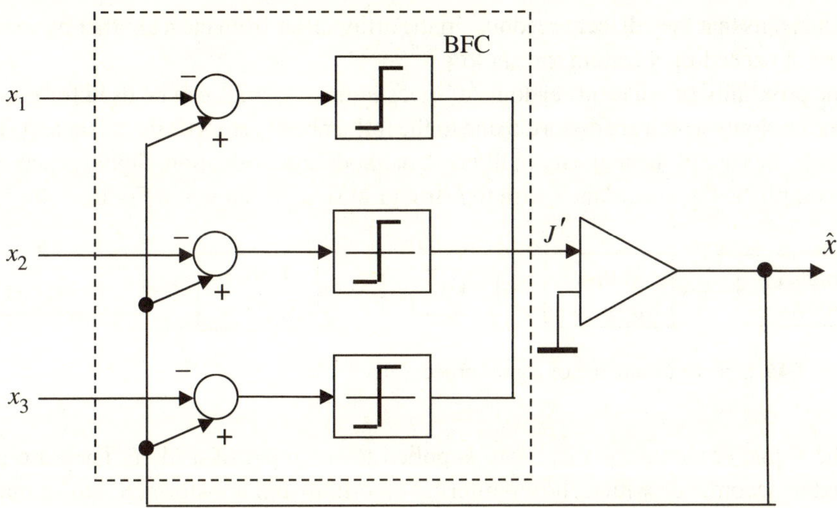

Figure 4.45. Block-diagram realizing the solution of equation (4.5.19).

the output of the amplifier an arithmetic mean estimate will be formed; if a linear characteristic with restriction is used, then estimate 37 from Table 4.11 will be obtained, and so on.

In conclusion let us mention some means which are not included in Table 4.11. Among these are chronological means, progressive means, factorial means, and a number of others.

4.6 Application of means for filtering problems

4.6.1 Digital filters with finite memory

The classical method of dealing with errors, noise, and mistakes in measuring systems consists of applying various filtering algorithms, taking into account some known properties of the signals and noise. If the data measurements, due to their physical nature, represent smooth continuous functions set by an array of readouts, then to process them filters based on averaging a number of adjacent readouts can be used.

In particular, such an approach can be used in airplane navigation devices, where the signal smoothness is caused by the fact that the location of an aircraft and its navigation parameters change smoothly.

In an airplane computer any signal is represented in the form of an array of readouts $x_1, x_2, \dots$, relating to equally spaced time moments. Therefore, the condition of smoothness can be written in the form of the inequality

$$|x_{i+1} - x_i| \leq \varepsilon, \tag{4.6.1}$$

which means that the adjacent readouts in the array differ from each another by a small value not exceeding a certain threshold ε.

The proximity of adjacent readouts of measurement signals can be used for removing the readouts which are distorted due to the influence of errors (fails, mistakes), This idea is at the base of the majority of filtering methods and evaluation. Signal processing is accomplished in accordance with the structural scheme shown in Figure 4.46.

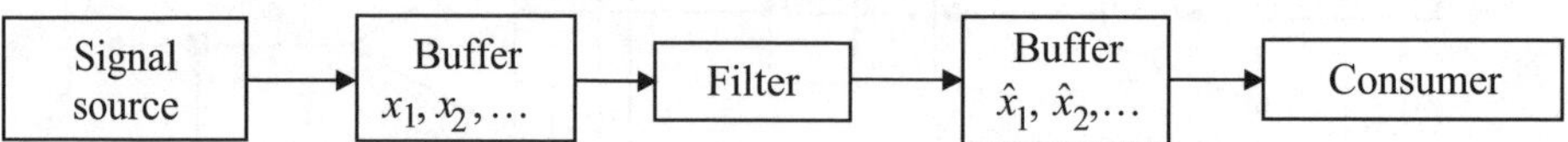

Figure 4.46. Structural scheme of signal processing.

The signal readouts $x_1, x_2, \ldots$ are supplied to the input of a filter. They are processed in accordance with a chosen filtering algorithm and transformed into an output sequence $\hat{x}_1, \hat{x}_2, \ldots$.

One simple filtering algorithm is the algorithm of a moving mean:

$$\hat{x}_i = \frac{x_i + x_{i-1} + \cdots + x_{i-k+1}}{k}, \tag{4.6.2}$$

where k is the width of a "window", moving over the signal readouts. This is the example of a filter with a finite memory or finite pulse–weight characteristic.

An arbitrary finite memory filter is described by the expression

$$\hat{x}_i = f(x_i,\, x_{i-1}, \ldots,\, x_{i-k+1}). \tag{4.6.3}$$

Therefore, the problem of filter synthesis satisfying the requirements set is reduced to the choice of the function f. From the mathematical point of view, this is close to the problem of the evaluation function selection, which was considered in the previous subsections of this volume. For example, the function of the moving average (4.6.2) at k =3 corresponds to the arithmetic mean function (4.2.4).

In the same way, any evaluation function indicated in Table 4.11 can be compared to the filtering algorithm defined by formula (4.6.3). In particular, using the harmonic mean formula the filter is obtained which is described by the formula

$$\hat{x}_i = 3\left(\frac{1}{x_i} + \frac{1}{x_{i-1}} + \frac{1}{x_{i-2}}\right)^{-1}.$$

Such an algorithm can be called the algorithm of a moving harmonic mean.

The simplest algorithm of secure filtering is supported by the formula given in point 47 of Table 4.11. It consists of the output readout of the filter being formed as a half-sum of the maximum and minimum readouts of three input readouts.

Thus, Table 4.11 automatically generates tens of filtering algorithms providing the processing of smooth signals. Let us consider in detail two kinds of these algorithms, median and diagnostic, which ensure removing single failures (mistakes, misses) in a signal.

4.6.2 Median filters

An algorithm of the median processing of three measurement values is described in Section 4.2 (see equation (4.2.8) and Figure 4.3). As applied to the signal filtering, this algorithm is turned into the algorithm of a moving median, which at $k = 3$ is described by the formula

$$\hat{x}_i = \operatorname{med}(x_i, x_{i-1}, x_{i-2}), \quad i = \overline{1, N}.$$

For reasons of symmetry this formula can be rewritten in the form

$$\hat{x}_i = \operatorname{med}(x_{i+1}, x_i, x_{i-1}). \tag{4.6.4}$$

This signifies that in the initial array data three adjacent readouts (the moving "window" of the width of three points) are looked through, and as an estimate $\hat{x}_i$ the readout, average in quantity, is accepted (Figure 4.47).

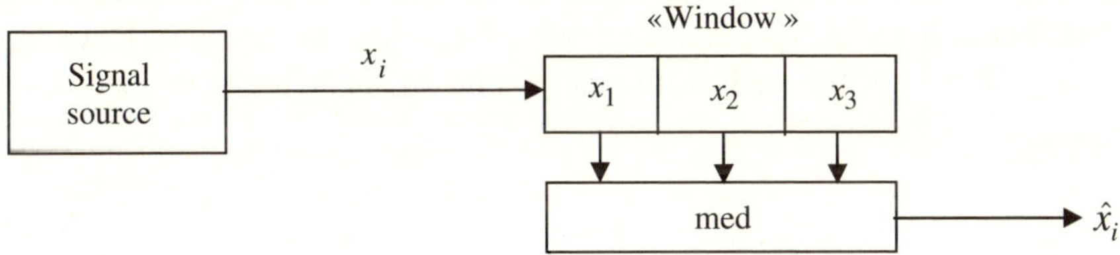

Figure 4.47. Three-point median filter.

An important property of the moving median algorithm is the fact that it rejects readouts distorted by single failures (misses, mistakes). Such readouts are replaced by an adjacent readout (by one of those whose value is closer). The algorithm thereby ensures the insensitivity to single failures (misses, mistakes), i.e., possesses the property of robustness.

The median filters are easily realized from the technical point of view and can be successfully used for avoiding failures (misses, mistakes) of any multiple factor. For example, selecting the width of the "window" $k = 5$, one obtains an estimate insensitive to double failures, etc.

The median filter possesses a number of interesting properties [18, 22–25, 29, 137, 167, 220, 221, 318, 319, 364, 378, 433, and others]. Let us consider some of them.

a) Criterion of filter optimal capabilities

In the same way as in performing the usual evaluation of measurement values, the median filter forms an estimate satisfying the module criterion of optimal capabilities. At the "window" duration $k = 3$ the criterion has the form

$$J = |x_{i+1} - \hat{x}_i| + |x_i - \hat{x}_i| + |x_{i-1} - \hat{x}_i| . \tag{4.6.5}$$

As follows from Figure 4.47, the estimate $\hat{x}_i$, calculated by formula (4.6.4), provides a minimum value for this criterion.

At the arbitrary duration of the "window" $k = 2n + 1$ formula (4.6.5) will contain k summands:

$$J = \sum_{j=-n}^{n} \left|x_{i+j} - \hat{x}\right|. \tag{4.6.6}$$

The minimum of such a criterion is achieved at $\hat{x}_i = \operatorname{med}(x_{i-n}, \ldots, x_{i+n})$.

b) Statistical properties of the filter

An analysis of statistical filter properties can be conducted by supplying to its input various random signals and studying the probabilistic characteristics of the output signal. One of the obvious and simple results of such an analysis is illustrated by Figure 4.48. In this figure the diagram of probability distribution density is shown for the output signal $p(\hat{x})$ of the median filter at the time of supplying to its input a signal uniformly distributed within the interval $[-1, 1]$.

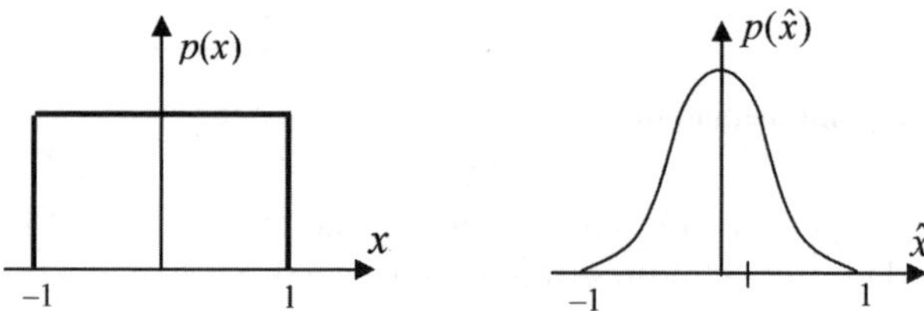

Figure 4.48. Transformation of the probabilities distribution density by the median filter.

The diagram of $p(\hat{x})$ has a bell-shaped form close to the Gauss curve. On an informal level the result given is explained by the fact that the median filter rejects the values with large amplitudes passing to the output the values with small and average amplitudes.

c) Nonlinear character of the filter

The median filter does not destroy and distort the pulse wavefronts, in contrast to any linear filter. Here the nonlinear character of this filter is revealed. Algorithmically, its nonlinearity is reflected in its use for realizing the function of med nonlinear sorting of signal values located in the filter "window". The calculation complexity of such an operation increases with the increase of the duration of the "window", which is evidence of a great efficiency of the algorithmic realization of the filter.

d) Root signals of the filter

One important characteristic of the filter is its behavior in the absence of noise. From this point of view all filters can be divided into two classes. The first class is formed by filters which do not introduce any distortions into input signals. Filters introducing some distortions into input signals are of the second class. For example, any linear filter inevitably changes the shape of the input signal.

The median filter belongs to the second class, i.e., in the general case it changes the form of input signals. First, this relates to the extreme points of a signal as well as to those what lie close to them. So, when a sinusoidal signal trough passes a three-point median filter, all the extreme points of the signal are "cut off". In case of a five-point signal not only extreme points are cut off, but also the points adjacent to them. As a result, the tops of sinusoid "waves" become flatter.

At the same time, there are signals which do not change when passing through the median filter. Such signals are called root signals, since they are the roots of the equation

$$\mathrm{med}(x) - x = 0,$$

where x is the input filter signal, and $\mathrm{med}(x)$ is the output filter signal.

Examples of root signals are a single "step", a linearly increasing signal, or any monotonic signal (Figure 4.49a,b,c). A rectangular pulse (Figure 4.49d) will be a root signal of the median filter if the pulse duration exceeds half of the filter "window" duration. These signals as well as their combinations pass through the median filter without change.

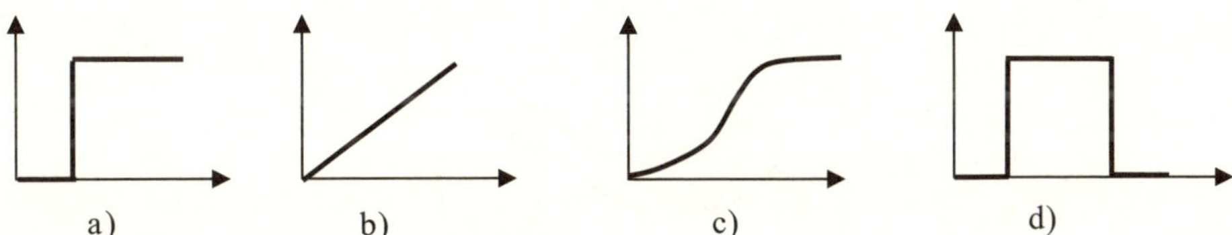

Figure 4.49. Root signals of the median filter.

e) Generalized median filters

A classical median filter optimizes the criterion defined by formula (4.6.6). This criterion permits the generalization in at least two directions. First, it can be written in the form

$$J = \sum_{j=-n}^{n} \left| a_j x_{i+j} - \hat{x}_i \right|,$$

where a_j means some weight coefficients. With the help of these it is possible to take into account the "importance" of various readouts. A separate class of weighted moving median filters will correspond to this formula.

Second, each summand in formula (4.6.6) can be raised to a positive power γ. This leads to the criterion

$$J = \sum_{j=-n}^{n} |x_{i+j} - \hat{x}_i|^{\gamma}. \tag{4.6.7}$$

Note the most important cases of the values γ: at $\gamma = 1$ one gets a classical median filter; at $\gamma = 2$ it will be a quadratic criterion, for which the optimal estimate is the moving arithmetic mean (4.6.2); at $\gamma \to \infty$ criterion (4.6.7) turns into a Chebyshev criterion with the optimal estimate in the form of a half-sum of minimum and maximum readouts located in the "window". This kind of estimate is secure and minimizes the maximum permissible error.

The case where $\gamma < 1$ is of interest. There the function J becomes multiextreme: all its minima are at points coinciding with readouts of an input signal. At $\gamma \to 0$ the optimal solution is achieved at the point x_j, minimizing the product

$$J = \prod_{\substack{i \neq j \\ j=-n}}^{n} |x_{i+j} - \hat{x}_i|.$$

In Figure 4.50 the character of criterion diagrams (4.6.7) is shown for different values of γ for a five-point filter. The values of the input signal readouts, fallen into the "window", are indicated on the x-axis, and the criterion values are on the y-axis.

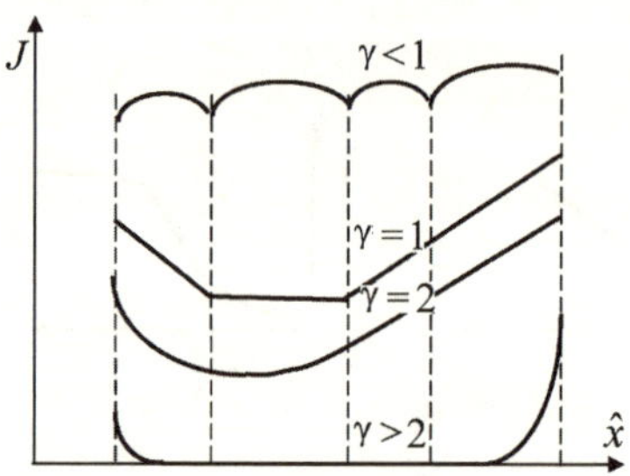

Figure 4.50. Character of the five-point filter diagrams for various values γ.

The choice of any value γ, and as a consequence the filter type, is defined by the physical essence of the problem, in particular by the signal character and noise.

4.6.3 Diagnostic filters

By a diagnostic filter we mean the filter intended to signal processing, the algorithm of which complies with the estimates corresponding to points 45 and 46 in Table 4.11.

The simplest diagnostic filter is obtained when considering adjacent pairs (i.e., using the "window" of two readout durations and checking for the fulfillment of the condition

$$\Delta = |x_i - x_{i-1}| \leq \varepsilon, \tag{4.6.8}$$

where ε is the fixed threshold determined by a power of "smoothness" of an input signal.

If condition (4.6.8) is met, then a current input readout is passed to the output of the filter: $\hat{x}_i = x_i$. In the opposite case the filter output is blocked. Blocking goes on until condition (4.6.8) is again met.

Within the blocking interval the output signal of the filter can be subjected to substitution with the help of different methods. For example, as an estimate within this interval it is possible to take the last true value: $\hat{x}_i = x_{i-1}$, i.e., to use the extrapolation of the zero order.

The corresponding algorithm has the form

$$\begin{aligned}
&\Delta := \text{abs}\,(x(i) - x(i-1))\\
&\text{if } \Delta \leq \text{eps then}\\
&\hat{x}(i) := x(i)\\
&\text{else}\\
&\hat{x}(i) := \hat{x}(i-1).
\end{aligned}$$

It is also possible to apply the extrapolation of the first or second order, but to do so will require an increase of the filter "window" duration. The algorithm, using the extrapolation of the first order, has the form

$$\begin{aligned}
&\Delta := \text{abs}\,(x(i) - x(i-1))\\
&\text{if } \Delta \leq \text{eps then}\\
&\hat{x}(i) := x(i)\\
&\text{else}\\
&\hat{x}(i) := 2\hat{x}(i-1) - \hat{x}(i-2).
\end{aligned}$$

Thus, the diagnostic filter at minimum duration of the "window" ($k = 2$) makes it possible to fight the failures (mistakes, misses) of an arbitrary repetition factor. This circumstance makes the diagnostic filter more beneficial than the median filter. But it should be noted that with an increase of the repetition factor the filter correction capability deteriorates.

Another advantage of the diagnostic filter consists of the fact that it does not introduce distortions into a signal with no fails. A signal of any shape which satisfies the condition of smoothness (4.6.8) is the root one for the diagnostic filter.

Both of these advantages above are provided by use of additional information about the threshold ε, which is not needed in case of the median filter. This is a disadvantage of the diagnostic filter considered above.

This disadvantage can be avoided by transferring to a *nonthreshold* diagnostic filter.

Let us describe the idea of the nonthreshold diagnostic filter for the "window" duration $k = 3$, and let the current readouts of an input signal x_i, x_{i-1}, x_{i-2} be denoted as y_1, y_2, y_3. Then let us form three control signals:

$$\Delta_1 = |y_2 - y_3|, \quad \Delta_2 = |y_1 - y_3|, \quad \Delta_3 = |y_1 - y_2|.$$

In the absence of fails, all three signals will be of small values. The appearance of a single fail will distort one of the readouts and lead to an increase in the values of two control signals. The index of the control signal which remains at a smaller value coincides with the number of the distorted readout. This readout is rejected, and an estimate is formed on the basis of two remaining readouts, for example, as their arithmetic mean, harmonic mean, etc.

Thus, the thresholdless diagnostic filter with the duration $k = 3$ in the same manner as the median filter gives us the possibility of helping to prevent single fails. The algorithm of such a filter has the form

$$\begin{aligned}
&y_1 := x(i);\\
&y_2 := x(i-1);\\
&y_3 := x(i-2);\\
&\Delta_1 := \text{abs}(y_2 - y_3);\\
&\Delta_2 := \text{abs}(y_1 - y_3);\\
&\Delta_3 := \text{abs}(y_1 - y_2);\\
&\text{if } \Delta_1 = \min(\Delta_1\Delta_2\Delta_3) \text{ then}\\
&\hat{x}(i) := (y_2 + y_3)/2\\
&\text{else if } \Delta_2 = \min(\Delta_1\Delta_2\Delta_3) \text{ then}\\
&\hat{x}(i) := (y_1 + y_3)/2\\
&\text{else if } \Delta_3 = \min(\Delta_1\Delta_2\Delta_3) \text{ then}\\
&\hat{x}(i) := (y_1 + y_2)/2.
\end{aligned}$$

Unfortunately, in the absence of fails this filter in the same manner as the median filter introduces some distortions into the original signal, i.e., is of the second class of filters.

4.6.4 Example of filtering navigational information

Let us illustrate the application of median filtering by an example of a procedure of processing the data received from an airplane navigation system. The block diagram of flight navigation information is shown in Figure 4.51.

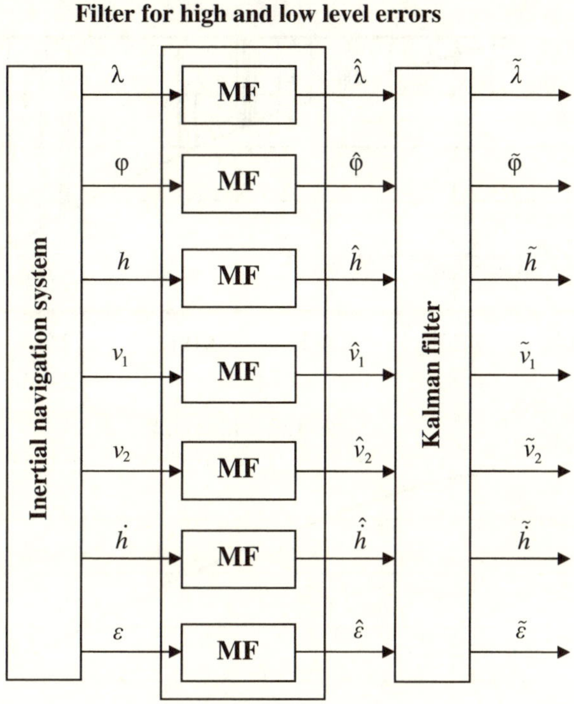

Filter for high and low level errors

Figure 4.51. Processing of signals of the inertial navigation system.

One source of primary information is the inertial navigation system, the output of which are estimates of the current values of seven aircraft parameters, i.e., longitude λ, latitude φ, altitude h, horizontal components of speed v_1, v_2, vertical speed $\dot{h}$, and azimuth ε are formed.

They contain noise of two types, distortions of a high level (overshoots, signal misses) and low level (errors caused by faulty hardware or some other errors).

To eliminate the influence of noise a two-stage procedure of data processing is applied. At the first stage the high-level errors are excluded. For this purpose a set of the median filters (MF) is used.

At the second stage of dealing with low level errors the Kalman filter is applied. As its output estimates of the navigational parameters "cleaned" of high-level noise are obtained.

As an example, in Figures 4.52 and 4.53 the results of the median filtering of signals h (altitude) and $\dot{h}$ (vertical speed) are given. For their construction real signal records of the inertial navigation system received in the process of flight tests were used. It can be seen that the upper diagrams in these figures contain a great number of high-level errors caused by regular signal misses caused by vibration and by the modest reliability of contacts.

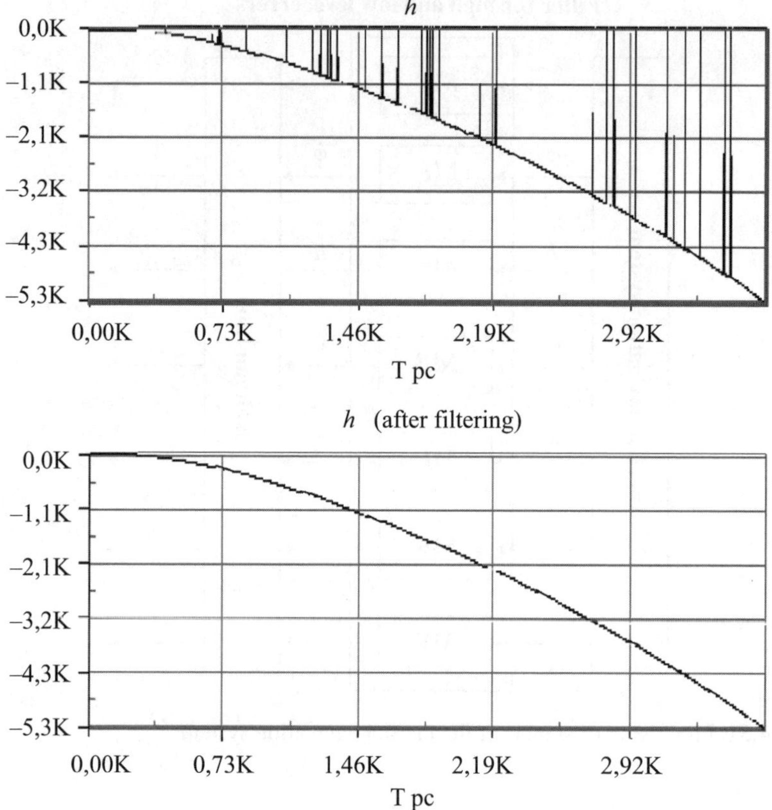

Figure 4.52. Results of the median filtering of real signals h (altitude).

The bottom diagrams in Figures 4.52 and 4.53 show the results of the median filtering. The median filter uses the "window" with a width equal to three readouts and eliminates single fails.

As a result the signal of the altitude channel appeared to be completely "cleaned" from high-level errors, and in the signal of the vertical speed channel all errors but two are removed. To eliminate the remaining errors it is necessary to use an additional filtering or to apply a five-point median filter.

The example considered shows a rather high efficiency of the median filtering used for preliminary signal processing of signals supplied from measuring sensors.

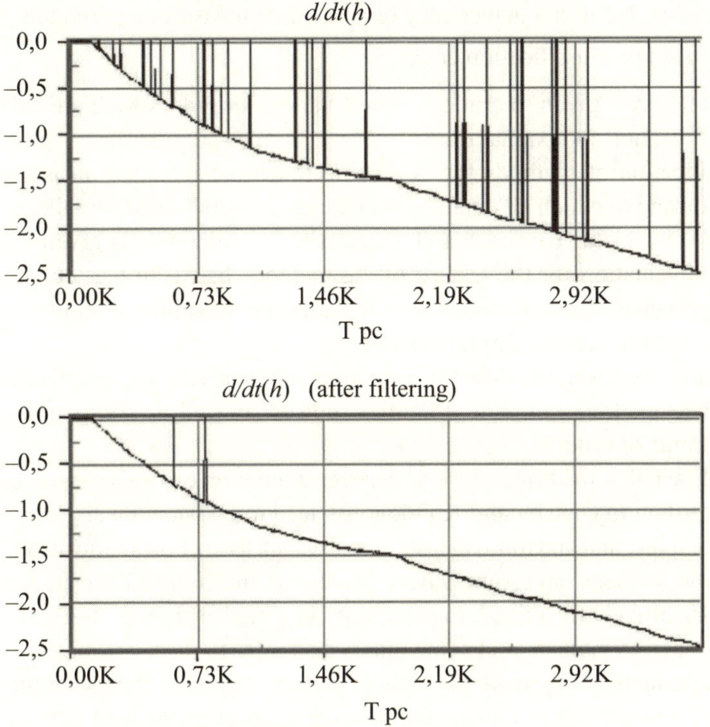

Figure 4.53. Median filtering of the signal of the vertical speed channel.

4.7 Summary

For many practical problems the situation is typical where it is required to obtain an estimate of an unknown quantity on the basis of a rather small number of measurements. In this chapter the evaluation algorithms relating to the classical problem of three or two measurements were collected, systematized, and studied. Along with this, some cases of a greater number of measurements were considered.

In accordance with the classification of measurement errors given in this book, the errors are classified as

- personal (subjective) errors, i.e., operator errors;
- hardware errors, i.e., errors of measuring instruments used;
- external errors caused by the influence of physical quantities which are not measurement objects;
- methodical errors, i.e., errors arising due to imperfection of the selected measurement method;

- errors arising due to the inadequacy of an applied measurement model;
- errors caused by classification errors.

Special attention is given to crude errors of various origins, which are called misses, mistakes, fails, noise, or signal misses.

A classification of methods for evaluating results of repeated measurements was given, according to which all evaluation methods are subdivided into the groups probabilistic, deterministic, heuristic, and diagnostic. Consideration is given to algorithms of optimal evaluation for the groups of the methods listed. In particular, the *probabilistic* approach is characterized, which includes the maximum likelihood method, as well as the Markovian and Bayesian estimates.

Algorithms realizing the *deterministic* approach use various optimization criteria, in particular, quadratic, module, minimax, power, as well as compound, combined, and other kinds of criteria.

It was shown that the principles of *heuristic* evaluation on the basis of definitions of means according to Cauchy and Kolmogorov lead to classical mean estimates for the case of two, three, and and more measurements, such as the arithmetic mean, geometric mean, harmonic mean, and some others. Moreover, linear, quasi-linear, and nonlinear estimates are also characterized. In particular, weighted arithmetic and quadratic mean esitimates, sample median, and some others also refer to these estimates.

When treating the *diagnostic* methods of getting estimates, the use of the *method of redundant variables* for increasing the evaluation accuracy as well as the application of *algebraic invariants* were described. It was shown that the use of the redundant variables method leads to obtaining Markovian estimates, and using algebraic invariants it is possible to reject unreliable measurement values.

Estimates with the rejection of one or two measurement values by either a maximum or minimum discrepancy were treated in detail. The comparison of various estimates, which differ in *evaluation algorithm character*, *method used for setting an algorithm*, and *easiness of technical realization*, was conducted.

The principle of applying mean estimates for the noise-immune filtration of signals, the use of which leads to filters with finite memory, was described.

High emphasis is placed on median filters and their characteristics, including optimization criteria, statistical properties, and root signals of filters. Consideration was also given to threshold and thresholdless versions of a diagnostic filter.

Algorithms of scalar quantity evaluation by three or two measurements were presented in the form of two tables in which more than 100 different estimates are reflected. It should be noted that the list of these estimates can be continued, since the number of possible estimates is not limited and the choice of one of them has to be made taking into account the specific features of a particular practical measurement situation.

Chapter 5

Metrological traceability of measurement results (illustrated by an example of magnetic recording instruments)

5.1 General ideas

Since the accuracy and trustworthiness of measurement results become "economic parameters" which have an influence on saving material and technical resources as well as on the quality of industrial products, the development of the scientific, engineering, organizational, and legal fundamentals of the metrological reliability of information-measurement systems of various destinations is of current importance.

In many cases instruments for the precise magnetic recording of analogue electrical signals (MRI) of measurement information and management data is an integral component of information measurement systems, measuring complexes, and automated control systems for technological processes used in various sectors of a national economy.

When using instruments for precise magnetic recording as one of the components of a measuring information system, its measurement characteristics influence the resultant metrological properties of the whole system. Solution of problems of an IMS analysis, i.e., the evaluation of the IMS errors with respect to the known metrological properties of the units it contains, as well as problems of IMS synthesis, i.e., determination of requirements for metrological characteristics of the system units on the given parameters and structure of this IMS, requires the development of a system of metrological assurance of MRI as a component.

By its metrological status the MRI is an intermediate linear transducer that can be considered as a "black case" characterized by its "inputs" and "outputs". A measurement MRI channel is designated for recording, keeping, and reproducing electrical information signals without disturbing their shape and has certain specific peculiar features as compared to traditional quadripoles (of the type of measurement amplifiers, dividers, etc.) and channels of information transmission systems which do not permit existing verification means to be used in full measure for metrological assurance of MRI.

The specific features are

- a large and frequently uncertain (depending on the operator's actions) time gap between the moment of recording and of reproducing a signal in both the mode of recording and the mode of reproducing;

- time scale distortions of a reproduced signal due to oscillations, drift, and the non-nominal character of a magnetic carrier speed in both mode of recording and mode of reproducing;
- the possibility of using the transposition of a tape-drive mechanism speed, thanks to which a spectrum of a recorded signal can be transformed.

In order to overcome the difficulties caused by the features listed above, there is a need to develop and study new specific methods and measuring instruments. These methods and measuring instruments have to determine the metrological characteristics of MRI taking into account a huge variety of types (a laboratory MRI of the stationary and transportable types, MRI for hard conditions of maintenance, and others), and different modes of its usage.

Thus, the solution of the problem of investigating and developing the scientific and technical fundamentals of the metrological assurance of measuring instruments with magnetic recoding and reproducing analogue electric signals is believed to be of current importance.

Therefore, two main lines of investigations will be considered in this chapter. Firstly, we will look at the development of the theory of the systems of metrological assurance as applied to measuring instruments for two physical values: electrical voltage and time. Secondly, we will investigate the creation of methods and means for verifying the analogue equipment of precise magnetic recording. Within the framework of these lines of investigation two groups of tasks are distinguished.

(1) The development of metrological assurance systems:

- evaluation of a maximum attainable accuracy of joint electric voltage and time measurements as an example of dynamic measurement variety;
- derivation of a measurement correctness condition and obtaining estimates of the upper and bottom limits of measurement time duration;
- development of a method for constructing functional dependencies for jointly measured measurands when the data is incomplete;
- investigation of a new method of normalizing the dynamic characteristics of a measurement channel using Markov's parameters;
- mathematical formalization of the concept of the system quality of metrological assurance and the development of an algorithm of their quality evaluation under the conditions of incomplete and inaccurate data on its elements, links, and properties.

(2) The creation of methods and measuring instruments for verifying MRI:

- development and investigation of methods and instruments for measuring errors of signal recording in a channel of an analogue MRI, taking into account error components caused by changes of a signal delay time in the channel or excluding this error component;

- development and investigation of new methods and measuring instruments for determining in an experimental way the values of the separate metrological characteristics of MRI, such as a pulse weight function of the channel, nonlinearity of its relative phase-frequency characteristic, level of nonlinear test signal distortions of a "white noise" type taking into account distinguished features of an apparatus under verification.

5.2 Precise magnetic recording instruments (MRI) of analog electrical signals as a part of measuring systems; MRI metrological traceability

5.2.1 Application of recording/reproducing electrical signals on magnetic carriers in measurement technique and specific features of MRI as an object of metrological investigations

The history of the discovery of magnetic recording is connected with recording sounds and radio broadcasting [6, 81, 82, 227, 322, 330, 410, and others]. Sound recording made and makes specific demands on the magnetic recording equipment used for these purposes, according to which it has to take into account physiological characteristics of human acoustic perception. These demands have are reflected in the setting of characteristics of magnetic tape recorders [276] such as the coefficients of harmonic distortions, detonations, relative noise level in a silent interval, etc. At the same time, we see a lack of interest in measurements of phase correlations of spectral components of the voltage which the magnetic tape recorder registers, i.e., in the form of a reproduced signal or, in other words in the error of signal transmission over a channel in magnetic recording instruments.

In the process of development, the technique of electric signal magnetic recording has been widely used for registering information of quite different types [6, 27, 136, 138, 162, 185, 186, 189, 278, 294, 324, 355, 357, 376, 410, 485, 523, and others.] in a digital or analogue form.

In particular, magnetic reproducing-recording equipment is frequently a necessary component of information measurement systems [253], measurement-computer complexes, and automatic control systems (for example, automatic systems providing a technological process control). It can be also used as a variable delay line for correlation analyzers, as a device providing the input of signals with a changeable time scale for spectrum analyzers [138], as a communication channel for verifying with a contactless method [146], etc.

When registering signals of measurement information, magnetic recording and reproducing equipment has to meet the registration accuracy requirements. In this case magnetic recording equipment is considered to be one of the units of a multiunit information-measurement system, the main and additional errors of which makes a significant impact on all metrological characteristics of the system in whole.

In other words, the accuracy requirements for magnetic recording equipment are similar to those which are required of any measuring instrument. From this it follows that this equipment has to be characterized according to a complex of normalized metrological characteristics in accordance with [191], giving sufficient information about its metrological properties.

Magnetic recording equipment intended for recording and reproducing electrical signals of measurement information and which has normalized metrological characteristics will be referred to herein as a precise magnetic recording instrument (MRI).

The basic assignment of MRI applied as a part of IMS is to record (register), store, and reproduce (read off) electrical signals of measurement information without distortions. Frequently MRI is used to transform an amplitude and time scale of a signal in a linear way, to provide multiple reproduction for the purpose of the analysis of its parameters, etc.

A generalized structural scheme of MRI [6] is presented in Figure 5.1.

In this figure there are three constituent parts: (1) an electronic unit of recording, (2) type driver mechanism (TDM) and (3) electronic unit of reproducing. The unit of recording (1) includes input converters (4), modulation devices (5), amplifiers of recording (6), and generator of a reference or control pilot-signal (7).

Type driver mechanism (2) contains magnetic heads of recording (8) and play-back heads (11) (or universal magnetic heads), receiving and feed cassettes (9), and magnetic carrier (10). The unit of reproducing consists of reproduction (read off) amplifiers (12), demodulators (13), output converters (14), detector of time error (15), and control (time error compensation) unit (16).

It should be noted that there is a similarity between the generalized structural diagram of MRI and a block diagram of the communication channel consisting of transmitter (1), media of propagation of signals transmitted (2), and receiver (3). In this case units (7), (15), and (16)operate as synchronizing devices.

Such a similarity is caused by the fact that both the communication channel and MRI are intended for transmitting electrical signals: the former in space, the letter in time. This causes us to conclude that the tasks of the metrological assurance of the communication channels (of information transmission) and the precise magnetic recording instruments are similar in many respects.

A large park of MRI of various types, working and being newly developed, can be classified on the basis of different attributes:

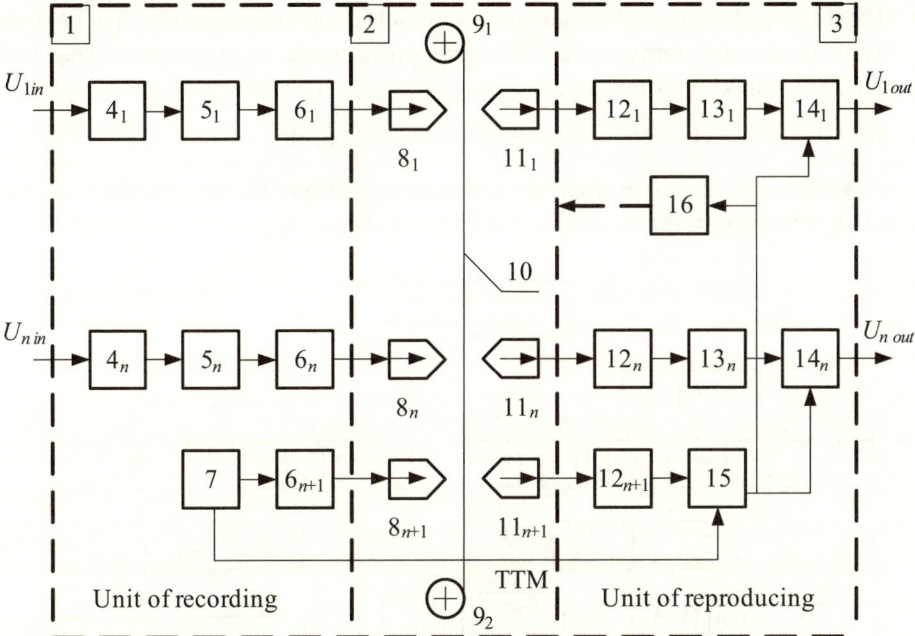

Figure 5.1. Generalized structural scheme of the precise magnetic recording instrument. Here: 1 – an electronic unit of recording; 2 – a type driver mechanism; 3 – an electronic unit of reproducing; 4 – input converter; 5 – a modulator; 6 – a recording amplifier; 7 – a generator of a pilot-signal; 8 – a magnetic head of recording; 9 – a cassette; 10 – a magnetic carrier; 11 – a magnetic head of reproducing; 12 – an amplifier of reproducing; 13 – a demodulator; 14 – an output converter; 15 – a detector of a time error; 16 – a unit of TTM control; $U_{i\ \text{in}}$ – an input registered signal; $U_{i\ \text{out}}$ – an output electric signal.

- type of applied signal modulation;
- volume of information stored, which depends on the number of channels (or tracks on a magnetic carrier), their pass band, speed of tape feed and carrier length (i.e., time of recording a signal) and amplitude range of electric voltage to be registered;
- possibility to apply the write and playback modes consecutively (for example, only the write mode on board and only the playback mode when an flying vehicle is landing; the mode that allows recording and reproducing to be performed simultaneously with a fixed or variable delay of a signal reproduced with respect to a signal recorded, and so on);
- possibility to transpose the speed of carrier movement in playback mode, which leads to the transformation of the recorded signal spectrum;
- type of an applied magnetic carrier (a magnetic or metal tape, wire, drum, disc, etc.);

- designation, conditions of operation, and special features of design (flight or ground-based instruments, form of a tape feed path: open, closed, U-shape, closed ring, coil type, or compact cassette, etc.;
- mass dimension characteristics, energy consumption, etc.

An example of MRI classification according to the modulation used (methods of precise magnetic recording) [6] is shown in Figure 5.2.

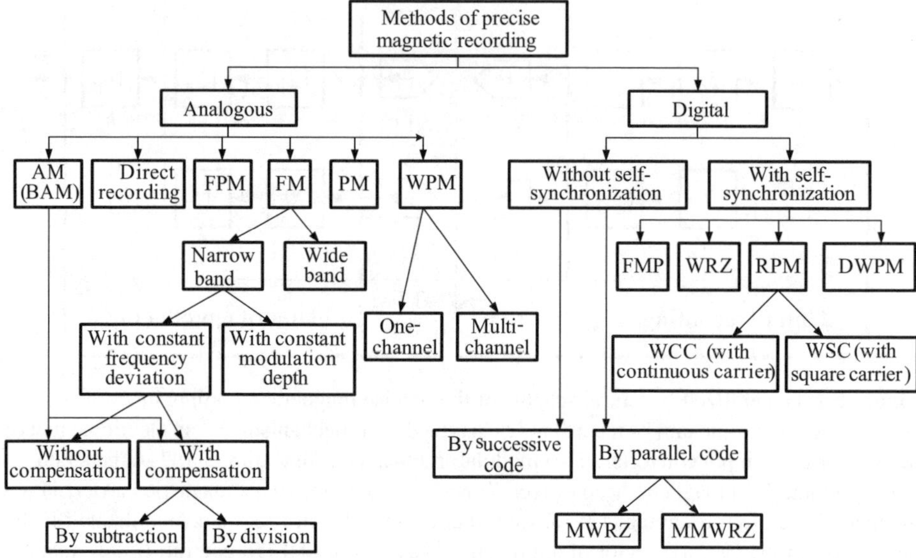

Figure 5.2. MRI classification according to the kind of used modulation (methods of precise magnetic recording). Here: AM – amplitude modulation; BAM – balance-amplitude modulation; FPM – frequency-pulse modulation; FM – frequency modulation; PM – phase modulation; WPM – width-pulse modulation; FMP – frequency manipulation; WRZ – with returning to zero; RPM - relative phase manipulation; DWPM – discrete width-pulse manipulation; MWRZ – method without returning to zero; MMWRZ – modified method without returning to zero.

Equipment for precise magnetic recording from the point of view of the modulation used can be divided into two classes, analogue or digital. Digital MRI are based on the application of the code-pulse modulation, and the error of registering signals in its channels depends mainly on two factors: the probability of failure (i.e., a signal miss or the appearance of redundant pulses) and the structure of a code word (or of a code pattern).

Analogue MRI has become widespread due to its higher information capacity (in spite of the fact that its volume is similar to that of the digital MRI) and comparatively simple technical realization. Various types of modulation, such as harmonic carrier modulation (amplitude modulation – AM; balance-amplitude modulation – BAM; frequency modulation – FM; phase modulation – PM), and pulse sequence modulation (amplitude-pulse – APM; frequency-pulse – FPM; phase-pulse – PPM; time-pulse – TPM; in particular, width-pulse – WPM), are also referred to.

The most frequently used modulation are frequency modulation (FM) and width-pulse modulation (WPM) (one-sided and double-sided types of modulation, and modulation of the first or of the second kind), since they provide a higher value of an index such as the product of a specific volume of a stored information on a registration accuracy.

An analogue MRI was chosen as the object of investigation. This is caused by the fact that, firstly, this equipment is widely applied as a part of IMS and, secondly, the problem of its verification, i.e., the experimental determination of an error of registering signals over the channels of the analogue MRI has not yet been solved.

As shown in Section 5.3, in accordance with the metrological status of the analogue MRI, the latter is an intermediate linear measurement transducer. The solution of tasks to determine metrological characteristics of the traditional linear electric transducers has been shown in numerous publications, e.g., [31, 144, 154, 281, 303, 368, 372, 412, 413, 483, 510, 535, and others]. At the same time, the analogue MRI, as an object of metrological investigations, has the following peculiarities.

1) Moments of recording and reproducing the signals in the MRI are spaced with respect to each other. When the write and playback modes exist simultaneously, a certain fixed delay takes place between recording and reproducing signals which is equal to the ratio obtained as a result of dividing the distance between a write head and a playback head by a value of the drawing speed of the magnetic carrier (so, at a distance between the heads of 4.76 cm and tape speed of 9.53 cm/s the interval of the signal delay will be 0.5 s).

During alternating activation of the write mode and playback mode, the delay interval between the signals becomes uncertain and depends on an operator's activity (for example, a signal was been written today, but it will not be reproduced for a year). Therefore, the traditional methods of verifying linear electric measuring transducers (for example, measurement amplifiers, dividers, attenuators, transformers, etc.) which consist of comparing variable test signals at the output and input of a transducer are not suitable for MRI, because they require the creation of a reference signal delay line with a wide range of delay time adjustment.

2) Owing to the instability of the magnetic carrier speed in the write and playback modes, the time scale of a reproduced signal is distorted. This leads to the appearance of an additional registering error due to the drift and the oscillations of the signal delay time in the MRI channel. This results in the necessity to compensate changes of the delay time in the process of an experimental determination of the main registration error, as well as to measure parameters of the drift process and oscillations of the delay time of the signal in the channel.

3) Many types of MRI allow the transposition of the magnetic carrier speed to be used, i.e., to realize the recording of a signal at one speed and reproducing at another, resulting in the transformation of the signal spectrum and the exclusion of the possibility of verifying MRI by means of traditional methods.

4) When selecting a method for verifying MRI channels (by a reference test signal or reference channel), the preference should be given to the reference test signal method because of the fact that the analogue equipment of precise magnetic recording is a multiparameter system with a great variety of types and modes of use. Moreover, to get MRI characteristics of a high quality, this equipment is developed and manufactured "on the cutting-edge of industrial engineering capabilities".

Therefore, to create a reference MRI (and many of its varieties) with an accuracy reserve of a 2–3 times higher at the minimum than that of a verified equipment is considered impossible as well as unreasonable from an economical point of view. It should be noted that such an approach allows the MRI to be considered to be a "black box", i.e., a linear dynamic system with known inputs and outputs and an unknown "content", providing the possibility of overcoming difficulties in the process of MRI verification arising due to the variety of its types and the design complexity of its internal components.

5.2.2 Main sources of distortions of measurement information signals in magnetic channels, and methods of their measurements

A great number of works performed both by Russian and by other authors are devoted to the reasons for signal distortion in a magnetic recording/reproducing channel [61, 90, 91, 131, 236, 237, 249, 293, 294, 388, 423, 443, 452, 479, 491, and others].

The main distortions are

- frequency distortions arising in a magnetic head-tape-head system and in electronic units of recording/reproducing;
- phase distortions arising in electronic units of recording/reproducing and in a magnetic head-tape-head system;
- nonlinearity of the amplitude characteristics of electronic units and of the process of recording a signal in a head-tape-head system;
- parasitic amplitude modulation;
- tape noise;
- coupling noises;
- penetration effect;
- effect of copying;
- effect of self-demagnetization;
- defects and damage to a magnetic carrier;

- noise of electronic units of recording and reproducing;
- oscillations, drift, and the nonrated speed of magnetic tape movement;
- static and dynamic magnetic tape skew relative to the working gap of a head;
- noncoordination of input and output impedances of a channel of recording/reproducing with output impedance of a load, correspondingly.

Frequency distortions in magnetic recording equipment arise because of the inequality of the amplitude frequency characteristics of the electronic units of recording and reproducing, as well as of the frequency and wave losses in the magnetic head-carrier-head system.

It is possible to separate out the wave losses caused by the remoteness of slices of a working tape layer from the surface of the reproducing magnetic head: nonzero gap width and finite dimensions of the magnetic reproducing head; relative slope of the recording and reproducing heads; frequency losses introduced by a magnetic material of the head; differentiating the action of the induction magnetic reproducing head; and others.

The sources of the **phase distortions** arising in recording electric signals are the nonlinearity of the phase-frequency characteristics in electronic recording and reproducing units, as well as the distortions in the head-tape-head system.

In this system the phase shifts are summed up from shifts that occur when

- electric current, running through a winding, is transformed into a magnetomotive force in the working gap of the magnetic chain of the recording head;
- magnetomotive force is transformed in the working gap of the recording head into a strength of the recording head field;
- strength of the recording head field is transformed in the residual magnetization of the carrier in the process of recording;
- residual magnetization of the carrier is transformed into a magnetomotive force in the working gap of the reproducing head;
- magnetomotive force in the working gap is transformed into a magnetic flux in the core of the reproducing head;
- magnetic flux in the core of the reproducing head is transformed into a electromotive force, and so on.

The nonlinearity of the transfer characteristic of the channel of recording/reproducing leading to nonlinear distortions of the signal reproduced is invoked by the nonlinearity of the modulation characteristic of the electronic recording units used with the frequency-modulation method (or some other modulation method) of recording; by the nonlinearity of the process of recording a signal in the magnetic head-tape-head system, which has a principally irremovable nature, as well as by the nonlinearity of the demodulation characteristic of the electronic units of reproducing, which is caused, in general, by the nonlinearity of an amplitude characteristic of low frequency filters which frequently play the role of demodulators.

One of the main reasons for the **parasitic amplitude modulation** of a signal reproduced from a magnetic tape is the so-called "noncontact", i.e., an imperfect contact of the magnetic head with the tape. The signals recorded on the tape both during direct recording and recording with saturation cannot be reproduced with high amplitude accuracy. The coefficient of parasitic amplitude modulation is, as a rule, of the order of 10 %. At the same time the signals with a lesser wave length of recoding are subject to an impact of faster and stronger amplitude fluctuations compared to signals with a greater wave length.

The main reason of the appearance of **signalogram selfnoise** is the magnetic nonuniformity of the tape. In signal carriers used in practice, a working layer is not magnetically uniform, since it consists of separate magnetic crystals of different dimensions spread differently in the layer. When takes are manufactured the absolute accuracy of the magnetic layer dimensions cannot be achieved. As a result of the layer width and thickness oscillations, a mass of a magnetic substance is transferred per unit of time through a writing field of the recording head changes. The extent to which perfection of the contact between the head and the carrier strongly is achieved depends on the surface roughness (lack of smoothness) of the magnetic layer. This results in changes of the head and carrier interaction with time, when the latter is moving. This leads in turn to fluctuations of the "effectively acting" magnetic mass of the signal carrier in the recording and reproducing processes.

There are two types of **noise arising during reproduction,** the source of which is the tape: noise of a demagnetized tape and modulation noise. **Modulation noise** is an additional source of distortions which can be detected when a signal is present and depends on the level of tape magnetization.

Transient disturbances are caused by signal penetration from one channel of the magnetic recording equipment to another one in the recording and reproducing heads as well as in the signalogram. Necessary gaps in the head cores inevitably create leakage fluxes, and a complete interhead screening is practically impossible because of the location of the tape on the head surface. Therefore, it is clear that the screens cannot extend through the tape.

With rapid variation of the magnetizing field in the surface layers of a ferromagnetic body eddy currents protecting internal parts of the signal carrier from influence of an external alternating field arise. This phenomenon is called the **effect of penetration.** It consists of the fact that when the signal frequency increases, the magnetization is unevenly spread over the whole signal carrier width, concentrating in its surface layers.

The **effect of copying** consists of the generation of additional distortions of a signal reproduced from the signal tape. The presence of an external field of the magnetic signalogram leads to the formation of a field near the magnetized parts of the tape wound in a roll, which impacts adjacent magnetic tape layers located in this roll. The signalogram magnetic field strength diminishes according to the exponential law the

the increase in distance. This is why at a given wave length the most intensive field only influences those tape layers located just near the layer carrying a record of a signal. Layers of the wound tape adjoining this part of the signalogram are partly magnetized by a field of the recorded signalogram. As a result, "magnetic prints" of the recorded signal are formed on these parts.

Self-demagnetization of the signalogram is caused for the following reasons. When a longitudinal signalogram is being recorded, the length of the forming magnets is inversely proportional to the frequency of the signal recorded. Constant magnets formed during recording are directed relative to each other by their opposite poles, i.e., they mutually demagnetize each other. During the process of recoding the signalogram is located just near the core of the head with a high magnetic permeability, and therefore it is possible to assume that there are practically no fields of self-demagnetization in it. When the signalogram moves away from the head, the fields of self-demagnetization reach their maximum, the output of the latter decreases, and the frequency characteristic of the signalogram in the field of high frequencies become restricted.

Defects in the magnetic tape occur as a result of wear during the process of operation as well as due to accidental damage and incorrect storage of the tape. The wearing out of the tape in the process of operation happens because of the scraping the working layer and the depositing of wear byproducts onto the surface of the tape. The clumps form a layer between the tape and heads. Because of this layer the probability of signal dropout increases and the frequency characteristic of the equipment in the field of small wave lengths becomes worse. Accidental damage to the tape, including ruptures and patches, take place due to either the incorrect behavior of an operator with the tape or equipment failures. At small tape tension at the moments of switching on and off the tape drive mechanism, the roll laps slip relative to each other, forming riffles. Scratches on the working layer of the tape arising due to traces of unmovable component pieces of the tape drive mechanism lead to "columnar deflection" of the tape. Storage of magnetic tapes at higher temperature and humidity causes adhesion of the roll laps and mechanical deformation of the tape in the form of ripples and waviness, as well as of bending and distortion of the roll itself.

Some of the reasons for **noises of the recording and reproducing units** are the noise of circuit elements, the instability of response thresholds of the shapers used in modulation recording, and also a background of supply unit.

The causes for **carrier speed variations** in magnetic tape recording devices can be divided into two groups. The first group includes the properties of the tape drive mechanism; the second one consists of the mechanical properties of the tape interacting with TDM elements. In a tape drive mechanism, the sources of fluctuation are the elements whose movement is characterized by low frequency irregularities and which influence the tape movement. Moreover, a relatively long part the unwound tape is in frictional contact with parts of the mechanism. As a result of this it is subjected to perturbation action of random forces which cause the high frequency fluctuations of the tape speed.

Another source of speed fluctuation is the magnetic tape. It is made of an elastic material which makes contact with the TDM elements at a continuous stretch. The noninequality of the stretch and instability of rotation moments arising in TDM are transferred from one element of the mechanism to the other through the tape. Due to internal irregularities of TDM work, the tape moves in spurts. These spurts also happen because of the friction between the tape and magnetic heads which do not have ideally smooth working surfaces. The mechanical properties of the tape complicate the picture (by this is meant the transformation of longitudinal vibrations of the tape into transverse ones and vice versa). The accidental vibrations of the tape arise due to its graininess and because when the tape rubs against unmovable surfaces, for example against the heads, the tape experiences the influence of many force pulses similar to the shot effect of electronic lamps. The tape response to these pulses is rather complicated because of the a great number of possible resonances within the range between subsonic frequencies and hundreds of kHz.

Thus, the spectrum of magnetic carrier vibrations in recording and reproducing is wide, and as a result of their impact the noise of a complicated spectrum is imposed onto the recorded signal. However, it should be taken into consideration that a finite pass band of the MRI channel significantly reduces the impact of speed variation components whose frequency exceeds the upper boundary frequency of the channel pass band.

A **skew of the working gap of the magnetic head** leads to constant and alternative time and phase shifts between signals recorded by one unit of heads on different paths. This phenomenon is also caused by skewing of the tape as it passes through the head unit. A varying skew caused by speed fluctuations of the tape leads to varying time and phase shifts which are of a significant importance for some applications of multichannel MRI. A dynamic skew can be also be due to the constant and alternative stretch stress across the tape, as well as due to some deformation or other dimensional nonuniformities peculiar to the tape.

To an MRI, as to any other electric four-port network used in the capacity of a measurement transducer, the amplitude scale distortions of signals recorded are inherent due to **finite-valued input and output impedances** of its channels. The number of these distortions can be decreased by coordinating the mentioned impedances with an output impedance of the signal source as well as with an input impedance of a load. The input impedance of the channel has to be much greater than the output impedance of the signal source and, in an ideal case, approaches infinity. The output impedance of the channel has to be much less than the output impedance of the load and, in an ideal case, must be equal to zero.

The above analysis of signal distortions which are characteristic for a magnetic channel of recording/reproducing shows that they can be divided into four groups:

- dynamic distortions;
- nonlinear distortions;

- noises;
- distortions of a signal time scale.

At the same time the first three groups can be considered as particular components of the basic error and the last one as an additional error of a signal transfer over a MRI channel.

Dynamic distortions, i.e., the distortions that appear during the registration of signals varying in time, contain frequency and phase distortions as well as distortions caused by the effects of self-demagnetizing and penetration.

Noises distorting an output signal of a recording made by a MRI include parasitic amplitude modulation, self-noises of the signalogram, transient noises between the channels, combinational components at the modulation methods of recording, copying effect, noises due to defects of the magnetic tape, and noises of electronic recording/reproducing units. Signal distortions of an amplitude scale due to finite-valued impedances of the channel also belong to this list of noises.

Distortions of a signal time scale are caused by variations of the magnetic carrier speed in the recording and reproducing processes, by its drift, by the difference between the average speed at recording and of reproducing, and by a skew of the magnetic head gap, and are characterized by statistical parameters of the process within which the delay time of a signal reproduced relative to a signal recorded changes.

There are many methods and means that can be applied when measuring metrological parameters of MRI [6, 185, 186, 293, 294, 423, 452, and others]. Among them there are the following:

- inequality of AFC;
- dynamic range;
- nonlinearity of an amplitude d.c. characteristic;
- harmonic coefficient;
- zero drift and transfer coefficient drift;
- depth of PAM;
- level of combinatorial distortions and transient noises;
- time shift of signals between channels;
- deviation of carrier speeds from nominal values and sliding;
- tape speed vibration coefficient, and others.

At the same time there are a number of the, the experimental determination of which requires the creation of new unconventional methods and measuring means due to peculiarities of MRI as an object of the metrological investigations considered in Section 5.1. These metrological MRI characteristics are as follows:

- nonlinearity of phase-frequency characteristics;
- pulse weight function of a channel;
- level of nonlinear distortions of a signal constituting a white noise, restricted by MRI band width;

- changes of a signal delay time in a channel;
- signal registration error.

Therefore, the main attention is given below to an analysis of the methods suggested for analyzing just these parameters.

A detailed and thorough review of methods for measuring the nonlinearity of PFC MRI is given in [443]. On the basis of the critical analysis of more than 60 sources including monographs, articles in Russian and foreign scientific and technical journals, papers presented at conferences, inventions and patents, the author of the analytical review has come to the following general conclusions.

(1) To estimate phase relations in a MRI channel it is only possible to apply indirect methods of measurements which are based on the use of both two-frequency and multi-frequency (group type) test signals with recoding on one track. However, neither of the measurement methods considered provides complete information about PFC MRI since knowledge of at least one point of an actual phase-frequency characteristic of the channel is required.

(2) The absence of repeating results of measuring the PFC channels of MRI is explained by many authors as due to a great number of factors which influence the obtained results as well as to the lack of a unified procedure of performing measurements which can be recommended when investigating the MRI channel.

The latter point is the actual task of developing a method and means, as well as of a procedure of fulfillment measurements of the nonlinearity of phase-frequency characteristics of MRI channels.

The possibilities to determine other full dynamic characteristics of the MRI channel such as the weight function or pulse weight function presenting a response of a system to an input signal in the form of a single jump of voltage or a Dirack's δ-function, correspondingly, have been investigated up to now to a lesser extent. A sufficiently complete review of the existing methods for determining such characteristics is given in [457].

In determining a weight function of a magnetic recording/reproducing channel, rectangular pulses of long duration are frequently used as test signals. Visualization of a reproduced transient process is realized with the help of a cathode ray oscillograph. The accuracy of a weight function estimate obtained with this method is determined by the errors of the oscillographic measurements of time and amplitude relations.

For cathode ray oscillographs these errors are nearly units of percents. In some works it is noted that experiments with short test pulses, the purpose of which was to obtain an estimate of a pulse weight function of a channel, had not been successful because of the impossibility of obtaining a reaction at the output which would be significantly higher than a noise level and thus meeting the condition of providing the system linearity.

An attempt to measure the values of the pulse weight function of the MRI channel with recording narrow pulses on an irremovable tape and then reproducing their response did not provide a sufficiently high accuracy in the results obtained. This fact is explained by the small signal-to-noise ratio at the output of the equipment channel.

From that it is possible to conclude that it is needed to develop indirect methods of measuring values of the weight function of a MRI channel which will take into account the peculiarities of the equipment under study and provide the required accuracy.

Special methods for measuring the **level of nonlinear distortions** in magnetic recoding equipment are considered in detail in [236]. The methods for measuring nonlinear distortions can be classified as single-, double-, and multi-frequency ones, according to a form of the test signal spectrum .

For a single-frequency signal, a harmonic distortion coefficient equal to the ratio of the square root for the sum of dispersions of the highest harmonics to the amplitude of the first harmonic, or a coefficient of nonlinear distortions, connected with it in a one-by-one manner, are measured. For measuring a power of nonlinear distortions, either selective action devices of the spectrum-analyzer or selective voltmeter type, or devices for measuring nonlinear distortions with a rejection filter for suppressing the first harmonic, or devices for measuring nonlinear distortions with a filter of high frequencies, are used.

One main disadvantage of such a method of measurements is the fact that a selected model of a test signal in the form of a harmonic voltage underestimates the nonlinear distortion level for real signals with a more complicated spectral content.

Among the dual-frequency methods for determining nonlinear distortions the difference frequency method is widely applied. When in recording mode, sinusoidal voltages of two frequencies that are close to each other are simultaneously supplied to the input of the magnetic recording channel under study, and in of reproducing mode these voltages are measured with the help of a spectrum analyzer or selective voltmeter amplitudes of combinatorial frequency voltages. Then a nonlinear distortions coefficient is calculated with a formula. Analysis of the methods which use a dual-frequency test signal (methods of difference frequency, mutual modulation, and others) has shown that the method of difference frequency can only find restricted application when the depth of PAM does not exceed 20 %, and the method of mutual modulation is not suitable for measuring nonlinear distortions in magnetic recording equipment.

Application of multifrequency methods for investigating nonlinear distortions such as the method of noise bands [249], dynamic, correlation, with use of periodic pulses of different forms, etc., is attractive thanks to the possibility of selecting a test signal with a spectrum which is close to the real registered signals. However, it is hereby necessary to overcome difficulties connected with a time break between the processes of recording and reproducing as well as with the complexity of hardware realization. Nevertheless, development of a perspective method for measuring the level of nonlinear distortions with a test signal of the "white noise" type which would result in an acceptable degree of accuracy is considered.

The problems of measuring parameters of the process of magnetic carrier speed variations during recording and reproducing are widely covered in the literature (for example, [6, 61, 90, 91, 185, 186, 293, 294, 423, and others]). However an error component of signal registration, caused by oscillations, drifts, and nonnominal value of the carrier speed, depends on the variations of the signal delay time in the MRI channel. These variations, in turn, are connected with the carrier speed oscillations by integral transformation, which does not result in an unambiguous value, even in the case when the process parameters of tape vibrations are known.

Therefore an current problem is the development of methods and instruments for measuring changes of the time delay of a signal in the MRI channel.

With regards to determining the error of signal registration with precise magnetic recording equipment, in spite of the many attempts to theoretically and experimentally estimate it [6, 131, 185, 186, 236, 237, 276, 294, 452, and others] which were done before 1973, no practical results were obtained in this direction. At the same time the lack of methods and means to verify MRI results in the the problem of its metrological assurance not being solved. Therefore the creation of such methods and measuring instruments is one of central tasks in developing the fundamentals of the metrological assurance of measuring instruments using the magnetic recording/reproducing of electric signals.

Considering investigations of MRI metrological characteristics described by existing Russian and foreign scientific schools, we need to mention firstly the Moscow school headed by M. V. Gitlits, A. I. Viches and A. I. Goron, which include such researchers as V. A. Smirnov, V. B. Minukhin, V. G. Korolkov, N. N. Slepov, V. N. Filinov, Yu. L. Bogorodsky, R. M. Belyaev, V. A. Aksenov, V. I. Rudman, R. Ya. Syropyatova, A. A. Fridman and others. Moreover, significant results have been obtained by the Leningrad school headed by V. A. Burgov and Yu. M. Ishutkin (M. A. Razvin, K. M. Matus, A. S. Zaks and others), the Kiev school headed by M. V. Laufer and V. K. Zheleznyak (V. A. Geranin, N. A. Korzh, A. G. Machulsky and others), as well as the Lithuanian school headed by K. M. Ragulskiss and R. P. Yasinavichus and Kishinev school (L. S. Gordeev). Among foreign researchers we mention the American school of scientists such as G. I. Devis, S. C. Chao, Ch. Mi, Ch. Pear and others.

5.2.3 Problems of the metrological traceability of MRI

By metrological assurance we mean the "establishment and use of scientific and organizational fundamentals, technical means, rules and norms required to achieve the uniformity and needed accuracy of measurements" [195].

At the same time, by scientific foundation of the metrological assurance of measurements we mean metrology as "the science of measurements, methods, and the means of ensuring their uniformity and the methods for producing the required accuracy" [195].

The technical foundations of metrological assurance are:

- system of state measurement standards of physical quantity units;
- system of transferring dimensions of physical quantities from the measurement standards to all measuring instruments using working measurement standards and other means of verification;
- system of government tests of measuring instruments intended for series or mass production or for importation from abroad in lots, which produces the uniformity of measuring instruments in their development and putting them to circulation;
- system of the national and provincial verification or metrological certification providing the uniformity of measuring instruments in their manufacturing, operation, and repair, and others.

One of the main goals of metrological assurance is the increasing of the quality of industrial production.

Based on these basic statements, let us formulate the tasks to be solved in creating the scientific and technical fundamentals of metrological assurance of measuring instruments with magnetic recording/reproducing of analogue electric signals.

Among these tasks are

- the study and analyzation of the particular error components of a signal over a MRI channel caused by various specific factors inherent in the process of recording/ reproducing a signal on a magnetic carrier;
- the investigation of the possibility of taking them into account by a grounded complex of normalized metrological characteristics;
- the development of a procedure for the theoretical calculation of the resultant error of registration according its components, which is described in [457].

Among the scientific tasks still needed to be solved for creating the scientific and methodical grounds of the metrological assurance of MRI, we need to note the following.

Since the transfer or registration of an analogue electric voltage is an intermediate transformation in carrying out joint measurements of two physical quantities, the voltage of electric current and time, then it is useful to perform an analysis of peculiarities of joint measurements; to estimate a maximum attainable accuracy of joint measurements of electric current voltage and time as one of dynamic measurements variety, as well as the upper and lower boundaries of measurement time duration; to develop a method of constructing functional dependencies for jointly measured quantities at incomplete initial data.

The problem of creating the scientific fundamentals of the metrological assurance of measuring instruments includes, as one of its important components, the development of a project for a verification scheme. For that we see a number of problems: its technical and economical validation, the degree of centralization or decentralization, ties to existing measurement standards of physical quantity units and government verification schemes, the required number of steps of this scheme, the relationships of WMS accuracies with respect to steps of the verification scheme, estimation of a quality of the system of metrological assurance under construction, and others [9].

It should be noted the problem of metrological assurance of precise magnetic recording equipment arose and the possibility for their solution were first seen in the late 1960s. As with any comparatively new matter, the solution of this problem called for an improvement of the metrological status of MRI, analysis of an equation of connection between input and output signals, revealing a complex of metrological characteristics to be normalized, grounds of a metrological model of forming a resultant error of signal registration, improvement of methods for normalizing separate metrological characteristics of MRI, etc.

The current importance of realizing the metrological assurance of MRI which is determining part of modern calculation, communication, measurement and other systems of transmitting information is explained by its wide application. Production of MRI is one of leading branches of the radioelectronic industry.

The engineering output of MRI in countries with a well developed industry amounts to 5–8 % of the industrial and 15–20 % of household radio-electronics yield [355]. In many countries tens of companies produce hundreds of MRI models of the four basic groups: laboratory stationary devices (for example, SE-5000, Great Britain; ME 260I, France), laboratory transportable devices (the typical examples here are SE 3000 and SE 7000 M, Great Britain; MT 5528, France; H067 and H068 produced by "Vibropribor", Kishinev), portable devices (for example, MP 5425, France; MR-10 and MR-30, Japan), MRI for hard conditions of operation: ME 4115, Schlumberger, France; 5600 C, Honeywell, USA) and others. Russian industry produces a set of MRI models of general industrial designation (types N036, N046, N048, N056, N057, N062, N067, N068), and also a great number of varieties of special equipment for accurate magnetic recording (for example, "Astra-2V", "Astra-144N", and others). At the same time, according to an estimate of the American company "Ampeco" the costs of all precise MRI produced in the world in 1969 was more than 70 % of the total costs of magnetic recording equipment [6].

Such an abundance of MRI equipment types, as well as the incomparability of data with regard to their parameters measured by different methods and types of measuring instruments, which can be frequently seen, have put into the forefront of urgent problems the mathematical formalization of the concept of the quality of a system of MRI metrological assurance, and the estimation of such a system under conditions of incomplete and inaccurate data about its elements, links, and properties.

Moving on to the problem of creating the technical grounds for the metrological assuarance of measuring instruments with magnetic recording/reproducing of electric signals, it is necessary to consider its basis, namely the development and study of methods and equipment for determining experimentally the main registration error of a signal in the channel of an analogue MRI, as well as the additional error caused by variations of the delay time of the signal in this channel. Moreover, from the scientific and practical point of view, a strong interest in developing methods and instruments for measuring some specific metrological characteristics of MRI such as the pulse weight function, nonlinearity of a relative phase frequency characteristic of a channel, the level of distortions of a test signal in the form of "white noise", and others, which could previously not be accurately estimated by means of experiments, is demonstrated (see Section 5.2.2).

The scope of problems of creating the means of metrological assurance of MRI is wider than the above list of them. For example, sometimes in practice there is a need to know not merely the metrological characteristics of the through channel (from input to output) of a magnetic recording/reproducing equipment, but also the metrological characteristics of a specific recording and reproducing units (of the board and ground types, correspondingly), as well as those of separate constituents of the signal registration error (in the development and adjustment of MRI). However, at present the central problem, for which no solution has previously been found,, is to create methods and instruments for measuring the resultant error of transmitting a signal over the trough channel of MRI. Here, the determination of metrological characteristics separately for a recording unit and reproducing unit can be realized with the help of comparative (relative) measurements. For example, one of the best recording units, with regard to its parameters, is accepted as the reference unit. This unit is used for adjusting all reproducing units and determining metrological characteristics of the through channel, as well as vice versa, when a selected reproducing unit is chosen as the reference unit. In essence, while solving the question of determining the error of signal transmission over the through channel of MRI, the problem of metrological assurance of its separate units is reduced to ensuring their interchangeability.

Characterizing the interrelation of investigations done Russian and foreign scientific schools in this direction, we mention, firstly, the school of Soviet metrologists, such as: M. F. Malikov, V. O. Arutyunov, Yu. V. Tarbeyev, S. V. Gorbatsevich, E. F. Dolinskiy, K. P. Shirokov, M. F. Yudin, A. N. Gordov, P. N. Agaletskiy, P. V. Novitskiy, E. D. Koltik, I. N. Krotkov, O. A. Myazdrikov, B. N. Oleynik, P. P. Kremlevskiy, P. P. Ornatskiy, L. I. Volgin, V. Ya. Rozenberg, I. B. Chelpanov, F. E. Temnikov, N. V. Studentsov, M. P. Tsapenko, S. A. Kravchenko, R. R. Kharchenko, E. I. Tsvetkov, V. A. Ivanov, V. V. Skotnikov, V. A. Balalaev, V. I. Fomenko, K. A. Krasnov, I. F. Shiskin, Yu. I. Alexandrov, V. N. Romanov, A. N. Golovin, V. A. Granovsky, A. E. Fridman, L. A. Semenov, T. N. Siraya, V. S. Alexandrov, I. A. Kharitonov, G. N. Solopchenko, and others. Among foreign scientists whose works are the most widely known we should mention such scientists aut as L. Finkelstein, J. Pfanzagl, D. Hofmann, K. Berka, Ya. Piotrovsriy, J. Jaworski, L. Zadeh, and many others.

5.3 Methods of determining MRI metrological characteristics

5.3.1 Problems in developing metrological traceability systems

A wide use of the system approach in various areas of activity has led to the introduction of the term "metrological system" (MS) (see Chapter 2). By this term is meant a class of organizational and technical systems with a specific structure and content, which are represented in the form of a hierarchy of subsystems having their own general targets, directions and tasks of development, and connected with performing accurate measurement procedures and their metrological ensuring [505].

Generally, the basic properties of metrological systems are large dimensions, complexity, imperfection and fuzziness of information concerning their elements and links, reliability of functioning, stability, efficiency, etc.

The large dimensions of metrological systems are determined by the number of elements and their links, of which there are tens and thousands, by the dissemination scale, i.e., from a branch of industry to a region of a country or a group of countries, and by the degree of influence on external systems such as the instrument industry, scientific investigations, high and critical technologies, industrial production, and others.

The complexity of metrological systems consists in their complicated structure (the multilevel nature of the system, non-uniformity of components and links between them), their complicated behavior and the nonadditivity of properties, and the complexity of formalized description and control of such systems.

The fuzziness of information about the metrological systems is caused by obscure ideas concerning about what an "ideal" system is, the conditions of its operation, the impossibility unambiguously predicting how the trends of external media development will influence the metrological systems, the multicriterion character of describing the systems, and the ambiguity of the estimates of whether or not the systems are optimal.

The development of metrology as a metrological system is intended to increase the accuracy and reliability of measurement results, to extend an assortment of measurement problems to be solved. The main problem in the further development of metrology consists in the necessity of resolving the contradictions between the growing demands of society for a product of a given science, i.e., for accumulated reliable measurement information, and the limited ability of society to bear the costs for its development.

So far two sufficiently complete subsystems have been formed in metrology: a system of fundamental metrological research [505] and a system of the metrological assuarance of a national economy.

The basic task of the system of measurement assurance (SMA) is the most complete satisfaction of the perspective demands of a national economy in the SMA, the main purpose of which is to accelerate scientific and technical progress by means of providing various branches of the national economy with reliable quantitative information about the totality of quantities and parameters in use.

In accordance with [195] metrological assurance is the "establishment and application of scientific and organizational grounds, rules, and norms needed for achieving the required uniformity of measurements and accuracy", i.e., it includes both the interconnected and open subsystems, such as scientific metrology and technical metrology (including a system of measurement standards, a system of transferring units of physical quantities from measurement standards to all measuring instruments, and others).

At present, **metrology**, as the "science of measurements, methods, and means for ensuring their uniformity, and the methods for achieving the required accuracy" [195] restricts its object of study essentially to the problems of accurate and reliable measurements of physical quantities with the help of special technical means.

At the same time, attempts [320, 382, 493, 496, and others] to expand its object of study to include so-called "natural" measurements carried out without the use of special technical means, as well as to extend the concept of measurements to include not only physical quantities. Providing proof of the appropriateness of such an "extension" is a matter for future. However, it is not useful to ignore this possibility or the trends towards the further development of metrology.

If this possibility is realized, it will inevitably result in a revision of the conceptual apparatus of metrology, which should be prepared for this to happen.. One step in this direction has been made [407] in the form of considering the principles of defining the fundamental concepts of metrology as well as the requirements for the system of concepts as principles of defining the basic concepts of metrology. Among these are such concepts as commonness, ambiguity, internal logic and consistency, compliance with other concepts (i.e., the ability of a concept to be self-derived from more general concepts and to be used for deriving less general concepts from it), simplicity, comfort, historical portability and the possibility of carrying out verification, i.e., a validity check of a defined concept. When considering the system of fundamental concepts it is necessary to follow such principles as completeness (enclosure), consistency (compliance), mutual independence, and utility.

Let us consider the concepts "measurement", "measurand", and "measurement accuracy" as the basic concepts. Ignoring or reducing the number of these concepts would most likely not be very helpful, since this could result in unnecessary complications when defining the remaining concepts.

The concept "measurement", being the basic concept in metrology, is methodically important for philosophy (within the framework of epistemology), natural and technical sciences, as well as its use in a number of social sciences. At present there are tens of versions of the definition of this concept [54, 144, 202, 216, 242, 297, 304, 313, 323, 382, 390, 437, 438, 481, 487, 489, 509, and others], using in particular a concept of the isomorphism (homomorphism) of analyzed properties of an object and multitude of numbers.

For example, in philosophy "measurement" is defined as the interaction between something physical and something mental.

The most successful definition of the concept "measurement" was given by prof. M. F. Malikov [323]: "By measurement we mean a cognitive process consisting of comparing a given quantity with a certain value, accepted as a unit of comparison, by means of a physical experiment".

To give a sufficiently general and substantial definition of the concept "measurement", let us use the following chain of "generating" concepts: "reflection" as the perception of the environment → "experiment" as a cognitive activity done with a definite (qualitative, quantitative, mental, or virtual) purpose → "measurement".

Thus, the definition of the concept "measurement" has to reflect the following aspects: a cognitive activity, activity characterized by a purpose or goal, providing for an interaction of an object and subject of cognition; obtaining information of a quantitative character as a result of carrying out a procedure of measurement by means of comparison with a measure.

Taking into account these aspects it is possible to assume the following definition of this concept: "Measurement is the form of cognitive activity providing for an interaction of a subject and object with the aim of obtaining information about it expressed in numerical form".

For the concept "measurand" the chain of "generating" concepts has the form: "objects (processes)" of a real world → "properties (characteristics)" of objects (processes) → "measurands (parameters)".

From this it follows that a "measurand" is the property (characteristic) of an object (process) allowing a presentation in a numerical form. Sometimes such a presentation is called the "numerical model".

For the concept "measurement accuracy" the chain of "generating" concepts is as follows: "quality" → "reliability" → "accuracy". Therefore, the "measurement accuracy is the degree of coincidence of the numerical presentation of a property with a true property of the object".

On the basis of these definitions for the basic concepts it is possible to define other derivative metrological concepts, e.g., the "reproduction" of a dimension of a unit of physical quantity, its "transfer" as a type of measurement, and so on.

The definitions of basic metrological concepts are based on a number of postulates. Attempts to formulate them have often been made [144, 201, 243, 246, 304, 496, and others]; however, there is still no universally recognized system of postulates. The requirements of the system of postulates are similar to those of the system of concepts. Clearly, the conceptual basis of metrology, as the science of measurements, together with the basic concepts should contain the following postulates:

(a) Objects (processes) of a real world can be measured, i.e., any properties of them can be measured).

At the same time, it should be noted that obtained information is used for constructing models of the object (the process).

(b) Measurement establishes the correspondence between the property of an object and its (numerical) model.

The above postulates and concepts have advantages such as compliance with the possible development perspectives for modern metrology, the suitability for application in all measurement scales, and equal defining the same concepts in different fields of scientific knowledge.

Describing the structure of the measurement process in greater detail [37], it is possible to present the following sequence of necessary actions: influence (interaction), discrimination, comparison (confrontation), registration (formation of a "stable connection", i.e., obtaining a reflection of the property of an object or, in other words, its image). In terms of materialistic dialectics, this process takes place between two objective realities entering into an interaction with each other, since only because of this efficient interaction can they obtain "knowledge" about each other as the final result of the measurement process. From experience it is known that not all interactions lead to knowledge about the existance of an object. For example, a human being (as distinct from a bat) does not sense ultrasound oscillations, i.e., does not obtain any knowledge about them, although they really exist. Therefore, the introduction of a "discrimination" (recognition of) procedure into the structure of the measurement process is justified.

To perform measurement it is necessary to formulate a concrete measurement problem, indicating its components: a measurand (a property), the object to be studied (its carrier), conditions of measurement (external influencing quantities), error given (a required one) (accuracy), the form for presenting the measurement result, the space–time coordinates (answering the questions of when, where, and for what time the measurement has to be performed), and others.

At the second stage a plan for the measurement experiment is developed, which answers the question of how it is to be done. The components of this plan may be chosen, which is not possible for the measurement problem. It is possible to assign to them the following: the chosen unit of a measurand, the method of measurement, the type of a measuring instrument, the operator who makes the measurement experiment plan, the means of processing the measurement results, auxiliary means, etc. The development of the plan is implemented on the basis of a priori information, i.e., metrological information accumulated before the beginning to solve the measurement problem.

At the third stage a process of real transformations takes place. These transformations are connected with the physical interaction of the chosen measuring instrument and the object, the external conditions, and the operator (observer) who carries out the procedures of distinction, comparison, and registration of the result.

At the fourth stage the procedure of processing measurement information is performed on the basis of the available a priori information with or without the help of computers and auxiliary means [32, 33].

In these stages, in essence, the process of measurements is presented in the form of a certain algorithm for finding the value of the measurand. Since our knowledge is principally limited, the distinction of a real algorithm for any process of finding a measurand value from the ideal (required) algorithm is inherent in any process of measurement. This is expressed by the presence of an inevitable error in a measurement result, which is the most important property of any measurement [32, 33].

The sources of measurement errors are all components of the measurement problem and the plan of the measurement experiment, as listed above, i.e., the object of measurement being a source of the measurand, the measuring instrument with its inherent properties, the measurement conditions with the quantities which influence them, the subject performing the measurements (operator/observer), etc.

Moreover, the following limitations, which are principal for any measurement, should be also taken into account:

- the presence of a final (nonzero) time interval $\Delta t = \tau_{\text{meas}}$, needed in order to realize the measurement algorithm;
- the necessity of the a priori establishment of requirements for the limits of a measurement error (then and only then is the measurement of any real practical value).

Taking into account all the principal properties of any measurement enables us to formulate a general condition of measurement correctness. For the simplest case of direct single measurements (which are the most important ones in practice, since all remaining cases reduce to them) this condition can be written in the form

$$\delta_0 + \delta_\varphi(\tau_{\text{meas}}) + \delta_s(\tau_{\text{meas}}) + \delta_\psi(\tau_{\text{meas}}) + \delta_v(\tau_{\text{meas}}) \leq \delta_n. \tag{5.3.1}$$

Here
δ_0 is the relative measurement error of an "instantaneous" value of a measurand φ;
$\delta_\varphi(\tau_{\text{meas}})$ is the relative error caused by a change of the measurand φ for the time of measurement τ_{meas};
$\delta_s(\tau_{\text{meas}})$ is the relative error caused by a change of metrological characteristics of a measuring instrument under normal conditions for the time τ_{meas};
$\delta\psi(\tau_{\text{meas}})$ is the relative resultant error caused by changes of all influencing quantities ψ (parameters of the environment) beyond the limits of normal conditions for the time τ_{meas};
$\delta_v(\tau_{\text{meas}})$ is the relative error caused by changes of parameters of perception organs of an operator/observer for the time τ_{meas};
δ_n is the normalized value (given, required, admitted) of the relative error of measurement.

All errors in equation (5.3.1) are taken with regard to the module and reduced to the input measurand.

It should be noted that since "instantaneous" measurements (measurements within infinitesimal time intervals) cannot be realized in practice and are purely an mathematical idealization (abstraction), the error component δ_0 in equation (5.3.1) should be considered as the "static" component of the measurement error, while the four remaining members of the left part of inequality (5.3.1) should be considered as the "dynamic" components of the measurement error.

From the general condition of measurement correctness (5.3.1) it is possible to deduce boundaries of a required time interval τ_{meas}.

Let t_0 (Figure 5.3) be the moment of "switching on" of the measurand φ source. It is clear that the measurement time interval cannot be greater than the limit value at which the change of the measurand ($\Delta\varphi_n = \varphi_0 \cdot \delta\varphi$) itself relative to the value measured φ_0 oversteps the limits of the admitted measurement error $\Delta\varphi_n = \varphi_0 \cdot \delta_n$, i.e.,

$$\tau_{\text{meas}} \leq (\tau_{\text{meas}})_{\text{lim}} = t_{\text{max}} - t_{\text{min}}. \tag{5.3.2}$$

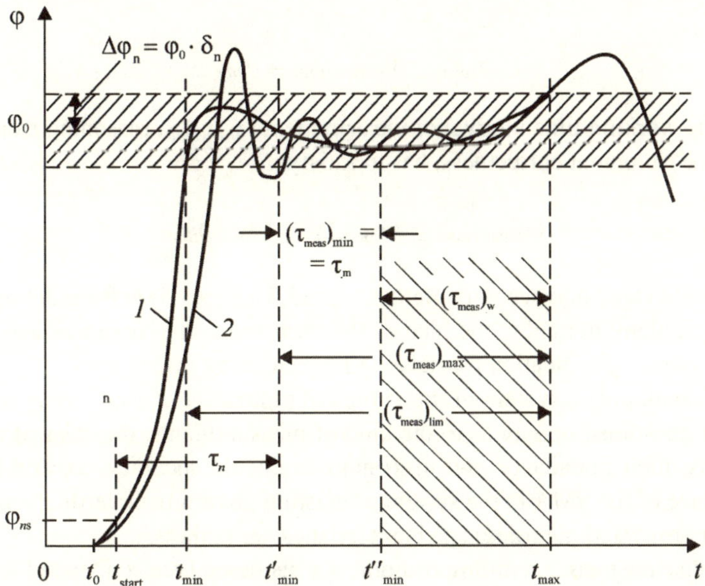

Figure 5.3. Analysis of the time relations in the measurement process. Here: 1 is the measurand variation in the process of measurement; 2 is the transient process in the measuring instrument.

The interaction of the measurand (of the object measured) and measuring instrument starts just at the moment t_{start} determined by the speed of the φ variation as well as by the value of the sensitivity threshold φ_{ns} of the applied measuring instrument. Here, the inertial properties of the measuring instrument determining its response speed provoke a transient process of τ_n duration (curve 2), which pushes back the possible start until the moment $t'_{\min}$, when the error component δ_s becomes less than δ_n.

Thus, taking into account the components δ_φ and δ_s,the maximum time of measurement is determined by the condition

$$\tau_{\text{meas}} \le (\tau_{\text{meas}})_{\max} = t_{\max} - t'_{\min}. \tag{5.3.3}$$

A similar argumentation with the use of components δ_ψ and δ_v leads to the maximum time interval of measurement becoming even more narrow. However, believing that these improvements are not principal from the point of view of the picture accuracy and that in metrological practice they aspire to provide normal measurement conditions beforehand, let us consider the lower boundary of the time interval of measurement.

Within the limits of $(\tau_{\text{meas}})_{\max}$ this boundary will be determined by the response speed of a readout device of the measuring instrument and the operator/observer organs of perception. If the inertial properties of these measurement components are expressed through a "dead time" τ_m, then the following relationship will have to be fulfilled

$$\tau_{\text{meas}} \ge (\tau_{\text{meas}})_{\min} = \tau_m. \tag{5.3.4}$$

Thus, the working time interval of measurement $(\tau_{\text{meas}})_w$, reasoning from the condition of measurement correctness, has to satisfy

$$(\tau_{\text{meas}})_{\max} \ge (\tau_{\text{meas}})_w \ge (\tau_{\text{meas}})_{\min}. \tag{5.3.5}$$

It is useful to take into account relationships (5.3.1)–(5.3.5) in formulating the measurement problem, itemizing the plan of the measurement experiment and preparing the requirements specification for developing measuring instruments.

Before developing a **system of metrological assurance** it is necessary to answer a number of questions, mainly as to the kind of measurements, the class of measuring instruments, their abundance and maximum attainable accuracy, as well as quality indices, place of the SMA in the system of existing government verification schemes, the set of normalized metrological characteristics, and others.

The precise magnetic recording equipment considered here is intended for registration (recording, reservation and reproducing) analogue (time continuous) electric signals of measurement information, i.e., for the registration of the dependence of electric current voltage on time, $U(t)$. Among four kinds of measurements which are known at present (direct, indirect, combined, and joint measurements) [438], the measurements of $U(t)$ are related to the joint ones.

According to the definition given in [195], by joint measurements is meant the "simultaneously performed measurements of two or more different quantities for finding dependence among them". The comparison characteristics of various measurements are given in Table 5.1.

Table 5.1. Comparative characteristics of the kinds of measurements (a method of obtaining a numerical measurand value serves as a criterion for classifying)

Features (kind of measurement)	Number of measured physical quantities n	Number of means in use m	Number of measurement equations P	Law of linking physical quantities	Goal of measurements
Direct	$n = 1$	$m = 1$ $(n = m)$	$P = 1$	—	Comparison of a measurand with a measure (scale)
Indirect	$n > 1$	$m > 1$ $(n = m)$	$P > 1$	Known	Calculation of the derivative value of a physical quantity against a known dependence on measurands
Agregated	$n \geq 1$	$m \geq 1$ $(n = m)$	$P > 1$	Chosen	Calculation of values of the cosen law of linking physical quantities
Joint	$n > 1$	$m > 1$ $(n = m)$	$P > 1$	Unknown	Finding out a dependence (law of linking) among measured quantities

A distinctive feature of measuring instantaneous values $U(t)$, as opposed to the joint measurements of other kinds, is their clear dynamic character, since one of the different measurands is the time serving as the argument of functional dependences.

In accordance with the stated classification of the kinds of measurements [456], the measurements of this kind are related to the average d.c. and a.c. electric currrent voltages of low and average frequency and are designated as 03.02.01.04.010.13 and 03.02.01.04. (011.+02323.+02324.).13, correspondingly.

The structure of the classifier is based on taking into account the interactions taking place both between the physical quantities themselves and between the physical quantities and the measurement process:

- correspondence of measurands to the fields of physics with selecting groups and individual physical quantities;
- chacteristics of the measurand or parameters (levels and ranges);
- specific character of the measurement problem to be solved connected to the field of its use.

In accordance with these classification signs the following rubrics of the classifier are selected.

1. Field of measurements (a division or field of physics): for $U(t)$ it is rubric 03, "electricity and magnetism".

2. Group of measurements (a subdivision of the field of physics): for $U(t)$ it is rubric 02, “electric circuits”.

3. Kind of measurements (a measurand or parameter to be measured): for $U(t)$ it is rubric 01, “electric voltage”.

4. Range of measurements: for $U(t)$ it is rubric 04, “average values” (from 10^{-6} to 10^{3} V).

5. Conditions including:
   5.1. Character of the time dependence: for $U(t)$ it is rubric 010, “static measurements” (a constant quantity) and rubric 011, “dynamic measurements” (a variable quantity), +02323 and 02324 mean that electrical voltage frequency (0232) has low values (02323. – less than 1 kHz) and average values (02324. – from 10^{3} to 10^{6} Hz);
   5.2. Dependence of influencing quantities;
   5.3. Characteristic of the environment (an aggregate state);

6. Field of application: in providing the metrological assurance of measurements for $U(t)$ it is rubric 13, “measurements for scientific investigations and problems of ensuring the uniformity of measurements”.

As it is shown in [457] the MRI as a measuring instrument (i.e., “a technical means used in measurements and having normalized metrological properties”) is related to the class of measurement transducers, since it is “designated for generating a signal of measurement information in the form suitable for transmitting, further transformation, processing, and/or reservation, but resistant to direct perception by an observer” [195].

The abundance of MRI in country used as units of measuring information complexes and computer systems is too great.

Suffice it to mention measurement magnetographs of the type “NO-…”, which are produced by the Kishinev plant “Vibropribor”, special equipment for magnetic recording of various types which were produced in small numbers by the Kiev NPO “Mayak”, import measurement magnetophones of the companies “Brüel & Kjær”, “Schlumberger”, “Ampex Corporation”, “Honeywell”, “Philips”, “Lockheed”, “Winston”, “Telefunken”, “Teac” and others, as well as a tendency to adapt inexpensive domestic magnetophones for registering signals of measurement information of biomedical destination.

Taking into account such a great demand for MRI and the need for its metrological assurance, we need emphasize the urgency for creating fundamentals of SMA in the form of the means for providing metrological certification and verification of precise magnetic recording equipment. To create such a means it is necessary to consider the MRI quality indices, in particular the accuracy requirements for the means of their verification. To begin with, let us try to assess the maximum attainable accuracy of joint measurements of instantaneous values of electric voltage $U(t)$. Relying on the results obtained in Section 3.3.3 and reasoning from the known relationship of uncertainties for energy and time, we obtain

$$\Delta E \cdot \Delta t \geq h/2, \tag{5.3.6}$$

where ΔE and Δt can be considered as the errors of the energy and time measurements, respectively, and h is the Planck constant.

Let us present the energy in the form of the product $q \cdot U$. Having chosen an elementary charge (an electron charge $\mathrm{e} \cong 1.6 \cdot 10^{-19}$ C) as q, formula (5.3.6) can be written as

$$\Delta U \cdot \Delta t \geq h/2\mathrm{e} \cong 2 \cdot 10^{-15}\ \mathrm{V \cdot s}. \tag{5.3.7}$$

Using the time of measurement from equation (5.3.5) instead of Δt, we obtain the maximum accuracy of instantaneous voltage value measurement:

$$\Delta U \geq h/(2\mathrm{e} \cdot \tau_{\mathrm{meas}}), \tag{5.3.8}$$

i.e., for example, at $\tau_{\mathrm{meas}} = 1 \cdot 10^{-6}$ s we obtain $\Delta U \geq 2 \cdot 10^{-9}$ V. This correlates with the theory described in detail in [368], where it is shown that the energy threshold of a measuring instrument sensitivity is proportional to its error γ, power P used by the object of measurement, and to the time τ needed to establish the result of measurement, and which cannot be less than $C = 3.5 \cdot 10^{-20}$ J:

$$\gamma^2 \cdot P \cdot t \geq C. \tag{5.3.9}$$

In the majority of MRI types, the dynamic range of their channels (i.e., the signal-to-noise ratio) does not exceed 60 dB and at the upper limit of the amplitude range on the level of a few volts, the question relating to the maximum attainable accuracies of registering, naturally, does not arise.

In accordance with [427] the **quality indices** of MRI, as those of measuring instruments, include indices of designation, reliability, economical use of raw materials, energy, and human resources, the ability to be ergonomic, aesthetic, technological, transportable, unified, standardized, and in compliance with legal regulations, as well as to meet industrial and environmental safety requirements.

The **indices of designation**, which characterize the main functions of the equipment and its field of application, are subdivided into indices of equipment functions and technical efficiency as well as design indices.

The **indices of MRI functions and technical efficiency** includes the most important technical and metrological characteristics, in particular those of amplitude and frequency range of signals to be registered, indices of accuracy, speed, versatility, automatization, compatibility with the measured object (from the viewpoint of consuming its power), adaptability to verification, and others.

While assessing a technical level of production and organizational-technical systems (for example, MRI as a measuring instrument or SMA as a variety of large systems), the need arises for a generalized (integral) quality index as distinct from the particular indices listed above. In this case qualimetry suggests various algorithms of "combining" particular quality indices, among which the method of their "weighted" summing up is most widely used. The difficult problem of a well-grounded choice of weight coefficients is solved, as a rule, by using the expert method.

A similar situation occurs when multicriteria optimization takes place and objects (systems) optimal for to one criterion are far from being optimal for other criteria (criterion). The problem becomes more complicated in the case, when information about object parameters (or of the system) is incomplete (fuzzy, not definite or clear enough), which in practice is frequently met with, just as in making a decision with incomplete initial information.

The concept of quality for the complicated organizational-technical systems connected with the performance of precise measurements and their metrological assurance (of "metrological systems"), can be related to both the system as a whole, i.e., its structure, operation, adaptability for use, and separate properties of system: reliability, stability, efficiency, and is defined differently [30, 165]. By the quality of a metrological system we mean its optimality in any way, from the point of view of reaching a definite goal, solution of a measurement (metrological) problem, and by the quality criterion – the index of the optimum. For any definition the quality of a system is expressed through the totality of quantitative (objective) and qualitative (subjective) factors which resist direct metrization, e.g., simplicity, suitability, beauty, etc. Quality assessment is connected with ranking systems against a certain set of objective and subjective features which depend on the goal or problem to be solved [405–407].

Let us define "quality threshold" as a guaranteed quality characterizing a definite level of attaining a goal or solving a problem and include into this concept the following indices (factors): accuracy, expressed by means of the error δ_0; trustworthiness (reliability), characterized by the probability P_0; time response, in the form of measurement time τ_0; resource T_0, determined by the time within which the system can be applied to solve the problem; and completeness α_0 reflecting a degree of satisfying demands of the subject. Each of these factors can have objective and subjective components.

To provide a definite quality (attaining the goal) for a set of systems it is necessary to comply with the conditions for realization in the form of a totality of inequalities:

$$\delta \leq \delta_0, \quad P \geq P_0, \quad \tau_{\text{meas}} \leq \tau_0, \quad T \geq T_0, \quad \alpha \geq \alpha_0. \tag{5.3.10}$$

From equation (5.3.10) it follows that in the general case the assessment of quality is the problem of multicriteria optimization. The process of formalizing the quality concept is connected with the choice of a certain totality of axioms (conditions) depending on the structure of a problem and the kind of system. The choice of axioms determine the general restrictions applied to the method of describing the quality concept. For example, let us require thtat the following axioms be complied with.

(1) If $\Sigma_i \subset \Sigma$, then $K(\Sigma_i) \leq K(\Sigma)$.

This means, in essence, that the quality is "measured" in scale of the order.

(2) If $\Sigma = \Sigma_1 \oplus \Sigma_2 \oplus \cdots \oplus \Sigma_n$, then $K(\Sigma) \geq \max_i K(\Sigma_i)$, $1 \leq i \leq n$ (parallel connection of systems), where $K(\Sigma)$ is the quality of system Σ.

(3) If $\Sigma = \Sigma_1 \oplus \Sigma_2 \oplus \cdots \oplus \Sigma_n$, then $K(\Sigma) \geq \min_i K(\Sigma_i)$, $1 \leq i \leq n$ (series connection of systems).

The second and third axioms allow the quality of a system to be calculated by their components in the simplest cases.

(4) If there are two interacting systems, Σ_1 and Σ_2, then $K(\Sigma_1 \oplus \Sigma_2) \geq \max\{K(\Sigma_1 \leftrightarrow \Sigma_2), \min[K(\Sigma_1), K(\Sigma_2)]\}$, where $K(\Sigma_1), K(\Sigma_2)$ is the quality of each of these two systems, and $K(\Sigma_1 \leftrightarrow \Sigma_2)$ is the feedback quality (interaction).

The totality of axioms given above is not the only possible one, since choice of axiom depends on the goal (of the problem to be solved) and the kind of systems. For example, for quality characteristics such as accuracy (error) and time of solving the problem (time of measurement), the laws of (max-sum)-composition are true, and for the reliability the laws of summing up and multiplying of probabilities are in force, and so on. Since a quality estimate always contains subjective components and, moreover, their impact on the final result is unknown, the use of common methods for describing the quality concept in a formalized form is restricted. It can be performed with the help of a mathematical apparatus of the fuzzy-set theory [205, 269, 371].

Let a set of metrological systems be denoted as m. As its elements we can have separate measurements, measuring instruments, verification schemes, as well as other objects of the type of a system of measurement uniformity assurance or a system of metrological assurance.

Let $\{m^i\}$ be the set of uniform systems of the i-th class: $\{m^i\} \in m$, and $m^i_k \in \{m^i\}$ of a k-th system of the i-th class; $1 \leq i \leq n$ (where n is the number of classes of metrological systems).

Let us take the subset $\{m^i\}$ and consider a set of its reflections, $L_i^{\{mi\}}$ in L_i, where L_i is a certain set of the array type ordered or quasi-ordered, the elements of which have lower and upper boundaries [205]. The set L_i is connected with a group of quality estimates of elements of the subset $\{m^i\}$ and, generally speaking, can have a structure differing for each subset $\{m^i\}$. Then let us determine on $L_i^{\{mi\}}$ a fuzzy subset $m^i = \{m^i_k, \lambda_{mi}(m^i_k)\}$ characterizing a system of a definite quality, which poses for each element $m^i_k \in \{m^i\}$ an element $\lambda_{mi}(m^i_k) \in L_i$ (the element λ_{mi} is an estimate of the system quality m^i_k). Thus, one poses to a set m in conformity the set of fuzzy subsets (their product) $\prod_{i=1}^n L_i^{\{m^i\}}$, characterizing systems of definite quality, the membership function of which correspondingly possesses a value in $L_1, L_2, \dots L_n$. The structure of the set L_i depends on the totality of axioms characterizing the quality concept. For example, the array L_i can be a vector one formed by the product of sets corresponding to the criteria of "threshold of quality":

$$L_i = L_i^{(1)} \cdot L_i^{(2)}, \dots, L_i^{(5)}, \tag{5.3.11}$$

where $L_i^{(1)}$ is the set of accuracy estimates, $L_i^{(2)}$ is the set of reliability estimates, and so on.

In a particular case each $L_i^{(k)}$ (it means that L_i too) can be a numerical interval [0, 1].

As an example, let us consider the quality estimate of a possible verification scheme for MRI (Figure 5.4a), the structure of which contains three levels: the first level constitutes initial (for the given verification scheme) WMS taken from existing national verification schemes for the government primary measurement standard of EMF unit [192] (or the government special standard of the voltage unit within the voltage range of $0.1 \div 10$ V and the frequency range of $20 \div 3 \cdot 10^7$ Hz) and the government primary standard of time and frequency unites [198]; the second level is presented by working measuring standards and the third one by working measuring instruments (precise magnetic recording equipment of various types).

Its representation L is demonstrated in Figure 5.4b, where $\lambda_{j(r)}$ is the quality estimate of subsystems, where $\lambda_{j(r)} \in L_i$.

The estimation of the quality of such a verification scheme can be determined by the formula

$$K = \left.\begin{matrix}\lambda_{1(1)} \\ \lambda_{1(2)}\end{matrix}\right\} \wedge \left\langle \begin{matrix}\{\lambda_{2(1)} \wedge [\lambda_{3(1)} \vee \dots \vee \lambda_{3(i)}]\} \vee \{\lambda_{2(2)} \wedge [\lambda_{3(i+1)} \vee \dots \\ \dots \vee \lambda_{3(k)}]\} \vee \dots \vee \{\lambda_{2(l)} \wedge [\lambda_{3(k+1)} \vee \dots \vee \lambda_{3(p)}]\}\end{matrix} \right\rangle, \tag{5.3.12}$$

where the symbols $\vee$ and $\wedge$ denote the procedures of taking the upper and lower boundaries determined on the array L_i.

From equation (5.3.12) it is possible to make some decisions. The system quality is determined mainly by the totality of $\lambda_{1(1)}$ and $\lambda_{1(2)}$, i.e., by an estimate of borrowed IMMI, as well as by quality estimates of a lower level (of the third level) of the system. The quality of each branch of the system also does improve because of any additional levels: increasing the number of parallel branches, i.e., an increase in the number of WMS the quality of the system.

To simplify expression (5.3.12) it is necessary to know the ratio of the order between $\lambda_{j(r)}$. If ranking $\lambda_{j(r)}$ is successful, so that they have the ratio of the order as

$$\begin{aligned}\left.\begin{matrix}\lambda_{1(1)} \\ \lambda_{1(2)}\end{matrix}\right\} > \lambda_{2(1)} \geq \lambda_{2(2)} \geq \dots \geq \lambda_{2(l)} > \lambda_{3(1)} \geq \dots \\ \dots \geq \lambda_{3(i)} \geq \lambda_{3(i+1)} \geq \dots \geq \lambda_{3(k)} \geq \lambda_{3(k+1)} \geq \cdot \geq \lambda_{3(p)},\end{aligned} \tag{5.3.13}$$

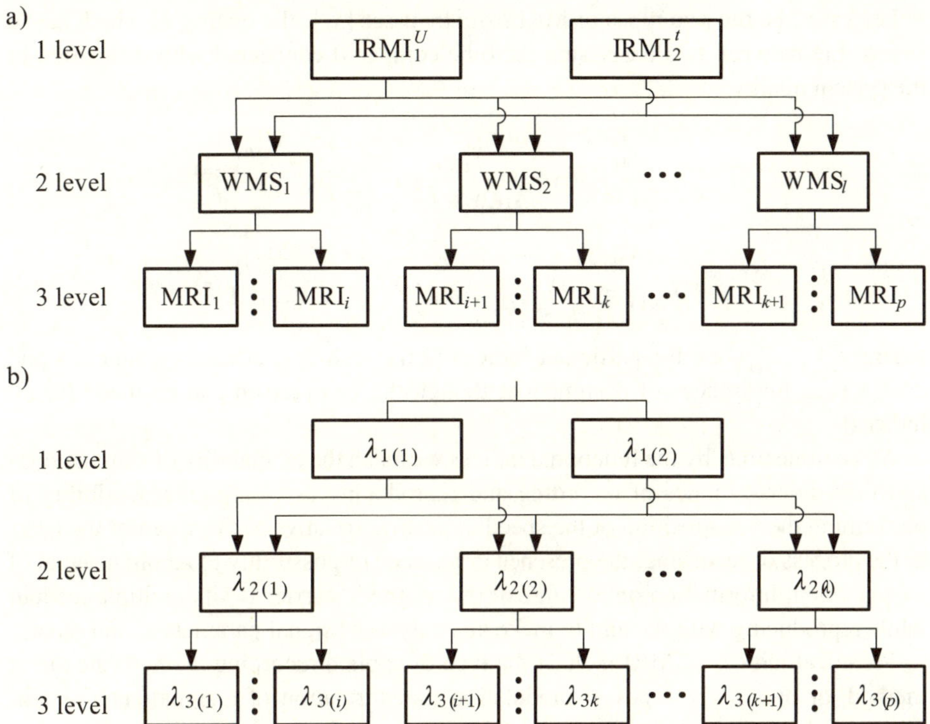

Figure 5.4. Example of getting a quality estimate of a possible verification scheme for the equipment of precise magnetic recording. Here: IRMI $_i^U$ is the initial reference measuring instrument from the state verification scheme of the state primary measurement standard of electromotive force (or of the state special measurement standard of the voltage unit); IRMI $_i^t$ is the initial reference measuring instrument from the state verification scheme for the state primary measurement standard of the time and frequency units; WMS $_i$ is the working measurement standard; MRI $_j$ is the magnetic recording instrument.

then from (5.3.12) it follows that

$$K = \begin{Bmatrix} \lambda_{1(1)} \\ \lambda_{1(2)} \end{Bmatrix} \wedge \lambda_{3(1)} = \lambda_{3(1)}. \tag{5.3.14}$$

In this case the system quality depends on the quality of the best representatives from the lowest link.

As another example take an algorithm for comparing the quality of uniform systems, whereby a certain type of precise magnetic recording equipment is assumed (it should be noted that a similar comparison is possible for heterogeneous systems, the interaction of which cannot be properly described and presented by a simple structural scheme).

Let x_1^i, x_2^i be the two types of MRI from their set $\{x_i\}$, the quality of which has to assess. Let us write two fuzzy subsets to be compared connected with estimation of the system quality:

$$x_1^i = \frac{x_{1(1)}^i}{\lambda_{1(1)} \in L_i^{(1)}} \; \frac{x_{1(2)}^i}{\lambda_{1(2)} \in L_i^{(2)}} \; \frac{\cdots}{\cdots} \; \frac{x_{1(n)}^i}{\lambda_{1(n)} \in L_i^{(n)}}$$

$$x_2^i = \frac{x_{2(1)}^i}{\lambda_{2(1)} \in L_i^{(1)}} \; \frac{x_{2(2)}^i}{\lambda_{2(2)} \in L_i^{(2)}} \; \frac{\cdots}{\cdots} \; \frac{x_{2(n)}^i}{\lambda_{2(n)} \in L_i^{(n)}}, \qquad (5.3.15)$$

where $x_{1(k)}^i, x_{2(k)}^i$ are the particular indices of the technical efficiency, function and design (i.e., the indices of designation, though the comparison can be made for all indices).

At the same time, by the functional indices we mean the availability of simultaneous and time-diverse modes of recording and reproducing as well as the possibility of performing the transposition of the speed of a carrier relative to the speed of the latter in the process of recording, the presence or absence of possibility to record a signal of measurement information on a "ring" of the magnetic carrier for its multiple readout while reproducing with the aim to make an analysis of signal parameters, and others.

Technical indices of MRI include the type of applied recording method (the direct method, or one of the types of modulation transformation of a signal such as amplitude, balance amplitude, frequency, phase, amplitude phase, amplitude pulse, frequency pulse, phase pulse, pulse width and delta modulation), speed of feeding the carrier, time of recording, number of channels (tracks), the type of the carrier (magnetic tape, metal tape, wire, and others), density of recording (longitudenal, transverse, surface type), coefficient of oscillations and drift of the carrier speed, power used, parameters of the supply network, etc. The design indices include the MRI mass, dimensions, capacity of a cassette and tape width, element base, form of executing (in the air, stationary, transportable, executed in one case or in the form of separate units of recording and equipment of reproducing), and others. The metrological characteristics of MRI are considered in Section 5.3.2.

The components $x_{1(k)}^i$, $x_{2(k)}^i$ can be estimated on the basis of various criteria, and their estimates $\lambda_{1(k)}\lambda_{2(k)}$, respectively, will accept the values in the sets $L_i^{(k)}$ having different structures. To obtain a comparative estimate of quality, the following scheme can be applied.

Let us represent the structure of each $L_i^{(k)}$ in the form of a simple graph and determine the order levels.

Distances, corresponding to different criteria (components) in each graph, will be as follows:

$$D_k^{(i)} = \left|N_{1(k)}^i - N_{2(k)}^i\right|, \qquad (5.3.16)$$

where $N_{1(k)}^i$, $N_{2(k)}^i$ are the levels of the estimate order for $x_{1(k)}^i, x_{2(k)}^i$.

Let the relative distances between elements be calculated:

$$\Delta^i_{(k)} = \frac{D^{(i)}_k}{N^{(k)}_{0i}}, \tag{5.3.17}$$

where $N^{(k)}_{0i}$ is the number of levels of the order in $L^{(k)}_i$.

Let us assign "+" or "−" to each $\Delta^i_{(k)}$ in accordance with the sign of the difference $N^i_{1(k)} - N^i_{2(k)}$:

$$\lambda^i_{(k)}(x^i_1, x^i_2) = \pm\Delta^i_{(k)}. \tag{5.3.18}$$

As a result the "table" is obtained:

$$\lambda^i_{1,2} = \begin{matrix} x^i_{(1)} & x^i_{(2)} & \cdots & x^i_{(n)} \\ \lambda^i_{(1)1,2} & \lambda^i_{(2)1,2} & \cdots & \lambda^i_{(n)1,2} \end{matrix}. \tag{5.3.19}$$

The value of $\lambda^i_{(k)1,2}$ can be considered as the relative difference of the quality estimates of the components $x^i_{1(k)}, x^i_{2(k)}$, and its sign shows which of the estimates is greater.

Let us find the relative generalized distance between the fuzzy sets $x^i_{1(k)}, x^i_{2(k)}$:

$$\delta(x^i_1, x^i_2) = \left| \frac{1}{n} \sum_{k=1}^{n} \lambda^i_{(k)}(x^i_1, x^i_2) \right|. \tag{5.3.20}$$

Let the sign be assigned to each δ in accordance with the sign of the sum $\sum_{k=1}^{n} \lambda^i_{(k)}(x^i_1, x^i_2)$:

$$\delta^* = \pm\delta. \tag{5.3.21}$$

The value δ^* can be considered as the relative difference of the quality estimates of two compared systems. It should be noted that $\delta^* = 0$ corresponds to their equal quality, and $\delta^* = \pm 1$ to a maximum difference of their qualities.

From the analysis it follows that it is easy to compare the quality estimates of systems, but not those of the estimates themselves, i.e., it is handy to estimate the quality in the scale of intervals. The scheme considered can be generalized for the case of a number of systems [406]. It is reduced to solving a problem of the multicriteria optimization by the quality criteria on a set of systems which consists of the choice of a system having as far as possible the maximum estimates according to all criteria and satisfying the conditions of realization of the form shown by equation (5.3.10).

The solution of the problem is done in two stages: constructing the relations of the fuzzy preference on the set of systems according to the quality criteria and aggregating the relations obtained. In a real case the complexity of its solution is connected with the fact that the threshold values of the quality criteria have not been been accurately preset and become "fuzzy", just like the estimates of quality of the systems considered. The width of a "fuzzy area" is determined by the incompleteness of the initial information concerning the goal or problem solved.

5.3.2 Metrological characteristics of MRI and their normalization

By metrological characteristics of MRI we mean those characteristics which directly influence and determine the accuracy of the registration of measurement information signals and are related to the "through" channel of MRI containing electronic transducers used in recording, the magnetic head-carrier-head system and electronic transducers used in reproducing.

An analysis of the properties of MRI considered as a measurement transducer is based on its mathematical description representing an abstract model of the transducer. At the same time, the quality of functions implemented by the transducer is the main interest, and here it should be pointed out that in this case the quantitative characteristic of quality is the error of transformation.

The basic function of MRI as a unit of information measurement systems is recording (storage), preservation and reproducing information without distortions. This assumes that MRI is a linear stationary system realizing similarity transformation or, in a more general case [457], transformation in the scale of intervals, satisfying the following equation of connection of output $y(t)$ and input $x(t)$ signals:

$$y(t) = A\{x(t)\} = K \cdot x(bt - \tau) + d, \tag{5.3.22}$$

where A is the operator of transformation; $K$is the transfer coefficient (coefficient of transformation of the amplitude scale); b is the coefficient of transformation of the time scale; τ is the time delay (lag) of a signal; d is the value of the constant constituent which is needed in registering a variable signal centered with respect to the channel, the amplitude range of which is asymmetrical with respect to zero (for example, from 0 to +6.3 V).

In practice the real parameters of model (5.3.22), K_r, b_r, τ_r, d_r, differ from the ideal and nominal (indicated in a certificate) parameters K_n, b_n, τ_n, d_n, which leads to a transformation error:

$$\Delta y(t) = y_r - y_n = [K_r \cdot x(b_r t - \tau_r) + d_r] - [K_n \cdot x(b_n t - \tau_n) + d_n]. \tag{5.3.23}$$

From (5.3.23) it is seen that the registration (transformation) error $\Delta y(t)$ depends on both the disagreement of real and nominal parameters of MRI and type (characteristics) of an input signal of measurement information $x(t)$.

Here, the difference of transfer coefficients

$$\Delta K = K_r - K_n \tag{5.3.24}$$

"comes apart", forming the following constituents:

- nonlinearity of amplitude characteristic leading to nonlinear distortions;
- difference between the angle of slope of the amplitude characteristic and that of the nominal one, which corresponds to the nonnominality of the channel sensitivity and results in distortion of the amplitude scale of an output signal and appearance of a multiplicative noise;

- "zero shift", percieved in the same manner as $\Delta d = d_r - d_n$ and forming an additive noise;
- lack of nominality of the complete dynamic characteristic of the channel in the frequency or time domain, causing the dynamic distortions.

The difference $\Delta b = b_r - -b_n$ and $\Delta\tau = \tau_r - \tau_n$ leads to distortions of the time scale of an output signal, Δb corresponding to the speed drift of the carrier and $\Delta\tau$ – to oscillations of the time delay of the signal reproduced.

The difference $\Delta d = d_r - d_n$ corresponds to the noise level in the channel and contains the noise of electronic units of recording and reproducing, parasitic amplitude modulation, tape noise caused by its granularity as well as by defects and damages, noises such as background noise, transient noise due to the effects of copying, penetration, self-demagnetization, etc.

Moreover, it is necessary to take into account the limits of the input and output impedance of the channel, since a noninfinitely great input impedance gives rise to distortions in state of the object whose properties are being measured, and the output resistance of the channel, whose value can not be infinitely small, may spoil the normal interaction of MRI with the next unit of the measurement system. All this taken together (including the nonlinearity of sensitivity) leads to distortion of the amplitude scale of output signals.

Taking into account the above, the metrological model of resultant error formation in the MRI channel can be shown in the form of the scheme in Figure 5.5 and containing a stationary link represented by the linear operator L (the operator of the linearized channel taking into account the disagreement between the slope angle of the AC and the nominal one, nonnominality of full dynamic characteristic and limits of input and output impedances of the channel), and by the nonlinear operator H (corresponding to the nonlinearity of the AC), the link of variable time delay D (reflecting the influence of Δb and $\Delta\tau$) and the source of additive noise e (containing the "zero drift" and Δd).

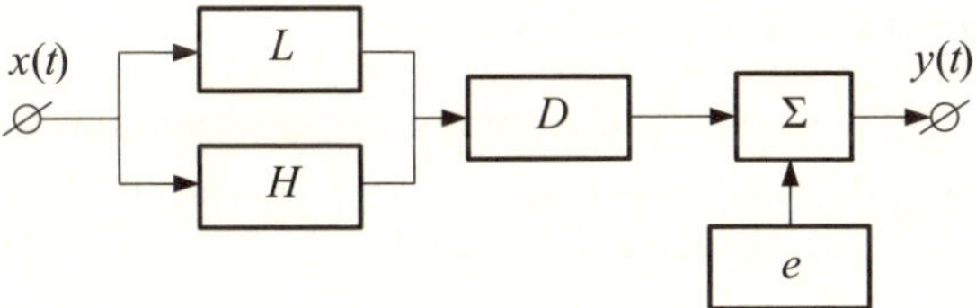

Figure 5.5. Model of resultant error formation in registration of a measurement information signal in a MRI channel. Here: $x(t)$ is the input signal; L is the operator of a linearized channel; H is the nonlinear operator; D is the link of a variable time delay; e is the source of additive noise; $\sum$ is the linear summator; $y(t)$ is the output signal.

Thus, in addition to the function of transformation, amplitude and frequency ranges of the signal registered, the list of metrological characteristics [191] of MRI, subjected to normalization must include

(1) nonlinearity of the amplitude characteristic;
(2) nonnominality of full dynamic characteristic;
(3) level of noises;
(4) variations of the time delay;
(5) limits of the input and output impedances of the channel;
(6) resulting basic and additional errors of transformation without their systematic and random constituents, as well as the functions of the influence of environmental conditions overstepping the limits of normal value areas.

At a known shape or characteristics of the input signal $x(t)$ by metrological characteristics (1)–(5) it is possible to estimate the corresponding particular constituents of the resultant error (6). All listed metrological characteristics and characteristics of the MRI error are of the instrumental type, i.e., they relate to a measuring instrument. In conducting measurements, in addition to them, it is also necessary to take into account the methodical errors and operator errors (including crude errors, or so called "blunders".

The degree of the interest in these parameters (see Table 5.2) depends on the categories of persons participating at different stages of the "life-cycle" of precise magnetic recording equipment such as: (a) developer, (b) manufacturer, (c) consumer (it is useful to distinguish between the consumer who implements "technical measurements" in industry, and the consumer who carries out scientific researches with the help of MRI), (e) personnel engaged in equipment repair, (d) verifier (a metrologist, specialist on quality control).

To estimate the error characteristics of both the resultant error and its particular constitutiens, one has to give or choose a kind of test signal. The problem of choosing an optimal test signal is solved by means of the theory of planning an experiment [163] and intends to obtain parameters of a dynamic object model least sensitive to random noises.

At the same time, it should be mentioned that

- D-optimal test signal is a signal that minimizes the determinant of the covariance matrix of estimates of unknown dynamic model coefficients (the volume of an ellipsoid of such estimates dispersion is minimal;
- A-optimal signal is a signal minimizing a trace;
- E-optimal signal is a maximal proper number of the covariance matrix.

From the point of view of metrology, we have following requirements for choosing the kind of test signal:

- signal parameters have to be as far as possible similar to those of real registered signals if the latter are known in advance and their parameters are given and do

Table 5.2. Interest (+) of various specialists in getting knowledge of normalized metrological characteristics of MRI.

Metrological characteristics / Category of specialists	Error: basic	Error: additional	Function of influence	NMC (1)–(5)	Particular instrument error constituents	Methodical error	Error of an operator	Notes
Developer	+	+	+	+	+	–	–	stipulated in "Specification of requirements"
Manufacturer of MRI	+	+	+	+	–	–	–	indicated in "Technical conditions for a product"
Consumer in industry	+	+	–	–	–	+	+	depends on conditions of MRI application
Consumer in reasearch	+	+	+	+	–	+	+	
Repairman	+	+	–	–	–	–	–	in accordance with the requirements of an agreement for repair
Verifier (quality controller)	+	+	+	+	+	–	–	determined by the availability of adequate verification equipment

not change; otherwise the test signal has to be "worse", i.e., it should provide the estimation of a registration error "from top";

- test signal parameters have to be easily reproducible, time stable and with a low dependence on an impact of influencing quantities (factors), providing the possibility to synchronize it in reading off, and at the same time to be suitable for equipment implementation;
- "binding" of test signal parameters to corresponding measuring instruments, contained in verification schemes, has to be implemented quickly enough for transferring the dimension of a unit from measurement standards to generators of the test signal, serving as a "reference measure" at metrological certification and verification of MRI.

In [302] it is shown that a D-optimal test signal for the objects, at the output of which there is an additive noise (a "white" or "colored" one with a known correlation function), is the signal with a final and constant power, representing the approximation of white noise in the frequency pass band of the object. The pseudo-random binary sequence of maximum length (Huffman codes or M-sequences) with the δ-form autocorrelation function relates to the class of signals satisfying the requirements listed above.

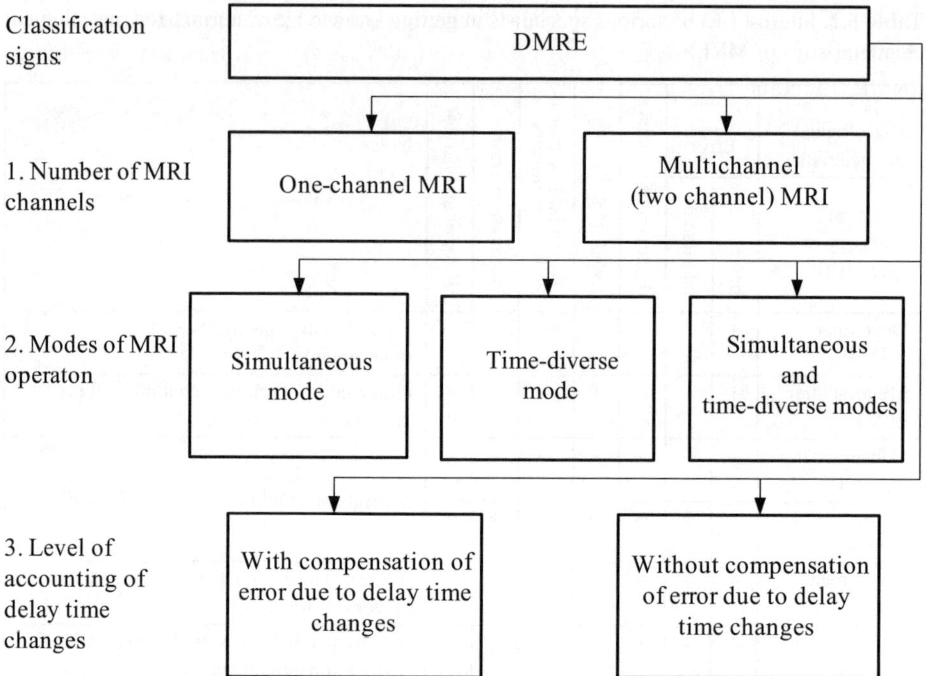

Figure 5.6. Classification of devices for measuring magnetic registration errors (DMRE).

The choice of such a signal as the test signal corresponds to the approach developed in [457] for calculating the resultant registration error with respect to its particular constituents.

The essence of the calculation procedure consists in summing up powers of dynamic (linear), nonlinear and noise distortions which have been assumed to be independent centered ergodic random processes, for receiving a dispersion of the resultant error.

To evaluate the resultant error of the MRI channel "from the top" when forms of the signal registered and parameters of the normalized metrological characteristics of the channel are not known, the calculation of particular error components is realized using the Cauchy-Bunyakovsky inequality.

At the same time normalization [452, 453, 455, 457, 459] has been performed with respect to particular constituents which depend on both the above metrological characteristics of MRI and the chosen parameters of the test signal. Such an approach is "integral" and does not conflict with normalization of particular metrological characteristics without taking into account the test signal parameters.

The process of normalizing the metrological characteristics of measuring instruments (in particular, MRI) is most complicated when the metrological characteristics represent a function (dependence on an argument), for example, amplitude- and phase-frequency characteristics, pulse weight function, etc. [289, 450, 556].

In [457] a method is proposed for normalizing dynamic characteristics of measurement transducers with the help of moments of its pulse weight function. It is shown that the dynamic error is

$$\Delta_D(t) = \sum_{i=2}^{\infty} \mu_i \cdot x^{(i)}, \tag{5.3.25}$$

where $x^{(i)}$ is the i-th derivative of the input signal $x(t)$,

$$\mu_i = \frac{m_i}{(-1)^i \cdot i!}, \tag{5.3.26}$$

μ_i is the i-th modified moment of PWF, $i = 0, 1, \ldots,$

$$m_i = \int_0^\infty t^i \cdot q(t) dt, \qquad i = 0, 1, 2, \ldots, \tag{5.3.27}$$

$q(t)$ is the pulse weight function of the investigated measurement channel, characterizing response of $y(t)$ on the pulse impact $x(t) = \delta(t)$:

$$y(t) = \int_0^\infty q(t) \cdot x(t - \tau) \mathrm{d}\tau. \tag{5.3.28}$$

It is known [191, 557] that the full list of dynamic characteristics, in addition to pulse weight function, include the transfer function $G(p)$, connected with PWF $q(t)$ by Laplace transform

$$G(p) = \int_0^\infty c^{-pt} q(t) dt\,. \tag{5.3.29}$$

In addition to these "external" (input–output) mathematical descriptions of finite-measured linear systems there are applied with increasing frequency "internal" descriptions, i.e., descriptions of states in space, taking into account internal variables of the system $s_1, \ldots, s_n$:

$$S = A \cdot S + b \cdots s; y = c \cdot s, \tag{5.3.30}$$

where S is the vector of a system state, A is the squared matrix of the n-order with constant elements a_{ij}, $b = [b_1, \ldots, b_n]^T$ is the constant vector-column, and $c = [c_1, \ldots, c_n]$ is the constant of a vector line.

The internal description is more detailed than the external one, since it contains information about the structure of the system (the number of parameters for the internal description is $N = n^2 + 2n$, while for the external description only $2n$ parameters are used). The connection between the pulse weight function $q(t)$ and transfer function $G(p)$ with the equation of a system state parameters is characterized by the following relations:

$$q(t) = c \cdot e^{At} \cdot b, \tag{5.3.31}$$

$$G(p) = c(pE - A)^{-1} \cdot b, \tag{5.3.32}$$

where E is the unitary matrix.

Alternative parameters, with respect to the PWF moments, which are equivalent to the latter in the sense of completeness of the description of linear systems properties, are Markov's parameters (MP) h_i [341]:

$$h_i = q^{(i)}(0); \quad i = 0, 1, 2, \ldots, \tag{5.3.33}$$

where $q^{(i)}(0)$ is the i-th derivative of the pulse weight function at $t = 0$,

$$q(t) = \sum_{i=0}^{\infty} \frac{h_i \cdot t^i}{i!}. \tag{5.3.34}$$

Markov's parameters are the elements of the Gankel's matrix H, representing a product of a transposed observability matrix of the system D^T and its controllability matrix R:

$$H = D^T \cdot R. \tag{5.3.35}$$

Connection of Markov's parameters with the transfer function and with parameters of state equation (5.3.29) is characterized by the relations

$$G(p) = \sum_{i=0}^{\infty} \frac{h_i}{p^{i+1}}; \tag{5.3.36}$$

$$h_i = c \cdot A^i \cdot b; \quad i = 0, 1, 2, \ldots. \tag{5.3.37}$$

The interaction of the equations, which, allowing the known modified moments (5.3.26) of pulse weight functions μ_i and their Markov's parameters to be found (and vice versa), is shown in Table 5.3.

Table 5.3. Interaction of the modified moments and their Markov's parameters.

$\cdots$	$c \cdot A^{-3} \cdot b$	$C \cdot A^{-2} \cdot b$	$c \cdot A^{-1} \cdot b$	$c \cdot b$	$c \cdot A \cdot b$	$c \cdot A^2 \cdot b$	$\cdots$
$\cdots$	μ_2	μ_1	μ_0	h_0	h_1	h_2	$\cdots$

Moments of pulse weight functions well describe the system in modes close to the established ones (i.e., in the low frequency range), while Markov's parameters do this in the transient modes (i.e., in the high frequency range). In the infinite sequence of moments of PWF, μ_i (the same way as of Markov's parameters h_i), corresponding to finite-measured linear system of the n-th order, only $2n$ terms are independent. In other words, the system of the n rank is completely characterized by $2n$ moments of PWF (or $2n$ Markov's parameters). Here, the coincidence of the moments (those of Markov's parameters) ensures the coincidence of the pulse weight functions and vice versa. It is interesting to point out that Markov's parameters of a system with discrete time are simply the values of its pulse weight function at successive time moments:

$$h_i = q(i); \quad i = 0, 1, 2, \ldots. \tag{5.3.38}$$

Since Markov's parameters of discrete systems are numerically equal to "step-by-step" values of PWF, it is of interest to answer the question, whether or not it is possible for a continuous (analogue) finite-measured linear system to be characterized by similar "step-by-step" values of PWF (which can be conventionally assumed as Markov's parameters).

Taking into account equations (5.3.33) and (5.3.37), the condition of coincidence of Markov's parameters for discrete and continuous systems can be written in the form

$$q(i) = q^{(i)}(0). \tag{5.3.39}$$

As an example, let us consider PWF of a resonance circuit with infinitely high Q-factor. For such a system $q(t)$ represents an undamped oscillation with a period corresponding to an frequency of the resonance circuit, i.e.,

$$q(t) = \sin \omega_0 t. \tag{5.3.40}$$

Successively differentiating (5.2.19), we obtain

$$\left.\begin{array}{ll} q'(t) = \cos \omega_0 t, & q'(0) = h_1 = 1, \\ q''(t) = -\sin \omega_0 t, & q''(0) = h_2 = 0, \\ q'''(t) = -\cos \omega_0 t, & q'''(0) = h_3 = -1, \\ q^{(4)}(t) = \sin \omega_0 t, & q^{(4)}(0) = h_4 = 0, \\ q^{(5)}(t) = \cos \omega_0 t, & q^{(5)}(0) = h_5 = 1, \\ \cdots\cdots\cdots & \cdots\cdots\cdots \end{array}\right\}, \tag{5.3.41}$$

and so on, the values of h_i being equal to four values $[1, 0, -1, 0]$. Thus, for a full characteristic of an ideal resonance circuit it is enough to have four Markov's parameters with a "step" equal to a quarter of the period of its eigen frequency, i.e., $2n$ Markov's parameters are equal to 4. This corresponds to the order $n = 2$ of the differential equation describing the circuit.

Extrapolating such an approach to continuous systems with dissipation, which are described with more complicated differential equations (of the finite order), it is useful to take PWF readouts every quarter of an upper boundary frequency period of its band pass, and the order of the differential equation describing the system can be estimated as half the sum of a number of extremes of the PWF and number of its intersections (without the point $t = 0$) with the x-axis (right up to the level when the PWF values stop to exceed the level of error of their measurement, or until the beginning of cyclic MP repetition), which corresponds to [277].

For a finite linear system the infinite series in formulae (Sections 5.3.25–5.3.27, 5.3.33, 5.3.34, and 5.3.37–5.3.39) are transformed into finite sums. Taking into account the mutually univocal correspondence of PWF moments and Markov's parameters, following from (5.3.34), it is possible to conclude that to normalize the dynamic characteristics by these parameters is similarly convenient and more efficient, since to give $2n$ parameters is easier than performing the "point-to-point" normalization of values of the weight or transfer function (AFC and PFC) from a continuous argument.

The final result of joint measurements is the construction of functional dependencies of measurands $U = f(t)$. The solution of such a measurement problem assumes, generally speaking, the use of a number of models: the model of the object or the process to be studied, theoretical model representations, describing the interconnection of the quantities analized, and the model (of the plan) of an experiment, as well as their synthesis on the basis of initial data for getting a model of the measurement process satisfying necessary requirements. Thus, the process of constructing a general model is reduced to solution of a multicriteria optimization problem.

The known procedures for optimizing, as a rule, are based on the use of the maximum likelihood method or its modifications (the least-squares method, least-module method etc.). The next choice of the best model is accomplished reasoning from the results of comparing the sum of squares (sum of modules) of the rest deviations [306].

Such an approach can only be used in the presence of sufficiently complete a priori information about the kind of the functional dependence of the measurands. If it is not known in advance and the class of possible models is broad enough, as in the case criteria mentioned for performing joint measurements, then the application of the criteria is of a low efficiency. Such an approach also does not provide any possibility to totally take into account all the initial data of different levels of formalization, the quantitativc and qualitative factors, objective and subjective estimates.

A more general approach to the solution of such a problem which eliminates these defects is based on the application of the theory of fuzzy sets [205, 269, 371].

Let the problem of choosing a model from the set Ω_f be considered [35]. It is possible to use available data concerning smoothness of the function desired. For example, assume with a certain degree of approximation that it belongs to the set L_{φ}^{m0} of linear combinations m_0 of the "basis" functions $\varphi_1, \varphi_2, \ldots, \varphi_{m0}$. The subset L_{φ}^{m} of this set consists of the linear combinations $\sum_{i=1}^{m} a_i \cdot \varphi_i$ of basis functions. Then there arises the possibility of considering other models from Ω_f, interpreting them as alternatives $f_i (f_i \in \Omega_f)$, and the sets L_{φ}^{m} $(m = 1, 2, \ldots, m_0)$ as the signs to which they are compared.

Comparison of the alternatives can be made by various methods, giving, for example, a ratio of preference, establishing a degree of permissibility of alternatives, and so on. Let us consider one of these versions, i.e., when fuzzy relations of preference are given by the functions of usefulness $U_j(f_i)$. In this case the values $U_j(f_i)$ provide getting a numerical estimate of the alternative f_i with respect to the sign j $(j = 1, 2, \ldots, n_0)$. Each of the functions of usefulness describes the relation of preference on Ω_f of the kind:

$$R_j = \{(f_k, f_i) |\ f_k,\ f_i \in \Omega_f,\ u_j(f_k) \geq u_j(f_i)\} \tag{5.3.42}$$

The choice of u_j depends on the measurement problem, i.e., on a set of signs considered. If the signs are not equal, then the degrees of importance $(0 \leq \lambda_j \leq 1)$ of each relation R_j are introduced. Expressions for λ_j also depend on the measurement problem, and they can be determined on the basis of objective and subjective estimates (by an expert method).

The task consists of a choice of an alternative (model) having, as far as possible, maximum estimates with respect to all signs. It should be noted that in the general case the requirements for alternatives are contradictory (due to fuzziness of the initial data concerning a type of the model), and there is no ideal solution.

To construct efficient models, let us apply a scheme based on theory of fuzzy sets, where we construct the set $Q_1 = \bigcap R_j{}_{j=1}^{n_0}$ with the membership function:

$$\mu_{Q_1}(f_k, f_i) = \min\{\mu_1(f_k, f_i), \dots, \mu_{n_0}(f_k, f_i)\}, \tag{5.3.43}$$

where, for example,

$$\mu_j(f_k, f_i) = \begin{cases} 1, \ (f_k, f_i) \in R_j, \\ 0, \ (f_k, f_i) \notin R_j. \end{cases}$$

The subset of effective alternatives in the set (Ω_f, μ_{Q_1}) is determined by the membership function

$$\mu_{Q_1}^{ef}(f_k) = 1 - \sup_{f_i \in \Omega_f} \left[\mu_{Q_1}(f_i,\ f_k) - \mu_{Q_1}(f_k,\ f_i)\right]. \tag{5.3.44}$$

Now it is possible to form a set of the other kind, namely the set $Q_2 = \bigcup_{j=1}^{n_0} \lambda_j \cdot R_j$ which allows the difference of the degrees of importance of the initial relations to be taken into account, with the membership function

$$\mu_{Q_2}(f_k,\ f_i) = \sum_{j=1}^{n_0} \lambda_j \cdot \mu_j(f_k, f_i). \tag{5.3.45}$$

The subset of effective alternatives in a set is determined in the same manner as above, i.e., by the membeship function:

$$\mu_{Q_2}^{ef}(f_k) = 1 - \sup_{f_i \in \Omega_f} \left[\mu_{Q_2}(f_i,\ f_k) - \mu_{Q_2}(f_k,\ f_i)\right]. \tag{5.3.46}$$

It is rational to choose the alternatives (models) from the set

$$\Omega_f^{ef} = \left\{ f'_k \middle| \ f'_k \in \Omega_f, \quad \mu^{ef}(f'_k) = \sup_{f_k \in \Omega_f} \mu^{ef}(f_k) \right\}. \tag{5.3.47}$$

It is necessary to point out that the quantity $\mu^{ef}(f'_k)$ can be interpreted as a quality criterion of the model determined on the set Ω_f.

As an example, let us assume that there are three classes of models for describing a file of experimental data:

- polynomials of the form of expansion in Taylor's series in the neighborhood of a certain fixed value of the argument – f_1;
- orthogonal polynomials constructed by the totality of experimental data – f_2;
- approximation model constructed on the basis of a priori (theoretical) ideas about the dependence considered – f_3.

The following signs are of interest:

(1) closeness of a model to the totality of experimental data;

(2) closeness of a model to a sought functional dependence;

(3) plan of an experiment;

(4) value of a mean square deviation of model parameters;

(5) number of model parameters

(if necessary, this list of signs can be extended). For further formalization of the initial data it is necessary to choose a quantitative measure of each sign (it is also possible to use subjective estimates). With respect to the signs listed it is possible to use, respectively:

(1) the estimate of the dispersion (S_0^2);

(2) nonparametric statistics (Kendall concordance coefficients, Wilcoxon coefficient);

(3) determinant of the information matrix detM (for D-optimal plans);

(4) maximum relative value of the mean square deviation of model parameters, (max S_{a_k}/a_k);

(5) number of model parameters of a type given, (K).

Setting our mind on preference relations from (5.3.42) in the form:

S_0^2	f_1	f_2	f_3	K	f_1	f_2	f_3
f_1	1	0	0	f_1	1	1	0
f_2	1	1	1	f_2	0	1	0
f_3	1	0	1	f_3	1	1	1

(5.3.48)

and forming from them the relations Q_1 and Q_2, we find μ_{Q_1} and μ_{Q_2} with respect to equations (5.3.43) and (5.3.45) in a form similar to equation (5.3.48), respectively, and then a set of nondominated alternatives $\mu_{Q_1}^{ef}$ and $\mu_{Q_2}^{ef}$ by the formulae in equations (5.3.44) and (5.3.46), respectively, in the form

$$\mu_Q^{ef}(f_i) = \frac{f_1 \;\; f_2 \;\; f_3}{0 \;\;\; 1 \;\;\; 1}. \tag{5.3.49}$$

On this basis we obtain the set (5.3.47) of effective alternatives Ω_f^{ef} in a form similar to equation (5.3.49), from which it is possible to choose the model f_i with a maximum value of the membership function.

In addition to jointly taking into account the objective and subjective (quality) signs and incompleteness of initial information about the models being compared, such an approach makes it possible to reveal "noninformative" data (when the exclusion of corresponding preference matrices does not influence the result). Moreover, this approach can be modified for normalizing functional dependencies of the type, e.g., the amplitude characteristic of a measurement channel.

To solve the problem of metrological assurance of analogue MRI, which contemplates the organization of metrological certification and the verification of MRI, in addition to determining a complex of normalized metrological characteristics, particular error components and the creation of a procedure for calculating the resultant error of registration (transformation) in accordance with its particular components, it is also necessary to develop methods and measuring instruments, which allow the metrological characteristics of MRIs to be estimated and are the technical basis for carrying out their certification and verification.

At the same time, it is useful to give our main attention to those characteristics whose parameters cannot be determined with the traditional measuring instruments which are presently available, namely: registration error, dynamic characteristics, and change (variation) of the time delay of the signal in the MRI channel. The problem of creating instruments for measuring the error of magnetic registration is especially urgent, since this problem has not yet been solved anywhere.

5.3.3 Methods for experimental evaluation of the basic error of measurement information signals registration in MRI channels

According to [191], one of the main parameters of MRIs used as a unit of information measurement systems or measurement computation complex, representing a measurement transducer, from the point of view of metrology [195], is the characteristics of the basic and additional errors.

The process of measuring the distortions of a signal which has passed through a traditional stationary four-pole network does not present any principal difficulties at a given signal form [413] and consists of obtaining a current (instantaneous) difference between the signal passed through the four-pole network and the test (reference) signal which is time-delayed and scaled in conformity with the nominal coefficients of transfer and time delay of the four-pole network under study.

Attempts to use this method for estimating the error of the magnetic registration of analogue signals of measurement information have been unsuccessfully completed, since the values of the reduced error vary over time within the limits from $+200$ to $-200\,\%$. This takes place because of the variations of the time delay of the signal in the MRI channel, caused by some imperfection of the MRI tape driving mechanism, leading to a deformation of the time scale of the signal reproduced.

Furthermore, it is necessary to keep in view that frequently the moment of recording and of reproducing can be separated by a rather significant time interval, and sometimes it is difficult to determine the difference between the distorted reproduced signal and the reference signal, since the latter can be absent (may not exist at a particular time), and it is necessary to create it anew (to restore it).

It should be added that usually the real signals, registered by MRI, have a complicated form and rich spectral content and can be considered to be the realization of a random process. It is not difficult to show that the level of distortions will be higher than that of the signals of a simple form (for example the distortion level of d.c. voltage, or harmonic oscillation of a given amplitude and frequency). For a more adequate estimation of the magnetic registration error it is useful to choose the test signal of a complicated form whose parameters approach the real registered signals. But the choice of the test signal of an irregular complicated form creates difficulties of its reconstruction in reproducing and synchronizing with the signal reproduced from the MRI channel under study.

These difficulties can be overcome by using quasi-random binary sequences instead of the test signal, in particular, the M-sequences (Huffman codes) [44, 45, 379, 383, 449, 455, 473, 480]. On the basis of the methods suggested there were developed measuring instruments which make it possible to realize the metrological certification and verification of MRI channels and devices for measuring the magnetic registration errors (DMRE). Their classification is shown in Figure 5.6, taking into account the variety of MRI used in practice.

The essence of the method used to measure an error of magnetic registration consists of reconstructing in the process of reproducing the test (reference) signal that was used in recording, synchronizing it with the signal reproduced from the MRI channel under study, adjusting the amplitude (scaling) of the reconstructed signal in accordance with a nominal transfer coefficient of the channel, subtracting it from the signal reproduced and measuring the difference obtained (or processing it according to a given algorithm). These operations have been successfully realized technically in the case where the test signal can be formed by some discrete technical means, has a regular structure (i.e., is a periodical one), and possesses some distinguished feature which allows the synchronization to be performed within any period.

One of the ways of realizing the proposed method is explained by the *functional scheme DMRE* [383, 449], shown in Figure 5.7.

From GRS 1 the pulses of the clock rate enter the input of GTS 2 and through UVD 5 are supplied to the input of similar GTS 9. GTS 2 (GTS9) is designed on the basis of a counter divider with weighted summing of its output voltages with weighted coefficients $2{-}n{+}i$, where i is the number of a trigger in the counter divider containing $n$triggers.

The shape of GTS output voltage resembles a sawtooth voltage [59, p. 199] with a sharp jump once per period. From the output of GTS 2 the quasi-random test signal through VR 3 and of FLF 4 is supplied to the input of MRI channel 8 to be recorded there. The FLF 4 (FLF 13) cutoff frequency just as the GRS 1 clock rate frequency are selected such that they are equal to the upper cutoff frequency of the band pass of MRI 8 channel for the spectrum width of the test signal does not exceed the channel band pass.

The reproduced signal from the output of the MRI 8 channel is supplied to one of the inputs of SU 11. To the other input of SU 11 the test signal reconstructed by GTS 9 which has been passed in advance through VR 12 (to be scaled correctly) and FLF 13. The level of difference voltage from the output of SU 11 is measured with V 14.

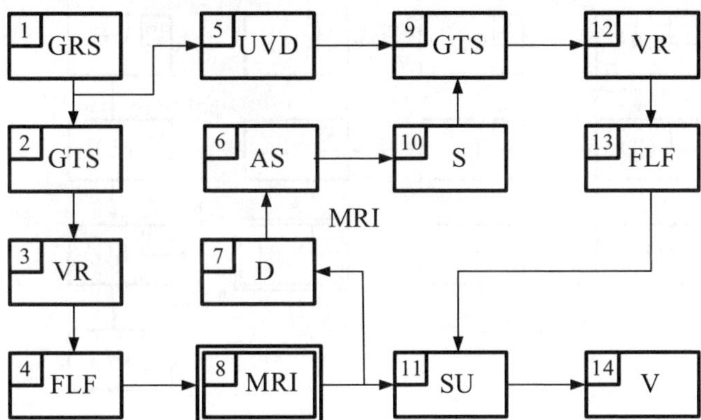

Figure 5.7. Functional scheme of the device for measuring the magnetic registration error of a one-channel MRI without compensation of the error caused by changes of the time delay. Here: GRS 1 is the generator of a reference signal; GTS 2, GTS 9 is the generator of a test quasi-random signal; VR 3, VR 12 is the voltage regulator; FLF 4, FLF 13 is the filter of low frequencies; UVD 5 is the unit of variable delay; AS 6 is the amplitude selector; D 7 is the differentiator; MRI 8 is the magnetic recording instrument channel under study; S 10 is the pulse shaper; SU 11 is the subtraction unit; V 14 is the voltmeter.

Furthermore, from the output of MRI 8 the signal reproduced is supplied through D 7, AS 6, and S 10 to set up inputs of the GTS 9 counter divider implementing "rough" synchronization of the output signal of this generator with the reproduced signal. D 7, AS 6, and S 10 separate the moment of a sharp change of the signal reproduced once per period and establish the GTS 9 counter divider into the required state. The "accurate" synchronization is accomplished by adjusting the time delay in UVD 5 with respect to the minimum of V 14 readouts. If the transfer coefficient of the MRI 8 channel is equal to 1, then VR 12 can be absent.

The test signal amplitude at MRI 8 input is selected, with the help of VR 3, to be equal to the nominal level of the MRI input signal. Thanks to this, as well as to the choice of the cutoff frequency of FLF 4, which is equal to the upper cutoff frequency of the band pass of the channel under study, the voltage value at the output of SU 11, which is proportional to the error, includes components caused by the inertial properties of the channel, the nonlinearity of its amplitude characteristic, the level of noises due to various reasons, and oscillations of the time delay.

The defect of such an implementation of DMRE is a noticeable inequality of a line spectrum of the test signal both with respect to the amplitude of its discrete constituents and the frequency scale. Moreover, it requires manual synchronization adjustment.

To remove these defects the **DMRE** [44] has been suggested and developed, which allows the **synchronizing process to be automated** as well as the MRI verification with both the **simultaneous and time-diverse modes** of recording and reproducing to be realized. The DMRE functional scheme is given in Figure 5.8.

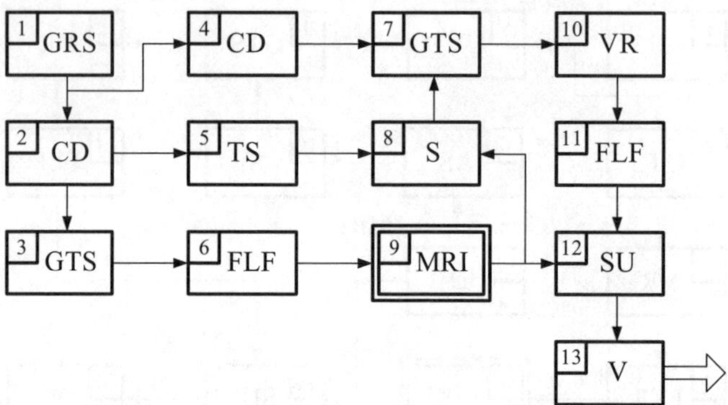

Figure 5.8. Functional scheme of the device for measuring the magnetic registration error of the one-channel MRI with the simultaneous and time-diverse modes of recording/reproducing. Here: GRS 1 is the reference signal generator; CD 2, CD 4 is the counter divider; GTS 3, GTS 7 is the generator of a test quasi-random signal; TS 5 is the time selector; FLF 6, FLF 11 is the filter of low frequencies; S 8 is the pulse shaper; MRI 9 is the studied channel of MRI; VR 10 is the voltage regulator; SU 12 is the subtraction unit; V 13 is the voltmeter.

The operation principle of DMRE consists of the following.

From the output of GRS 1 the square-wave voltage of frequency f_0 is supplied to the inputs of CD 2 and CD 4. These dividers are the counting trigger-circuits with a chosen counting coefficient K. The square-wave voltage from the outputs of CD 2 and CD 4 with the frequency f_0/K is supplied to the circuits of shift register of GTS 3 and GTS 7, respectively. A quasi-random binary sequence of a maximum length has been chosen (M-sequence) as the test signal. It is useful to choose the cutoff pulse frequency equal to the doubled upper cutoff frequency of the MRI 9 channel pass band. A further increase of the pulse shift frequency, although it improves the quality of the spectral content of the test signal in the pass band of the MRI 9 channel at the same period of the signal (the spectrum becomes closer to a more continuous one, uniformly distributed in the frequency band of the MRI 9 channel), but makes it more difficult to reconstruct its shape by S 8 due to a distortion of the test signal by FLF 6 and MRI 9 channel and, consequently, the operation of the time selector, TS 5, which provides synchronization of the signals at the inputs of SU 12. From the output of GTS 3 the test signal is supplied to FLF 6, which prevents the harmonics from overstepping the limits of the pass band of the MRI 9 channel. In the mode of reproducing the signal that has been subjected to distortions in the process of recording, preserving, and reading out, is supplied to one of inputs of SU 12. The voltage at the output of S 8 is a sequence of square wave pulses of different duration with different intervals (pauses) between them. Once per period of the test signal at the output of S 8 a pulse of the longest length will be present which is equal to the period of shift pulses multiplied by a number of register length of GTS 3.

This specific feature of the chosen test signal is used in constructing TS 5 which is intended for automatically synchronizing signals (once per test signal period) at the inputs of SU 12. The synchronization is accomplished by setting triggers of CD 4 and GTS 7 into the state at which the signals at the inputs of SU 12 appear to be synchronous. Setting is performed by a short pulse entering from TS 5 once per test signal period after the longest pulse at the selector output connected to the output of S 8 finishes.

The time selector TS 5 is based on a pulse counter supplied from an intermediate output of CD 2. Counting is performed only when the pulse passes from S 8, and pause is nulled during pauses. A logic circuit "AND" which is an element of TC 5 separates that state of the pulse counter which is only characteristic of the passing of the longest pulse. The short pulse of setting states of CD 4 and GTS 7 is formed by a desired front of a pulse at the output of the circuit "AND". The parasitic time shifts in units TS 5, GTS 7, VR 10, and FLF 11 are removed by the corresponding selection of the threshold level of S 8. A resolution of TS 5 is selected so as to distinguish the longest pulse among those which are closest to it with respect to their length, taking into account admitted oscillations of the signal-carrier speed.

From the output of GTS 7 the reconstructed quasi-random test signal, being of a shape identical to that of the GTS 3 and synchronized with the output signal of the MRI 9 channel, is supplied to VR 10, which provides for the equality of the amplitude scales of the reproduced and reconstructed (reference) signals and for which its transfer coefficient is set to be equal to a nominal (according to its specification) transfer coefficient of the MRI 9 channel. From the output of VR 10 the signal is supplied through FLF 11, the functions of which are identical to those of FLF 6, to the other input of SU 12. The output signal of SU 12, which is equal to the instantaneous difference of the reproduced and reconstructed signals, i.e., to the voltage proportional to the absolute error of magnetic registration, is supplied to the input of V 13.

As V 13 a voltmeter for mean, root-mean-square, or peak values can be used (depending on the problem at hand: what characteristic of error needs to be estimated). Moreover, instead of V 13 the analyzers of various kinds can be used: instrument measuring dispersion, correlation, and density of the probability distribution law, as well as spectrum analyzers, etc.

In DMRE considered above the MRI error, including a component caused by a drift and oscillations of the magnetic carrier speed, i.e., by oscillations of the time delay of the signal in recording and reproducing, at least within the period of the test signal, is estimated. To exclude this component an additional channel of MRI has to be used for registering a "pilot-signal" or putting information about a time scale in recording into the test signal.

Figure 5.9 illustrates the functional scheme of *DMRE for a multichannel MRI* [473].

Its operating principle consists of the following. In recording mode C 3 is in the low position and connects the "pilot-signal" from the output of GSS 1 through PC 6 and S 9 to the register shift circuit of GTS 12. Furthermore, this harmonic voltage is

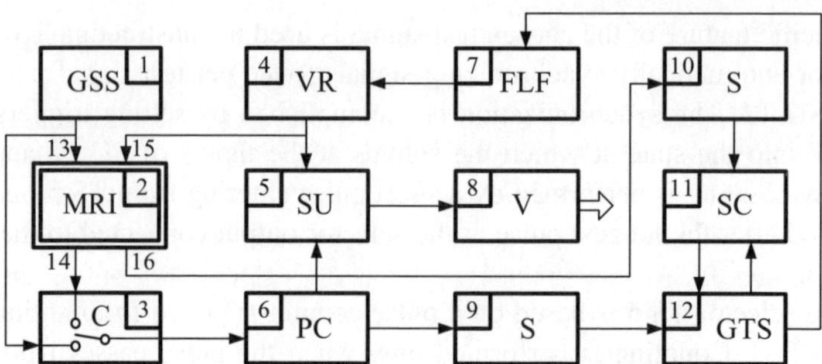

Figure 5.9. Functional scheme of the DMRE for multichannel MRI. Here: GSS 1 is the generator of a sinusoidal signal; MRI 2 is the precise magnetic recording instrument under study; C 3 is the commutator; VR 4 is the voltage regulator; SU 5 is the subtraction unit; PC 6 is the phase changer; FLF 7 is the filter of low frequencies; V 8 is the voltmeter; S 9, S 10 is the pulse shaper; SC 11 is the synchronizer; GTS 12 is the generator of the test signal.

recorded in the additional channel MRI 2 (input 13). An adjustment performed by PC 6 is provided for only in reproducing mode. The quasi-random signal from the output of GTS 12, having passed through FLF 7 and VR 4, is recorded in the channel of MRI (input 15). The cutoff frequency of FLF 7 is selected so as to be equal to the upper cutoff frequency of the channel studied of MRI 2 and makes the spectrum width of the test signal agree with the pass band of the channel. VR 4 allows the amplitude of the test signal to be set so that it becomes equal to the upper limit of the amplitude range of the channel being studied.

In reproducing mode the commutator C 3 occupies the upper position and connects the readout "pilot-signal" passing from the output of the additional channel of MRI 2 (output 14) through the phase changer PC 6 and pulse shaper S 9 to the input of the shift circuit of GTS 12. The generator GTS 12 reconstructs the previously existing (in recording) test signal which through the filter of low frequency FLF 7 and voltage regulator VR 4 is supplied to one of the inputs of the subtraction unit SU 5. To the other input of SU 5 a distorted signal is supplied from output 16 of the MRI 2 channel being tested.

The level of voltage obtained as the difference between the distorted and reconstructed signals is measured by the voltmeter V 8, or is treated according to the given algorithm. To produce a "rough" synchronization of the distorted and reconstructed signals, the pulse shaper S 10 and synchronizer SC 11, selecting the longest pulse once a period of the M-sequence and setting GTS 12 into corresponding state, are used.

The "accurate" synchronization is performed taking into account the oscillations of the signal delays in the channel being studied of MRI 2 inside the period of the test signal with the help of the "pilot-signal" reproduced from the additional channel of

MRI 2 (output 14). Any influence of the difference of the time delays in two channels of MRI 2 and units of DMRE is excluded by the phase changer PC 6.

Use of "rough" and "accurate" synchronization allows for the compensation of both the low- and high-frequency components of the signal time delay oscillations in the channel, which are caused by detonation and sliding (drift) of the signal carrier speed. Use of the same units od DMRE in both recording and reproducing modes, namely the phase changer PC 6, pulse shaper S 9, generator GTS 12, filter of low frequencies FLF 7, and votage regulator VR 4, allows the methodical error of measurements, which is caused by their nonidentity, to be excluded.

One disadvantage of DMRE as described above is the presence of a measurement result error which is caused by the dynamic shift of the signal carrier relative to the multitrack magnetic head, which leads to the fact that the time oscillations of the delay in the studied and additional recording channels of MRI do not coincide.

To prevent this disadvantage the *DMRE* has been suggested which provides measurement of the magnetic registration error *with compensation of the error component arising due to time-diverse modes of recording and reproducing in a single-channel MRI* [480], the functional scheme of which is illustrated in Figure 5.10.

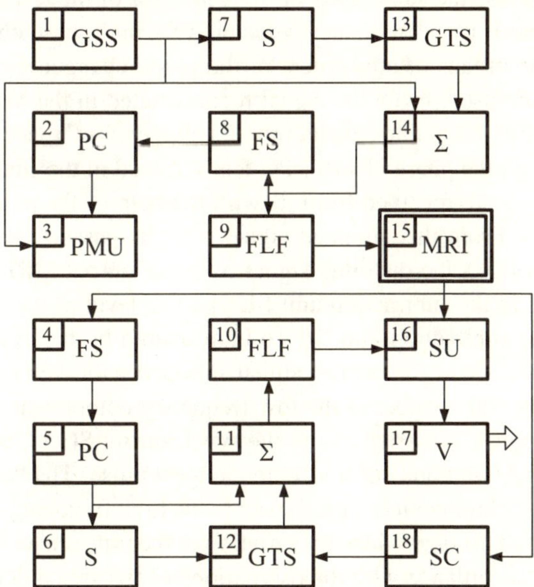

Figure 5.10. Functional scheme of the DMRE with the compensation of the error component arising due to time oscillations of the signal delay in the MRI channel. Here: GSS 1 is the generator of a sinusoidal signal; PC 2, PC 5 is the phase changer; PMU 3 is the phase measuring unit; FS 4, FS 8 is the frequency selector; S 6, S 7 is the pulse shaper; FLF 9, FLF 10 is the filter of low frequencies; $\sum$ 11, $\sum$ 14 is the summator; GTS 12, GTS 13 is the generator of test quasi-random signal; MRI 15 is the magnetic recording instrument channel under study; SU 16 is the subtraction unit; V 17 is the voltmeter; SC 18 is the synchronizer.

Its operating principle consists of the following. In adjusting and recording mode the reference harmonic voltage is supplied from the output of the generator GSS 1 through S 7 to the input of the shift circuit of the generator of the test quasi-random signals (M-sequence), GTS 13. The frequency of the reference voltage of G1 is chosen to be equal to the cutoff frequency of the MRI 15 channel being studied and is located at the point of the frequency spectrum of a quasi-random signal, in which a component of its discrete (line) spectrum is equal to zero. This harmonic voltage is summed with the test signal in the linear summator $\sum$ 14, and having passed through FLF 9 with the cutoff frequency equal to the upper cutoff frequency of the pass band of the channel under study, is supplied to MRI 15. From the summed signal at the output of $\sum$ 14 the harmonic voltage of reference frequency is separated by FS 8. The separated voltage will be phase-shifted relative to reference one due to time delays arising during the operations of summing and separating. Therefore, the separated voltage is made to be in-phase with the reference one, i.e., with the help of PC2 and PMU 3 the zero phase difference between them is achieved. After finishing the operation of phasing, the procedure of recording the summed up signal is realized in the MRI 15 channel under study.

In reproducing mode the same units of the generator of the test quasi-random signals (GTS), summator ($\sum$), frequency selector (FS), and phase changer (PC), as in recording mode the organs of adjustment of the phase changer remaining in the unchanged position are used. From the signal reconstructed in the MRI 15 channel the frequency selector FS 4 separates the harmonic voltage of reference frequency, which after passing through the phase changer PC 5 is summed in the summator $\sum$ 11 with the quasi-random signal, received from it (with the help of the pulse shaper S 6 and generator GTS 12). From this summed signal after its passing through the filter of low frequencies FLF 10, the distorted signal from the output MRI 15 channel under study is subtracted in the subtraction unit SU 16. The level of the difference voltage at the output of the subtraction unit SU 16 is measured by the voltmeter V 17. The "rough" synchronization of the control signal summed with the reproduced one, i.e., the synchronization with respect to the low frequency component, realized once per period of the test signal, is carried out by the synchronizer SC 18, which functions as a time selector identifying and separating the longest pulse. The "accurate" synchronization, i.e., the synchronization with respect to the high frequency component, takes place automatically. This is provided by variation of the time scale of the reconstructed (control) signal in accordance with the oscillations of the signal delay time just as in in the MRI 15 channel, since the phased reference harmonic voltage (being summed with the test quasi-random signal) was recorded in the channel being studied.

Thus, two methods are proposed for the experimental estimation of magnetic registration errors of analogue measurement information signals:

- a method taking into account the error component caused by oscillations of the signal delay time in the MRI channel [44, 383, 449, 455];

- a method of error estimation with compensation of the component caused by oscillations of a time delay [45, 473].

The proposed engineering solutions have been implemented at the level of inventions, and three of them are used in metrological practice [383, 473].

It is interesting to note that ideas which the above-mentioned inventions contain were evidently independently developed at the same time by NASA specialists in the USA. In US patent N° 4003084 [529], registered on 25 March 1975 and published on 11 November 1977, which is devoted to a method and means for testing systems with magnetic recording/reproducing, the technical solution suggested by the authors of [383] for 29 March 1973 is essentially repeated.

5.3.4 Methods for determining the dynamic characteristics of MRI channels

The channel of any analogue equipment of precise magnetic recording/reproducing electrical signals of measurement information can be considered as a finite linear system, the dynamic properties of which can be characterized by both the "external" (input–output) and "internal" mathematical descriptions.

The similarity to each other regarding their full dynamic characteristics, such as the pulse weight function $q(\tau)$, transfer function $Q(p)$ and differential equation, are related to the "external" mathematical descriptions. The description of system states in space, which is more detailed as compared to the input–output descriptions, since it carries in itself information about the system structure, is related to the "internal" descriptions.

The classical tasks of the analysis of linear systems are identifying, modeling, and "reducing" a model. For metrological analysis a finite result is the estimate of a dynamic error of the system.

The traditional approach to determining the dynamic characteristics of MRI channels was the evaluation of the component "nonnominality" of the transfer function of each channel, i.e., the nonuniformity of an amplitude frequency characteristic and the nonlinearity of a phase frequency characteristic in the working frequency pass band of the channel.

However the experimental determination of PFC of the MRI channels has some technical difficulties. Therefore, in [474] an alternative method is proposed which makes it possible to estimate the channel dynamic error on the basis of results of the experimental determination of its pulse weight function.

5.3.4.1 Methods for determining the pulse weight function of MRI channels

The measurement equation of weight function values is based on the use of the Wiener–Hopf equation, linking the cross-correlation function $R_{xy}(\tau)$ input and output signals of the channel with the pulse weight function

$$R_{xy}(\tau) = \int_{-\infty}^{+\infty} R_{xx}(t) \cdot q(t-\tau)dt, \tag{5.3.50}$$

which changes into the equality

$$R_{xy}(\tau) = m \cdot q(\tau) \tag{5.3.51}$$

(here m is the coefficient of proportionality) for the case when the autocorrelation function of the input signal $x(t)$ can be fairly accurately approximated by the Dirac δ-function.

One convenient test signal satisfying this requirement is the quasi-random binary sequence of a maximum length (the so-called M-sequence), generated by the shift register which is captured by feedback with respect to module 2 [59]. Such a signal represents a two-level voltage with the autocorrelation function (Figure 5.11) with the envelope proportional to $(\frac{\sin x}{x})^2$

$$R_{xx}(t) = \begin{cases} a^2\left(1 - \dfrac{|t|}{\Delta t}\right) & \text{at } |t| \le \Delta t \\ \dfrac{-a^2}{(2^n-1)} & \text{at } |t| > \Delta t, \end{cases} \tag{5.3.52}$$

and spectral power density

$$S(\omega) = 2\pi\left(\frac{a}{2^n-1}\right)\Bigg\{ -(2^n-1)\cdot\delta\cdot\omega + \sum_{k=-\infty}^{\infty} \times 2^n \cdot \left[\frac{\sin\left[k\pi/(2^n-1)\right]}{(k\pi/2^n-1)}\right]^2 \cdot \delta\left[\omega - \frac{2\pi k}{(2^n-1)\cdot\Delta t}\right]\Bigg\}, \tag{5.3.53}$$

where a is the upper level (amplitude) of the signal, Δt is the period of shift pulses, and n is the register capacity.

Such a signal is distinguished by a good reproducibility of its form, since to maintain the stability and accuracy of its statistical parameters it is necessary to set and maintain the frequency of shift pulses and levels of "zero" and "unit" with high accuracy, which is technically rather easily realized. Moreover, when the lower and upper limits of the channel amplitude range are equal to the levels of "zero" and "unit", this signal has the maximum power permissible for the channel, and since it is generated with elements of digital technique, its discrete time delay is easy realized by a blockage of passing a required number of shift pulses. One more advantage of the selected quasi-random binary test signal is the comparative simplicity of the technical realization of the multiplicating device of the correlation meter, which can be made on the basis of simple key circuits.

For various MRI modifications and modes of their use, a number of engineering solutions have been proposed which enable correlation measuring instruments to

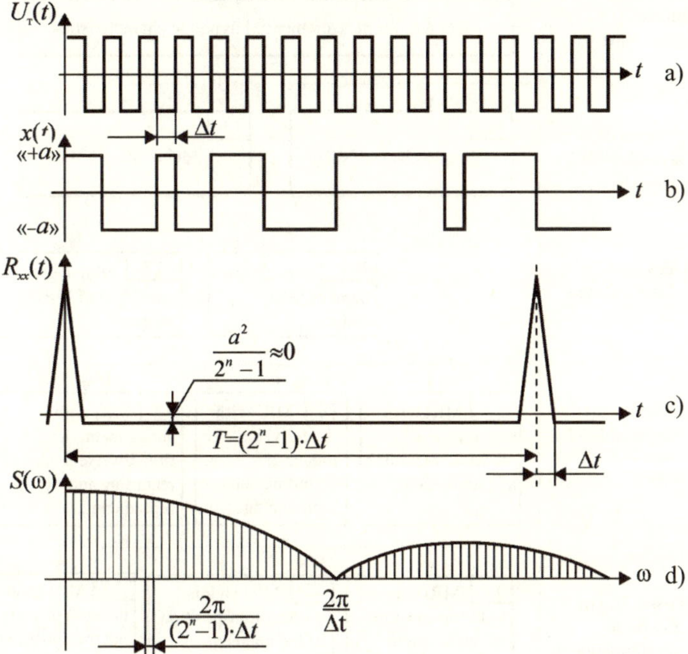

Figure 5.11. Voltage form of the reference signal of clock frequency (a), quasi-random test signal (b), its correlation function (c), and the spectral density of power (d).

determine the pulse weight function of the MRI channels with the help of experiments designed for this task [42, 43, 469–471, 474, 475].

The classification of such instruments, called devices for measurement of dynamic characteristic (DDC), is given in Figure 5.12.

The essence of the applied method of measuring the pulse weight function of the MRI channel is explained by the functional schemes of the multichannel MRI shown in Figure 5.13 and of the single channel MRI in Figure 5.14.

The operation principle of *DDC for the multichannel MRI* consists of the following [474]. The voltage of the clock frequency, supplied from the generator of reference signals GRS 2, is recorded in the additional channel of MRI 3 under study and serves as the "pilot-signal". At the same time this voltage of the clock frequency is used as shift pulses for the generator of the quasi-random sequence GTS 1. The M-sequence generated by GTS 1 is recorded in channel 8 of MRI 3. In the mode of reproducing the "pilot-signal" from the output of additional channel 11 passes to pulse shaper 5 and then through the adjustable time delay unit UVD 7 to the circuit of shift pulses of GTS 6, which reconstructs the M-sequence, similar to that which has been recorded in channel 8, with the only difference being the fact that the time scale of this sequence takes into account oscillations of the time delay in MRI channel 3. Two signals are

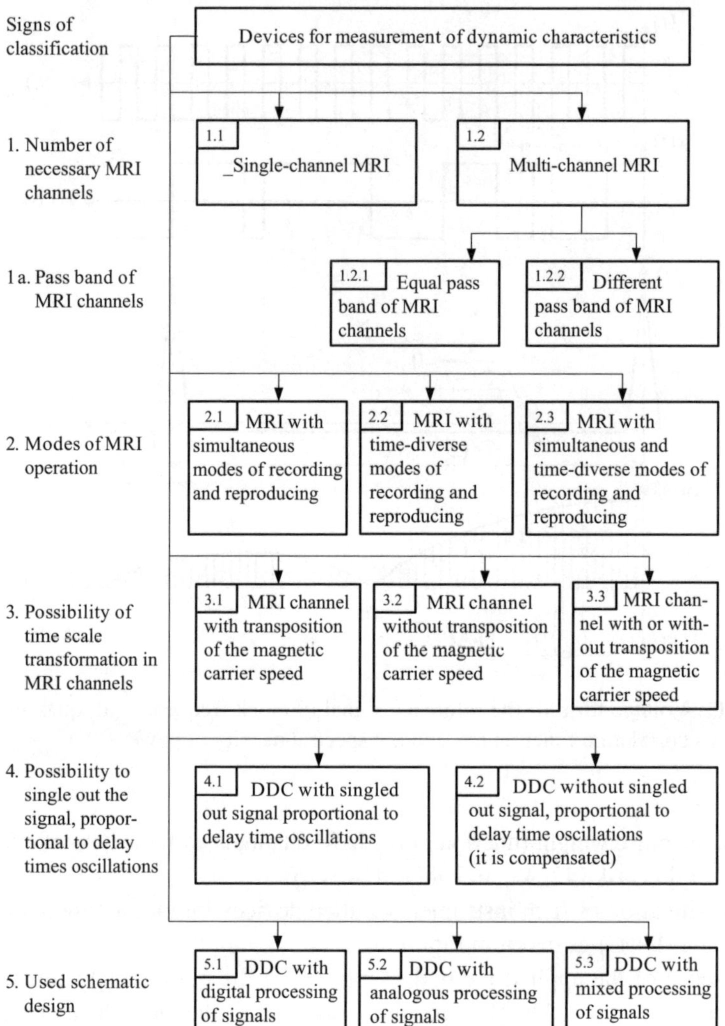

Figure 5.12. Classification of devices for measuring dynamic characteristics of MRI channels. Here: DDC are devices for measurement of dynamic characteristics of MRI channels; 1.1 is one-channel MRI; 1.2 is multi-channel MRI; 1.2.1 is for equal pass band of MRI channels; 1.2.2 is for different pass band of MRI channels; 2.1 is MRI with simultaneous modes of recording and reproducing; 2.2 is MRI with time diverse modes of recording and reproducing; 2.3 is MRI with simultaneous and time diverse modes of recording and reproducing; 3.1 is MRI channel with transposition of the magnetic carrier speed; 3.2 is MRI channel without transposition of the magnetic carrier speed; 3.3 is MRI channel with transposition of the magnetic carrier speed or without it; 4.1 is DDC with singled out the signal, proportional to delay time oscillations; 4.2 is DDC without singled out the signal, proportional to delay time oscillations (it is compensated); 5.1 is DDC with digital processing of signals; 5.2 is DDC with analogous processing of signals; 5.3 is DDC with mixed processing of signals.

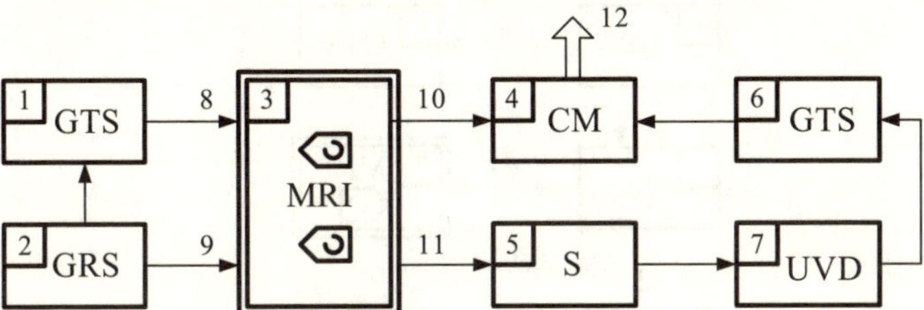

Figure 5.13. Functional scheme of DDC for measuring values of dynamic characteristics of the multi-channel MRI. Here: GTS 1, GTS 6 is the generator of a test quasi-random signal, GRS 2 is the generator of reference signal of clock frequency, MRI 3 is the equipment of precise magnetic recording, CM 4 is the correlation meter, S 5 is the pulse shaper, UVD 7 is the unit of variable delay.

supplied to the inputs of correlation meter CM4. One of them is the reproduced signal supplied from MRI 3 channel 10, and the second one is the reconstructed and delayed signal supplied from generator GTS 6. At output 12 of correlation meter CM 4 the cross-correlation function values that are proportional to the values of the pulse weight function of through channel 8–10 of MRI 3 are obtained.

A specific feature of the **device for measuring dynamic characteristics of the one-channel MRI** [42], the functional scheme of which is given in Figure 5.14, is that the signals of clock frequency from GRS 1 and the test M-sequence from GTS 6 are summed in a linear manner in $\sum 2$ and then are recorded passing over studied channel 11 of MRI 7. At the same time a specific feature of the quasi-random signal spectrum is used, namely the zero value of the spectral density of power corresponding to the clock frequency $2\pi/\Delta t$ (see Figure 5.11d).

This makes it possible for the pass band filter PBF 3 to identify and separate the voltage of the clock frequency from the reproduced signal coming from channel 12 of MRI 7 under study, and then to suppress this voltage in the reproduced quasi-random signal by rejection filter RF 8. A further treatment of signals does not differ from the one applied in DDC shown in Figure 5.13.

For the **multichannel MRI having the channels with different pass bands,** a device for measuring dynamic characteristics is used. Its functional scheme is given in Figure 5.15. The multichannel MRI with different pass bands of the channels is frequently used in practice, when in the equipment there is a channel of sound accompaniment or of direct overhead information recording. The pass band of such a channel significantly exceeds the pass band of the measurement channels in which one of modulation methods of measurement information signal recording (FM, FPM, WPM and others) is applied.

The operation principle of DDC, the functional scheme of which is given in Figure 5.15, consists of the following [471]. From the generator of reference signal GRS 3, the pulse sequence is supplied to the input of the counter divider CD 2, which has two

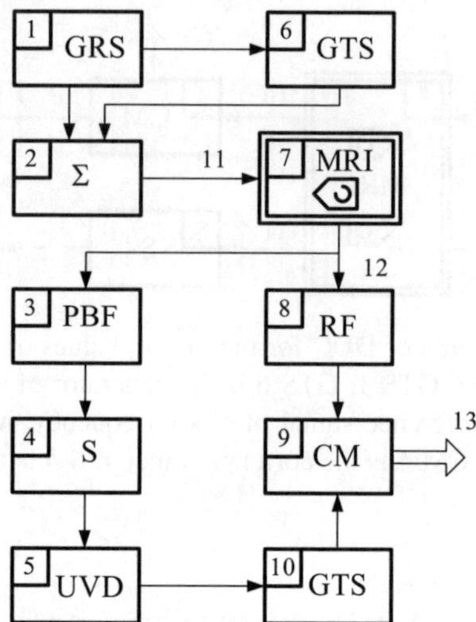

Figure 5.14. Functional scheme of DDC for measuring values of dynamic characteristics of the one-channel MRI. Here: GRS 1 is the generator of reference signal, $\sum$ 2 is the summator, PBF 3 is the pass band filter, S 4 is the pulse shaper, UVD 5 is the unit of variable delay, GTS 6, GTS 10 are the generators of test quasi-random signals, MRI 7 is the channel under study of the equipment of precise magnetic recording, RF 8 is the rejection filter, CM 9 is the correlation meter.

outputs from different triggers. From the one output of the counter divider the "pilot-signal" is supplied to additional channel 14 of MRI 4, which is intended for recording overhead information or voice accompaniment, the pass band of which exceeds the pass bands of the channels of recording/reproducing signals of measurement information by about one order, in particular that of channel 13. From another output of counter divider CD 2 the pulse sequence is supplied to generator of the quasi-random pulse sequence GTS 1 as the shift pulses. The test signal from the output of GTS 1 is recorded when passing through channel under study 13 of MRI 4. There, the frequency of shift pulses is selected so that it agrees with the upper cutoff frequency of the pass band of channel 13.

In the mode of reproducing the test signal, having been subjected to distortions, is taken from the output of studied channel 15 of MRI 4 and supplied to the first input of correlation meter CM 6, to the second input of which the reconstructed and delayed reference test signal in the form of the quasi-random binary sequence is supplied from the output of GTS 9, the parameters of which are identical to those of GTS 1. Units S 7, CO 8, CG 5, "+1" 12, CD 11, UVD 10 and GTS 9 are used to reconstruct this signal. The "pilot-signal" from the additional channel 16 of MRI 4 is reproduced by passing it through the pulse shaper S 7) to one of the inputs of the comparison operator CO 8. From GRS 3 the pulse sequence that has been passed through CG 5 and reversible

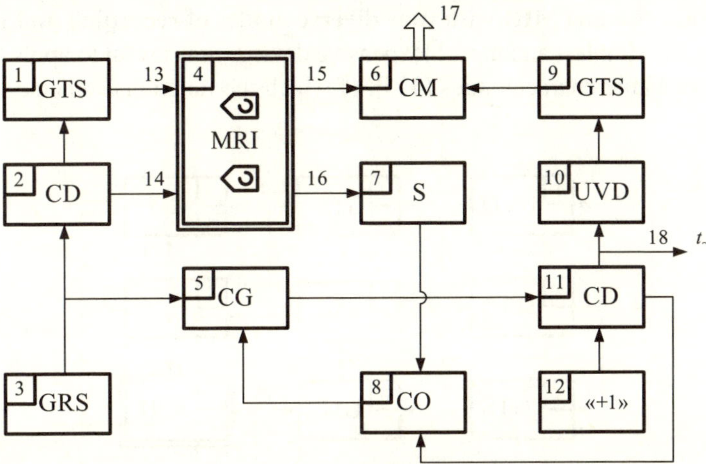

Figure 5.15. Functional scheme of DDC for measuring values of dynamic characteristics of the multi-channel MRI with a different pass bands. Here: GTS 1, GTS 9 is the generator of a test quasi-random signal, DU 2, DU 11 is the divider unit, GRS 3 is the generator of reference signal, MRI 4 is the equipment of precise magnetic recording, CG 5 is the controllable gate, CM 6 is the correlation meter, S 7 is the pulse shaper, CO 8 is the comparison operator, UVD 10 is the unit of variable delay, "+1" 12 is the unit of generating command "+1".

counter divider CD 11 is supplied to the second input of comparison operator CO 8. Comparison operator CO 8 carries out comparing the frequencies of the "pilot-signal" reproduced with MRI 4, as well as the reference "pilot-signal" obtained by dividing the frequency of the pulse sequence supplied from GRS 3. The coefficients of division at both outputs of divider CD 11 are similar to the coefficients of division at the similar outputs of divider CD 2. The frequency of the pulse sequence, taken off from the output of S 7, changes due to oscillations of the carrier speed in recording and reproducing. The frequency of the pulse sequence taken off from the output of GRS 3 is the reference one. If the frequency of the pulse sequence taken off from the output of pulse shaper S 7 appears to be higher than the frequency of the pulse sequence taken off from the output of CD 11, then comparison circuit CO 8 gives a permission to the input of unit of generating command "+1" 12, otherwise the comparison circuit gives a prohibition for pulses to pass through controllable gate CG 5. Thus, the pulse sequence that is supplied from the second output of divider CD 11 through unit of variable delay UVD 10 to the shift pulse circuit of quasi-random pulse sequence generator GTS 9, appears to be synchronized with the reproduced pilot-signal with the accuracy within one period of the reference signal at the output of GRS 3 and contains information about oscillations of signal time delay in the channel of MRI 4 at output 18 of divider CD 11.

It should be noted that application of such a DDC is useful in the MRI modes of recording/reproducing when they exist simultaneously without transposition of the carrier speed.

For the **one-channel MRI with time diverse modes of recording and reproducing**, i.e., for the simplest and most inexpensive devices, it is useful to apply the device measuring dynamic characteristics [475], of which the functional scheme is given in Figure 5.16.

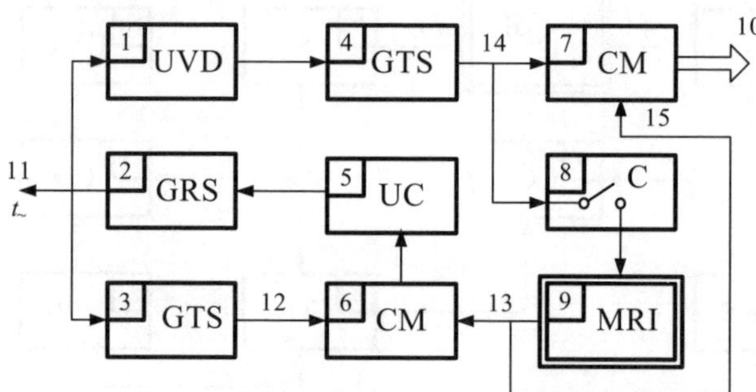

Figure 5.16. Functional scheme of DDC for measuring values of dynamical characteristics of MRI with time diverse modes of recording and reproducing. Here: UVD 1 is the unit of variable delay; GRS 2 is the controllable generator of reference signal; GTS 3, GTS 4 are the generators of test quasi-random signals; UC 5 is the unit of control; CM 6, CM 7 are the correlation meters; C 8 is the commutator; MRI 9 is the equipment of precise magnetic recording under study.

Its operation principle consists of the following. In the mode of recording the reference quasi-random binary pulse sequence, used as the test signal, is supplied to the input of the MRI 9 channel under study, where it is recorded on the magnetic carrier.

In the mode of reproducing the output signal of the MRI 9 channel under study is supplied to input 15 of correlation meter CM 7, to the second input of which the reference signal is supplied from GTS 4. When the frequencies of the reproduced and reference signals are equal and the phase of the reference signal is changed in continuous succession with the help of UVD 1, correlation meter CM 4 generates a voltage proportional to the pulse weight function values of the channel under study in time points determined by a value of the phase shift between the reproduced and reference signals.

To eliminate the influence of variations in the reproduced signal frequency which are caused by oscillations of the carrier speed it is required to provide a phase "binding" of the GRS 2 reference signal that controls generator GTS 4, to the phase of the reproduced signal.

For this purpose a loop of phase autotuning of the frequency, formed by generator GTS 3, correlation meter CM 6, and unit of control UC 5, is used. When the phase of the reference signal changes with respect to the phase of the signal reproduced, the envelope of output voltage of correlation meter CM 6 in the same way as that of correlation meter CM 7, repeats the form of the pulse weight function of the channel under study.

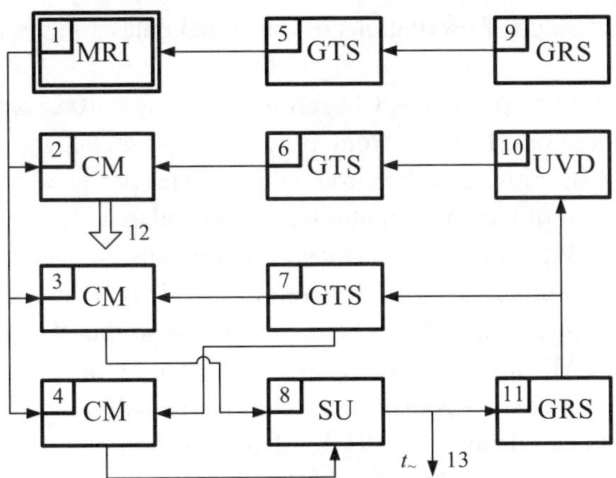

Figure 5.17. Functional scheme of DDC for measuring values of the dynamic characteristic of MRI with the transposition of the magnetic carrier speed. Here: MRI 1 is the studied equipment of precise magnetic recording; CM 2, CM 3, CM 4 is the correlation meter; GTS 5, GTS 6, GTS 7 is the generator of test quasi-random signals; SU 8 is the subtraction unit; GRS 9, GRS 11 is the generator of reference signal; UVD 10 is the unit of variable delay.

The output voltage of correlation meter CM 6, via unit of control UC 5, affects generator GRS 2 changing the reference signal phases so that it "traces" oscillations of the signal time delay in the channel of MRI 9. Thanks to this in the process of reproducing the reconstructed reference signal and reproduced signal appear to be synchronous with respect to each other. Shifting the reference signal with the help of variable delay unit UVD 1, it is possible to measure all values of the weight function of the MRI 9 channel under study at output 10 of correlation meter CM 7. At the same time, at output 11 of GRS 2 in the mode of reproducing the reference signal contains information about oscillations of the signal time delay in the channel. It should be noted, that such a device for measuring dynamic characteristics is not adapted for operation when the MRI magnetic carrier speed is transposed.

To measure values of the pulse transfer function of the MRI channel with transposition of the magnetic carrier speed, it is possible to use DDC [43], the functional scheme of which is shown in Figure 5.17. Its operation principle consists of the following. From the output of generator GTS 5, to the input of which the shift pulses are supplied from generator GRS 9, the reference test signal in the form of the quasi-random binary sequence, is recorded passing through the studied channel of MRI 1.

From the studied channel of MRI 1 the reproduced signal is supplied in parallel to the inputs of correlation meters CM 2, CM 3 and CM 4. Correlation meter CM 2 is the basic one for measuring the PWF values, and correlation meters CM 3 and CM 4 together with generators GRS 11, GTS 7 and subtraction unit SU 8 are intended to compensate the influence of oscillations of the signal delay time in the studied channel of MRI 1.

The signal from the output of GRS 11 through unit UVD 10 is supplied to the shift pulse circuit of generator GTS 6 and from its output to the second input of CM 2, where the values of the cross correlation function, which are proportional to the values of the pulse weight function of the channel studied, are calculated. To the second inputs of correlation meters CM 3 and CM 4, the quasi-random pulse sequence is supplied from different taps of the shift register of generator GTS 7. At that point, the time shift between these two quasi-random sequences is chosen so that the output voltages of correlation meters CM3 and CM4 are equal to $a^2/2$ (see Figure 5.11c) and symmetrically located in the middle of opposite slopes of the first half-wave of the pulse weight function of the studied channel of MRI 1. At such a choice of time shift the signals from the outputs of correlation meters CM 3 and CM 4 in subtraction unit SU 8 provide voltage 13, proportional to time delay oscillations in the studied channel, which is supplied to the control input of adjustable generator GRS 11. Thus, the frequency of shift pulses, supplied from the output of generator GRS 11 "traces" the time delay oscillations, excluding by their influence on the measurement results of the PWF values with correlation meter CM 2.

To provide measurements of PWF values in the mode of transposing the magnetic carrier speed, the frequency of pulses of generator GRS 11 is chosen so that its value is greater (less) than that of the pulse frequency of generator GRS 9 by the value of the transposition coefficient.

One of versions of constructing the **device for measuring dynamic characteristics of MRI with digital processing of information and automated searching for an initial point of the pulse weight function** [470] is given in Figure 5.18.

Its operation principle consists in the following. The test quasi-random reference signal from generator GTS 7 passes through MRI 12 channel under study where it is recorded. With the help of correlation meter CM 16 the cross-correlation function of the signal, reproduced from the MRI 12 channel under study, and of the reference reconstructed test signal from the output of generator GTS 4, the time scale of which "traces" oscillations of the time delay of signals in the MRI 12 channel, is determined.

The value of this cross-correlation function in the point determined by the time delay of the reference reconstructed test signal relative to the signal reproduced (t_{td}) is proportional to the value of the pulse weight function $q(t)$ of the channel under study in the point $t = t_{td}$. To obtain the required number of points of the pulse weight function the time delay t_{td} is automatically changed in a discrete manner by a command from the unit of control UC 10 with the help of the unit of variable delay UVD 9.

With the help of commutator C 6 either a "cyclic" or "one-time" mode of operation is established. In the "cyclic" operation mode, after a given number of time delays, with the help of unit of control UC 10 and circuit "AND" 21, the time delay t_{td} automatically returns to the value corresponding to the initial point of measurements.

Moreover, returning to the initial point of measurements can be formed by hand with the help of commutator C 11 in any mode of operation.

The synchronization signal ("pilot-signal") supplied from the output of counter divider CD 2 is recorded in the additional MRI 12 channel. In the process of reproducing, the time delay oscillations, caused by the nonstability of the signal carrier speed in recording and reproducing, are "traced" with the help of this signal and units of CG 8, CO 13, CD 14, S 17 and S 18.

The generator of test signal GTS 7 (the same way as generators GTS 4 and GTS 20 identical to this one) are constructed on the basis of a 10-bit shift register to the output of the first bit a sum of signals with respect to module 2, taken from outputs of the third and tenth bits. Logical circuit "AND" 1 (as that of "AND" 5, "AND" 15, "AND" 21) fixes a definite state of the shift register of generator GTS 7 (GTS 4, GTS 20) by generation of a pulse only one time within the period of the M-sequence.

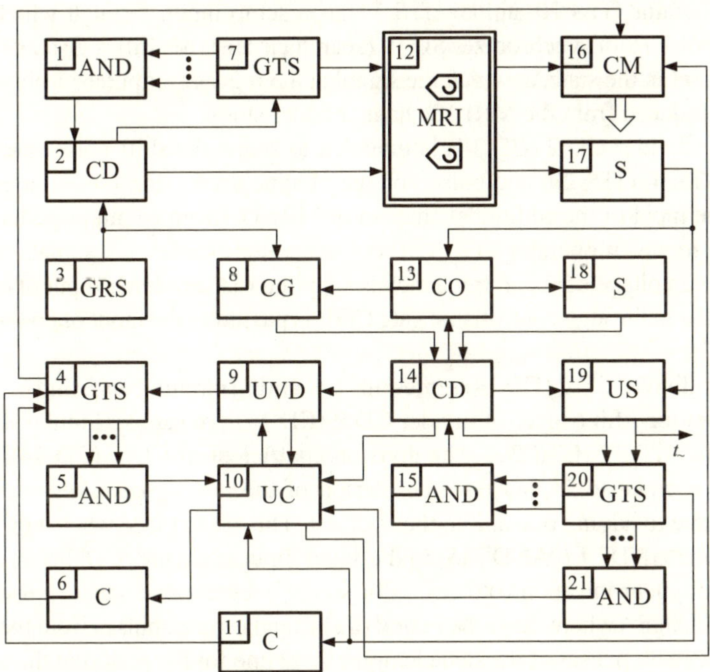

Figure 5.18. Functional scheme of DDC with digital processing of information and automated searching for an initial point of the pulse weight function. Here: AND 1, AND 5, AND 15, AND 21 is the logic circuit "AND"; CD 2, CD 14 is the counter divider; GRS 3 is the generator of reference signal; GTS 4, GTS 7, GTS 20 is the generator of test quasi-random signal; C 6, C 11 is the commutator; CG 8 is the controllable gate; UVD 9 is the unit of variable delay; UC 10 is the unit of control; MRI 12 is the equipment of magnetic recording under test; CO 13 is the comparison operator (comparator); CM 16 is the correlation meter; S 17, S 18 is the pulse shaper; US 19 is the unit of synchronization.

Test signal generator GTS 7 has a shift circuit input, to which the pulse sequence is supplied from the output of the fifth bit of counter divider CD 2. Test signal generator GTS 4, apart from the similar inputs and outputs, has two additional inputs, through which the connection of the feedback from the summator with respect to modulo 2 of generator GTS 20, at the same time the own feedback in shift register of generator GTS 4 being switched off.

The connection of such a feedback can take place either automatically in cyclic operation mode with unit of control UC 10, logical circuit "AND" 21 and commutator C 6 (having two positions corresponding to the "one time"and "cyclic" operation modes), or by hand with the help of commutator C 11 ("Start of measurements") at any mode of operation.

This way an in-phase operation of generators GTS 4 and GTS 20 is achieved which provides a delay of the reference reconstructed test signal at the output of generator GTS 4 relative to the signal reproduced from the output of MRI 12 channel under study. Generator GTS 20, unlike GTS 7, has a setup input, through which with the help of a pulse from synchronizer SC 19 (search circuit) the shift register of generator GTS 20 is set in the state, at which the signal at the register output is in-phase with the signal reproduced from the MRI 12 channel under study.

Counter divider CD 2 (CD 14) is intended to shape the shift pulses for generator GTS 7 (GTS 4, GTS 20) and pulse voltage of synchronization ("pilot-signal"), supplied to the input of the additional channel of MRI 12 (from counter divider CD 14 to one of the inputs of operator CO 13). From generator of reference signals GRS 3, the square wave voltage of stable frequency is supplied to the counter input of counter divider CD 2 and through controllable gate CG 8 is passed to the input of counter divider CD 14.

Counter divider CD 2 (CD 14) represents a counting circuit, consisting of 9 triggers. Pulses from the fifth trigger of divider CD 2 (CD 14) are supplied to the shift circuit of generator GTS 7 (GTS 20). The division coefficient of CD 2 (CD 14), generally speaking, is equal to 2^9, but once per period of the test signal it becomes equal to $(2^9 - 2^5)$ due to adding one unit to the sixth bit. This takes place when a pulse arrives from circuit "AND" 1 ("AND" 15) to the setup input of divider CD 2 (CD 14).

Thus the synchronization voltage ("pilot-signal") represents a square pulse sequence of the $8\Delta t$ length (where Δt is the period of shift pulses, i.e., pulses from the output of the fifth bit) with pauses of the same length except one for the period of the test signal, the length of which is smaller by Δt, namely $7\Delta t$. The moment when this shortened pause elapses corresponds to the definite state of the shift register of generator GTS 7 (GTS 20).

This specific feature is used to provide the automatic synchronization of the reconstructed signal at the output of generator GTS 20 with the reproduced signal supplied from the channel under study of MRI 12. Furthermore, counter divider CD 14 has three intermediate outputs and one additional input. The pulse sequence from one of these outputs is supplied to the signal input of unit of variable delay, UVD 9, and further on to the shift circuit of generator GTS 4.

From the other two intermediate outputs of counter divider CD 14 the pulse voltage is supplied to one of additional inputs of control unit UC 10 and to one of two inputs of synchronizer SC 19 (search circuit) and used to achieve the time agreement of UC 10 and SC 19 input signals. The signals from pulse shaper S 18 are supplied to the additional input of divider CD 14. With the help of these signals as well as with use of pulse shaper S 1, controllable gate CG 8, and comparison operator CO 13, the process of "tracing" oscillations of the "pilot-signal" time delay in the MRI 12 track of recording/reproducing is realized.

If the voltage of synchronization ("pilot-signal") reproduced from the MRI 12 channel and reconstructed by pulse shaper S 17, which is supplied to one of two inputs of synchronizer SC 19, is not synchronized with that of the output voltage of counter divider CD 14, which is supplied to the other input of synchronizer SC 19, then at the output of the latter a pulse is generated corresponding by time to elapsing the shortened pause in the pulse sequence at the output of pulse shaper S 17.

With the help of this pulse the shift register of generator GTS 20 is set into a state in which the signal at its output, similar to that of generator GTS 7, appears to be in-phase with respect to the signal reproduced from the MRI 12 channel under study. Moreover, this change in operation of generator GTS 20 through logical circuit "AND" 15 influences the operation of counter divider CD 14 in such a manner, that the synchronization signal at the final output of divider CD 14 appears to be synchronous with the signal at the output of pulse shaper S 17 with respect to its shape.

Tracing the time delay oscillations of the signal reproduced in the MRI 12 recording and reproducing channel under study is realizes at the arrival of the "pilot-signal" recorded in the additional MRI 12 channel with the help of units S 17, CG 8, CO 13, S 18, CD 14. Comparison operator CO 13 has two inputs. To one of these the signal is supplied from pulse shaper S 17 and to the other the signal of a similar shape is supplied from the final output of the divider CD 14. The phase comparison of these signals takes place in CO 13.

If the signal at the output of CD 14 takes the phase lead over the signal supplied from pulse shaper S 17, then at one of the outputs of comparison circuit CO 13 a pulse appears the length of which corresponds to phase mismatching. This pulse is supplied to the control input of controllable gate CG 8 and switches it off, providing a blockage for one or a number of pulses from the output of frequency stable generator GRS 3 to the input of divider CD 14. This results in a corresponding delay (time shift) of the output signal of this divider.

If the signal mentioned is phase lagged, then at the other output of the comparison circuit CO 13 a pulse is generated, which through pulse shaper S 18 is supplied to the set up input of divider CD 14, setting it into the state at which the phase of its output signal is shifted aside, as desired. The threshold sensitivity of comparison circuit

CO 13 to the phase mismatching corresponds to the length of one period of reference signal generator GRS 3.

Correlation meter CM 16 determines the values of the cross-correlation function of the signal reproduced from the channel under study of MRI 12 and the reconstructed reference signal supplied from the output of generator GTS 4 in the point, corresponding to the time shift between these signals. Correlation meter CM 16 contains three connected in series devices: a multiplication unit representing a key-circuit controlled with the reference reconstructed signal, an integrator, and a voltmeter.

Unit UC10 controls the operation of all units of DDC and provides the automatic and cyclic process of measurements. It forms the commands "stop", "read out", and "off", which are supplied to the control input of correlation meter CM 16. The command "stop" is supplied to the control input of UVD 9. At this command it is prohibited for one or several pulses to pass (depending on a selected "step" of the delay change).

Moreover, control unit UC 10 generates the command "cycle finish" which is supplied to the control input of unit of variable delay UVD 9 and provides a return of the delay to the initial point of measurements. It is generated at the moment of coincidence of the pulse supplied from UVD 9, corresponding to the moment of delay change, and logical circuit "AND" 21, fixing a definite state of the shift register of generator GTS 20.

It should be noted that the pulse sequence, supplied from the output of synchronizer SC 19, contains information about the signal time delay oscillations in the channel of the studied equipment of precise magnetic recording.

The devices for measuring dynamic characteristics which have been considered up to now are based on a direct method of measurement, in particular, on the method of immediate evaluation of values of the pulse weight function of the MRI channel under study. At the same time it is known [323, p. 342] that the differential (zero) method of measurements provides a higher accuracy of measurement results. However, use of the differential method of measurements supposes application of the reference analogue (model) of the channel under study with controllable parameters.

Creation of a "reference" MRI represents a technical problem, the realization of which is very difficult, firstly because of the real park of MRIs being maintained and developed is characterized by a great variety of its basic technical parameters: number of channels, their pass bands, synchronism of recording and reproducing modes, ability to transform a spectrum of signal, value of the coefficient of carrier speed oscillations, and so on, and secondly because the registration error of a "reference" MRI has to be less by 2–3 times than the registration error of the MRI under study, while the MRI designers make all maximum efforts for achieving a minimum error of MRI used as a part of IMS, wherever possible.

Taking into account these difficulties, an attempt was made to propose a **differential method of measuring values of the pulse weight function of the MRI channel** considered as a quadripole under study [469]. Here it was assumed that the speed oscillations of the carrier could be neglected and the MRI studied had simultaneous modes of recording/reproducing without any transposition of carrier speed.

The operation principle of the proposed DDC, the functional scheme of which is illustrated in Figure 5.19, consists of the following.

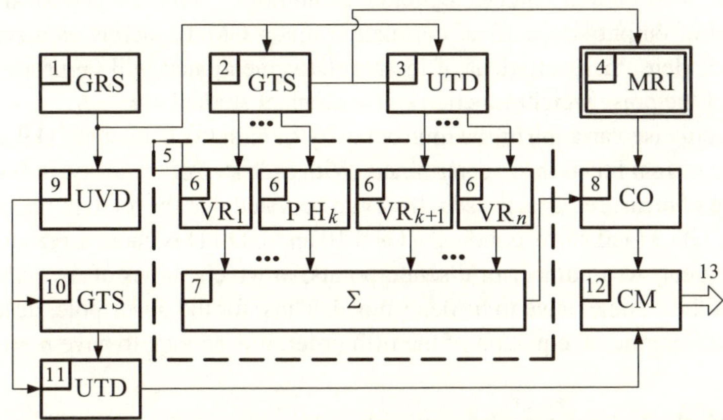

Figure 5.19. Functional scheme of DDC for measuring dynamic characteristics of MRI by the differential method. Here: GRS 1 is the generator of reference signal; GTS 2, GTS 10 is the generator of test quasi-random signal, UD 3, UD 11 is the delay unit; MRI 4 is the channel under study; Unit 5 is the reference quadripole with adjustable parameters; VR 6_i are the voltage regulators; $\sum 7$ is the summator; CO 8 is the comparison operator; UVD 9 is the unit of variable delay; CM 12 is the correlation meter.

A pulse sequence of clock frequency from the output of reference signal generator GRS 1 is supplied to the shift circuits of test quasi-random signal generator GTS 2 and delay unit UD 3, as well as through the UVD 9 to the shift circuits of GTS 10 and UD 11.

The output quasi-random pulse sequence from GTS 2 is supplied to the input of the UD 3 and to the input of the MRI 4 channel under study. At the outputs k of test signal k-bit generator GTS 2 and outputs m of m-bit UD 3 $n = k + m$ of quasi-random pulse sequences are generated, which differ in an initial phase. These sequences through corresponding VR 6 are supplied to a linear analogous summator $\sum 7$.

Voltage regulators VR 6 and summator $\sum 7$ form a reference quadripole with adjustable parameters, the output signal of which is subtracted by comparator CO 8 from the output signal of the MRI 4 under study. The difference of these signals is supplied to one of the inputs of correlation meter CM 12, to another input of which the delayed reference quasi-random pulse sequence is supplied from the output of delay unit UD 11.

Correlation meter CM 12 generates a signal proportional to the difference of pulse weight functions of the studied channel of MRI 4 and reference quadripole at a value of the time argument $t + t_i$, equal to the delay time, adjustable with the help of UVD 9, between the output pulse sequences of test quasi-random signal generators GTS 2 and GTS 10. The value of the output signal of correlation meter CM 12 depends on the transfer coefficient of that voltage regulator VR_i 6, the number i of which coincides with the number of delay time steps of the GTS 10 output pulse sequence.

With the help of this voltage regulator a minimum value (in the ideal case the zero value) of output signal 13 of correlation meter CM 12, acting as a zero organ, is achieved. Here, the coefficient of the regulator mentioned will be proportional to the value of the pulse weight function of the channel studied at $t = t_i$.

At the step-wise variation of the time delay by t_i, with the help of UVD 9, all values of the pulse weight function are determined. With each step of the delay, in essence, the static mode of measurements is used. It should be noted that the total number of bits of GTS 2 and UD 3 (and correspondingly GTS 10 and UD 11) is chosen, reasoning from the requirements to a number of discrete points, in which values of the pulse weight function of the channel have to be determined. Thus, for the quadripole, described by the linear differentional equation of the fifth order, it is enough to have $n = 20 \div 30$.

5.3.4.2 Methods for determining the phase-frequency characteristics of MRI channels

In [272, 273, 448, 457, 465] the method of coherent frequencies is proposed and described, which is intended to measure the nonlinearity of relative phase-frequency characteristics (PFC) of MRI channels which are characterized by uncertainty of delay time of the signals to be registered and, as a consequence, by an unknown PFC angle. In this method the sum of coherent voltages of multiple frequencies with known phase relationships is used as a test signal.

The operation principle of an instrument for measuring the PFC nonlinearity of MRI channels is explained by the example of the functional scheme [272, 273] given in Figure 5.20. This method consists of the following.

From the reference signal generator GRS 1, a sinusoidal signal of frequency f_0, chosen to be equal to the upper cutoff frequency of the pass band of the MRI 4 channel under study, is supplied through voltage regulator VR 2 to one of two inputs of linear summator $\sum$ 3. To its second input the sinusoidal voltage of a multiple frequency is supplied from the output of coherent frequencies unit UCF 13.

Unit UCF 13 includes a counter divider based on triggers with a counter input CD 5, ..., CD 10; voltage regulators VR 6, ..., VR 11; synchronizer SC 12 and adjustable divider AD 17.

From GRS 1 the pulse voltage of the same frequency f_0 is supplied to the input of coherent frequencies unit UCF 13. This pulse sequence is divided by D 5, ..., D 10, which contain filters of low frequencies for identifying and selecting the first harmonics of the frequency divided. If voltage U_0 from the output of generator GRS 1 has the form

$$U_0 = U_{m0} \cdot \sin 2\pi f_0 t, \tag{5.3.54}$$

then the voltage from the output of the i-th divider, which is supplied through the corresponding VR to one of the inputs of adjustable divider AD 7, will be of the form

$$U_i = U_{mi} \cdot \sin 2\pi \frac{f_0}{2^i} t. \tag{5.3.55}$$

Synchronizer SC 12 governs adjustable divider AD 7, which consecutively connects one of the inputs of adjustable divider AD 7 with the input of UCF 13. The synchronizer represents a counter divider, to the input of which the pulses from the output of the last CD 10 are supplied. The voltages from the outputs of adjustable divider AD 7 and GRS 1 (through VR 2) are linearly summed in summator $\sum$ 3. The voltage of a complicated form from the output of $\sum$ 3 is supplied to the input of MRI 4 for being recorded there. "Packs" of voltages of a complicated form are recorded on the magnetic carrier, each of them consisting of two spectral components with frequencies f_0 and $f_0/2^i$.

In the reproducing mode, the "packs" of voltages of a complicated form are supplied simultaneously from the output of MRI 4 to the inputs of high frequency filters FHF 8 and low frequency filters FLF 9. At the output of FHF 8 the voltage is

$$U_0' = U_{m0}' \cdot \sin (2\pi f_0 t). \tag{5.3.56}$$

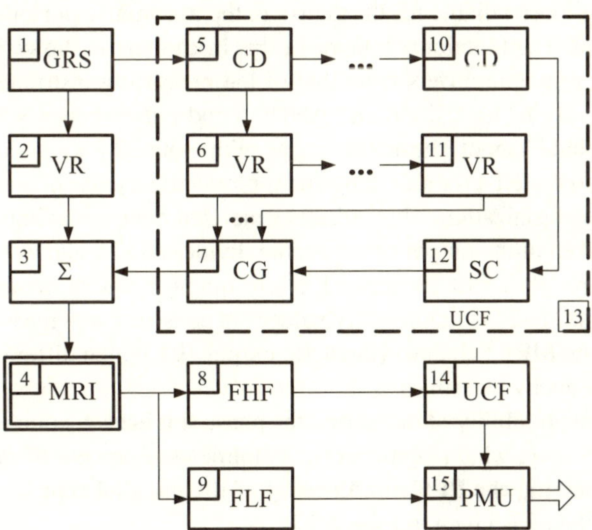

Figure 5.20. Functional scheme of the device for measuring the nonlinearity of PFC. Here: GRS 1 is the generator of reference signal; VR 2, VR 6, ..., VR 11 is the voltage regulator; $\sum$ 3 is the summator; MRI 4 is the studied channel of MRI; CD 5, ..., CD 10 is the counter divider; CG 7 is the controllable gate; FHF 8 is the filter of high frequencies; FLF 9 is the filter of low frequencies; US 12 is the unit of synchronization; UCF 13, UCF 14 is the unit of coherent frequencies; PMU 15 is the phase measurement unit (phase meter).

Its phase shift can be considered to be equal to zero, since it is used as the reference. This voltage is supplied to the input of UCF 14, the design of which is similar to that of UCF 13. At the output of UCF 14 the voltage is equal to

$$U_i' = U_{mi}' \cdot \sin\left(2\pi \frac{f_0}{2^i} t + \varphi_i\right). \tag{5.3.57}$$

From the outputs of UCF 14 and FLF 9 the voltages are supplied to two inputs of phasemeter PMU 15 (a phasemeter or device for measuring time intervals). While measuring time intervals τ_i, the values of the PFC $\Delta\varphi(f)$ nonlinearity are calculated by the formula

$$\Delta\varphi_i(f_i) = 2\pi \frac{f_0}{2^i} \tau_i. \tag{5.3.58}$$

Thus, the instrument measuring the PFC nonlinearities allows the phase shifts to be automatically measured in points of a frequency scale, which are uniformly distributed along the abscissa axis (of frequencies) in a semilogarithmic scale.

Using the described instrument for measuring the PFC nonlinearity in practice has shown that "fall-outs" of a signal in the channel of magnetic recording/reproducing of MRI under study (particularly in the case of a low accuracy class and quality) result in fails of the synchronization of the "packs" of voltage, having a complicated form, and also in an uncertainty of the results of separate series of measurements.

To eliminate this drawback it has been proposed an improved device [465], which provides getting more reliable results thanks to the **automatic periodic master-slave synchronization**. The functional scheme of this instrument is shown in Figure 5.21.

Its operation principle differs from that of the measuring instrument described in [273] by the ability of logical circuit "AND" 6 and pulse shaper S 5, as well as by using an additional channel (the auxiliary one with input 20 for recoding and input 21 for reproducing) of MRI 4 for recording and reproducing a pulse of automatic periodic master-slave synchronization. This pulse is generated once a maximum period of the test signal received at the output of adjustable divider AD 10 and, while reproducing it is supplied through S 5 to the network of cleaning (of setting the zero state) of the UCF 12 dividers. The PFC nonlinearity values are consecutively measured in the pass band points of the MRI 4 channel (input 18, output 19), which differ in frequency for two adjacent points by two times.

To provide the possibility of measuring the phase shit in each point of the frequency range simultaneously, which significantly simplifies the process of measurements, a **device for measuring the PFC nonlinearity of the parallel type** is proposed [448]. Its functional scheme is given in Figure 5.22.

Its operation principle consists of the following. The square-wave voltage of clock rate f_0 with on-off time ratio 2, selected so that it is equal to the upper cutoff frequency of the pass band of the MRI 4 channel under study, from the output generator of reference signal GRS 1 is supplied to the input of the counter divider based on triggers CD 7, ..., CD 16 and then through VR 2 to one of the inputs of linear summator $\sum$ 18. From the output of each trigger CD 7, ..., CD 16 the rectangular-wave voltages with

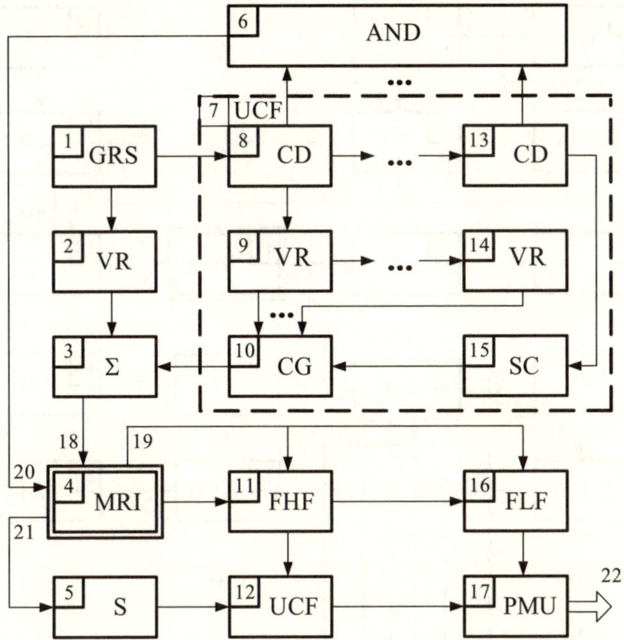

Figure 5.21. Functional scheme of the device for measuring the nonlinearity of PFCs of a consecutive type with the automatic periodic master-slave synchronization. Here: GRS 1 is the generator of reference signal; VR 2, VR 9, ..., VR 14 is the voltage regulator; $\sum$ 3 is the summator; MRI 4 is the equipment of precise magnetic recording; S 5 is the pulse shaper; "AND" 6 is the logical circuit "AND"; UCF 7, UCF 12 is the unit of coherent frequencies; CD 8, ..., CD 13 is the counter divider; CG 10 is the controllable gate; FHF 11 is the filter of high frequencies; US 15 is the unit of synchronization; FLF 16 is the filter of low frequencies; PMU 17 is the phase measuring unit (phase meter).

the frequency of $f_0/2^i$, where i is the trigger number of the counter divider, through VR 8, ..., VR 17 are supplied to the remaining inputs of summator $\sum$ 18. At the output of $\sum$ 18 a test signal, representing a weighted sum of square-wave voltages, is obtained.

While selecting transfer coefficients of voltage regulators VR 8, ..., VR 17 so that $U_i = U_0 \cdot 2^i$, where $U_0 \cdot$ is the amplitude of voltage at the output of VR 2, and U_i is the amplitude of the i-th voltage at the output of voltage regulators VR 8, ..., VR 17, at the output of summator 18 a quasi-random test signal is obtained, the shape of which is close to that of the sawtooth voltage and which has a great swing once per period. This peculiarity of the test signal is used for synchronizing within the process of reproducing.

With the help of differentiating circuit D 9, amplitude selector AS 3, and pulse shaper S 4 from the signal reproduced by MRI 4 a short pulse is identified and selected once per period. This pulse is supplied to the swing circuit of the counter di-

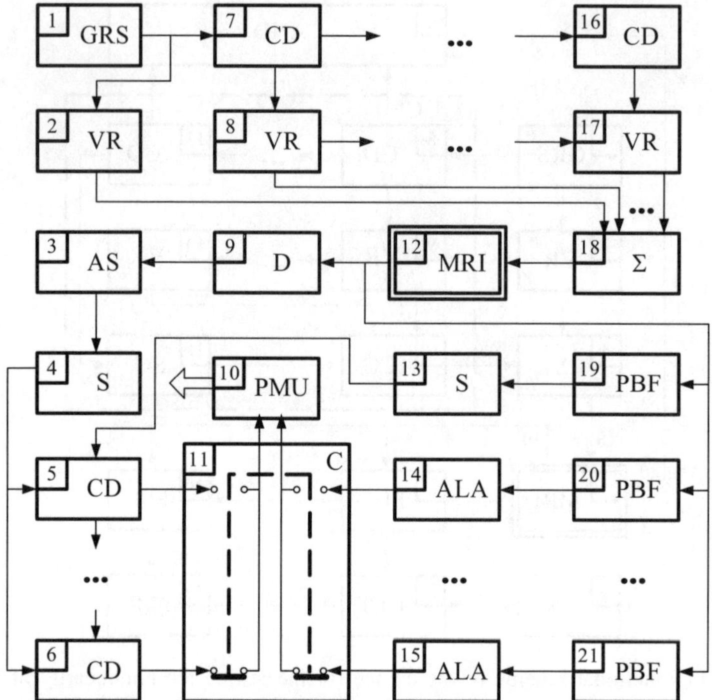

Figure 5.22. Functional scheme of the device for measuring the nonlinearity of PFC of the parallel type. Here: GRS 1 is the generator of reference signal; VR 2, VR 8, ..., VR 17 is the voltage regulator; AS 3 is the amplitude selector; S 4, S 13 is the pulse shapers, CD 5, CD 6, CD 7, ..., CD 16 is the counter divider; D 9 is the differentiator; PMU 10 is the phase measuring unit; C 11 is the commutator; MRI 12 is the MRI channel under study; ALA 14, ..., ALA 15 is the amplitude-limiting amplifier; $\sum$ 18 is the summator; PBF 19, ..., PBF 21 is the pass band filter.

vider based on triggers, CD 5, ..., CD 6, and realizes its master-slave synchronization with the signal reproduced. To a counting input of the counter divider based on trigger CD 5, ..., CD 6 from the output of S 13, the pulses are supplied which are selected by pass band filter PBF 19 and have the frequency f_0.

The remaining pass band filters, i.e., PBF 20, ..., PBF 21 select from the signal reproduced the harmonic voltages with the frequencies $f_i = f_0/2^i$, $i = 1, \ldots, n$. These harmonic voltages are transformed into the square shape by amplitude-limiting amplifiers ALA 14, ..., ALA 15 and then supplied to the inputs of commutator C 11. To the other inputs of this commutator the output pulses of counter divider triggers CD 5, ..., CD 6 are supplied.

Commutator C 11 is controlled by hand or automatically and allows any pair of voltages from the outputs of triggers CD 5, ..., CD 6 and amplitude-limiting amplifiers

ALA 14, ..., ALA 15 to be selected. If n phasemeters PMU are used, then the parallel mode of measuring the nonlinearity of the phase-frequency characteristic is provided. In such a way the fail protection of the device and possibility of parallel measurements of the PFC nonlinearity in the points selected are provided.

In all three modifications [272, 448, 456] of devices for measuring the PFC nonlinearity the ability of self-calibration has been embedded, i.e., such devices are able to reveal systematic errors (and corrections too) in each point of the frequency range, as well as mean-root-square deviations, when the channel of a MRI under study is replaced by an inertialess quadripole with the zero resistance.

Thus, two methods are proposed for determining pulse weight functions:

- correlation method of direct measurements of PWF values with five modifications for various types of MRI;
- correlation zero method of PWF determination,

as well as two methods of coherent frequencies for measuring nonlinearity values of phase-frequency characteristics of MRI channels:

- method of the successive type (with two modifications);
- method of the parallel type [42, 43, 272, 273, 448, 465, 469–471, 474, 475].The novelty of engineering solutions has been defended by intellectual property rights (11 author's certificates of the USSR for inventions among which some [42, 43, 272, 273, 465, 470, 471, 474, 475] have been implemented in metrological practice).

5.3.5 Methods for determining the nonlinear distortions and oscillations of a signal time delay in MRI channels

The method of experimental evaluation of a magnetic registration error [383] attracts attention by the fact that a result can be obtained on-the-fly and will characterize the quality of the MRI channel of magnetic recording/reproducing on the whole. However, in practice very often more detailed information concerning channel properties is needed. Therefore, in addition to an "integral" evaluation of the operation quality of the channel, it would be useful to develop methods and measuring instruments, which will allow particular components of a resultant error, caused by imperfect normalized metrological characteristics of the channel to be evaluated.

This experimental data is important from the theoretical point of view, since their use makes it possible to improve the model of forming the resultant error of magnetic registration, to ground the algorithm of summing up particular error components and solve the question concerning a degree of their independence upon each other. At the same time, it is desirable to determine the estimation of one type of distortions with elimination of the influence of other types of distortions.

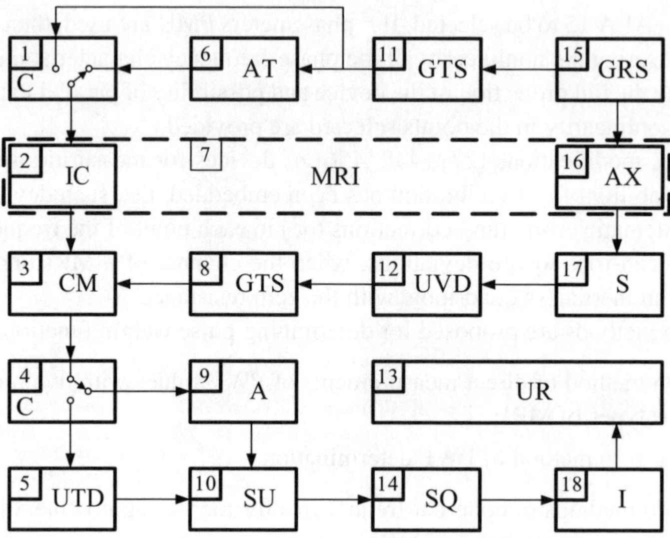

Figure 5.23. Functional scheme of the device for measuring nonlinear distortions of the "white noise" type in the MRI channel. Here: C 1, C 4 is the commutator; IC 2 is the investigated channel of MRI; CM 3 is the correlation meter; UTD 5 is the time delay unit; AT 6 is the attenuator; MRI 7 is the studied equipment of precise magnetic recording; GTS 8, CTS 11 is the generator of test quasi-random signal; A 9 is the amplifier; SU 10 is the subtraction unit; UVD 12 is the unit of variable delay; UR 13 is the unit of registration (voltmeter); SQ is the unit for raising to the second power; GRS 15 is the generator of reference signal; AX 16 is the auxiliary channel of MRI; S 17 is the pulse shaper; I 18 is the integrator.

The data is just as important for practical reasons, since it provides ways for improving the MRI characteristics, as well as for determining the factors contributing mostly to resultant errors with which we struggle.

In addition to the assessment of the dynamic characteristics and distortions, there is a great interest in methods of the experimental determination of the nonlinearity of amplitude characteristics and nonlinear distortions, interference, and noise levels in the MRI channel, as well as the time delay oscillations of a measurement information signal in the magnetic recording/reproducing channel, which can be considered as a reason for an additional registration error.

In [457, 265] some methods for evaluating the nonlinearity of the amplitude characteristic of a quadripoles are proposed and evaluated, including the MRI channel, as well as the total noise and interference level. In [476] the correlation method of measurements, which is based on determining the pulse weight function of the channel for high and low levels of the test signal, is proposed to evaluate nonlinear distortions of the "white noise" type signal in the MRI channel.

The functional scheme of the **device for measuring nonlinear distortions** [476] is shown in Figure 5.23.

The method, in its essence, consists of the following. Over investigated channel IC 2 of precise magnetic recording equipment MRI 7, the test signal of the form of a quasi-random binary pulse sequence of a maximum length is recorded. This signal is generated by GTS 11 with various amplitudes. The recording is made once with the level maximum for the channel (commutator C 11 occupies its left position), and again with the minimum level (the commutator is in its right position). To auxiliary channel AX 16 of MRI 7, the reference signal of clock frequency is recorded from the output of generator GRS 15.

In the process of reproducing, the reference signal from the output of AX 16 is shaped, time-delayed, using units S 17 and UVD 12, and then on this basis with the help of GTS 8, a quasi-random pulse sequence is reconstructed. Correlation meter CM 3 determines the cross-correlation function of the reconstructed and reproduced signals. In this way the values of the pulse weight function of investigated channel IC 2 are measured for the test signal of low and high levels.

Then the values of the pulse weight function, obtained using the test signal of low level, are muliplied with the help of amplifier A 9 by a coefficient, equal to the ratio of the high and low test signal levels (i.e., by the coefficient that is inverse with respect to the transfer coefficient of attenuator AT 6) in the process of recording.

In subtraction unit SU 10 the difference between the PWF values, delayed in UTD 5 and obtained with a high level of the test signal, and the PWF values multiplied by amplifier A 9, obtained with a low level of the test signal. The difference signal is squared, integrated, and measured with the help of units: SQ 14, I 18 and UR 13 for determination of its dispersion.

At a high level of the registered test signal, the whole amplitude range of the MRI channel under study is used, the amplitude characteristic of this channel having a certain nonlinearity. The output signal of the channel can be approximately represented in the form of a Duhamel's integral:

$$y(t) \cong \int_0^\infty q(t-\tau) \cdot x(\tau) d\tau \,. \tag{5.3.59}$$

At a low level of the test signal $x(\tau)/k$ only a small part of the amplitude characteristic of the channel is used, which can be considered as practically a linear one. Therefore the pulse weight function $q_0(t-\tau)$, determined at the low level signal, differs from PWF $q(t-\tau)$, measured at a high level signal. The output signal of the channel after dimensional scaling will be of the form

$$y_0(t) \cong k \int_0^\infty q_0(t-\tau) \cdot \frac{x(\tau)}{k} d\tau = \int_0^\infty q_0(t-\tau) \cdot x(\tau) d\tau \,. \tag{5.3.60}$$

Dispersion of nonlinear distortions D_n is evaluated as

$$D_n = \overline{(y-y_0)^2} = \overline{\left[\int_0^\infty (q-q_0) x(\tau) d\tau\right]^2} , \tag{5.3.61}$$

and at $x(\tau) = \delta(\tau)$, i.e., in the case when to the input of the channel a signal in the form of the Dirac δ-function with the frequency-constant spectral density of power ("white noise"), (5.3.61) can be written as

$$D_n = \overline{(q - q_0)^2}\,. \tag{5.3.62}$$

For practical use of the proposed method it is enough to experimentally determine the pulse weight function of the channel under study at two levels of the test signal. The remaining operations can be performed by calculation.

One of the basic metrological characteristics of MRI is the **oscillations of the signal delay time in the channel**, which are caused by a drift and oscillations of the speed of the signal-carrier in both recording and reproducing modes. Attempts to characterize MRI with the coefficient of speed oscillations or coefficient of detonation provide information about the quality of a tape driving mechanism of MRl, but do not provide for getting sufficient data concerning the resultant distortions of the time scale of the signal reproduced, which is of interest from a metrological point of view, since they are the reason for the appearance of an additional magnetic registration error.

This is explained by the fact that the speed oscillations$V(z)$ of the signal carrier in recording and reproducing modes are not identical, that they have a complicated form and represent a parametric process. The delay time by itself $\Theta(t)$ of the signal in the channel is determined from the integral equation

$$\int_0^{\Theta(t)} V(z)dz = 0, \tag{5.3.63}$$

where it is the upper limit of integration. The solution of such an equation is ambiguous and cannot be applied in practice. It is therefore important to fubd a solution for the problem of experimental measurements of time delay oscillations relative to the nominal value that frequently is unessential.

For this purpose the **correlation method** is suggested in [457, 480], makes it possible **to determine oscillations of the delay time**. The essence of this method consists of the following. The test signal in the form of a quasi-random binary sequence with the clock frequency, the period of which exceeds a maximum swing (peak-to-peak) of time delay oscillations, is recorded in the MRI channel under study. Further on, the values of the cross-correlation function of the reproduced and test signals are determined. To provide this, the test signal is delayed for an interval which allows the measurement to be performed on one of sloping sections of the cross-correlation function.

To make measurements a frequency of the shift pulses is selected significantly less than the upper cutoff frequency of the pass band of the channel under study. Two conditions are thereby fulfilled: first, the test signal passes through the channel with small dynamic distortions, hence the cross-correlation function of the input and output signals of the channel differs only a little from the autocorrelation function of the test signal, which has a triangular peak once a period (Figure 5.11c); second, the length of each of its sloping sections, equal to the period of shift pulse passing, deliberately exceeds the maximum peak-to-peak value of delay time oscillations.

When the value of the test signal time delay is correctly selected, then the mean time delay corresponds to the middle of the sloping section of the cross-correlation function. And since the sloping section of this function is linear, then such an device represents a linear transducer of time-delay oscillations into the output electric voltage of the correlation meter, which can be subjected to further analysis for determining its statistical parameters.

The extreme case of such an instrument is a device in which a test signal generator is the source of a square wave pulse sequence with an on-off time ratio equal to 2, i.e., of the meander having a time connection function (by analogy with the algorithm of determining the autocorrelation function) of the symmetrical triangular sawtooth form.

For getting results of measurement in a digital form, a **method of measuring the signal time delay in the MRI channel** is proposed [472], which is explained by the functional scheme in Figure 5.24.

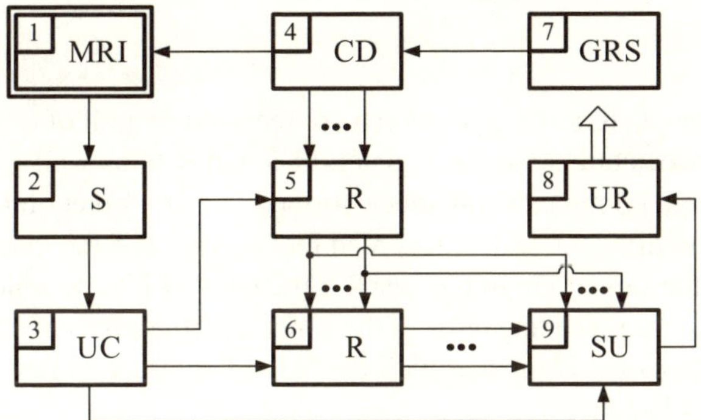

Figure 5.24. Functional scheme of the device for measuring time delay oscillations of signals in the MRI channel. Here: MRI 1 is the equipment of precise magnetic recording under study; S 2 is the pulse shaper; UC 3 is the unit of control; CD 4 is the counter divider; R 5, R 6 is the shift register; GRS 7 is the generator of reference signal; UR 8 is the unit of registration; SU 9 is the subtraction unit.

The essence of the method consists in the following. Over the MRI 1 channel under study a test periodic signal is recorded and reproduced (see Figure 5.25), before recording the reference signal frequency being divided, and in the process of reproducing the pulses are formed with a period of the signal read out. At the moments of their appearance the instantaneous values of the reference signal frequency division are memorized, the adjacent results are compared and on the basis of a current difference the oscillations of the delay time are determined.

The pulse sequence from generator of reference signals GRS 7 through counter divider CD 4 is supplied to the input if the MRI 1 channel under study in the mode of recording. In reproducing pulse shaper S 2 generates short pulses U_{repr}, passing with

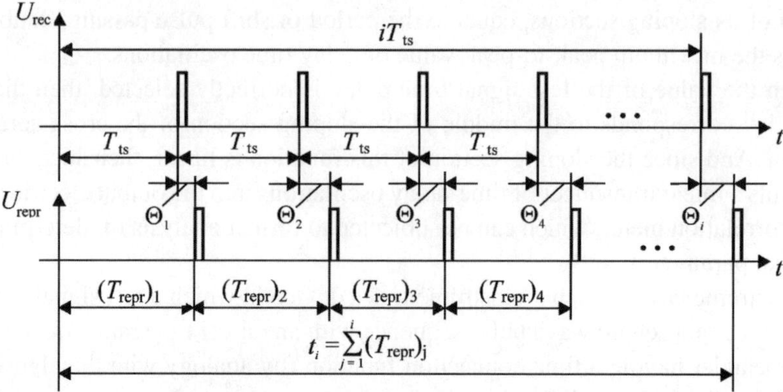

Figure 5.25. Time diagrams explaining the essence of the method for measuring the time delay oscillations of signals in the MRI channel under study. Here: U_{rec} is the signal at the MRI input in the mode of recording; U_{repr} is the signal at the MRI output in the mode of reproducing; T_{ts} is the period of the test signal; $(T_{repr})_i$ is the period of the reproduced signal; Θ_i is the time delay value.

a period of a readout signal $(T_{repr})_i$, which are received by unit of control UC 9. Unit of control generates commands for registers R 5 and R 6, as well as for subtraction unit SU 9. At these commands a number, formed at this moment by the states of all triggers of counter divider CD 4, is recorded into register R 5 (an instantaneous result of division). At the same time in register R 6 the number is kept which was recorded into register R 5 at the preceding pulse of the command (the adjacent result of division). Two adjacent results are compared with the help of subtraction unit SU 9, i.e., their difference is calculated as

$$\Delta\Theta_i = \Theta_i - \Theta_{i-1} \tag{5.3.64}$$

between the numbers in registers R 5 and R 6.

This difference corresponds to the variation of $\Delta\Theta_i$, the delay time for the preceding period of the readout test signal. Then at the next command of unit of control UC 3 the result of comparing is supplied to registration unit RU 8 and the number from register R 5 is recorded into register R 6. When the next pulse U_{repr} arrives, the process of measurements is repeated.

As distinct from the correlation method of measuring the oscillations of delay time, this method is efficient even at a possible shift in speed of the signal carrier, as well as at available small distinctions of the tape-driving mechanism speed in the modes of recording and reproducing.

Thus, to evaluate the nonlinear properties of the MRI channel, two methods are proposed:

- a method of automated determination of the nonlinearity of amplitude characteristics [265];

- a correlation method of evaluating nonlinear distortions of a signal of the "white noise" type [476].

To measure the oscillations the time delay of the signal in MRI channels two methods are also proposed:

- a correlation method with use of a quasi-random signal [467];
- a digital method that allows not only oscillations, but also drift of a time delay to be evaluated [472].

The novelty of the proposed engineering solutions has been confirmed by USSR invention patents of the USSR [265, 472, 476], two of which [472, 476] were implemented in metrological practice.

5.4 Hardware implementation of the methods for determining MRI metrological characteristics

5.4.1 Measuring instruments for experimentally determining MRI metrological characteristics

5.4.1.1 Devices for measuring magnetic registration errors

On the basis of materials indicated in Sections 5.3.2 and 5.3.3, there are three versions of initial measuring instruments which can be used to achieve verification of analogue equipment of precise magnetic recording. In these measuring instruments the inventions indicated in [44, 383, 473] have been realized. The first instrument is called a "device for measuring magnetic registration errors", the second is called a "device for verifying equipment of magnetic recording" and the third a "measuring instrument for determining MRI error".

In the "device for measuring magnetic registration errors" (DMRE) as the test signal generator a recurrent shift register is applied as described in Section 5.3.2 and according to recommendations given in [457]. This recurrent shift register is involved through the summator in such a modulo 2 feedback, which, when pulses are supplied to the register shift circuit, at the output of any bit of this register a quasi-random binary sequence of a maximum length (the M-sequence) is formed, the period of which is equal to

$$T = (2n - 1)\tau, \tag{5.4.1}$$

where n is the number of bits in the register and τ is the period of shift pulses.

Among the properties of the M-sequence, the form of which is shown in Figure 5.11, the following ones should be noted in particular [379]:

- in the M-sequence being generated there is no code combination consisting only of zero symbols, i.e., all bit cells of the register cannot keep their zero state simultaneously;
- the number of zero symbols passing one after another is limited and does not exceed $(2n - 3)$;
- the number of unit symbols also passing one after another is also limited and not greater than $(3n - 3)$.

At the same time, the pulse of the $(3n - 3)\tau$ length is present in the test signal only once per period T, which allows the problem of its synchronization with the distorted signal, reproduced from the channel of MRI under study, to be solved sufficiently easily.

In the developed measuring instrument of DMRE two modes of operation are provided for:

- mode I – without "tracing" variations of time delay;
- mode II – with "tracing" variations of time delay.

Moreover, the operation of the measuring instrument varies, depending on whether or not the equipment under study has the modes of recording and reproducing concurrently, or whether or not these modes are only alternately switched on.

The functional scheme of the device in mode I at the separate recording and reproducing processes is given in Figure 5.26.

In the process of recording (Figure 5.26a) the stable frequency voltage from generator GRS 1 is supplied to the input of pulse shaper S 2, which transforms the sinusoidal voltage into rectangular pulses. Then the frequency of this pulse sequence is divided by the seven-bit counter divider CD 3 with the division coefficient 128, which is based on the triggers with a counting input. The pulses from the output of CD 3 pass into the shift circuit of register R 4 being a part of generator of test signal GTS 7. The frequency of shift pulses f in the device is selected to be equal to the twofold upper cutoff frequency of the working range of the MRI channel under study ($f_h = 5\,\text{kHz}$):

$$F_s = 2f_h = 10\,\text{kHz}. \tag{5.4.2}$$

A variation of the frequency range of the test signal, depending on the pass band of the channel studied, is realized by way of the corresponding replacement of FLF 5 and selection of the frequency of shift pulses f_s.

The quasi-random sequence formed at the output of register R 4 is supplied to filter of low frequencies FLF 5, the cutoff frequency of which is equal to f_h. The test signal from the output of FLF 5 is recorded over the MRI 6 channel. FLF 5 has adjustable "zero" and "level", with the help of which the parameters of the test signal agree with the amplitude range of the channel under study.

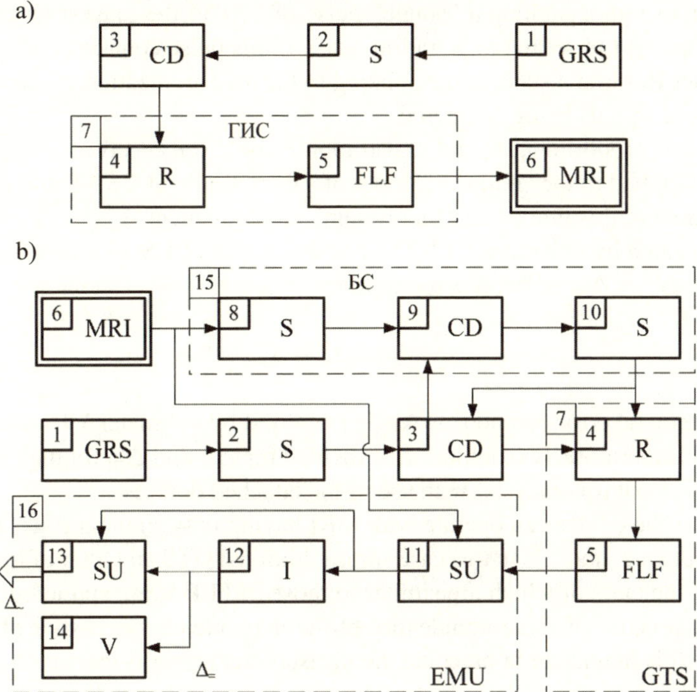

Figure 5.26. Functional scheme of DMRE for the separate modes of "recording" and "reproducing" in mode I. Here: GRS 1 is the reference signal generator; S 2, S 8, S 10 is the pulse shaper; CD 3, CD 9 is the counter divider; R 4 is the shift register; FLF 5 is the filter of low frequencies; MRI 6 is the channel under study of MRI; GTS 7 is the test quasi-random signal generator; SU 11, SU 12 is the subtraction unit; I 13 is the integrator; V 14 is the voltmeter; US 15 is the synchronization unit; EMU 16 is the error measurement unit.

In the process of reproducing (Figure 5.26b) the voltage of stable frequency from reference signal generator GRS 1 through pulse shaper S 2, counter divider CD 3, shift register R 4, and filter of low frequencies FLF 5 is supplied to the subtrahend input of subtraction unit SU 11 in the error measurement unit EMU 16 and serves as a reconstructed reference signal $U_{\text{ref}}(t)$. The reproduced distorted signal $U_{\text{repr}}(t)$ from MRI 6 is supplied to the input of decreasing subtraction unit SU 11, at the output of which the difference

$$\Delta(t) = U_{\text{repr}}(t) - U_{\text{ref}}(t) \tag{5.4.3}$$

is obtained, corresponding to the instantaneous values of the absolute error of magnetic registration.

Moreover, the reproduced signal from the MRI 6 output is supplied to the input of pulse shaper S 8 of synchronization unit US 15, containing the series connection to S 8, counter divider CD 9 (similar to CD 3), pulse shaper S 10, counter divider CD 3, for the purpose of separating a "single" pulse of 27 τ in the longest length over the whole period, and of setting by a trailing edge of this pulse counter divider CD 3 and shift register R 4 into a state, corresponding to the synchronization of the reproduced and reference signals being compared.

The signal, proportional to the registration error $\Delta(t)$, received at the output of US 11, is supplied to the "minuend" input of subtraction unit US 13, and to the "subtrahend" input of this unit a constant (systematic) component $\Delta_{\text{sist}} = \Delta_{=}$ of the error signal, separated by integrator I 12. Thus, at the output of US 13 a variable (random) component $\Delta^0 = \Delta_{\approx}$ of the Δ error appears:

$$\Delta_{\approx} = \Delta(t) - \Delta_{=}.$$

The systematic error component $\Delta_{=}$ is measured by voltmeter V14 with a pointer indicator. Furthermore, two outputs are provided for $\Delta_{\approx}$ and $\Delta_{=}$ for their more accurate measurement (or analysis) with use of the attached devices.

In mode I the device can operate with MRI having only simultaneous "recording" and "reproducing" modes. In this case some additional GTS and filter of low frequencies, the parameters of which are similar to those of FLF 5, are applied. It should be noted that, because of some nonidentity of the dynamic characteristics of these two FLFs, which is inevitable in practice, the measurement error is increased and will be equal to 4 % (according to data obtained by experiment in the mode of self-checking).

The functional scheme of DMRE, operating in mode II at separate recording and reproducing modes, is given in Figure 5.27.

The difference between the device operation in mode II and that in mode I (Figure 5.26) consists of th e use of the "pilot-signal", recorded simultaneously with the test signal onto the adjacent track of the magnetic carrier, and in special features of generating the shift pulses supplied to the registers of test signal generator GTS 7. Thus, when recording in mode II (Figure 5.27a) this voltage from the output of counter divider CD 3 is supplied to the counter input of trigger T 18. The pulse voltage of the rectangular shape (meander) from the output of T 18 (with the frequency $f_s/2 = f_h$) passes to the input of FLF 19, the cutoff frequency of which, as that of FLF 5, is equal to the upper threshold frequency f_h of the MRI 6 pass band.

In the process of recording (Figure 5.27b) the shift pulses entering the shift circuit of register R 4 for getting a reconstructed reference test signal at the output of GTS 7 are formed by S 21 from the "pilot-signal", reproduced and supplied by additional auxiliary channel II MRI 6.

Since the "pilot–signal" and test quasi-random sequence are recorded simultaneously on adjacent tracks of the same magnetic carrier, then the variations of their time scales are practically synchronous, due to oscillations of the carrier speed, sliding, and inequality of mean speed values in the processes of recording and reproducing. This causes the "tracing" variations of the signal time delay in the channel MRI 6 and elimination of their influence on the measurement results of the registration error.

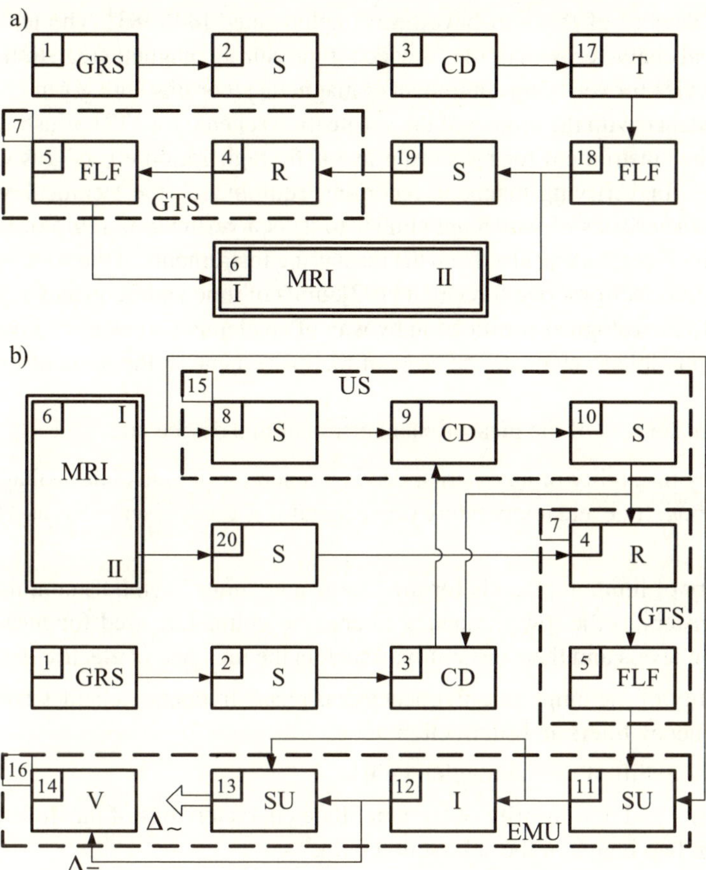

Figure 5.27. Functional scheme of DMRE for separate procedures of "recording" and "reproducing" in mode II. Here: GRS 1 is the reference signal generator; S 2, S 8, S 10, S 19, S 20 is the pulse shaper; CD 3, CD 9 is the counter divider; R 4 is the shift register; FLF 5, FLF 18 is the filter of low frequencies; MRI 6 is the MRI channel under study; GTS 7 is the test quasi-random signal generator; SU 11, SU 13 is the subtraction unit; I 12 is the integrator; V 14 is the voltmeter; US 15 is the synchronization unit; EMU 16 is the measurement error unit; T 17 is the trigger.

It should be noted that in measurements in mode II the reference test signal is reconstructed with high accuracy, since in both the recording reproducing processes the same generator, i.e., GTS 7, is used. One positive property of the synchronization unit, US 15, which is applied for this purpose, is the fact that setting into synchronization is periodically "confirmed", removing harmful any consequences of possible failures (signal misses) due to defects of the magnetic tape.

Several devices of this sort have been implemented [44, 383]. The technical and metrological characteristics of the "device for measuring magnetic registration errors" and the "device for verifying equipment of magnetic recording" are given in Table 5.4.

In accordance with the project of the verification scheme for MRI, illustrated in Figure 5.4a, the "instrument for measuring magnetic registration errors" (as well as the "instrument for verifying magnetic recording equipment", the technical and metrological characteristics of which are similar to it) is a verification setup of the highest accuracy, receives the unit size from the measuring instruments of the state verification scheme of the electromotive force unit [192], units of time and frequency [198], and is subjected to metrological certification by way of an element-to-element investigation, making it possible to evaluate the error of the instrument on the basis of information obtained.

The main sources of the measurement errors of this device are:

- frequency instability of an external driving generator of sinusoidal voltage with respect to both time and temperature (the generator is used, essentially, as a time measure);
- instability of limitation levels for the "zero" and "units" symbols coming from the shift registers to the low frequency filters (the voltmeter, used for measuring the limitation levels and their instability, transfers the unit size of electric d.c. voltage);
- nonidentity of the amplitude-frequency and phase-frequency characteristics of the low frequency filters and subtraction units;
- "zero" time drift of the subtraction units;
- nonidentity and nonlinearity of the amplitude characteristics of the differential amplifiers at two inputs of the subtraction units;
- distinction of the transfer coefficient of the subtraction units and low frequency filters from the nominal ones and their instability;
- inaccuracy of the time synchronization of the reference signal with the signal being reproduced.

An experimental study of the main metrological characteristics of the instrument has been performed. The results of this study are shown in Table 5.4 [452].

The frequency stability of the crystal-controlled generator of sinusoidal voltage which was used in the experiment was checked with the aid of a frequency meter. The frequency measurement error was equal to ±1 of the lower order bit, i.e., it was equal to ±1 Hz. The generator frequency corresponded to 1 MHz.

The frequency stability was not worse than $\pm 1 \cdot 10^{-6}$. Moreover, it is necessary to take into account that the frequency stability at the output of counter divider CD 3 is even higher by two orders, since its division coefficient is equal to 128. Hence, we can conclude that the error component caused by this factor is negligible and can be ignored.

Table 5.4. The technical and metrological characteristics of the "Device for measuring magnetic registration errors" and "Device for verifying equipment of magnetic recording"

Characteristic	Values area
Frequency range of test signal (Hz)	0–5000*
Number of spectral components of a test signal in the working frequency range	512
Interval between the frequency components (Hz)	10*
Level of a spectral components envelope at frequencies (Hz):	
0	1
5000	0,5
Attenuation of the test signal higher than the threshold frequency of the frequency range (dB/octave), no less than	54
Amplitude range (V)	0 – (+6)**
Input resistance (kOhm), no less than	80
Output resistance (kOhm), no greater than	8
Limit of the permissible fiducial error in the mode of self-checking (%), no greater than:	
for a systematic component	$\pm 0,1$
for a random component (root-mean-square deviation)	4
Particular components of the basic error:	
instability of the driving generator of reference sinusoidal voltage, no greater than	$\pm 1 \cdot 10^{-6}$
instability of levels limiting "zeros" and "units" at the shift register output (%) , no greater than	0,1
nonidentity of AFC of the low frequency filters (%)	from -2.2 to $+0.2$
nonidentity of PFC of the low frequency filters (degrees), no greater than	7
drift of "zeros" for subtraction units in interval 3 h. (%), no greater than	± 0.1
nonlinearity of AC for subtraction units (%), no greater than	± 0.2
nonidentity of differential amplifiers for subtraction units at different inputs (%), no greater than	$\pm 0,6$
nonidentity of AFC of subtraction units for two inputs (%)	from –1.4 to +0,2
Power supply (V)	$\sim 220(\pm 10\ \%)$
with frequency (Hz)	50
Power consumption (W), no greater than	6
Time of continuous work (h.), no less than	8

* It is possible to change the range by selecting the reference frequency and replacing the filters of low frequencies.

** Adjustment of "zero" and "level" is provided for.

The instability of the levels of the limited quasi-random binary signal is mainly caused by three things: the instability of d.c. voltage supplied to the shift registers, the instability of the "zero" and "units" levels of the elements forming a register bit, and the instability of the limitation levels in the circuit of level fixation.

The power supply of the registers is +5 V from a stabilized power source. The instability of its output voltage (according to its certificate) is no greater than 0.3 mV (due to rippling). The elements being used are characterized by instability of the "zero" and "units" levels of the order of 0.3 mV. The instability of limitation levels in the level fixation circuit is equal to 0.1 %, which is negligible for evaluating an instrumental error of a measuring device.

The AFC nonuniformity of the low frequencies filters was measured with a signal generator, from the output of which the voltage was supplied to the input of a corresponding filter of low frequencies, as well as to the input of a digital voltmeter. With the help of this voltmeter a root-mean-square value of voltage was measured at the input and output of the filter in each selected point of the frequency range. An analysis of the measurements has shown that the nominal transfer coefficient of both filters, which corresponds to the ratio of the output filter voltage to the input one, is equal to 1. The cutoff frequency of both filters is 5 kHz at the steepness of 54 dB per octave. The AFC nonuniformity of the lowest frequencies filter was within the limits from −1.5 % to +2.2 %, and that of the second lowest frequencies filter it was from −7 % to +4.1 %. The difference of the transfer coefficients of the filters was within the limits −2.2 % to +0.2 %. The calculations made by a formula given in [452] and formula (**??**) show that the measurement error due to the AFC nonidentity of the low frequencies filters is equal to some fractions of a percent:

$$D_D = \frac{P_0}{\pi(\omega_h - \omega_l)} \int_{\omega_l}^{\omega_h} [\Delta \mathrm{K}(\omega)]^2 \, d\omega \,, \tag{5.4.4}$$

where D_D is the dispersion of the dynamic error; P_0 is the power of the test signal with a spectral density uniform within the working frequencies range of the equipment, $[\omega_l, \omega_h]$; $\Delta(\omega)$ is the nonnominality of the transfer coefficients.

The values of phase-frequency characteristics of the low frequencies filters were measured using a frequency meter, which in measuring mode operated the length Δt of the interval A–B. From a signal generator the sinusoidal voltage of a selected frequency was supplied to the input of the corresponding low frequencies filter and to input A of the frequency meter. The input of the latter B was connected to the output of the low frequency meter. The phase was determined by the formula

$$\varphi = 360 \cdot f \cdot \Delta t \text{ (degree).} \tag{5.4.5}$$

Analysis of the measurement results has shown that the mean delay time of two FLFs is equal to 175 and 179 μs. The nonlinearity of the phase-frequency characteristics is ±17° (±5 %).

The difference of the mean intervals of the filter delay time is equal to 4 μs. The nonidentity of their phase-frequency characteristics does not exceed 7°. Thus, in accordance with (**??**) the error due to the nonidentity of phase-frequency characteristics of the filters is some fractions of a percent.

As the experimental results show the "zero" drift of the subtraction units at $U_{in1} = U_{in2} = 2.964$ V is not more than 2 mV. Thus, the error due to the "zero" drift of the subtraction units is about 0.1 %.

The nonlinearity of the amplitude characteristics of the subtraction units was measured for each of the inputs of a differential amplifier (here its other input was grounded). An analysis of the measurement results has shown that the transfer coefficient on two inputs is equal to 0.533 and the nonlinearity of amplitude characteristics does not exceed ±0.2 %. The nonidentity of the transfer coefficients of the subtraction units on two inputs does not exceed ±0.6 %.

The nonidentity of amplitude-frequency characteristics of the subtraction units on two inputs was within the limits −1.4 % to +0.2 %.The values of the amplitude-frequency characteristics of the subtraction units were measured when supplying the same signal with the level of 3.008V to both of its inputs. Within the limits from 100 Hz to 10 kHz the level of the output voltage did not exceed 2 mV (the error was less than 0.1 %).

To avoid a rather laborious and painstaking investigation to determine the law of summing up particular components of errors caused by the imperfection of metrological characteristics of listed elements, the following experiment was performed.

In the scheme illustrated in Figure 5.27, instead of MRI 6 a direct connection was made. Thus, the instrumental error, formed by the particular components of practically all the units of the functional scheme, was evaluated by the self-check method. The systematic component of the error did not exceed 0.1 %, and the fiducial root-mean-square deviation of the random error component was 4 %.

With the help of the "device for measuring the magnetic registration error" and other devices, investigations of the metrological characteristics of the MRI with a tape driving mechanism "Astra-V", MRI of the "Astra-2V" – "Astra-144N" types, and those of the complex of instruments for recording/reproducing reference signals, were conducted. Information on this is given in Table 5.5.

For the MRI with the tape-transporting mechanism "Astra-V" the results of investigations have shown that the fiducial root-mean-square deviation of the random error of magnetic registration is 2.5 %. For the purpose of comparison using this procedure, given in [457], a theoretical estimate of the error of the signal transfer over the MRI measurement channel has been calculated. It corresponds to 2.2 %.

Thus, the dispersion between the theoretical and experimental estimates is about 10 %, which confirms the correctness of the metrological model of forming the MRI error from Section 5.3.2 and the satisfactory accuracy of the developed procedure of its theoretical calculation, and it indirectly gives us confidence in the results of the DMRE metrological certification. In verifying the MRI of the "Astra-2V"–"Astra-

Table 5.5. Investigated MRIs

Name of the MRI equipment and its type	Measuring instruments used for investigations and references to inventions	Goal of investigation
Equipment of magnetic recording/reproducing, type "Astra-2V" – "Astra-144N"	Device for measuring magnetic registration errors [39, 275]. (Numbers of authors' certificates for inventions: 473125, 538310)	Experimental evaluation of the accuracy of magnetic recording/reproducing
Data input device for the equipment of statistical analysis, type KSICh	Device for measuring the nonlinearity of phase-frequency characteristics [197-1, 198, 330]. (Numbers of authors' certificates for inventions: 288141, 296054, 541124)	Evaluation of dynamic characteristics of MRI in the form of nonuniformity and of relative nonlinearity of PFC
Consumer recorder, type "romantic"	–"–	–"–
MRI with a tape transporting mechanism "Astra-V"	Device for measuring dynamic characteristics [37, 335, 336, 338–340]. (Numbers of authors' certificates for inventions: 442437, 474840, 501366, 513372; 547831; 562863)	Metrological certification of MRI
Magnetograph, type 4 MIZV	–"–	–"–
Analogue equipment of magnetic recording, type BPR-1	Device for determining MRI errors [275, 337]. (Numbers of authors' certificates for inventions: 585539, 473125)	Metrological assurance of measuring instruments belonging to customers
MRI, type "MIR"	–"–	–"–
MRI, type MO-22-01	Unit of control for a setup for verifying information-measuring systems, [336]. (Numbers of authors' certificates for inventions: 556492)	Determination of the law of signal time delay distribution in a channel of MRI

144N" types, the technical and metrological characteristics, the results of of which are given in Table 5.6, are shown in Table 5.7.

The analysis of the measurement results has shown that the fiducial root-mean-square value of deviation of the random error component of signal magnetic registration does not exceed 9.6 % at the confidence probability of 0.9, which entirely corresponds to the theoretical evaluation of the magnetic recording accuracy, which has been done according to the procedure described in [457].

The systematic component of the signal magnetic registration error (from −2.2 % to −0.1 %) is caused by the incorrectness of the channel calibration provided for in the equipment "Astra-2V"–"Astra-144N", as well as by the time and temperature drifts in the process of the experiment.

The "**DMRE for multichannel MRI**" was developed [452] for verifying the analogue MRI, in particular for the equipment of the BPR-1 and MIR-6 types. In this device a number of inventions have been realized [383, 473].

The functional scheme of the device is given in Figure 5.9, and the structural scheme is shown in Figure 5.28.

The special features of the device operation are as follows. The synchronization voltage represents a harmonic signal supplied from an external generator through the

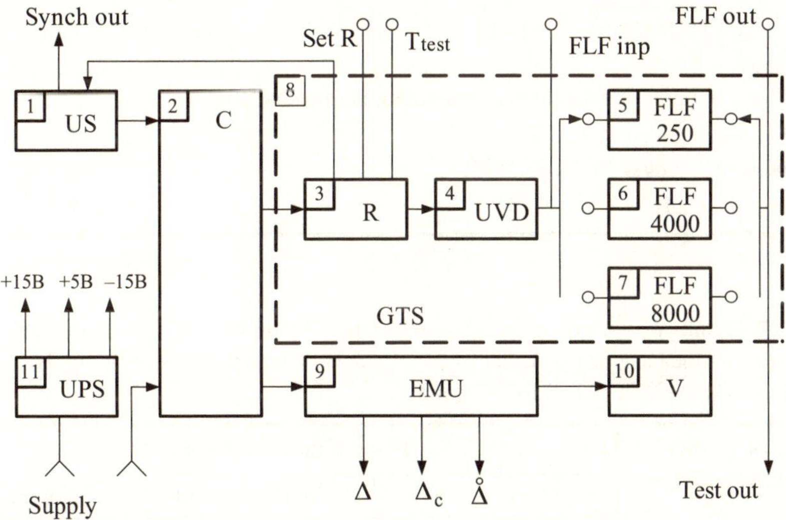

Figure 5.28. Structural scheme of the "Device for determining MRI errors". Here: US 1 is the synchronization unit; C 2 is the commutator "Kind of operation" ("Recording"–"Reproducing"–"Control"); R 3 is the shift register; UVD 4 is the unit of variable delay of the quasi-random binary sequence; FLF 5, FLF 6, FLF 7 is the filter of low frequencies with the cutoff frequency of 250 Hz, 4000 Hz, 8000 Hz, correspondingly; GTS 8 is the test signal generator; EMU 9 is the unit of selecting and measuring an error; V 10 is the voltmeter; UPS 11 is the unit of power supply.

Table 5.6. Technical and metrological characteristics of the multi-channel analogue equipment of precise magnetic recording of the "Astra-2V"–"Astra-144N" types.

Characteristic	Values area
Frequency range of the signal registered (Hz) at the speed of 76,2 cm/s	0–5000
Kind of modulation	FPM
Speed of magnetic tape transportation (cm/s)	4.7; 9.53; 19.05; 38.1; 76.2
Speed deviation (%), no greater than	1
Instability of the speed (%), no greater than	±0, 35
Total coefficient of detonation over the through tract at the speed of 4,76 cm/s, no greater than	1
Amplitude range of the input signal (V)	from 0 to +6.3
at deviation of carrier frequency (%)	40
Input d.c. resistance of the measurement channel (kOhm), no less than	80
Nonuniformity of AFC of the through tract (dB), no greater than	±1.5
Nonlinearity of the d.c. and a.c. amplitude characteristic within the whole pass band of the frequency range (%), no greater than	2
Coefficient of nonlinear distortions of the through tract (%), no greater than	4
Signal-to-noise ratio over the through tract (dB), no greater than:	
for the speeds (cm/s) 4.76; 9.53	35
for the speeds (cm/s) 19.05; 38.1; 76.2	40
Number of channels	14

Table 5.7. Results of the experimental determination of the fiducial systematic and random components of the basic instrumental errors of magnetic signals registration by the analogue MRI of the "Astra-2V"–"Astra-144N" types.

Number of the MRI channel, type of magnetic tape used	Place of the tape in the roll					
	Beginning of the roll		Middle of the roll		End of the roll	
	γ_{syst}, %	γ_{rand}, %	γ_{syst}, %	γ_{rand}, %	γ_{syst}, %	γ_{rand}, %
Channel 1, tape 6D	−1.6	9.6	−1.8	9.5	−2.2	9.6
Channel 7, tape 6D	−0.1	7.5	−0.6	975	−0.9	8.5
Channel 7, tape I 4403	−0.7	8.0	−1.4	7.6	−1.3	7.6

plug-and-socket "f_0" to the input device of synchronization unit US 1 for shaping the voltages: reference voltage and synchronization voltage. At a moment given in advance, namely at the moment of the pulse arrival from test pulse generator GTS 8, the amplitude of one of the half-waves of the synchronization voltage is shaped and becomes much greater than the remaining ones.

Thus, the moment of shaping this "marked" half-wave is determined by generator GTS 8, i.e., by the time position of the shaped test signal. The synchronization voltage is supplied to the plug-and-socket "output synch" and recorded in the channel that is adjacent to the MRI channel under test.

The reference voltage, formed from the harmonic signal of the external generator, represents the voltage of the rectangular shape (meander) with the pulse repetition rate equal to the input signal frequency. The reference voltage is supplied through commutator "kind of operation" C 2 to the GTS 8 input circuits.

The output signals of GTS 8 are as follows:

- the test signal representing a periodical quasi-random signal, agreeing with respect to the frequency and amplitude ranges with the corresponding characteristics of the MRI verified. It is supplied to the commutator "Test signal output" and recorded in the MRI channel under study;
- pulse voltage with the pulse repetition rate, equal to the test signal repetition rate. These pulses are supplied to synchronization unit US 1 for shaping "marked" half-wave in the synchronization voltage. Moreover, the pulse voltage is supplied to the control socket "T_{test}" and used for synchronizing a scan-out of an oscillograph, with the help of which the availability of a test signal is controlled in the process of recording.

When reproducing, two signals are simultaneously read out from two MRI channels: synchronization voltage and test signal. From the synchronization voltage, supplied through the plug-and-socket called the "input synchr", the pulses, setting the shift register "Set R", and reference voltage are shaped.

The reference voltage, representing the meander, is supplied to the input circuits of GTS 8 to shape the shift pulses from it. Since the test signal and synchronization voltage, which are read out simultaneously, take similar time distortions, then the test (reconstructed) signal at the output of GTS 8, received with the help of the shift pulses, shaped from the readout synchronization signal, will also take the same time distortions.

The pulses, setting register R 3 and shaped from the "marked" half-waves of the synchronization signal, set regiser R 3 into the state at which in the mode "recording" the pulses have been shaped for forming the "marked" half-wave, i.e., they provide the synchronization of the reconstructed and readout test signals.

The test signal, read out from the MRI channel, through the plug-and-socket "input test" is supplied to one of inputs of the unit for selecting and measuring the error

(EMU 9). To another input of EMU 9 the reconstructed test pulse is supplied from the output of GTS 8. The output signals of EMU 9 are:

Δ the registration error equal to an instantaneous difference of the input signals of EMU 9 (the plug-and-socket "Δ");

Δ_{syst} the systematic component of the error (the plug-and-socket "Δ_{syst}");

$\overset{0}{\Delta}$ the random component of the error (the plug-and-socket "$\overset{0}{\Delta}$").

Voltmeter V 10 with the pointer indicator measures the mean rectified value of the random error component:

$$\frac{1}{T}\int_0^T \left|\overset{0}{\Delta}\right| dt .$$

On the basis of a minimum of readings of this indicator a smooth adjustment of synchronization is performed with the help of variable delay unit UVD 4 for the constant time shift to be compensated between the channels, in which the test signal and synchronization voltage have been recorded, correspondingly.

In addition to the modes "recording" and "reproducing", in the device a control mode is provided ("control"). In this mode both inputs of EMU 9 are made free from internal contacts and are connected with external plug-and-sockets ("input test" and "output test"). This makes it possible to do the following:

- to check EMU 9 with respect to the known test signals, supplied from the outside;
- to use EMU 9 as the unit of subtracting analogue electric signals varying from -5 V to $+5$ V within the frequency range from 0 to 10 kHz.

The main technical and metrological characteristics of the "DMRE for multichannel MRI" c are given in Table 5.8.

The device was certified in accordance with the procedure for certification and periodical verification developed [452] for the "device for measuring the MRI errors". Moreover, two documents were created: "procedure for verifying the magnetic recording equipment" and "procedure for determining the delay interval between the channels of recording/reproducing of the equipment of the BPR-1 type" [452].

5.4.1.2 Devices for measuring dynamic characteristics of MRI channels

On the basis of material in Sections 5.3.2 and 5.3.4 two varieties of measuring instruments have been developed which provide the experimental determination of MRI dynamic characteristics: the "device for measuring dynamic characteristics", for measuring the values of the pulse weight function of a MRI channel, and the "device for measuring the nonlinearity of PFC", for measuring the nonlinearity of a relative phase-frequency characteristic of the channel of equipment under study [42, 43, 272, 273, 448, 465, 469, 470, 471, 474, 475, 477].

Table 5.8. Technical and metrological characteristics of the "DMRE for multichannel MRI".

Characteristic	Values
Frequency range of a test signal (Hz)	
Range I	0–250
Range II	0–4000
Range III	0–8000
Test signal attenuation above cutoff frequency of each frequency range (dB per octave), no less than	50
Amplitude range (V), no less than	±6
Test signal shift level (V) at the permissible deviation of ±1 %	+3
Input resistance (kOhm), no less than	100
Output resistance (Ohm), no greater than	100
Limit of a permissible absolute basic error for MRI (mV), with oscillations of the signal carrier speed up to 0.05 %, no greater than:	
for a systematic component	±3.5
for random component (mean square deviation):	
Range I	1
Range II	3
Range III	4
Limit of a permissible additional error (in fractions of the basic one) at variation of power supply voltage within the limits from –15 % to +10 %, no greater than	0.5
Power supply (V) from:	220 (±10 %)
Network of frequency (Hz)	50
Consumed power (W), no greater than	5
Time of continuous running (h.), no less than	8

The "**device for measuring dynamic characteristics**" is a further development of the underlying ideas of the first versions of a similar instrument [452, 457].

The functional scheme of the developed measuring instrument is given in Figure 5.17 for discrete tacking and in Figure 5.29 for analogue tracking of the delay time oscillations of the reproduced test signal from the MRI channel.

The "device for measuring dynamic characteristics" is a desk-top measuring instrument. Its front panel contains the following controls: commutators "intervals of averaging" and "generator frequency, kHz", seven high frequency plug-and-sockets, six toggles, two buttons, ten test jacks, the regulating devices "synchronization adjustment" and "zero indicator", ground terminal, and three light-emitting diodes and

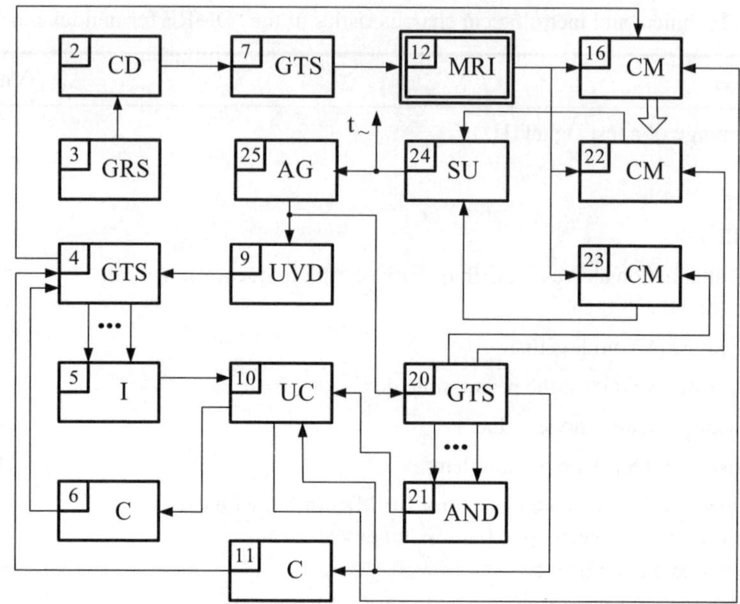

Figure 5.29. Fig. 5.29: Functional scheme of the "Device for measuring dynamic characteristics" with analogue tracking delay time oscillations of the signal reproduced. Here: CD 2 is the counter divider; GRS 3 is the reference signal generator; GTS 4, GTS 7, GTS 20 is the generator of test quasi-random signal; "AND 5", "AND 21" is the logic circuits "AND"; C 6, C 11 is the commutator; UVD 9 is the unit of variable delay; UC 10 is the unit of control; MRI 12 is the equipment of precise magnetic recording; CM 16, CM 22, CM 23 is the correlation meter; SU 24 is the subtraction unit; AG 25 is the adjustable generator.

display lamp, signalling power-up. Seven versions of this instrument [42, 43, 470, 471, 474–476] have been implemented.

The operation of the instrument, according to the functional scheme illustrated in Figure 5.17, is described in Section 5.3.4. The distinctive features of operation are in accordance with the functional scheme shown in Figure 5.29, are as follows. In the mode of analogue "tracing" of delay time oscillations of the signal in the MRI channel the commutator "synchronization", located on the front of the instrument, is installed in the position "analogue". From the output of channel MRI 12 the reproduced test signal is supplied to the first inputs of correlation meters CM 16, CM 22, CM 23.

The last two correlation meters, together with subtraction unit SU 24, generate a signal proportional to the value and sign of the time disagreement between the reproduced test signal from the output of channel MRI 12 and the voltage generated by GTS 20. The inputs of the 8-th and 10-th bits of the M-sequence generator GTS 20, are connected to the second inputs of correlation meters CM 22 and CM 23, respectively.

Output signals of these correlation meters are proportional to values of the cross-correlation function of the reproduced signal and M-sequence from the output of

GTS 20. They are calculated in points shifted relative to one another by two periods of the shift pulses of GTS 20. The difference between these signals has a control effect on controllable generator CG 25, the frequency of which changes according to the value and sign of this effect. The output sequence of pulses of CG 25 is supplied to the shift circuit of GTS 20 and then through unit of variable delay UVD 9 they are supplied to the shift circuit of GTS 4.

Thus, the variation of the time scale of the reproduced signal, caused by oscillations of the time delay, results in similar variation of the time scale of the M-sequences generated by generators GTS 4 and GTS 20.

The technical and metrological characteristics of the device were studied according to the procedure described in [457] and are shown in Table 5.9.

Table 5.9. Technical and metrological characteristics of the "device for measuring dynamic characteristics".

Characteristic	Values
Frequency of the reference signal generator, kHz	800, 400, 200, 100, 50, 25, 10
Number of shift register bits	10
Amplitude of the M-sequence (V)	From 0 to +5
Input resistance (kOhm), no less than	100
Output resistance (kOhm), no greater than	5
Limit of the permissible main fiducial error (%)	1
Mode of operation	Cyclic or one-shot
Length of the cycle (τ)	24 or 30
Shift step (τ)	1, 1/2, 1/4
Averaging interval (τ)	From 5 to 11 (in 1)
Power supply (V)	220 ($\pm$10 %)
from the network (Hz)	50
Consumed power (W), no greater than	17
Time of continuous running (h), no less than	8

Using of the "device for measuring dynamic characteristics" the values of the pulse weight functions of measurement channels of MRI with the tape driving mechanism "Astra-V" and magnetograph 4 MIZV (Table 5.5) were determined. In particular, for MRI with the tape driving mechanism "Astra-V" the channel PWF was obtained. This is shown in Figure 5.30.

The error of the PWF values obtained did not exceed 0.4 % at the confidence probability of 0.9. On the basis of these measurement results the incorrectly posed problem of calculating the transfer function from the PWF values, known with the error, obtained with the help of the Fourier transformation, was solved using a computer and

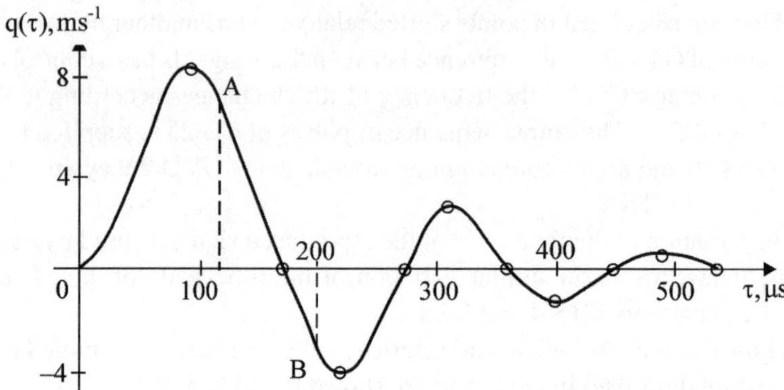

Figure 5.30. Results of the experimental determination of the pulse weight function of MRI channel.

obtained fractionally rational approximation function of the minimal order:

$$K(p) = \frac{\sum_{i=0}^{m} b_i p^i}{\sum_{j=0}^{n} a_j p^j}. \tag{5.4.6}$$

The values of calculated coefficients a_j, b_i are given in Table 5.10.

Table 5.10. Values of the coefficients of the approximating transfer function of the MRI channel, calculated on the basis of the pulse weight function.

Indices i, j	Coefficients	
	b_i	a_j
0	1.0000	1.3775
1	$-0.2974 \cdot 10^{-4}$	$0.1383 \cdot 10^{-3}$
2	$0.6470 \cdot 10^{-9}$	$0.7074 \cdot 10^{-8}$
3	$-0.6946 \cdot 10^{-14}$	$0.2100 \cdot 10^{-12}$
4	$0.5240 \cdot 10^{-19}$	$0.3957 \cdot 10^{-17}$
5	—	$0.4945 \cdot 10^{-22}$

The approximating transfer function has the fifth order and provides a satisfactory approximation of experimental data within the pass band to 10 kHz. This fact is the confirmation of the possibility of estimating the order of such a linear system as a half sum of a number of PWF extremums and that of its crossings with the x-coordinate (exept $t = 0$), accepting them for the values of Markov's parameters. In fact, a number of the dots marked in Figure 5.30 by circles are equal to 10, corresponding to a system of the fifth order.

The "**device for measuring the nonlinearity of PFC**" was developed on the basis of three inventions [272, 273, 465]. The functional schemes of the developed prototype

Table 5.11. Technical and metrological characteristics of the "Device for measuring the PFC nonlinearity".

Characteristic	Values
Frequency range of the test signal (Hz)	From 20 to 10000
Frequency of the external driving generator of sinusoidal signal (kHz)	20
Number of points of phase-frequency characteristic measurements	10
Slope of attenuation of high and low frequencies filters (dB per octave), no less than	46
Amplitude range of the test signal (V), no less than	3
Input resistance (kOhm), no less than	100
Output resistance (kOhm), no greater than	500
Limit of the permissible absolute basic error (angular degree):	
for systematic components	≈ 0 (with use of the self-checking mode)
for random components (root-mean-square deviation)	3
Limit of the permissible additional error (as a fraction of the basic one) at supply voltage changes within the limits from −10 % to +10 %	0.5
Time of measurements of the ten PFC values in automatic mode (min), no greater than	5
Supply voltage (V)	220 (±10 %)
from the network with the frequency (Hz)	50
Consumed power (W), no greater than	40
Mass (kg), no greater than	7
Time of continuous running (h.), no less than	8

of the measuring instrument in different modes of operation are given in Figure 5.21 and Figure 5.22. The operation principle of the device is described in Section 5.3.4 [452]. The main technical and metrological characteristics of the "device for measuring the nonlinearity of PFC" are illustrated in Table 5.11.

This instrument is a desk device. On the front the terminals "input 20 kHz" are located for connecting to an external driving generator of sinusoidal oscillations with the frequency of 20 kHz. In addition to these, there are terminals "MRI" for connecting the input and output of the measurement channel of the MRI under study and terminals "to phase meter" for connecting with an external phase meter or frequency meter.

Moreover, on the left side of the front plate there are 20 control sockets of the "trans." triggers of the transfer part of the device. Here the control sockets "control of power suppl." of the supply unit, sockets for control of input and output voltages of the high frequency filters, i.e., "input of FHF" and "output of FHF", respectively, as well as the socket of the control pulse of zeroing the receiving part of the instrument "pulse of zeroing receiv.".

On the right-hand side of the front plate there are 20 control sockets of the triggers of the receiving part of the "receiv." device, as well as sockets for controlling the output voltages of the filters of low frequencies with cuttoff frequency from 10 kHz to 20 Hz. In the middle of the front plate are the following regulators and attention devices: a toggle and control lamp "switched on", "zeroing" button, and dual commutator "MRI", which allows the self-check of the "device for measuring the PFC nonlinearity", i.e., the connection of the transfer part of the device to the receiving part directly, avoiding the MRI channel under study.

The "device for measuring the nonlinearity of phase-frequency characteristics" was used to determine by means of experiments the values of the nonlinearity of the relative phase-frequency characteristics of measurements channels of the "data input device for statistical analysis equipment" of the KSICh type, as well as of a magnetic recording/reproducing path of sounds of a consumer recorder of the "romantic" type (Table 5.5).

The latter is of interest from the point of view that in practice the phase-frequency characteristics of a consumer recorder path have been never determined, since its quality parameters (the coefficient of parasitic and frequency modulation, level of noise and of nonlinear distortions, etc.) are significantly worse than the parameters of MRI, which makes measurements of PFC values of magnetic recorders more difficult.

At the same time, taking into account a growing trend towards using modernized magnetic recorders of a low accuracy class as a part of measuring instruments (for example, in measuring instruments applied in medicine and biology), the task of finally solving the problem of determining PFC of their channels is urgent .

In Figure 5.31 the experimentally obtained dependence of the nonlinearity of the relative phase-frequency characteristic of the recorder "romantic" path is given.

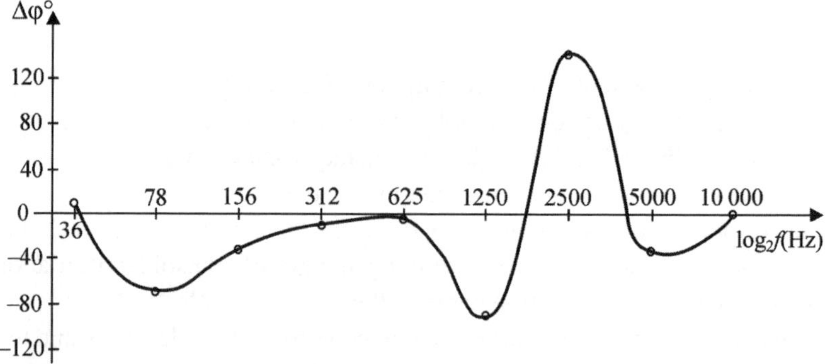

Figure 5.31. Nonlinearity of the relative phase-frequency characteristic of the magnetic recording/reproducing path of the recorder "Romantic-3".

Table 5.12. Basic technical characteristics of the “romantic”-type recorder.

Characteristic	Values
Speed of tape transporting (cm/s)	9.53
Frequency range (Hz)	63–12500
Dynamic range of the equivalent resistance of a loud-speaker (dB), no less than	42
Coefficient of nonlinear distortions at a harmonic signal of 400 Hz (%), no more than	5
Coefficient of detonation (%), no more than	0.4

It should be noted that its values strongly depend on positions of timbre controls for the low frequency and high frequency parts of the working frequency range. The dependence indicated was obtained at the middle position of the both timbre controls.

The basic technical characteristics of the recorder of the “romantic” type are indicated in Table 5.12.

5.4.1.3 Setup for verifying information measurement systems

The reference “Setup for verifying the information-measurement systems” includes “unit of control “GPS-2” that fulfils the following functions:

- comparison of the instantaneous values of a test signal before and after its passing through a channel of the equipment being tested;
- tracing the current delay time of a test signal in a channel with generation (according to the invention indicated in [472]) of the voltage proportional to its instantaneous values, as well as synchronization of moments. The test signal values are read out from the output of the equipment under testing;
- output of the analogue signals which are proportional to the error of transforming instantaneous signal values in the amplitude (ΔU) and time (Δt) domains, as well as their transformation into digital code.

The functional scheme explaining the method [452] of measuring oscillations of the signal time delay in the MRI channel is shown in Figure 5.24, and the basic technical and metrological characteristics of the unit of control “GPS-2” are listed in Table 5.13. The essence of this measurement method is described in Section 5.3.5.

The “unit of control GPS-2” has six cassette modules which are installed on a common frame. The modules are: two boards of “logic”, board of shapers, ADC, DAC, and a time-error separation board. In the process of testing, the IMS unit operates in two modes.

Table 5.13. Technical and metrological characteristics of the Unit of control "GPS-2".

Characteristic	Values
Frequency range of the test signal (kHz)	0–20; 0–10; 0–5; 0–2.5; 0–1.25; 0–0.625; 0–0.312
Number of spectral components of the test signal within the working frequency range	1048575
Amplitude range (V)	±1; ±10; 0–(+1); 0–(+6)
Constant shift of the test signal (V)	0; +0.5; +3
Input resistance (kOhm), no less than	
for the "pilot-signal"	20
for the test signal	600
Output resistance (Ohm), no greater than	200
Limit of the permissible basic reduced error (%)	1.6
Root-mean-square deviation of the random error component (%), no greater than	0,7
Supply voltage (V)	220 ($\pm$10 %)
from the network (Hz)	50
Consumed power (W), no greater than	30
Mass (kg), no greater than	10
Dimensions (mm)	$480 \times 350 \times 220$
Time of constant running (h.), no less than	8

Here the test and reference signals are shaped in the mode "recording", which are supplied to IMS under testing through the plug-and-sockets "Output to MRI, Channels 1, 2".

In the mode "reproducing" the test and reference signals, having passed through IMS under testing, are supplied to the above unit. The instantaneous values of these signals are compared with the corresponding instantaneous values of the signals from GPS-2. The voltages obtained, which are proportional to the errors ΔU and Δt, are supplied through the plug-and-socket "output to PC" to the computer for further processing in accordance with the appropriate software, and then for printing.

With the help of the unit of control of the "setup for verifying information-measurement systems", the measurement channels of MRI of the M022-01 type were investigated with the purpose of getting the law of distribution of signal time delay in a MRI channel.

The results of the investigations are arranged in the form of a printout that can be used to construct a histogram of the distribution probability density law of time delay oscillations. Moreover, the software allows the minimum ($\Theta_{\min}$) and maximum ($\Theta_{\max}$) delay time to be separated within the given time interval Δt, and their estimates to be calculated:

$$\gamma_T = \frac{\Theta_{\max} - \Theta_{\min}}{T} \cdot 100\,\%$$

and

$$\gamma_{\Delta t} = \frac{\Theta_{\max} - \Theta_{\min}}{\Delta t} \cdot 100\,\%,$$

the maximum peak-to-peak of signal time delay oscillations, which is reduced to the period T of the upper threshold frequency of the pass band of the channel under study and to the selected time interval Δt.

5.5 Summary

An overview of the literature, both theoretical and practical, on the determination of metrological characteristics of precise magnetic recording equipment for recording analogue electric signals, used as a part of information-measurement systems and measurement-calculation complexes, was presented.

On the basis of this generalization a number of new approaches to creation of methods and means for measuring metrological characteristics of MRI have been developed.

The results provide the groundwork for a solution of the problem of metrological assurance of precise magnetic recording equipment, which is of great importance in connection with a wide use of MRI in IMS of various destinations.

Within the framework of the solution of this problem and the concrete tasks which follow from it, the results indicated below have been obtained.

An algorithm for evaluating the quality of systems of metrological assurance under the conditions of incomplete and inaccurate data on their elements, connections, and properties, was developed which makes use of the fuzzy-set theory. This algorithm was illustrated by an example of the project of a MRI verification scheme.

Taking into account the metrological status of MRI as an intermediate measuring transducer, on the basis of an analysis of the equation of connection between the input and output signals of its measurement channel, a complex of its metrological characteristics which need to be normalized was presented. These characteristics include the error of transferring the signal over a channel, the dynamic characteristics in the form of the transfer or pulse weight function, the level of nonlinear distortions, and the time oscillations of the signal delay. Moreover, the metrological model of forming the resultant error was also grounded.

A new method for normalizing the dynamic characteristics of a measurement channel was proposed, based on the application of Markov's parameters. The proposed method is alternative and equivalent with respect to the method based on applying moments of the pulse weight functions. In addition to this method, one more method was developed which can be used to construct functional dependencies for jointly measured quantities under the condition of incomplete initial data. This method is based on solving the problem of multicriteria optimization.

For the first time, a method for the experimental evaluation of the registration error of a signal in the MRI channel has been developed which is based on the use of the peculiar features of the test signal in the form of a quasi-random binary sequence of the maximum length, and limited with respect to its spectrum according to the pass band of the channel under study.

Correlation methods and instruments for direct and differential measurements of values of the pulse weight functions of the MRI channel, taking into account the following special features of the equipment verified, were presented: the number of channels, the width of their pass band, the different time sequence of recording and reproducing modes, the availability of transposing the speed of the magnetic carrier movement, and the possibility of separating a signal proportional to oscillations of the time delay. An analysis of this new method of coherent frequencies for the experimental determination of the nonlinearity of phase-frequency characteristics of MRI channels was made.

A correlation method was developed for measuring nonlinear distortions of the signal of the type of "white noise" in the MRI channel. This method is based on determining the pulse weight functions of the channel on high and low levels of the test signal. The development of methods and instruments for measuring oscillations of the time delay of signals in the channel also allows both the oscillations and time delay drift to be estimated.

Various versions of the technical realization of instruments for measuring the MRI metrological characteristics were developed. Their novelty is confirmed by 18 author's invention certificates.

Eight types of measuring instruments of eight types, developed researchers can be successfully used for experimental determination of the metrological characteristics of any equipment of precise magnetic recording and provide its verification and certification.

Chapter 6

Validation of software used in metrology

6.1 General ideas

In recent years the use of software for solving problems in metrology has significantly increased. The underlying reason of this trend is connected with the increasingly extended application of computer methods and means, firstly for data acquisition and processing, transmitting, storing and presenting measurement results, which requires auxiliary infrastructural information, and, secondly, for the metrological maintenance and simulation of measurement experiments.

The EU Measuring Instruments Directive (MID) [327] has demonstrated a new approach to the technical harmonization of standards in the field of commercially important measuring instruments (MI), subjected to legal metrological control.

For the purpose of the execution of this MID, the European Cooperation in legal metrology, WELMEC, has developed Document WELMEC 7.2 Software Guide (Measuring Instruments Directive 2004/22/EC) [541], and at the same time the Organisation Internationale de Métrologie Légal has prepared the Document OIML D 31 General Requirements for Software Controlled Measuring Instruments [181].

The Regional Organization COOMET (Euro-Asian Cooperation of National Metrological Institutions) has issued the recommendation "Software in Measuring Instruments: General Technical Specifications" [123].

Moreover, many countries have national normative documents relating to this matter. In particular, in Russia these issues are touched upon in normative documents such as [196, 199, 200, 332, 335–337, and others], as well as in a number of scientific and technical publications [115, 284, 307, 482, 502, 537, and many others]. In the countries of Western Europe, America, and in Japan a number of normative documents and publications [1, 56, 123, 145, 180, 181, 224, 231–235, 244, 327, 541, and others] are also devoted to these issues.

The software applied in metrology, which expands the functional capabilities of program-controlled measuring instruments, as well as the solution for providing automation of other no less important problems of metrology, are of interest, above all connected with the accuracy characteristics of software.

In connection with the publication of the ISO "Guide to the Expression of Uncertainty in Measurement" (GUM) [243] (also translated and edited in Russian) which has become a document recognized by the international metrological community; the accuracy characteristics are determined in terms of uncertainty (the standard, combined and extended ones) together with the previous characteristics of errors, systematic and random, which are more habitual, whereby the terminology of the International vocabulary on metrology (VIM) [246] was also used.

The task of this chapter is to acquaint all those who work in the manufacturing measuring instruments industry, the producers and users of software for MI, and the entire metrological community with the normative documents now in force, and also with publications relating to quality estimates of the software used for measurement data processing, laying special stress on software accuracy characteristics. This chapter can be used as a reference manual for carrying out work on the metrological certification of measuring instruments and measuring systems with embedded software control facilities.

The software (S) used in metrology includes, firstly, software for data processing facilities hardwired into measuring instruments and/or computing units of measuring systems, software tools automating the processes of the acquisition, maintenance, transmission and presentation of measurement data, as well as the treatment of measurement results in accordance with the appropriate divisions of the certified procedures of performing measurement methods (MM).

The question of software quality evaluation naturally arises. The task of describing this quality, in itself, is unusual, since it is rather difficult to propose a unified set of indicators of this quality. It is necessary to take into account the inevitable division of labor in developing, creating, debugging, studying, applying, and accompanying the software, on the one hand, and on the other hand, the difference in ideas of both designers and customers of various levels with respect to the software.

An adequate set of quality indices for software products depends on the functional designation and properties of each software product. Software and software complexes for computers and microprocessors, as objects of designing, developing, testing, and quality evaluation are characterized by the following *generalized indices* [307]:

- problem-oriented field of application;
- technical and social destination of a given complex;
- particular type of functional problems to be solved with a sufficiently defined field of application, which is determined by their respective users;
- volume and complexity of the totality of software and data bases for solving a single target problem of a given type;
- required composition and desired values of quality characteristics of software operation, as well as the permissible amount of damage caused by insufficient quality;
- degree of connection of problems solved in a real time scale or the permissible latency time for getting a result of the problem solution;
- predicted values of maintenance and a the perspective of creating a great number of versions of software and software complexes;

- expected edition of production and application of software;
- degree of providing software with required documentation.

To determine the quality of software operation, the availability of technical capabilities which provide software compatibility, interaction, improvement and development, it is necessary to use documented standards in the field of the evaluation of the quality indices of software. Analysis of modern Russian and other normative documents and publications in this field shows that documented standards [181, 200, 541] are, in our view, fundamental for regulating the quality indices of software.

In addition to the general requirements for software quality, special requirements arise in each particular field of application. In this connection it would be correct to consider separately the problem of metrological accompaniment of the software for measurement results processing. A documented standard [195] briefly formulates the basic metrological requirements for a software, related to the availability of a detailed documentation, protection of the software, its univocal identification and adequacy for application under conditions in measurement, test, and calibration laboratories. Adequacy in the metrological sense is understood, firstly, as the possibility of achieving the required accuracy of a final result of measurement when a particular software is used.

When we speak of the accuracy characteristics of the algorithms and software used in metrology, we need to pay attention to the uncertainty components of errors and toe methods for their evaluation. The accuracy of final measurement results is influenced by various factors, such as the accuracy of the input data, the algorithm of the input data processing, the algorithm of the evaluation of a final result, as well as the realization of the listed algorithms in the software.

In 1980s, at the D. I. Mendeleyev Institute for Metrology (St. Petersburg, Russia) the methodology for the certification of algorithms for data processing in carrying out measurements was developed [332]. The certification of an algorithm (software) of data processing is an investigation of the algorithm properties with the help of models of initial data which results in the determination of properties and he evaluation of quantitative characteristics of the algorithm (software). However, there is a difference between the procedures of the general and metrological certification. As a result of the general algorithm (software) certification, evaluation of the algorithm (software) characteristics regarding accuracy, stability and complexity are obtained, using various models of input data.

With regard to metrological certification, the estimates of error (uncertainty) components characteristics are obtained when the processing of the results is performed under particular conditions of applying this algorithm. The concept "metrological certification" is close to the concept "validation", which is now widely used. Validation is a confirmation with the help of presenting objective evidence that the requirements concerning a particular expected application or use are met.

In essence, the procedure of software validation consists of certifying the algorithm of data processing realized by this software. "Good" software will not introduce any significant errors into the total errors of the result. In other words, errors of the software itself have to be insignificant as compared to the transformed errors of input data and to the methodical errors of the algorithm. When algorithm validation results are available, testing the software is an effective means for confirming the fact that this software does not introduce any significant additional errors.

When performing a test of the software, one important question is the proof that testing is complete and includes the development of test tasks and the formation of sets of "standard" data. The answer to this question is based on the specification of the software and the results of the algorithm certification, which allow "weak" features of the software to be indicated, i.e., parameters of models.

This chapter is devoted to various aspects of solving the practical task of evaluating the accuracy of measurement results obtained with the help of software for measurement data processing. In this chapter the factors influencing the accuracy of a final result of measurement are discussed. Great attention is paid to issues of the metrological validation of modern software-controllable measuring instruments.

6.2 Tasks of the metrological validation of software (MVS) used in metrology

6.2.1 Classification of tasks for MVS used in metrology

The field of applying software products in metrology is rather wide. This leads to the fact that the requirements for software and problems of its metrological validation differ, depending on the problem being solved. Therefore, a need arises to make an attempt to classify available problems of metrological validation of software used in metrology.

To construct a classification "tree" is not easy, since its form depends on selected classification signs. The following signs can be included into the hierarchy of this classification:

- degree of tie to particular measuring instruments and measuring systems;
- degree of autonomy and capability of the software developer;
- degree of use of computer aids (CA);
- degree of protection or a risk class of data falsification;
- possibility to separate an important (from the point of view of metrology) and legally relevant part of a software being in use;
- possibility to download modified versions of software, and others.

The following categories of software can be distinguished **by the degree of tie to particular measuring instruments** and measuring systems:

- firm tie to particular samples and/or types of measuring instruments and/or measuring systems;
- practical lack of any conformity. As an example of this category are the software products, intended to be used for data processing of measurement results obtained in key, regional, and interlaboratory comparisons of measurements standards. Moreover, the software for processing data obtained before from various measuring instruments and/or measuring systems, as well as for providing a normal operation of the infrastructure of any measurement, testing or calibration laboratory belong in this category;
- simulation technique of a measurement experiment, as well as test "standard" input signals and response to them (for example, see [56, 244, and others]).

Depending on the **degree of autonomy and capability of a developer**, the software can be divided into:

- software firmly tied to measuring instruments and/or measuring systems, which is developed specially for solving posed measurement problems and cannot be separated as an object of metrological validation (and/or for sale), i.e., so-called built-in or user (ordered) software;
- software that can be entirely or partly separated and subjected to metrological validation (and sail) as an independent object. Modified commercial software with a changed source code (for example, software developed at LabView, and others) are also of this category;
- autonomous commercial software purchased as a finished product and used without any modification (for example, Microsoft Excel, MATLAB, and other application software packages).

By the **degree of use of computer aids**, the software support is classified as:

- software, developed as a task-oriented product for computer aids applied in a particular measuring instrument and/or measuring system (built-in or of the P type);
- software used for a personal universal multipurpose computer (in particular, in a measuring instrument of the U type).

By the **degree of immunity or risk class of data falsification**, the software can be of the following types:

- software protected from accidental use;
- software protected from unauthorized use, including cracked software, use of undocumented commands entered through a user interface, protection of controlled parameters, etc., which are protected by physical (hardware) as well as electronic and cryptographic means.

Reasoning from the **possibility of separating the metrologically important from the legally relevant part of a software** being used, one distinguishes:

- software that can be divided into metrologically important and legally relevant parts;
- software that cannot be divided into metrologically important and legally relevant parts.

Reasoning from the **possibility of downloading modified versions of the software** (software update procedure), we distinguish

- a verified update;
- a traced update.

In accordance with [197] it is possible to divide t **software types** into **5 different categories**, which initially divide software support into:

- COTS (commercial off-the-shelf), i.e., finished purchased software;
- COTS (modified off-the-shelf), i.e., modified purchased software;
- CUSTOM, i.e., the software manufactured by request or "homemade".

The special features of these categories are given in Table 6.1.

Table 6.1. Categories of software.

Category	Type	Software groups	Software examples
1 (COTS)	Operation systems	Operation systems	Windows, LINUX
2 (COTS)	"Hard" software	Embedded software	Devices, voltmeters, testing machines
3 (COTS)	Standard program packages	E-mail programs, word-processors	Word, Excel (only as tables), Outlook, Internet Explorer, Acrobat and others
4 (MOTS)	Configured and modified software packages	Software as a means of programming and configuring an environment. An appliance is needed before use.	Equation, Excel, LabView, LabWindows, Labtech Notebook, Mathcad
5 (CUSTOM)	Software for users, manufactured by request or "homemade"	Software developed for a user and applying software tools. This includes documents from Word/Excel with macrocodes	Applications written in C++, SOL+, Java Visual Basic, XML, LabView, LabWindows, and other languages

One version of software for measuring instruments is given in the WELMEC 7.2 Software Guide [541], which represents a structured set of blocks of requirements. The total structure of the guide follows the classification of measuring instruments, which rests on the basic configurations and classification of so-called configurations of measurement technologies. The set of requirements is supplemented with special requirements for measuring instruments.

As a result the following three types of requirements exist:

(1) requirements for two basic configurations of measuring instruments (called type P and type U);

(2) requirements for four configurations of measurement technologies (called extensions L, T, S and D);

(3) special requirements for measuring instruments (called extensions I.1, I.2 ...).

The first type of requirements can be applied to all measuring instruments with software (measuring instruments of the P type: built-for-*purpose* measurement instruments; measuring instruments of the U type: measurement instruments using *universal* computers).

The second type of requirements relates to the following functions of information technology:

- memory of long-term storage of measurement data (L – *long*-term storage);
- transmission of measurement data (T – *transmission*);
- download of software (D – *download*);
- separation of software (S – software *separation*).

Each set of these requirements is applied only in cases where a corresponding function exists.

The last of the four types represents a collection of further special requirements for measuring instruments. At the same time, the numeration in the guide [541] follows the numeration used in Supplements of Special Requirements for Measuring Instruments in the Directive MID [327].

The set of requirement blocks that may be applied to a given measuring instrument is schematically shown in Figure 6.1.

The schemes in the following Figure 6.2 show what sets of requirements exist.

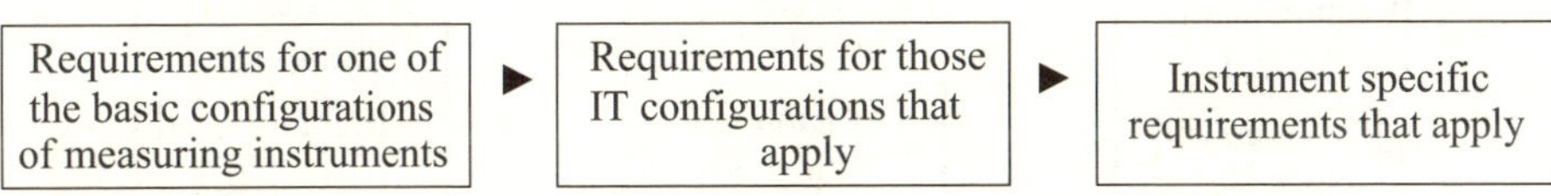

Figure 6.1. Type of requirement sets that should be applied to a particular measuring instrument.

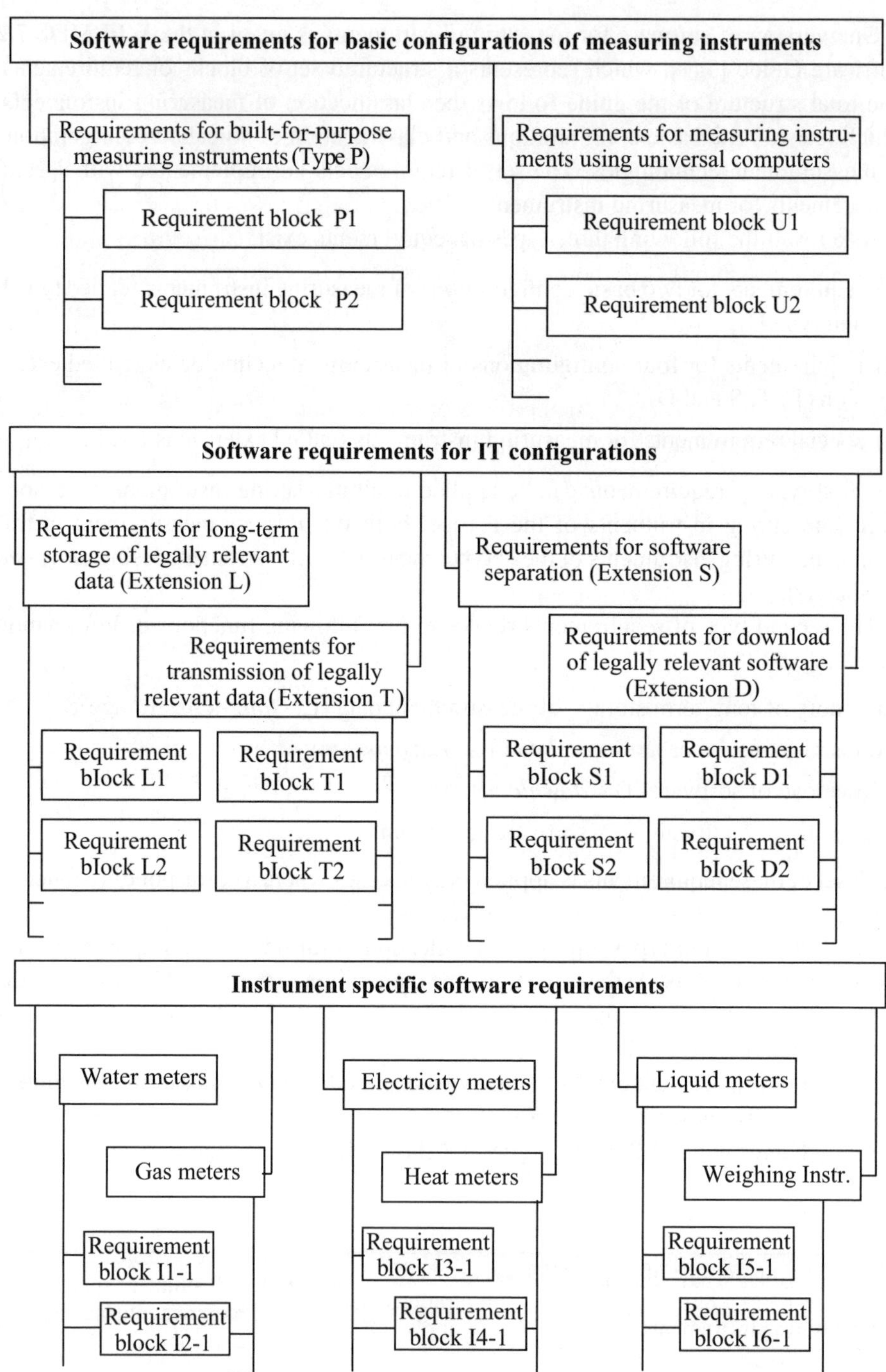

Figure 6.2. Overview of requirement sets.

In addition to the structure described, the guide requirements [541] differ in risk classes. In the guide six risk classes are presented, which are numbered from A to F with an assumed increase of the risk. At present the A class of the lowest risk and the F class of the highest risk are not applied. They have been formed as a reserve for possible cases which might make them necessary in the future.

The remaining risk classes, from B to E, cover all the classes of instruments which are related to the sphere of the MID regulation [327]. Moreover, they ensure a sufficient interval of possibilities for the case of computations in evaluating a changed degree of risk. The classes are determined in the guide, which has an exclusively informative character.

Each measuring instrument has to be assigned to a risk class, since special software requirements that should be applied, are regulated with the risk class to which the measuring instrument belongs.

The guide [541] can be applied to a great variety of measuring instruments, and is set up in modules. The appropriate sets (blocks) of requirements can be selected if the following algorithm is fulfilled.

Step 1. Selection of the basic configuration (P or U).

Only one of two sets for basic configurations has to be applied.

Step 2. Selection of the IT to be applied (extension L, T, S and D).

The configurations of the IT include: memory for long-term storage of legally relevant data (L), transmission of legally relevant data (T), separation of software (S), and download of legally relevant software (D). The corresponding sets of requirements, called module extensions, do not depend on each other. The sets are selected depending only on the IT configuration. If the extension set is selected, it has to be applied in full. It is necessary to determine which of the existing module extensions give the best fit, and then they can be used correspondingly (Figure 6.2).

Step 3. Selection of special requirements for measuring instruments (extension I).

It is necessary to select extensions I.x, using those corresponding to the requirements for MI, which fit to a particular measuring instrument, and then to apply them correspondingly (Figure 6.2).

Step 4. Selection of the applied class of risks (extension I).

It is necessary to select the class of risk as determined in the corresponding extension I.x for the special requirements for measuring instruments in subdivision I.x.6. Thus, the class of risk can be identically determined for the class of MI or even divided into categories, fields of application, etc. As soon as the class of risk has been selected, only those requirements and and the validation guidance have to be considered which correspond to this class of risk.

Blocks of requirements

Each requirement block (see Figure 6.2) contains a title and well-defined thesis of the requirement. Moreover, it consists of defining comments (a text containing explanatory and prescribed notes, additional explanations, exclusive cases, and so on), documentation demanded for presentation, validation guidance and examples of acceptable decisions (if they are accessible). The content inside the block of requirements can be divided according to the classes of risk. This leads to a schematically presentation of the block of requirement shown in Figure 6.3.

<table>
<tr><td colspan="3">Title of a requirement</td></tr>
<tr><td colspan="3">Main statement of therequirement (eventually differentiated between risk classes)</td></tr>
<tr><td colspan="3">Specifying notes (scope of application, additional explanations, exceptional cases, etc.)</td></tr>
<tr><td colspan="3">Documentation to be provided(eventually differentiated between risk classes)</td></tr>
<tr><td>Validation guidance for one risk class</td><td>Validation guidance for another risk class</td><td>...</td></tr>
<tr><td>Example of an acceptable solution for one risk class</td><td>Example of an acceptable solution for another risk class</td><td>...</td></tr>
</table>

Figure 6.3. Structure of requirement block.

The requirement block represents a technical content of a requirement, including validation guidance. It is addressed to both a manufacturer and a notified body two purposes: (1) to consider the requirement as a minimal condition, and (2) to not place demands beyond this requirement.

Notes for a manufacturer:

- it is necessary to adhere (keep) to the main thesis of a requirement and defining comments;
- it is necessary to present the documentation required;
- acceptable decisions are examples of fulfillment of the requirement. Following them is not obligatory;
- the validation guidance is of an informative character;

Notes for notified bodies:

- it is necessary to adhere (keep) to the main thesis of a requirement and defining comments;
- the validation guidance has to be followed;
- the completeness of the documentation presented should be confirmed.

6.2.2 State of affairs in this field in leading countries of the world, immediate tasks and ways for their solution

At present the problems of metrological validation of software for controllable measuring instruments and devices, measuring systems, and measurement-computation or apparatus-program measurement complexes, are the most studied ones.

In addition to efforts of national metrology institutes, where specialists have prepared publications and developed normative documents in this field (the most successful ones in Germany, Great Britain, USA, Canada, Japan, Russia, and others), a powerful impulse for further development in this direction is the EC Directive on Measuring Instruments (MID) 2004/22/EC [327], issued in 2004.

In this document meticulous attention is paid to special software requirements for the following types of measuring instruments (see MID, Supplement III):

- water meters;
- gas meters and volume conversion devices;
- active electrical energy meters;
- heat meters;
- measuring systems for the continuous and dynamic measurement of quantities of liquids other than water;
- weighing instruments;
- taximeters;
- material measures;
- dimensional measuring instruments;
- exhaust gas analyzers.

The work in the field of validation of software, important from the point of view of metrology, began in national metrology institutes in the 1980s. To estimate the state of matters in leading countries of the world, as well as to reveal the problems, which have been solved or not solved, and arising difficulties, it is necessary to trace the history of development of this direction in each of the countries mentioned and in the world metrological community as a whole.

6.2.2.1 The International Organization of Legal Metrology (Paris, France)

The Organisation Internationale de Métrologie Légale (OIML) is an intergovernmental organization which includes 59 member-states and 57 countries as participants [238, S. Just, U. Grottker].

The OIML is the world-wide intergovernmental organization, the main goal of which is the harmonization of rules of metrological control applicable to national metrological services or organizations connected with them.

The tasks of OIML are as follows:

- to provide governments with the requirements and procedures which can be used in national legislation and regulation;
- to harmonize national requirements and procedures;
- to establish mutual confidence, acceptance, and recognition;
- to sett up international systems for estimation and certification of measuring instruments;
- to compensate for the insufficient competence and disinterestedness of users; error and falsification protection.

The OIML publications are divided into the following basic categories:

- International Recommendations (OIML R), which are model (frame) rules establishing metrological characteristics for definite measuring instruments, as well as indicating methods and equipment to control compliance with the requirements;
- International Documents (OIML D), which have an informative character and are intended to improve the activity of the above-mentioned metrological services;
- International Guides (OIML G), which are also informative and intended to apply particular definite requirements on legal metrology;
- International Basic Publications (OIML B), containing the rules according to which different structures and systems of the OIML operate.

By 2008 the OIML system had published 137 International Recommendations related to 107 categories of measuring instruments, which are, in a certain sense, the international standards in accordance with the Agreement on Technical Barriers to Trade of the World Trade Organization. These Recommendations are openly available on the OIML web site.

Among the OIML and some other organizations such as the ISO or IEC the joint agreements have been set up with the purpose of avoiding contradictory requirements. Hence, the designers, manufacturers, and users of measuring instruments, testing laboratories, and so on, can simultaneously apply both the OIML publications and publications of other organizations.

In the meantime, the OIML has issued the Document OIML D 31:2008 "General Requirements for Software-controlled Measuring Instruments" [181].

An doubtless advantage of this document, which carefully takes into account the specific character of metrology and terminology, is its logical statements, use of the experience of various countries in this field and connection with other documents of the OIML and international organizations on standardization, as well as with the requirements for measuring instruments in case of approving their type.

This work was carried out by experts from various countries and the experience gained by such organizations as the BIPM (in particular, documents [243–245, and others were prepared there], ISO and IEC [231–235, and others], as well as the WELMEC [541] and COOMET [123].

6.2.2.2 Physikalisch-Technische Bundesanstalt (PTB, Germany)

In 1991 at the PTB a department for metrological information was established. Its activities are devoted to the following fields: testing software and ensuring its quality, information technologies in legal metrology, exchange of data and safety, game machines, and voting systems [238].

As a result of the activities of this PTB department and of the Working Groups (WG) of the WELMEC, a number of new documents have been issued. Thus, there was issued a normative document by the WELMEC entitled "Software Guide" [541], which was developed, firstly, to render assistance to manufacturers of software-controlled measuring instruments and authorized bodies in assessing the quality of software and its compliance with the requirements of the EC Directive on Measuring Instruments MID [327].

The WELMEC, a cooperation of the EC and European Free Trade Association member-states services on legal metrology, believes that this guide is the best practical manual.

At the same time, we note the guide has only a consultative character and does not force any restrictions or additional technical requirements beyond those which are indicated in corresponding EC Directives. Moreover, alternative approaches are allowed.

It should be noted that the WELMEC Guide represents the most detailed and currently up-dated document, and provides examples of control tables for the fulfillment of requirements.

A drawback of the guide is the fact that it embraces the requirements for software of measuring instruments and measurement systems and does not cover software for solving other problems in the field of metrology. This refers to the software not strictly intended for definite measuring instruments and measurement systems and, for example, those which are intended for treatment of measurement data of key, regional, and interlaboratory comparisons of measurement standards, data obtained earlier from various measuring instruments and/or measurement systems, etc., as well as used for simulating a measurement experiment, although many requirements for software given in the guide [541] can also be applied to the solution of such problems.

At the PTB the analysis of the Profiles of Protection for Measuring Instruments (V. Hartmann, N. Greif, and D. Richter) was performed in accordance with ISO/IEC 15408. Profiles of Protection means a set of protection requirements, independent of application, for the categories of measuring instruments which correspond to particular needs, namely to safety, i.e., protection from an unauthorized access (authenticity), modification (integrity), and loss of access (accessibility).

It was taken into account that the *probability of attacks* increases with the increased complicated nature of systems, a greater number of software components, and links among the components, etc., as well as the fact that future measuring instruments will undoubtedly be part of communication networks.

The general criteria for assessment the safety (*protection*) of information technology included a protection structure (safe environment and protection objects), functional protection requirements, and requirements for confirming the confidence in safety (with introduction of their assessment levels).

On the basis of this analysis it was concluded that general criteria can be applied in metrology and are suitable tools for metrological software validation. At the same time, the general criteria do not take metrological aspects into consideration. Therefore their adaptation is needed, i.e., "resealing a defensive line" between the protection requirements and metrological functional requirements.

6.2.2.3 National Physical Laboratory (NPL, UK)

Among numerous NPL publications [36, 56, and others] there are some which we especially ought to mention, namely: Best Practice Guide No 1. Validation of Software in Measurement Systems. Version 2.1, March 2004. Software Support for Metrology. B. Wichmann, G. I. Parkin, R. M. Barker, NPL DEM-ES 014, January 2007.1 [56].

In addition, proper attention should be given to the joint publication of the PTB and NPL representatives [1, R. Parkin, N. Greif], where the authors intend to prepare a new guide for development and assessment of the measurement software, motivated by the lack of any widely recognized basis for assessment and comparison, as well as by the absence of a comprehensive international guide in this field for scientists and experts of measurement practice and science.

At the same time both the NPL achievements and PTB system of the prepared guiding documents on developing and evaluating software will be used, in spite of the following drawbacks: they have not become international guides or standards; they do not cover all the aspects of the life-cycle of software and are not sufficiently resistant to various revisions.

The structure of the new guide is to contain, apart from the main text of the guide, some additional guidelines which will elucidate separate aspects of the type of static analysis of a code. The main part of the guide is to be built on the basis of a risk approach. It is intended to be practical, brief, and self-sufficient, covering all the types of software intended for measuring instruments: purchased software packets, embedded software, instrument control, mathematical computation, and graphical user interface.

Moreover, it is be connected, where necessary, with international standards, following their procedures, definitions, requirements, and recommendations. For examples of this see [232–235].

This guide needs to consider these two interrelated aspects from the point of view of both the programming process and software product. By the aspect of the programming process we mean the acquisition of proofs in developing the software through preventive audits of this process. By an aspect of the software product we mean an analytic testing of a final (or some intermediate) software product or system.

A "standard" model of the programming process which takes into account the stages of the life cycle has to be taken from ISO/IEC 12207 as a base for the guide. At the same time only significant key fields of programming should be considered. By significant processes of the software development is meant the analysis of requirements, design of software, its application, testing of software, and its maintenance in the operation state.

Structurization of software life-cycle helps to divide the whole variety of the recommended technique and requirements for the process into categories, as well as to make simpler their selection. The guide will provide the procedures for risk assessment, based on [233, 235], using a widely recognized approach for determining the risk category (factors and levels of the risk) and the means to make these risks minimal.

The content of the guide has to fulfill the following:

- to determine risk categories on the basis of the risk factors specific to such a software with acceptable risk levels;
- to provide the characteristics for measurement software for each risk category;
- to trace risk factors according to standardized risk indices;
- to indicate what technique has to be used for each risk category and what degree of activity in using this technique has to be applied for each process of a life-cycle of software.

The authors' suggestions concerning the risk classes are discussed in detail in Section 6.6.

6.2.2.4 Canadian agency of industry "Measurements in Canada"

The report "Metrological Software. The Canadian Experience" [238, D. Beattie] includes the history of software development, requirements for software, experience of the agency of industry "Measurements in Canada", the drawbacks of existing requirements, and activity on further development.

In the section "History of development" five periods are distinguished: before software, period of early electronics, period of early devices based on software, period of the appearance of personal computers, and the period of the process of development.

Within the period that preceded the appearance of software, all instruments were of mechanical nature and fulfilled only basic functions. They were assessed on the basis of their response to known inputs.

During the period of early electronics the instruments fulfilled only basic measurement functions. They were assessed by the same method as mechanical instruments. In the course of time their models had changed insignificantly and there was no hardware for connection with other equipment.

During the period of early devices based on software, all the instruments were specialized (of the P type). They fulfilled only basic metrological functions and there was no connection with other equipment. No software updating was applied. For such instruments it was typical to have rather limited resources.

For the period when personal computers appeared, the following features became characteristic: appearance of computers on high levels of trade; as before measurements were realized with specialized instruments; measurement data were sent to a personal computer for further processing, and complexes, on the whole, were retailed as advanced technologies.

The period of development included: internal discussions, work with other government bodies, arrangement of an open forum on metrological software, formation of working groups on software (government working groups, working groups of developers and manufacturers, final users, and owners of instruments), development of a project of requirements, improvement of the project by legal experts up to its final version.

Tthe following conclusions were made:

- the functions of software can be divided into three large categories: measurement, calculation, and control, each function making its own contribution to measurement accuracy;
- the type approval cannot increase the significance of all software types;
- requirements specific to software are necessary for all disciplines;
- these requirements are applicable only for those software versions that are used for measuring instruments in which a personal computer (of the U type) is applied.

Measurement functions of software: software fulfills all steps necessary to handle a signal into readable information; no facilities are used to determine the accuracy of indicated values; availability of advantages due to checking and evaluating the type; possibility of providing output data for other devices for further use; measurement functions are completed at the first indication of measurement quantities; all instruments require approving their type.

Computational functions of software: the software is a receiver of basic information from measurement software; executor of the basic calculations, for example, the calculation of a general price or tare allowance; executor of more complicated calculations, e.g., determination of a liquid temperature transition, the accuracy of which can be easily checked. The computational functions of software are not necessarily of a measurement character; for example they include stock-taking and checking. The computational functions of software are subjected to changes, which happens quite often. The type approval procedure is less important. These functions can be realized in the process of type approval, if a given software is combined with some other measurement software or embedded into a specialized measuring instrument.

Functions of programmed control to receive basic measurement information from some measurement software and apply data obtained for performing a control of a measurement process. Frequently, software this kind is created by a user for a specific location or task. The functions of such a are critical to the process of accurate measurements; they are not easily evaluate in a laboratory situation and frequently change; the type approval is of less significance for such functions; and initial verification is important for them. At present all these functions are inherent to specialized measuring instruments.

Canadian requirements for software: a a general nature; no procedures developed for their support; inapplicable for specialized measuring instruments; measurement functions are evaluated through a procedure of type approval; computational and control functions are evaluated through primary and subsequent checks in the process of operation.

Requirements for measurement software: compatibility; adequacy of system resources; integrity of code and configuration parameters. The code has to be protected against changes. Integrity tests of input/output signals have to be performed and in the event of a computer failure there should be no loss of measurement data. A built-in event recorder (audit trail) and an indicator of some other modes apart from the normal operational mode are required. Moreover, video indication facilities are necessary. Identification numbers can be displayed or printed out. Software this kind can be modified by an applicant for correcting problems that have appeared without any subsequent evaluation.

Requirements for computational software: measurement information has to be identified by an approved device. All operations should be recorded; records have to include all information needed for confirming calculations. Data and information which accompany this data have to be protected against loss in case of disconnection from a supply source or a malfunction of a computer.

Requirements for control software: measurement data and results should not be lost in case of computer failure or computer components trouble. All parameters of products which should be measured are measured. A model test under operating conditions has to be provided for confirming the correctness of performing functions.

Knowledge for today: it is impermissible to approve measurement software standing apart. The definition principles and principles of performing checks have been accepted and need to be approved.

Disadvantages of the Canadian requirements: they are not applicable for any specialized measuring instruments (devices of the U type which are "repackaged" as the type P devices, many of which are used by open networks for communications). There are some problems connected with updated software (unexpected consequences that are difficult to be detected and/or to give off an alarm).

Prospective development activity: for the time being there are no firm plans. The requirements for software of specialized measuring instruments have to become a part of long-term documents of OIML R. It is possible to "redefine" the type P instruments in order to exclude "repacked" personal computers. Addressing of updating software should be performed only for specialized devices. A revision and enhancement of existing requirements should take place. Generally recognized methods of checking should be accepted.

6.2.2.5 National Institute of Standards and Technologies (NIST, USA)

In [238] A. Thompson. describes the trends of software of the National Conference on Weights and Measures. In the US the structure of regulation is characterized by the following: almost all executive authorities are based on federal and local jurisdictions. The NIST does not possess any executive authority but is responsible for advancing unified measurement standards of weights and measures to promote the trade. The National Conference on Weights and Measures is the main developer of standards for the legal metrology for the US.

In 1901 at the *National Conference on Weights and Measures* (NCWM) a decision was made to found an institute (at present the NIST) whose purpose is to provide technical support for the NCWM in developing requirements and practical unified measurement standards on weights and measures. The NIST consists of official representatives in the field of weights and measures, manufacturers of instruments, as well as of representatives of industry and the federal government. In particular, it contains National Technical Committees (NTC) based on sectors such as the weighing, measuring, software ones, and others.

Zhe operational requirements for software of measuring instruments were first fixed in Handbook 44 (HB 44). Then Publication 14 (Pub 14) on type approval was issued. HB 44 and Pub 14 are annually revised and republished.

History of NCWM participation in providing software for industry and science: In 1989 the audit logs for electronic/software regulation and configuration parameters were accepted. In early 1995 in the NTCs general criteria for independent software were estimated and installed. Between 1995 and 1997 the First Working Group on software was active, and from 1997 until 1999 the Second Group continued this work. In 2005 the National Technical Committee on type approval and the sector on software were established.

Tasks of the sector on software: to understand the use of software applied in weighing and measuring devices; to enhance the requirements of HB 44 for software and verification facilities in the process of operation and requirements on protection and identification; to develop the criteria given in Pub 14 by including marking, protection, and functions important from the point of view of metrology; education of officials.

Audit logs (1989) contain: requirements based on accessibility; properties of sealing and parameters influencing metrological integrity (adjustment capabilities acting upon accuracy; choice of operations that influence the compatibility with HB 44); maintenance of recording sealed parameter changes.

Class of an audit log depends on: the easiness at which an HB 44 swindle can be detected; soundness of the capability not to detect the swindle for sure.

In the USA the basic risk is considered based on the fact f whether or not it is possible to remotely reconfigure an instrument. The requirements are based on needs for checking in the process of operation and are divided into three categories:

- *Category* 1 (physical sealing/counters): not capable of for performing remote configuration.
- *Category* 2 (physical sealing/counters): limited capabilities for performing remote configuration; a sealed hardware controls access to remote communications.
- *Category* 3 (registrator of events): unlimited distance capabilities; unlimited access to configuration parameters and adjustment facilities.

Sector of software support: beginning from 2005 a meeting of this sector takes place twice a year where OIML and WELMEC achievements are considered. This sector is concerned with the capabilities of national technical laboratories to perform checks of validated software under operating conditions and type approval procedures. At present the program code analysis does not refer to its functions.

Sector of discussions and recommendations is engaged in solving the following problems. The sector provides third persons with permissions to use the general software criteria in addition to a validated hardware part; devices of the U and P types; there has to be a possibility for the metrological version to be shown on a display or printed out. Moreover, the activity of this sector includes issues of software separation; identification of certified software; software protection as well as protection of software interfaces; enhancement of control audit for software, software authentication and updating.

Additional considerations: all requirements have to be applied to devices of the U type; the range of options for devices of the P type has to be applied according to risk. Concerning components, systems, and outside software, it should be said that differences between the USA and OIML normative documents need to be reduced to a minimum.

6.2.2.6 Japan (The National Metrology Institute of Japan and the National Institute of Advanced Industrial Science and Technology)

The report "Current state and perspectives of developing software expertise of taximeters, weighing devices and other specialized measuring instruments in Japan" [238, S. Matsuoka] is devoted to the following issues: fundamental features of the measurement legislation in Japan; reasons why software expertise is presently important for Japan; some rather practical issues on the expertise carried out for software of weighing devices.

The Measurement Law in Japan: rules for carrying out an examination and verification of measuring instruments subjected to legal control according to the Japan Industrial Standards (JIS) for these measuring instruments. To revise laws is difficult, but to revisions of standards is comparatively easily done.

Examples of specific measuring instruments:

(1) for trade: nonautomatic weighing instruments; taximeters; fuel and oil gauges (including fuel dispensers for automobiles);

(2) municipal meters: water meters, gas meters, heat meters, and active electricity gauges;

(3) measurements of environmental parameters: vibration level gauges, brightness meters, and sound level meters.

(4) clinical measurements: clinical thermometers, blood pressure measuring instruments.

The measuring instruments listed above have to be checked before putting them into operation.

Reasons why at present software examination is important in Japan: the OIML Recommendation P76-1 which recently appeared includes: software examination; necessity to take part in MAA (Message Authentication Algorithm) for OIML P76; revision of Japanese industrial standards to bring them into agreement with OIML P76-1; introduction of the scale type expertise for trucks and railway platforms, including weight indicators. The software examination is also important for other specialized measuring instruments, although this is not a very urgent task.

Summary: Not long ago in Japan practical software examination was initiated with regard to measuring instruments subject to the legal control.

At present the software examination of weighing devices is extremely current because of the revision of Japanese industrial standards for nonautomated weighing measuring instruments.

However, there are some problems concerning what should be done with modern computer systems (Windows, data bases, etc).

6.2.2.7 Russia (D. I. Mendeleyev Institute for Metrology – VNIIM, St. Petersburg; and the "All-Russian Research Institute of Metrological Service" – VNIIMS, Rostechregulirovaniye, Moscow)

The first science and technical publications devoted to the problems of software validation of measuring instruments and measurement systems appeared in Russia beginning in 1985 [482, 502, etc.]. The first normative document was the recommendation document "Government system of ensuring measurement uniformity. Attestation of algorithms and software for data processing" [332], developed at the VNIIM. Additional normative documents in this field [196, 199, 200, 333–337, and others] were also issued.

In metrology the organizational frames of software validation fulfillment can be:

- testing and validation required for software controlled measuring instruments that fall under governmental metrological control and supervision in accordance with the Russia Federation law on "Ensuring the uniformity of measurements". This refers mainly to measuring instruments of great importance for protecting the life and health of citizens, preservation of the environment, defense of the country, and social order;
- the validation of measuring instruments software is also necessary in those cases where in a specification for development and customer acceptance of a measuring instrument there is a list of normative documents, in accordance with which tests of measuring instruments should be carried out and which contain requirements for the validation of their software (in particular, [200]);
- !in other cases the validation of software used in measuring instruments is voluntary and performed within the framework of the existing system for the voluntary certification of measuring instruments.

At the same time, in this field there is a set of unsolved problems. In particular, at present in accordance with established procedure, the risk and degree of test rigidity is performed by an organization carrying out the validation of software in conformity with a customer order. Nevertheless, it is evident that to choose the level of requirements it is necessary to apply an expert method and to acquire a group of specialists competent in the corresponding field of measurements.

Further development of the generally accepted normative base in the field of technique and technology requires solving all existing problems of fulfillment the work on validation of the software used in metrology.

There is do doubt about the basic necessity for solving a number of tasks mentioned in Sections 6.2.2.2–6.2.2.6.

Reasoning from the above, in this chapter it is useful to give the contents and requirements of the OIML Document [181], partly specifying them on the basis of Guides of the WELMEC [541] and using other normative documents in this field which are at present in force.

6.3 Approaches to evaluating precision parameters of software used in metrology

6.3.1 Sources of uncertainty and methods of their evaluation when applying data processing software for obtaining a measurement result

6.3.1.1 Specification of software

By a software specification [56] we mean the mathematical description of a problem which a software produces. The problems being solved can be different. Among them, for example, there can be computation of a mean or mean-root–square deviation of a number of values, evaluation of regression dependence parameters, computation of a function integral, etc.

In addition to formulating a problem, an approach and method for its solution has to be described. For example, in evaluating parameters of regression dependence, the ordinary method of least squares (OLS) is applied, which has many modifications, such as classical OLS, weighted OLS, and generalized OLS. It is desired, when the opportunity presents itself, to present a formalized algorithm, with which the method selected for solving the problem can be realized.

In other words, it is necessary to determine: the vector of input data $\boldsymbol{X}$, vector of output data $\boldsymbol{Y}$, function dependence establishing the compliance between the input and output in an explicit or implicit form: $\boldsymbol{Y} = f(\boldsymbol{X})$ or $g(\boldsymbol{X}, \boldsymbol{Y}) = 0$. Hereinafter, for the purpose of simplification, the explicit method of setting the functional dependence of a scalar output quantity on the input quantity vector will be considered:

$$Y = f(X_1, \dots, X_n). \tag{6.3.1}$$

To such a form the rather great number of problems of processing results of measurements are reduced, among which it is possible to single out the following ones.

- Equation (6.3.1) coincides with the equation of indirect measurements, which is used in the GUM [243]. Within the GUM this equation is also the algorithm of finding a numerical value of the output quantity, which is, generally speaking, not always correct. In particular, in Appendix 1 of the GUM [244] a different algorithm for finding the numerical value of a measurand is considered, also based on the measurement equation, but the numerical estimates of the output quantity will slightly differ.
- The problem of evaluating coefficients of calibration dependence of measuring instruments is also reduced to the form in (6.3.1) for evaluating each coefficient. In this case the input data are the results of measurements of the input and response of measuring instruments within a certain range.

- The problems of processing of time series: for example, the evaluation of covariation functions and function integrals, presented by the results of value measurements in discrete time moments.

The software specification does not include any description of concrete methods of the software realization of the algorithm indicated in the software description. The formulae, equivalent from the point of view of mathematics, are indistinguishable in respect to the software specification. Therefore, in testing the functional abilities, the software is frequently represented in the form of a "black (or opaque) box".

In actual fact, it would be more correct to represent the software as a "gray (or semiopaque) box" where the "opaque color" is defined by a degree of refining the description of the algorithm realized with the software. Thus, the attempts to make the "box" completely "transparent" are natural. This would allow analytical methods to be applied for investigating the functional properties of the software, the influence of the latter on the accuracy of a final result, and thereby to increase the reliability of the investigation results.

In practice this would be implemented by performing an analysis of the initial code of the software, significantly complicating the problem of software investigation. Therefore, the analysis of the initial code is applicable only in particularly important situations of software use. In most practical cases the reliability of software validation is achieved due to the joint use of methods of analytical investigation of algorithms realized in the software and functional checks of the software.

6.3.1.2 Sources of the uncertainty; evaluation; control of the combined uncertainty

In evaluating the uncertainty of a measurement result obtained with software for data processing of a measurement experiment it is necessary to take into account the following sources of uncertainty:

- uncertainties of measurement experiment data which are the input quantities for an applied software, as well as uncertainties of other input quantities, i.e., calibration coefficients, reference data, and others;
- uncertainties related to the algorithm selected for solving a problem. The algorithm influences the accuracy of the final result in two ways. On the one hand, in accordance with the algorithm, the transformation of the input quantities into the output quantities takes place and, correspondingly, the uncertainties of the input quantities are transformed according to a definite rule. On the other hand, some additional uncertainties can arise. These uncertainties are caused by approximate estimates, inherent in the algorithm, or by assumptions related to the distribution law, which is important for the statistical algorithms of data processing;

- uncertainties related to the realization of the algorithm selected with a particular algorithm;
- uncertainties related to postulates, models being accepted when formalizing a measurement problem and making up a measurement equation.

These four sources exert a joint influence on the accuracy of a final measurement result. With the exception of the influence of the initial data, it is not practically possible to completely separate the influence of the remaining sources of uncertainties caused by models accepted, algorithm, and its realization in a software. Therefore, the combined standard uncertainty cannot be represented in the form of four independent summands caused by the above reasons.

Also, in its evaluation, some sources of uncertainty begin to form intersecting groups, which leads to a pessimistic estimate of the combined uncertainty.

Equation (6.3.1) is the main expression for calculating the uncertainty of the output quantity depending on the uncertainties of the input quantities. On the other hand, equation (6.3.1) itself is a certain approximation of the interconnection of the output and input quantities. The choice of this approximation is dictated by a measurement problem and the requirements for the accuracy of a measurement result.

The corresponding component of uncertainty is mentioned in the Guide [243], but to obtain this component is rather difficult, since it "falls out" from the procedure proposed in the Guide and based on the postulated model. In a number of cases this component can be evaluated from "above", for example, by replacing a function integral with a sum.

In this case it is possible to write the uncertainty caused by the discretization step and selected method of integration in an explicit form. In other cases attempts to obtain so uniquely an estimate of the approximation accuracy are unsuccessful, for example, in postulating a linear form of the calibration dependence which in reality is a nonlinear function. In this case after performing the computation in accordance with the model accepted, a check of the experimental data compliance with the measurement model is performed according to the criterion χ^2.

The positive results of the check indirectly confirm that the uncertainty of setting the model is insignificant as compared with the uncertainties of input data. This example illustrates the general approach to the evaluation of the combined uncertainty, i.e., the contribution of a part of the main sources is evaluated quantitatively and as to the remaining sources efforts are made to show that their contribution is not significant. This means that the software functions correctly and corresponds to the problem being solved.

By the example of the problem of integrating the continuous function according to the results of measuring its values "in knots", let us illustrate the evaluation of the uncertainty components caused by the uncertainties of initial data and selected algorithm of their processing.

In Table 6.2 the estimates of transformed uncertainties and upper boundaries of the methodical error of integration are given for two methods: the method of rectangles and Simpson's method (see Chapter 4). At transfer from the boundaries to the standard uncertainty, the uniform law of distribution was assumed: $u^2(S-I) = \frac{\Delta_i^2}{3}$.

Table 6.2. Evaluation of the uncertainty components of the algorithms for integrating according to function values measured in "knots".

	Method of rectangles	*Simpson's method*
Quantity to be evaluated	$I = \int_a^b y(t)dt$	
Model of input data	$y_i = y(t_i) + \varepsilon_i \quad \varepsilon_i \in N(0,\sigma^2) \quad i = 1,\dots,m \quad u^2(y_i) = \sigma^2$	
Quadrature formula	$S_1 = \frac{b-a}{m}\sum_{i=0}^{m-1} y_i$	$S_2 = \frac{h}{3}[y_0 + 4y_1 + 2y_2 + {}$ $+4y_3 + \cdots + 4y_{m-1} + y_m]$
Boundaries of the methodical error	$\Delta_i^{(1)} = \left\lvert S - \int_a^b y(t)dt \right\rvert \le \frac{(b-a)^2}{2m} \cdot \max_{[a,b]} \lvert y'(t) \rvert$	$\Delta_i^{(2)} = \left\lvert S - \int_a^b y(t)dt \right\rvert \le \frac{(b-a)^5}{90m^4} \cdot \max_{[a,b]} \lvert y^{(4)}(t) \rvert$
Standard uncertainty of the output quantity	$\frac{2\sigma(b-a)}{\sqrt{m}}$	$\frac{2\sqrt{2}}{3}\sigma(b-a)\sqrt{\frac{5m-1}{m^2}}$

The obtained estimates of the methodical component and component caused by transformation of the measurement result uncertainties are the a priori estimate of the software accuracy. Testing the corresponding software realization of the method of rectangles and Simpson's method will be made with the purpose of showing that the software realization itself does not significantly contribute to the combined uncertainty.

6.3.1.3 Computation of the transformed uncertainty

The uncertainty component caused by the uncertainties of input data can be evaluated analytically in accordance with the "law of transforming (or propagating) uncertainty" [243], or this can be done by statistical simulation in accordance with the "law of transforming distributions" [244].

The process of evaluating the uncertainty of the output quantity (magnitude, value) is schematically, i.e., in the form of a sequence of steps, as shown in Table 6.3.

Let us compare the advantages and disadvantages of both approaches to the computation of measurement uncertainties.

Table 6.3. Evaluation of measurement uncertainty in accordance with [173] and [174].

Law of propagation (transforming) uncertainties	*Law of transforming distribution pdf – probability density function)*
1. Input data introduction: $\{x_i, u(x_i)\}_{i=1}^{N}, \{\mathrm{cov}(x_i, x_j)\}_{i,j=1}^{N}$ $u(x_i) = u_i, \mathrm{cov}(x_i, x_j) = u_{ij}$	
2. Evaluation of the first derivatives of the function $Y = f(X_1, \ldots, X_N)$, describing the transformation of input data into the output data: $\left\{ \frac{\partial f}{\partial X_i} \lvert x_i \right\}_{i=1}^{N}$	2. Modeling *pdf*(x) on the basis of input data. *Example*: the totality of independent random quantities having a normal distribution: $pdf(X_i) = \frac{1}{\sqrt{2\pi} u_i} \exp\left\{ -\frac{(x - x_i)^2}{2u_i^2} \right\}$
3. Computation of the combined standard uncertainty of an output quantity: $u^2(Y) = \sum_{i=1}^{N} \left(\frac{\partial f}{\partial X_i} \right)^2 u_i^2 + \sum_{i \neq j} \frac{\partial f}{\partial X_i} \frac{\partial f}{\partial X_j} u_{ij}$	3. Obtaining the distribution of an output quantity *pdf*(Y) by the Monte Carlo method: • modeling sets of input data $\left\{ x_1^{(k)}, \ldots, x_N^{(k)} \right\}_{k=1}^{M}, \quad M \approx 10^6$; • obtaining output data $\{y\}_{k=1}^{M}$ with the help of the software processing of input data sets; • construction of *pdf*(Y)
4. Computation of the expanded uncertainty: $U = ku$	4. Computation of the standard and expanded uncertainty on the basis *pdf*(Y): $u(Y),\ U(Y)$
5. Representation of a result: $y = f(x_1, \ldots, x_N), U$	5. Representation of the result: $E(Y), U(Y)$

The merits and ease of applying the "law of propagation uncertainties" are evident. This is the analytical method which allows the output quantity uncertainty to be expressed in an explicit form as the function of output quantity values at a change of these values in a certain field. On the one hand, this method is rather simple, but on the other hand, it covers the whole field of variation of input quantity values.

The necessity of linearizing the measurement model should be considered as a disadvantage of this method. But strictly speaking, this disadvantage can be eliminated if an additional uncertainty component, caused by the linearization of the measurement equation, is introduced. It is not difficult to evaluate it using the remaining terms of expansion in a Taylor series.

But as a matter of fact, here a disadvantage which is difficult to overcome is the impossibility of obtaining in an analytical manner the law of output quantity distribution, which is required for correctly calculating the expanded uncertainty. Therefore to calculate the expanded uncertainty, a "will" solution has been accepted, i.e., the multiplication of the combined uncertainty by coefficient 2 for confidence level 0.95 and by coefficient 3 for confidence level 0.99, respectively. It is believed that this simplified approach to calculating the uncertainty for the most part of metrological problems will be the dominant one.

The advantage of the distribution transformation approach with modeling according to the Monte Carlo method is the fact that it does not make any assumption concerning the linearity, does not require any linearization of the function f, and allows an actual distribution of the output quantity to be obtained.

It should be noted that when applying the approach of "distribution transformation" the evaluation of the uncertainty of a measurement result obtained using the software realizing (6.3.1) is made rather than that of the uncertainty of a quantity obtained in accordance with equation (6.3.1). With such an approach some additional sources of uncertainty related to the necessity to use auxiliary software arise:

- software providing the formation of "standard" input and output quantities;
- software providing the computation of output quantity values at given input quantities in accordance with equation (6.3.1);
- software for processing an output quantity data array with the purpose of determining its distribution law, from which a measurement result and corresponding uncertainty, as well as standard deviation and expanded uncertainty, respectively, are determined.

In connection with the above, one more argument can be added in favor of the approach "transformation of distributions". At first glance, it seems to be attractive that the process of evaluating the transformed uncertainties of a measurement result can be combined with testing the software used for getting this result.

6.3.1.4 Testing of the software

The purpose of testing the software is to check whether it functions correctly. In the metrological aspect this is mainly the check of its adequacy for obtaining a result with the required accuracy. In practice the task of checking the software is frequently a constituent part of the task of testing measuring instruments, measurement systems, or MM certification. Therefore the generalized problem of determining the accuracy characteristics of a measurement result, obtained with the help of the software validated, is of current importance. This problem will be considered in this subsection.

Stating the problem thusly ("generalized testing") has a broader scope than the traditional notion of testing software as a confirmation of whether or not a software is suitable for solving a particular measurement problem, i.e., it does not introduce a significant contribution into a measurement result error ("testing in the narrow sense").

The statement of the problem of "generalized testing" is quite correct, since in the long run it is just this evaluation of the combined uncertainty of measurement that is required. An alternative to this approach is a two-stage approach, if it can be implemented in practice.

In the first stage the metrological validation of the algorithm of experimental data processing is performed, at the second stage the software is tested, the estimates of accuracy obtained at the first stage being the checkpoint for determining an admissible divergence between the "standard" results and results obtained using the software under test. The approach to validation of the algorithms of data processing is considered in Section 6.3.2.

Let us consider in more detail the procedure of the "generalized testing". To perform such a test it is necessary to have:

- "standard" input and output data (perhaps, some software is needed to generate such "standard sets" which also have to be validated);
- "standard" software for calculating a measurement result in accordance with the specification of tested software;
- software for processing output data arrays in order to obtain in the general case the output quantity distribution law and its characteristics, i.e., a mean and mean square deviation.

Testing is performed using the "standard" data. The basic recommendations concerning the formation of "standard" data are formulated in [56].

Mechanisms of getting the "standard" sets have to provide: detection of defects in a software under test; the design of test sets with properties known in advance (for example, a degree of noisiness); the possibility of getting the test sets of input data for which "standard" outputs are known and which, in a certain sense, are "close" to the input data sets arising in the process of practical use of the software; the possibility of getting a great number of test sets providing a sufficient coverage of ranges in which the inputs of software support can change.

The input data parameters are: the volume of measurement information (the number of elements of a sample), range, sampling increment in evaluating parameters of processes, level of noise, law of noise distribution, and others.

In testing the software the "standard" input and output data, determined in advance, are used, or to obtain the "standard" output data the "standard" software is applied. In the documents developed by NPL (Great Britain) the method of "zero-space" of "standard" data forming is proposed. The input "standard" data are formed in such a way that the "standard" output data remain unchanged. This method is called "zero-space". According to this approach, a single output set (vector) of "standard" data corresponds to different sets of input "standard" data.

Let us consider this in more detail. Assume that a problem is being solved, according to which the vector $y = [y_1\ y_2\ \cdots\ y_m]^T$ of the results of the observations carried out at discrete time moments $x_i, i = \overline{1,m}$ is given. With all this going on, let us consider that the results y_i are defined by a linear dependence $y_i = b_1 + b_2 x_i + r_i\ i = \overline{1,m}$, where $b_{1,2}$ the parameters to be evaluated are, r_i is the random errors arising in the process of measurement.

Let the system of equations be rewritten in the matrix form

$$y = Ab + r, \tag{6.3.2}$$

where A is the matrix of observations, y is the vector of observation results, b is the vector of model parameters, amd r is the vector of remainders:

$$A = \begin{bmatrix} 1 & x_1 \\ 1 & x_2 \\ \vdots & \vdots \\ 1 & x_m \end{bmatrix}, \quad y = \begin{bmatrix} y_1 \\ y_2 \\ \vdots \\ y_m \end{bmatrix}, \quad r = \begin{bmatrix} r_1 \\ r_2 \\ \vdots \\ r_m \end{bmatrix}, \quad b = \begin{bmatrix} b_1 \\ b_2 \end{bmatrix}.$$

It is known that the solution of the linear regression problem according to the least square method is characterized by the fact that

$$A^T r = 0.$$

Consequently,

$$\sum_{i=1}^{m} r_i = 0, \quad \sum_{i=1}^{m} x_i r_i = 0.$$

Let N be the matrix, the columns of which are the base vectors of the zero-space of the matrix A^T. Then $A^T N = 0$. Let $r = Nu$, then for different vectors of observation results y and $y + r$, similar values of the parameters b_i are obtained: $A^T A b = A^T y$ and $A^T (y + r) = A^T y + A^T N u = A^T y$.

Thus, the algorithm of constructing test sets of data is as follows:

(1) the vector of observation results $y_0 = Ab$ is calculated;

(2) a matrix N, the columns of which are a basis of zero-space for the matrix A^T, is constructed;

(3) the vector of remainders $r = Nu$, is formed, where the vector elements u are generated with the help of a transducer of random numbers;

(4) the normalization of the vector r is performed in such a way that it and its components have a distribution with a given mean and mean square deviation;

(5) the vector of observation results y is formed in accordance with the formula $y = y_0 + r$.

This approach allows the correctness of the applied software to be controlled. To express the influence of the software on the accuracy of a result quantitatively, the following measures [56], which are evaluated using the "standard" data or compared with the results obtained with the "standard" software, are used.

Absolute measures of accuracy
Let us have

$$\Delta y = y^{(\mathrm{test})} - y^{(\mathrm{ref})}, \tag{6.3.3}$$

where $y^{(\mathrm{test})}$ means the test output results and $y^{(\mathrm{ref})}$ means the "standard" output results corresponding to the "standard" input vector x.

Then $d(x) = \|\Delta y\| / \sqrt{n}$ represents the absolute measure of the test result deviation from the "standard" result at the "standard" input vector x.

Relative measures of accuracy
Let the number of accurate significant digits in the "standard" results corresponding to the "standard" input vector x be denoted as $M(x)$. Then the number of coinciding digits in the test computation results and "standard" results are calculated using the formula [115]

$$N(x) = \min\left\{ M(x), \log_{10}\left(1 + \frac{\|y^{(\mathrm{ref})}\|}{\|\Delta y\|}\right)\right\}, \quad \text{provided } M(x) \neq 0,$$
$$\text{and } N(x) = M(x) \text{ otherwise.} \tag{6.3.4}$$

Executing characteristic is the quantity

$$P(x) = \log_{10}\left(1 + \frac{1}{k(x)\,\eta}\frac{\left\|y^{(\mathrm{ref})}\right\|}{\|\Delta y\|}\right), \tag{6.3.5}$$

where η is the computation accuracy and $k(x)$ is the coefficient of the problem stipulation, which represents the quotient of division of a relative output by a relative input change

$$k\,(x) = \frac{\|\delta y\|}{\|y\|} \Big/ \frac{\|\delta x\|}{\|x\|}.$$

The executing characteristic is a number of accurate significant digits, "lost" due to test computations as compared to the results obtained with software realizing the optimally robust algorithm. For the "standard" software it is assumed that $\delta y \approx J(x)\delta x$, where $J(x) = \{\partial f_i / \partial x_j\}$ is the Jacobian of the function $f(x)$.

In other words, the "standard" software is interpreted as a completely "transparent" box, for which the dependence of output data on input ones is described by the function $f(x)$. The additional errors caused by the "standard" software are only the errors of rounding $\|\Delta, x\| \approx \eta \|x\|$, where η is the relative accuracy, and consequently

$$\begin{aligned}\|\Delta y_{\mathrm{ref}}\| &= J(x)\|\Delta x\| = J(x)\eta_{\mathrm{ref}}\|x_{\mathrm{ref}}\| \\ &= k(x)\eta_{\mathrm{ref}}\|y_{\mathrm{ref}}\| \end{aligned} \tag{6.3.6}$$

If the software tests are executed on the basis of the "standard" data (input and output quantities) without any use of the "standard" software, then at a known (6.3.6) computer accuracy of the software under test, the limit accuracy of output data, which is possible at steady realization of the algorithm by the software under test, can be evaluated with the analytical method

$$\begin{aligned}\|\Delta y_{\mathrm{lim}}\| &= J\,(x)\,\|\Delta x\| = J\,(x)\,\eta_{\mathrm{test}}\,\|x_{\mathrm{ref}}\| \\ &= k(x)\eta_{\mathrm{test}}\,\|y_{\mathrm{ref}}\|\,. \end{aligned} \tag{6.3.7}$$

Comparing this quantity with the really observed deviations of the output data of the software under test and with the "standard" output data it is possible to evaluate the accuracy lost due to realization of the algorithm by the software under test:

$$\frac{\|\Delta y\|}{\|\Delta y_{\mathrm{lim}}\|} = \frac{\|\Delta y\|}{k\,(x)\,\eta_{\mathrm{test}}\,\|y_{\mathrm{ref}}\|}.$$

Consequently, in (6.3.5) it is necessary to use an index of the computation accuracy of the "standard" software or the one which is being texted, depending on the method of testing.

It is also necessary to note that the accuracy characteristics given above differ from the expression of the measurement uncertainty which has been accepted for metrology.

Therefore, they cannot be used directly as the estimates of the uncertainty component introduced by the software. Moreover, they are not intended for this purpose. In international documents and reports on software validation these characteristics are called "measures of behavior (functioning)" of software which are used exclusively for making a conclusion with regard to suitability of the software for solving a particular measurement problem.

Nevertheless, if it is necessary to characterize the software contribution into the combined uncertainty quantitatively, then it is possible to suggest the following method. As a result of testing, a set of divergences $\{\Delta_i\}$ are obtained which correspond to different values of input quantities. These values are selected in such a way that ranges of their change overlap. The standard uncertainty, caused by realization of the software, can be estimated as $u_{\mathrm{soft}} = \frac{\max \Delta_i}{\sqrt{3}}$.

It is necessary to stress that this estimate is "from the top", namely the pooled estimate of a number of influencing factors, since it is impossible, for example, to single out the component caused by forming the "standard" data. Furthermore, if the "standard" sets of data have been generated without any dependence on the algorithm of data processing used in the software, then the divergences between the "standard" output values and the values obtained with the help of the software under study reflect not only the uncertainty contributed by the software, but also the methodical errors of the algorithm of data processing.

If the software is tested in the process of testing measuring instruments for which the software is an integral part, then the divergences considered result from a great number of uncertainty components. Therefore, it is necessary to analyze the divergences in each particular case in order to answer the question of which components they really characterize.

The basic sources of uncertainty, requiring a quantitative estimate, are the uncertainties caused by the input data and algorithm (method) used. In the problem of "generalized testing" the software, the "zero-space" method is insufficient, since it is necessary not only to reveal the software influence on the accuracy of a final result, but to evaluate exactly the combined uncertainty of the final result.

In the most general case this uncertainty is expressed by the law of output quantity distribution. To obtain the law of output quantity distribution, the input data have to simulate an error of measurements just at the algorithm input. At the same time, the volume of modeling has to be of the order of 10^6 iterations in order to provide the reliability of the estimate of the output quantity distribution.

In Figure 6.4 the scheme of testing the software is shown. In this scheme the "standard" software (SSW), software under test and simulation input data are represented. Use of SSW allows the output quantity uncertainty to be evaluated with the Monte Carlo method and the output data of SSW and those of the software under test to be compared with the purpose to check the last one. Such a check can be performed on the basis of real data, but this is insufficient for obtaining the output quantity distribution. The comparison of distributions of the output quantity for SSW and the software under test allows a systematic bias to be evaluated in the results of the last one, and scale of the corresponding distributions to be compared.

Thus, it can be summarized that in order to obtain the most efficient solution of the problem of evaluating the accuracy of a measurement result obtained with software, it should be divided into three stages in accordance with the stages of evaluating the basic components of uncertainty.

1. In the first stage a check of measurement equation correctness (6.3.1) and its correspondence to a particular measurement problem is carried out. In other words, the validation, i.e., the determination of the suitability, of the measurement model is performed. If some simplifications of the model are accepted, then the final result uncertainty caused by these approximations is evaluated. It should be noted that if measurements are carried out according to a certified procedure of performing measurements, then the problem of the first stage has already been solved.

2. In the second stage the evaluation of the "transformed" uncertainty of a measurement result is performed on the basis of input data uncertainties. The method of evaluation is selected based on the form (of nonlinearity) of the measurement equation.

When the Monte Carlo method is used ("law of distribution transformation") the software inevitably influences the uncertainty computation results. But an a posteriori

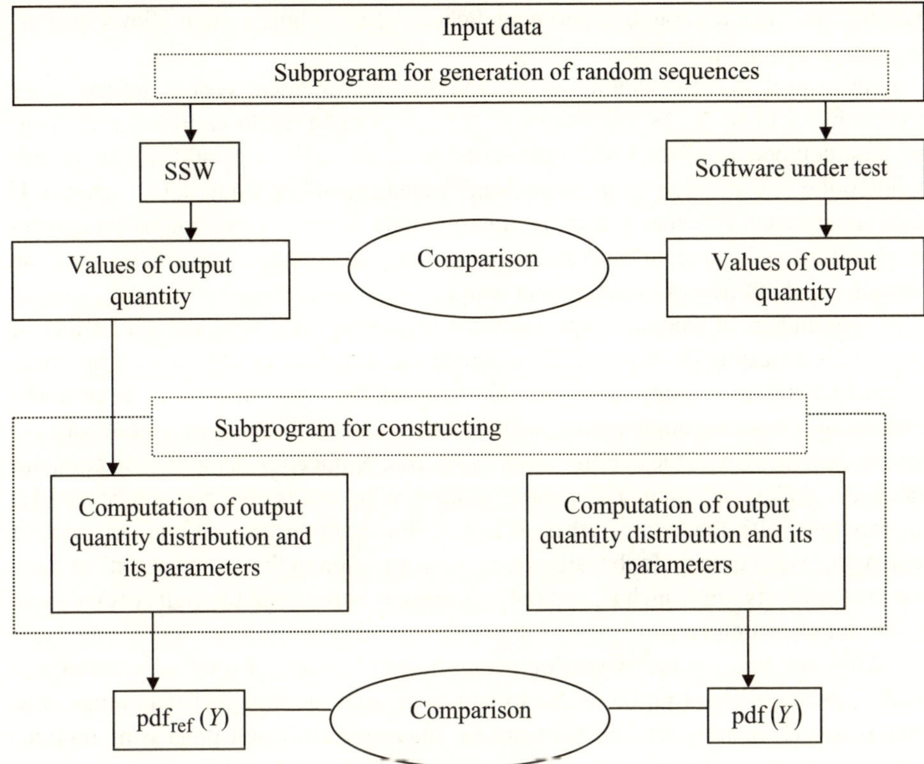

Figure 6.4. Scheme of "generalized testing".

check of the software using the "standard" data is needed (with use of the "standard" software), since application of the Monte Carlo method does not allow any systematic biases of the output quantities to be revealed. The content of the second stage was a constituent part of the MM certification if it was performed.

3. The third stage is testing the software, comparing the results obtained on the basis of the "standard" input sets with the "standard" output results. Testing allows unsuspected uncertainty sources to be revealed which are connected not only with the software but also with other influencing factors which for several reasons could not be evaluated at the a-priori stages. If SSW is used, then the second and third stages are naturally combined into one stage and the Monte Carlo method is applied.

Let us once again stress the fact that the testing of software should not be considered to be the evaluation of the uncertainty component contributed by the software, but as the procedure of checking its compliance with the requirements of a particular measurement problem. In the cases where the software is embedded, testing it can be carried out within the framework of the testing (validation, calibration, verification) of measuring instruments or measurement systems with the help of common measuring instruments. Such a through experimental check of the whole system allows the correctness of the applied software support to be defined.

With respect to the evaluation of information measurement system accuracy, the approach described in [482] is worthy of note. According to this approach each computation component of the IMS is suggested to be provided with subsoftware of calculating the uncertainty (characteristics of "transformed" or "inheritable" errors). If such an approach is realized, then the measurement system in addition to a measurement result would give an estimate of its accuracy in complete accordance with the present-day definition of measurement result.

At present a great amount of specialized software for calculating the uncertainty of measurement results is available. To evaluate the influence of software on the measurement result uncertainty, it is necessary to check the conformity of the uncertainty computation method, implemented with this software, with the requirements of normative documents and the conditions of its specific application, as well as to compare the results with some "standard" data or, at least, with results obtained independently, for example, with the help of other software, similar in destination and comparable with respect to accuracy. In the latter case, it is a question of the "comparison" of various kinds of software, which is, certainly, of current importance for high accuracy and important measurements.

Taking into account the expanding perspectives of using software in metrology, it seems that it will later on be useful to add additional metrological requirements to such software: namely, that all software for measurement result processing no considered to be "mickey mouse" programs (trivial software) must contain a subprogram of computation of measurement uncertainty. Evaluation of the quality of such a generalized program can be carried out in accordance with the procedure described above.

6.3.2 Methodology of algorithm validation for measurement data processing and its practical implementation

Testing the software according to the "black box" method is the basic method of checking its functional capabilities in the process of metrological validation. In realizing this method two cardinal interrelated questions appear:

(1) How must a set of test data be formed? What are the parameters of a useful signal have to be considered? What level of noise has to be known? What are the points of input signal ranges where it is useful to work?; etc.

(2) How can the completeness of testing be provided, since testing is a check in separate points and the results of testing are the propagation of estimates obtained in testing over the whole range of parameter variation?

Some recommendations on the formation of test "standard" data are given in Section 6.3.1. However, to reveal substantially possible "weak spots" of the software, an analysis of the algorithm of processing measurement results, realized by this software, is needed. Analytical estimates of algorithm accuracy as the function of the parameters of input data models allow the most critical values of parameter ranges of test signal models to be evaluated and, consequently, test data sets for testing software to be formed using the "black box" method.

In this case testing with the "black box" method is a check for confirming the estimates obtained in the analytical manner. Thus, an alternative to "generalized testing" of software in evaluating the accuracy characteristics of the measurement result is to carry out the metrological validation of the realized algorithm, allowing the estimates of algorithm accuracy in the analytical form in the range of input data parameter variation (and not in separate point as in testing). On the basis of metrological validation the selection of test data parameter values and then testing software in points selected are performed.

6.3.2.1 Scheme for validating experimental data processing algorithms in measurements

In the following a scheme for metrological validation of algorithms and software for data processing in measurements will be given. The methodology of validation of algorithms and software for data processing was developed at the D. I. Mendeleyev Institute for Metrology about 20 years ago [332]. The study of a data processing algorithm on the basis of unified models of initial data with the purpose to determine characteristics of its accuracy, reliability (stability), and complexity is called here validation.

Frequently validation is carried out for one group of algorithms, and the results are a basis for comparing the algorithms, in order to select the best one. By best we mean the algorithm that has a consistently high accuracy for a wide class of models, since in practice there is no information about the authentic properties of input data. For this the validation ideology differs from that of the optimization of statistic methods of evaluation, where the best algorithm for a selected criterion and postulated model is revealed.

For the group of algorithms $A = \{a\}$, the sequence of stages are:

- the establishment of indices of the algorithm $\Pi_1, \dots, \Pi_n$ which will be used for comparison and caused selection of algorithms in group A;
- the selection of test models of initial data entering the algorithm input: $\mu_1, \dots, \mu_m$, which correspond to a measurement problem considered;
- the evaluation of algorithm characteristic values on type models of input data:

$$\Pi_{ij}(a) = \Pi_i(a, \mu_j), \quad i = 1 \cdots n, \quad j = 1 \cdots m.$$

The values of the indices $\Pi_{ij}(a)$ are either the numbers or the dependencies of algorithm characteristics on the parameters of the input data models (including analytical expressions, interpolation formulae, tables, or diagrams).

The algorithm characteristics can be divided into three groups: accuracy, steadiness (stability), and complexity.

The accuracy indices of an algorithm characterize the accuracy of the measurement results obtained in its realization. They reflect both the methodical errors of the algorithm and the transformed errors (uncertainties) of measurements (caused by errors or uncertainties of initial data).

At present, the basic characteristic of measurement accuracy is the standard uncertainty or density probability function of distribution of measurand values, i.e., *pdf*. In evaluating the systematical biases caused by the nonlinearity of algorithms it is convenient to use their boundaries. In addition to the accuracy characteristics, in the methodology of algorithm validation the steadiness and complexity characteristics are also considered.

Models of initial data, i.e., *test models*, are the combination of models of useful signals and models of noises. Selection of useful signal models is based on the measurement equation. Models of noise are formed separately for random and for systematic components.

The random noise is described by random sequences (above all by uncorrelated sequences with a mean, equal to zero, and constant dispersion σ^2) which have particular standard distributions (in particular, Gaussian, uniform, double exponential, or "mixed" Gaussian distributions). For describing systematic components they more often use functional dependencies which are constant, changing in a linear manner, and harmonic.

In practice, when realizing the algorithm validation methodology we see a *general* or *research* validation of algorithms, which is the most complete study of such algorithm properties such as their accuracy, steadiness and complexity, and *metrological* validation. Results of the former have a reference character and are directed to choice algorithm for solving a particular measurement problem.

The *metrological* algorithm validation has a more concrete character and is carried out, as a rule, according to a definite procedure of fulfillment measurements or to a measurement system, and it is governed by normative and/or technical documents. The validation of this kind corresponds to the concept *validation* to the greatest degree.

6.3.2.2 Validation of algorithms for determining information parameters of analytical signals (an example)

To illustrate the use of the validation methodology for algorithms of data processing in measurements, let us consider the validation of algorithms of determining information parameters of the quantitative chemical analysis (QCA) signals, which often are called analytical. The algorithm for determining an information QCA signal parameter is a sequence of operations made with respect to an output signal of a primary measurement transducer with the purpose to determine the value of the information parameter and to evaluate its error.

The QCA signal is the output signal of the primary transducer which carries information, in particular, about the qualitative and quantitative composition of a substance analyzed. By information parameters we mean the parameters of a QAC signal which allow the qualitative composition of a sample under analysis to be identified, such as:

- of arithmetic operations, linearization of functions, QAC discretization, etc.);
- transformed error components, which caused by input data errors well as parameters which are functionally connected with the concentration values of the components being determined.

Usually the following algorithm error components can be distinguished:

- methodical error components caused by a particular realization of an algorithm of processing, e.g., errors caused by an approximate performance and their transformation in the process of algorithm fulfillment. The transformed errors are divided into systematic and random errors. For example, errors, caused by overlapping peaks when processing chromatograms and spectrograms, as well as errors caused by the nonlinearity of operations performed, are systematic transformed errors; and errors caused by random errors of initial data and changing in a random manner in the process of a repeated processing of some other realization of the QAC signal.

The algorithm error components are characterized by:

- boundaries in case of methodical error components;
- boundaries or standard uncertainty, evaluated according to type B, in the case of systematic error components;
- standard uncertainty, evaluated according to type A, in the case of random error components.

When using the metrological validation of algorithms for determining an information parameter, the models of a useful signal, background signal (zero drift) and random noise are established. A set of input data models at the validation of algorithms of determining information parameters is determined, reasoning from general requirements such as sufficient simplicity, small number, and variety needed, adequate for the initial data.

In specifying a set of models, it is necessary to take into account the following:

- the characteristics of an analytical signal peak, symmetrical or asymmetrical; single or imposing on the other peak; peak of Gaussian, Lorentz, or arbitrary form;
- a priori information about a background signal;
- information about the characteristics of a random noise (level, spectral characteristics, correlation interval, etc.).

The model of a useful signal can be presented in the form

$$s(t) = \sum_{1}^{m} A_i g(t - t_i),$$

where $g(t)$ is the standardized pulses.

Usually considered to be models $g(t)$ are:

- Gaussian model:

$$s(t) = \frac{1}{\sqrt{2\pi}\omega} \exp(-t^2/2\omega^2) \qquad \text{for symmetrical peaks;}$$

$$s(t) = \begin{cases} \frac{1}{\sqrt{2\pi}\omega_1} \exp(-t^2/2\omega_1^2) & \text{at } t < 0 \\ \frac{1}{\sqrt{2\pi}\omega_2} \exp(-t^2/2\omega_2^2) & \text{at } t > 0 \end{cases} \qquad \text{for asymmetrical peaks;}$$

- square-law model: $s(t) = t^2 - t/\omega$;
- Lorentz model: $s(t) = 1/(1 + t^2/\omega^2)$, and others

As the models of a background signal they usually use:

- linear model: $n(t) = a_0 + a_1 t$;
- square-law model: $n(t) = a_0 + a_1 t + a_2 t^2$.

As models of noise, the following random sequences and signals are used:

- random sequences with independent terms, distributed in accordance with the normal law of probability distribution:

$$p(x) = \frac{1}{\sqrt{2\pi}\sigma} \exp(-x^2/2\sigma^2);$$

- random sequences with independent terms, distributed in accordance with the "mixed" normal law with probability density:

$$p(x) = (1 - \varepsilon) p_1(x) + \varepsilon p_2(x),$$

where $p_1(x) = \frac{1}{\sqrt{2\pi}\sigma} \exp(-x^2/2\sigma^2)$,

$$p_2(x) = \frac{1}{\sqrt{2\pi}10\sigma} \exp(-x^2/200\sigma^2);$$

- random sequences with independent terms, distributed in accordance with the Poisson law with the probability density

$$p(k) = e^{-k}\frac{\lambda^k}{k!};$$

- random stationary processes with correlation functions of the form

$$K_1(\tau) = \sigma^2 \exp(-\alpha\tau^2),$$
$$K_2(\tau) = \sigma^2 \exp(-\alpha\,|\tau|), \quad \text{and others.}$$

This list of standard models can be changed and expanded at any time, reasoning from the purpose of validation and possible kind of initial data. Below, we give the results of validating a group of algorithms for determining peak values with the analytical method, where accuracy characteristics are obtained in the form of functions of initial data model parameters.

The number of algorithms under study includes the following algorithms of determining the position of the peak (t^*) and peak value (x^*) on the basis of the sequence of experimental data (t_i, x_i):

A: x^* is determined as the maximum term of the sequence (x_i), i.e., as the corresponding value of the sequence argument

$$x^* = \max(x_i), \quad t^* = \arg(x^*);$$

B: x^* and t^* are determined according to local square-law approximation $x_{sq}(t)$ nearby the maximum term of the sequence (x_i)

$$x^* = \max x_{sq}(t), \quad t^* = \arg\max x_{sq}(t).$$

Analytically the algorithm is written as

$$x^* = x_j - \frac{(x_{j+1} - x_{j-1})^2}{8(x_{j+1} + x_{j-1} - 2x_j)}, \quad t^* = t_j - \frac{h(x_{j+1} - x_{j-1})}{2(x_{j+1} + x_{j-1} - 2x_j)},$$

where x_j and t_j are the estimates obtained with the algorithm **A**; h is the discretization step.

C: x^* and t^* are determined on the basis of the square-law approximation function $s(t)$ by the least square method against (t_i, x_i):

$$x^* = \max x_{sq}^{lsm}(t), \quad t^* = \arg\max x_{sq}^{lsm}(t).$$

Analytically the algorithm is written as

$$t^* = \bar{t} - \frac{d_1}{2d_2}, \quad x^* = d_0 - \frac{T^2(N+2)}{12N} d_2 - \frac{d_1^2}{4d_2},$$

where d_0, d_1, d_2 are the coefficients of the orthogonal polynomials $P_0(t)$, $P_1(t)$, $P_2(t)$ for the zero, first and second power, correspondingly.

D: x^*, t^* are determined according to the local cubic approximation $x_{\text{cub}}(t)$near the maximum of the sequence (x_i):

$$x^* = \max x_{\text{cub}}(t), \quad t^* = \arg\max x_{\text{cub}}(t).$$

Analytically the algorithm is written as

$$\begin{aligned} t^* &= t_j - \frac{h(x_j - x_{j-1})}{x_{j+1} - 2x_j + x_{j-1}}, \\ x &= M_3(t^* - t_{j+1})(t^* - t_j)(t^* - t_{j-1}) + M_2(t - t_j)^1(t^* - t_{j-1}) \\ &\quad + M_1(t^* - t_{j-1}) + x_{j-1}, \end{aligned}$$

where

$$M_3 = (x_{j+1} - 3x_j + 3x_{j-1} - x_{j-2})/6h^3,$$
$$M_2 = (x_{j+1} - 2x_j + x_{j-1})/2h^2, \quad M_1 = (x_j - x_{j-1})/h,$$

t_j, x_j are the estimates obtained according to the algorithm **A**.

G: t^* is determined on the basis of cubic plain-interpolation near the maximum of the sequence (x_j):

$$\begin{aligned} t^* &= \frac{t_{j-1} + 4t_j}{5} - \frac{M_2}{5M_3}\left\{1 - \sqrt{(1 - h\frac{M_3}{M_2})^2 - \frac{M_3}{M_2}h - \frac{25M_1M_3}{M_2^2}}\right\}; \\ x^* &= x_j + \frac{(x_{j-1} - x_j)t_j}{h} - 0{,}4h^2M_2 + (M_1 - 1{,}2h^2M_3)t^* \\ &\quad + \frac{2}{5h}(M_2 - 4hM_3)(t_j - t^*)^3 + \frac{2}{5h}(M_2 + hM_3)(t^* - t_{j-1})^3. \end{aligned}$$

F: x^* is determined as the maximum term of the smoothed sequence z_i, where $z_i = (1-k)z_{i-1} + kx_i$ and t^* is the corresponding value of the argument

$$x^* = \max(z_i), \quad t^* = \arg\max(z_i)$$

As models of a useful signal $s(t)$ the following models were selected:

$$
\begin{aligned}
s_1(t) &= a_2 t^2 + a_1 t + a_0, \\
s_2(t) &= A \sin(\omega t + \varphi), \\
s_3(t) &= -k\,|t| + A.
\end{aligned}
$$

As models of random errors ε_i, the random sequences, distributed according to the Gaussian law with the parameters $(0, \sigma)$ or according to the uniform law with the parameters $(0, \lambda)$, were chosen.

It was assumed that the systematic errors had been excluded in advance from the measurement results, and for the nonexcluded remaining systematic errors in many cases the models of random errors give the best fit.

As a result of the metrological validation of algorithms of data processing the following indices were determined:

- boundaries of the methodical error component;
- boundaries of the systematic bias of a result;
- standard uncertainty caused by random factors.

The results of the analytical investigation are given in Table 6.4.

The parameters γ and μ in accordance with the models $s(t)$ have the following values:

$$
\begin{aligned}
&\gamma_1 = |a_2|\, h^2/4\sigma, \; \gamma_2 = A\omega^2 h^2/8\sigma, \;\; \gamma_3 = kh/8\sigma, \\
&\mu_1 = a_2 h^2, \qquad \mu_2 = -A\omega^2 h^2/2, \; ; \mu_3 = kh.
\end{aligned}
$$

The parameter ρ characterizes the degree of suppressing errors using the algorithm C:

$$
\rho = \sqrt{\frac{180N^3}{(N^2-1)(N+1)(N+3)}},
$$

and q is the initial data parameter equal to the ratio of a number observations at moments $t_i > 0$ to the number of all observations.

6.4 Requirements for software and methods for its validation

6.4.1 General requirements for measuring instruments with regard to the application of software

Measuring instruments installed and used in accordance with supplier specifications comply with the following requirements, but at the same time do not contradict the remaining technical and metrological requirements of the OIML Recommendations.

Table 6.4. The results of analytical investigation.

Algorithms	*Model*	*Boundaries of the methodical component*	*Transformed uncertainty*	
			Boundaries of the systematic bias	*Upper limit of the standard uncertainty*
A	1	$-h/2 < \Delta_t < h/2$ $0 < \Delta_x < a_2h^2/4$	$\Delta_t = 0$ $0 < \Delta_x < 0,45\sigma$	$s_t < 0,84\sigma$ $s_x < 4,8\sigma$
	3	$-h/2 < \Delta_t < h/2$ $-kh/2 < \Delta_x < 0$		
B	1	$\Delta_t = 0$ $\Delta_x = 0$	$\|\Delta_t\| < \dfrac{0,75h\sigma^2}{\mu^2}$ $0 < \Delta_x < \dfrac{\sigma^2}{2\|\mu\|}$	$s_t < 0,7\sigma h/\|\mu\|$ $s_x < \sigma$
	2	$\|\Delta_t\| < A\omega^2h^2/12$ $-0.02A\omega^4h^4 < \Delta_x < 0$		
	3	$\Delta_t < h/3$ $-0.5kh < \Delta_x < 0$		
D(G)	1	$\Delta_t = 0$ $\Delta_x = 0$	$\|\Delta_t\| < \dfrac{0,75h\sigma^2}{\mu^2}$ $0 < \Delta_x < \dfrac{4,5\sigma^2}{\|\mu\|}$	$s_t < 0,7\sigma h/\|\mu\|$ $s_x < \sigma$
	3	$\Delta_t < h/3$ $-0,5kh < \Delta_x < 0$		
D	2	$\|\Delta_t\| < A\omega^2h^2/12$ $\Delta_x < -8\cdot 10^{-9}A\omega^6h^6$	$\|\Delta_t\| < \dfrac{0,75h\sigma^2}{\mu^2}$ $0 < \Delta_x < \dfrac{4,5\sigma^2}{\|\mu\|}$	$s_t < 0,7\sigma h/\|\mu\|$ $s_x < \sigma$
G		$\Delta_t < h/12$ $\|\Delta_x\| < 0.06A\omega^2h^2$		
C	1	$\Delta_t = 0$ $\Delta_x = 0$	$\Delta_t = (t^* - \bar{t})\dfrac{\sigma^2\rho^2}{\mu^2(N-1)^4}$ $\Delta_x = \dfrac{-\sigma^2\rho^2}{(N-1)^4\mu^2}$ $\times((\bar{t} - t*)^2 + T^2/60)$	$s_t = \dfrac{\sigma\rho}{(N-1)^2\|\mu\|}$ $\times\sqrt{(\bar{t} - t*)^2 + T^2/60}$ $s_x = \sigma^2\left\{\dfrac{1}{N} + \dfrac{12}{N}(\dfrac{\bar{t} - t^*}{T})\right\}$ $+\rho^2\left\{(\dfrac{\bar{t} - t*}{T})^2 - \dfrac{N+2}{12N}\right\}$
	3	$\Delta_t = 167.5(q-0.5)^3$ $-8.3(q-0.5)$ $\Delta_x = 245.9(q-0.5)^4$ $-24.2(q-0.5)^2$ -0.9		

Since in this chapter we have used material from [181], it is necessary to explain that this international document OIML establishes the general requirements applicable to software for the functional capabilities of measuring instruments and provides a guide for checking the compliance of measuring instruments with the general requirements. We shall take this into account as a base for establishing particular requirements for both software and procedures which need to be indicated in international recommendations applied to particular categories of measuring instruments (hereinafter short: in the corresponding Recommendations).

At the same time the document does not cover all technical requirements specific for measuring instruments of a particular kind. These requirements will be indicated in a respective Recommendation document, for example, for weighing instruments, water meters, and others. Since instruments with software control, as a rule, are always electronic, it is necessary to take into account the document OIML D 11 General Requirements for Electronic Measuring Instruments.

At present the basic requirements reflect the state of matters in the sphere of information technology. In principle, they are applicable to all kinds of software-controlled electronic devices and subsystems, and have to be taken into account in any recommendations document. Unlike these elementary requirements, the special requirements in Section 6.4.2 refer to functional peculiarities that are not general for some kinds of measuring instruments or some fields of their application.

Note: The following are the conventional designations for the rigidity levels of tests:

(1) technical solution to be applied in case of a normal rigidity level of tests;

(2) technical solution to be applied in case of an increased rigidity level of tests.

6.4.1.1 Software identification

Legally relevant software of a measuring instrument/electronic device/subsystem must l be clear and unambiguously identified. An identifier may consist of more than one part, but one part shall be obligatory which is devoted only to a legally relevant purpose.

The identifier shall be permanently connected with software and presented or printed when on demand or reflected on a display in the process of operation or when switching on a measuring instrument which can be switched on or off. If a subsystem/electronic device has neither a display or printer, the identification number must be sent through a communication interface for representing/printing with some other subsystem/electronic device.

As an exception, for continuous measurements the solution may be a stamp with the identification number on the device under the following conditions:

- a user interface is not available l for activating an identifier indication on a display or has no display at all to show the identifier of software (analog indicating device or electro-mechanical counter);
- instrument/electronic device has no interface for transmission of the identifier;
- after manufacturing of an instrument/electronic device a change in software is impossible or can only take place only hardware or some component of it is changed.

A producer of hardware or of a corresponding component is responsible to correct the indication of an identifier on a corresponding instrument/electronic device.

The identifier of software and means of identification must be stated in a certificate of the measuring instrument type approval.

Note: All measuring instruments in use must comply with the type; a combined software identifier allows service staff and people obtaining measurement results to determine this compliance with the used measuring instrument.

Example: (I) A software contains a test series or number, unambiguously identifying the loaded version. This series is transmitted to a display of the instrument by pressing a button when the instrument is switched on, or cyclically under a timer control.

A number of the version may have, e.g., the following structure: A, Y, Z. If a flowmeter computer is considered, then A will represent a version of the basic software which counts pulses, Y is a version for recounting (none at 15 °C and 2 °C) and Z means an interface user language.

(II) The software calculates a checksum of the executable code and presents the result as the identification instead of or in addition to the string in (I). The checksum algorithm shall be a normalized algorithm e.g., the CRC16 algorithm is an acceptable solution for this calculation. Solution (II) is suitable if increased conformity is required, e.g.,, the identity of the whole executed code (see Section 6.4.2.5).

6.4.1.2 Correctness of the algorithms and functions

The measuring algorithms and functions of an electronic device smust be appropriate and functionally correct for the given application and device type (accuracy of the algorithms, price calculation according to certain rules, rounding algorithms, etc.).

The measurement result and accompanying information required by specific OIML recommendations or by national legislation must be displayed or printed correctly.

It must be possible to examine algorithms and functions either by metrological tests, software tests, or software examination.

6.4.1.3 Software protection

Prevention of misuse

A measuring instrument is to be constructed in such a way that the possibility of unintentional, accidental, or intentional misuse are minimal. In the framework of this OIML document, this applies especially to the software. The presentation of the measurement results should be unambiguous for all parties concerned.

Note: Software-controlled instruments are often complex in their functionality. The user needs adequate guidance for their correct use and for achieving correct measurement results.

Example: The user is guided by menus. The legally relevant functions are combined into one branch of this menu. If any measurement values might be lost by an action, the user should be warned and requested to perform another action before the function is executed.

Fraud protection

a) Legally relevant software must be secured against unauthorized modification, loading, or changes by swapping of the memory device. In addition to mechanical sealing, technical means may be necessary to secure measuring instruments which have an operating system or an option for loading software.

Note: When the software is stored on an inviolable memory device (on which data is unalterable, e.g., a sealed ROM “Read Only Memory”) the need for technical means are accordingly reduced.

Example: (I)/(II) The component containing the memory devices is sealed or the memory device is sealed on the PCB.

(II) If a rewritable device is used, the write-enable input is prevented by a switch that can be sealed. The circuit is designed in such a way that the write protection cannot be cancelled by a short-circuit of contacts.

(I) A measuring system consists of two subassemblies, one containing the main metrological functions incorporated into a component that can be sealed. The other subassembly is a universal computer with an operating system. Some functions such as the indication are located in the software of this computer. One relatively easy manipulation – especially if a standard protocol is used for communication between both software parts – could be swapping the software on the universal computer. This manipulation can be prevented by simple cryptographic means, e.g., encryption of the data transfer between the subassembly and the universal computer. The key for decryption is hidden in the legally relevant program of the universal computer. Only this program knows the key and is able to read, decrypt, and use the measurement values. Other programs cannot be used for this purpose, as they cannot decrypt the measurement values.

b) Only clearly documented functions can be activated by the user interface, which shall be realized in such a way that it does not facilitate fraudulent use. The presentation of information shall comply with Section 6.4.2.2.

Note: The examiner decides whether all of these documented commands are acceptable.

Example: (I)/(II) All inputs from the user interface are redirected to a program which filters incoming commands. It only allows and lets pass the documented commands and discards all others. This program or software module is part of the legally relevant software.

c) Parameters that fix the legally relevant characteristics of the measuring instrument must be secured against unauthorized modification. If necessary for the purpose of verification, the current parameter settings must be able to be displayed or printed.

Note: Device-specific parameters may be adjustable or selectable only in a special operational mode of the instrument. They may be classified as the ones that should be secured (unalterable) and the ones that may be accessed (settable parameters) by an authorized person, e.g., the instrument owner or product vendor. Type-specific parameters have identical values for all specimens of a type. They are fixed at the type approval of the instrument.

Example: (I)/(II) Device specific parameters to be secured are stored in a nonvolatile memory. The write-enable input of the memory is prevented by a switch that can be sealed.

d) Software protection comprises appropriate sealing by mechanical, electronic, and/or cryptographic means, making unauthorized intervention impossible or evident.

Example: (1) (I) Electronic sealing. The metrological parameters of an instrument can be input and adjusted by a menu item. The software recognizes each change and increments an event counter with each event of this kind. This event counter value can be indicated. The initial value of the event counter must be registered. If the indicated value differs from the registered one, the instrument is in an unverified state (equivalent to a broken seal).

(2) (I)/(II) The software of a measuring instrument is constructed such that there is no way to modify the parameters and legally relevant configuration except via a switch-protected menu. This switch is mechanically sealed in the inactive position, making modification of the parameters and of the legally relevant configuration impossible. To modify the parameters and configuration, the switch has to be switched, inevitably breaking the seal by doing so.

(3) (II) The software of a measuring instrument is constructed such that there is no way to access the parameters and legally relevant configuration except by authorized persons. If a person wants to enter the parameter menu item he has to insert his smart card containing a PIN as part of the cryptographic certificate. The software of the instrument is able to verify the authenticity of the PIN by the certificate and allows the parameter menu item to be entered. The access is recorded in an audit trail including the identity of the person (or at least of the smart card used).

Level (II) of the examples for acceptable technical solutions is appropriate, if increased protection against fraud is necessary.

6.4.1.4 Support of hardware features

Support of fault detection
The relevant OIML recommendation may require fault detection functions for certain faults of the instrument. In this case, the manufacturer of the instrument shall be required to design checking facilities into the software or hardware parts or provide means by which the hardware parts can be supported by the software parts of the instrument.

If software is involved in fault detection, an appropriate reaction is required. The relevant OIML recommendation may prescribe that the instrument/electronic device be deactivated or an alarm/record in an error log be generated in case a fault condition is detected.

The documentation submitted for type approval shall contain a list of faults that are detected by the software and its expected reaction and if necessary for understanding, a description of the detecting algorithm

Example: (I)/(II) On each start-up the legally relevant program calculates a checksum of the program code and legally relevant parameters. The nominal value of these checksums has been calculated in advance and stored in the instrument. If the calculated and stored values do not match, the program stops execution.

If the measurement is not interruptible, the checksum is calculated cyclically and controlled by a software timer. In case a failure is detected, the software displays an error message or switches on a failure indicator and records the time of the fault in an error log (if one exists).

Support of durability protection
It is up to the manufacturer to realize the durability protection facilities addressed in OIML D 11:2004 (5.1.3 (b) and 5.4) in software or hardware, or to allow hardware facilities to be supported by software. The relevant OIML recommendation may recommend appropriate solutions.

If software is involved in durability protection, an appropriate reaction is required. The relevant OIML recommendation may prescribe that the instrument/electronic device be deactivated or an alarm/report be generated in case durability is detected as being jeopardized.

Example: (I)/(II) Some kinds of measuring instruments require an adjustment after a prescribed time interval in order to guarantee the durability of the measurement. The software gives a warning when the maintenance interval has elapsed and even stops measuring, if a certain time interval has been exceeded.

6.4.2 Requirements specific for configurations

The requirements given in this section are based on typical technical solutions in IT, although they might not be common in all areas of legal applications. Following these requirements, technical solutions are possible which show the same degree of security and conformity as a type as instrument which is not software controlled.

The following specific requirements are needed when certain technologies are employed in measuring systems. They have to be considered in addition to those described in Section 6.4.1.

In the examples, where applicable, both normal and raised severity levels are shown.

6.4.2.1 Specifying and separating relevant parts of software and their interfaces

Metrologically critical parts of a measuring system – whether software or hardware parts – must not be inadmissibly influenced by other parts of the measuring system.

This requirement applies if the measuring instrument (or electronic device or subassembly) has interfaces for communicating with other electronic devices, with the user, or with parts of the software other than the metrologically critical parts within a measuring instrument (or electronic device or subassembly).

Separation of electronic instruments and subsystems

a) Subassemblies or electronic devices of a measuring system which perform legally relevant functions must be identified, clearly defined, and documented. They form the legally relevant part of a measuring system.

Note: The examiner decides whether this part is complete and whether other parts of the measuring system may be excluded from further evaluation.

Example: (1) (I)/(II) An electricity meter is equipped with an optical interface for connecting an electronic device to read out measurement values. The meter stores all the relevant quantities and keeps the values available for read-out for a sufficient time span. In this system only the electricity meter is the legally relevant device. Other legally nonrelevant devices may exist and may be connected to the interface of the instrument provided corresponding requirement is fulfilled. Securing of the data transmission itself is not required.

(2) (I)/(II) A measuring system consists of the following subassemblies:

- a digital sensor calculating the weight or volume;
- a universal computer calculating the price;
- a printer printing out the measurement value and the price to pay.

All subassemblies are connected by a local area network. In this case the digital sensor, the universal computer and the printer are legally relevant subassemblies and are optionally connected to a merchandize system which is not legally relevant. The legally relevant subassemblies have to fulfill the corresponding requirement and – because of the transmission via the network – also the requirements contained in Section 6.4.2.3. There are no requirements for a merchandise management system.

b) During type testing, it must be demonstrated that the relevant functions and data of subassemblies and electronic devices cannot be inadmissibly influenced by commands received via the interface.

This implies that there is an unambiguous assignment of each command to all initiated functions or data changes in the subassembly or electronic device.

Note: If "legally relevant" subassemblies or electronic devices interact with other "legally relevant" subassemblies or electronic devices, see Section 6.4.2.3.

Example: (1) (I)/(II) The software of an electricity meter (see example (1) above) is able to receive commands for selecting the quantities required. It combines the measurement value with additional information – e.g., time stamp, unit – and sends this set of data back to the requesting device. The software only accepts commands for the selection of valid allowed quantities and discards any other command, sending back only an error message. There may be securing means for the contents of the data set but they are not required, as the transmitted data set is not subject to legal control.

(2) (I)/(II) Inside a component which can be sealed there is a switch which defines the operating mode of the electricity meter: one switch setting indicates verified mode and the other nonverified mode (securing means other than a mechanical seal are possible; see the examples above). When interpreting commands received, the software checks the position of the switch: in nonverified mode the command set which the software accepts is extended compared to the mode described above; e.g., it may be possible to adjust the calibration factor by a command which is discarded in verified mode.

Separation of software parts

OIML TCs, and SCs may specify in the relevant recommendation the software/hardware/data or part of the software/hardware/data which are legally relevant. National regulations may prescribe that a specific software/hardware/data or part of the software/hardware/data be legally relevant.

a) All software modules (programs, subroutines, objects, etc.) which perform legally relevant functions or which contain legally relevant data domains form the legally relevant software part of a measuring instrument (electronic device or subassembly). The conformity requirement applies to this part (see Section 6.4.2.5) and it must be made identifiable as described in Section 6.4.1.1.

If the separation of the software is not possible or needed, the software is legally relevant as a whole.

Example: (I) A measuring system consists of several digital sensors connected to a personal computer which displays the measurement values. The legally relevant software on the personal computer is separated from the legally nonrelevant parts by compiling all procedures realizing legally relevant functions into a dynamically linkable library. One or several legally nonrelevant applications may call program procedures in this library. These procedures receive the measurement data from the digital sensors, calculate the measurement result, and display them in a software window. When the legally relevant functions have finished, control is given back to the legally nonrelevant application.

b) If the legally relevant software part communicates with other software parts, a software interface must be defined. All communication must be performed exclusively via this interface. The legally relevant software part and the interface shall be clearly documented. All legally relevant functions and data domains of the software must be described in order to enable a type approval authority to decide on correct software separation.

The interface consists of a program code and assigned data domains. Defined coded commands or data are exchanged between the software parts by storing to the assigned data domain by one software part and reading from it by the other. Writing and reading program code is part of the software interface. The data domain forming the software interface, including the code that exports from the legally relevant part to the interface data domain and the code that imports from the interface to the legally relevant part, must be clearly defined and documented. It must be impossible to circumvent the declared software interface.

The manufacturer is responsible for respecting these constraints. Technical means (such as sealing) for preventing a program from circumventing the interface or programming hidden commands must be impossible. The programmer of the legally relevant software part as well as the programmer of the legally nonrelevant part should be provided with instructions concerning these requirements by the manufacturer.

c) There be be unambiguous assignment of each command to all initiated functions or data changes in the legally relevant part of the software. Commands which communicate through the software interface must be declared and documented. Only documented commands are allowed to be activated through the software interface. The manufacturer must state the completeness of the documentation of commands.

Example: (I) In the example described above the software interface is realized by the parameters and return values of the procedures in the library. No pointers to data domains inside the library are returned. The definition of the interface is fixed in the compiled legally relevant library and cannot be changed by any application. It is not impossible to circumvent the software interface and address data domains of the library directly; but this is not good programming practice, is rather complicated, and may be classified as hacking.

d) Where legally relevant software is separate from nonrelevant software, the legally relevant software must have priority over nonrelevant software in using resources. The measurement task (realized by the legally relevant software part) must not be delayed or blocked by other tasks.

The manufacturer is responsible for respecting these constraints. Technical means for preventing a legally nonrelevant program from disturbing legally relevant functions must be provided. The programmer of the legally relevant software part as well as the programmer of the legally nonrelevant part should be provided with instructions concerning these requirements by the manufacturer.

Examples: (1) (I) In the above examples, the legally nonrelevant application controls the start of the legally relevant procedures in the library. Omitting a call of these procedures would of course prohibit the legally relevant functioning of the system. Therefore the following provisions have been made in the example system to fulfill the corresponding requirement. The digital sensors send the measurement data in encrypted form. The key for decryption is hidden in the library. Only the procedures in the library know the key and are able to read, decrypt, and display measurement values. If the application programmer wants to read and process measurement values, he is forced to use the legally relevant procedures in the library which perform all the legally required functions as a side effect when being called. The library contains procedures that export the decrypted measurement values, allowing the application programmer to use them for his own needs after the legally relevant processing has been finished.

(2) (I)/(II) The software of an electronic electricity meter reads raw measurement values from an analog-digital converter (ADC). For the correct calculation of the measurement values the delay between the "data ready" event from the ADC to finishing buffering of the measurement values is crucial. The raw values are read by an interrupt routine initiated by the "data ready" signal. The instrument is able to communicate via an interface with other electronic devices in parallel served by another interrupt routine (legally nonrelevant communication). Interpreting the corresponding requirement for such a configuration, it follows that the priority of the interrupt routine for processing the measurement values will be higher than the communication routine.

The examples mentioned above are acceptable as a technical solution only for a normal severity level (I). If increased protection against fraud or increased conformity is necessary, software separation alone is not sufficient and additional means are required, or the entire software should be considered to be under legal control.

6.4.2.2 Shared indications

A display or printout may be employed for presenting both information from the legally relevant part of software and other information. The contents and layout are specific for the kind of instrument and area of application and have to be defined in the relevant recommendations. However, if the indication is realized using a multiple-window user interface, the following requirement applies.

Software which realizes the indication of measurement values and other legally relevant information is part of the legally relevant software part. The window containing these data must have highest priority, i.e., it must not able to be deleted by other software, or overlapped by windows generated by other software, or minimized, or made invisible, as long as the measurement is running and the presented results are needed for the legally relevant purpose.

Example: (I) On the system described in the above examples the measurement values are displayed in a separate software window. The described means guarantee that only the legally relevant program part can read the measurement values. On an operating system with a multiple-window user interface an additional technical means is employed to meet the requirement in Section 6.4.2.2:

The window displaying the legally relevant data is generated and controlled by procedures in the legally relevant dynamically linkable library. During measurement these procedures check cyclically that the relevant window is still on top of all the other open windows; if not, the procedures put it on top.

If increased protection against fraud is necessary (II), a printout as an indication alone may not be suitable. There should exist a subassembly with increased securing means able to display the measurement values.

The use of a universal computer is not appropriate as part of a measuring system if increased protection against fraud is necessary (II). Additional precautions to prevent or minimize the risk of fraud, both in hardware and in software, should be considered when increased protection is necessary, such as when using a universal computer (for example PC, PDA, etc.).

6.4.2.3 Storage of data, transmission via communication systems

If measurement values are used at a place other than the place of measurement or at a later time than the time of measurement they possibly have to leave the measuring instrument (electronic device, subassembly) and be stored or transmitted in an insecure environment before they are used for legal purposes. In this case the following requirements apply.

• The measurement value stored or transmitted must be accompanied by all relevant information necessary for future legally relevant use.

Example: (I)/(II) A data set may include the following entries:

- measurement value, including unit;
- time stamp of measurement (see Section 5.2.3.7);

- place of measurement, or identification of the measuring instrument used for the measurement;
- unambiguous identification of the measurement, e.g., consecutive numbers enabling assignment to values printed on an invoice.

• The data must be protected by a software means to guarantee the authenticity, integrity, and, if necessary, correctness of the information concerning the time of measurement. The software displaying or further processing the measurement values and accompanying data must check the time of measurement, authenticity, and integrity of the data, after having read them from the insecure storage or after having received them from an insecure transmission channel. If an irregularity is detected, the data must be discarded or marked unusable.

Software modules which prepare data for storing or sending, or which check data after reading or receiving, belong to the legally relevant software part.

Note: It is appropriate to require a higher severity level when considering an open network.

Example: (I) The program of a sending device calculates a checksum of the data set (algorithm such as BCC, CRC16, CRC32, etc.) and appends it to the data set. It uses a secret initial value for this calculation instead of the value given in the standard. This initial value is employed as a key and stored as a constant in the program code. The receiving or reading program has also stored this initial value in its program code. Before using the data set, the receiving program calculates the checksum and compares it with the one stored in the data set. If both values match, the data set has not been falsified. Otherwise, the program assumes falsification and discards the data set.

• For a high protection level it is necessary to apply cryptographic methods. Confidential keys employed for this purpose must be kept secret and secured in the measuring instruments, electronic devices, or subassemblies involved. Means must be provided whereby these keys can only be input or read if a seal is broken.

Example: (II) The storing or sending program generates an "electronic signature" by first calculating a hash value and then encrypting the hash value with the secret key of a public key system. The result is the signature. This is appended to the stored or transmitted data set. The receiver also calculates the hash value of the data set and decrypts the signature appended to the data set with the public key. The calculated and the decrypted values of the hash value are compared. If they are equal, the data set has not been falsified (integrity is proven). To prove the origin of the data set the receiver must know whether or not the public key really belongs to the sender, i.e., the sending device. Therefore the public key is displayed on the display of the measuring instrument and can be registered once, e.g., together with the serial number of the device when it is legally verified in the field. If the receiver is sure that he used the correct public key for decryption of the signature, then the authenticity of the data set is also proven.

- **Automatic saving**

a) When, considering the application, data storage is required, and measurement data must be automatically stored when the measurement is concluded, i.e., when the final value used for a legal purpose has been generated.

The storage device must have sufficient permanency to ensure that the data are not corrupted under normal storage conditions. There must be sufficient memory storage for any particular application.

When the final value used for a legal purpose results from a calculation, all data necessary for the calculation must be automatically stored with the final value.

Note: Cumulative measurement values such as, e.g. electrical energy or gas volume must be constantly updated. As the same data domain (program variable) is always used, the requirement concerning the storage capacity is not applicable to cumulative measurements.

b) Stored data may be deleted if either

- the transaction is settled, or
- thes data is printed by a printing device subject to legal control.

Note: Other general government regulations (for instance for tax purposes) may contain strict limitations for the deletion of stored measurement data.

c) After the requirements in b) are fulfilled and when the storage is full, it is permitted to delete memorized data when both of the following conditions are met:

- data is deleted in the same order as the recording order and the rules established for the particular application are respected;
- deletion is carried out either automatically or after a special manual operation.

Note: The use of additional access rights should be considered when implementing the "special manual operation" prescribed above in the second bullet.

- **Transmission delay**

The measurement must not be influenced by a transmission delay.

- **Transmission interruption**

If network services become unavailable, measurement data must not be lost. The measurement process should be stopped to avoid the loss of measurement data.

Note: Consideration should be given to distinguish between static and dynamic measurements.

Example: (I)/(II) The sending device waits until the receiver has sent a confirmation of the correct receipt of a data set. The sending device keeps the data set in a buffer until this confirmation has been received. The buffer may have a capacity for more than one data set, organized as a FIFO (first in–first out) queue.

- **Time stamp**

The time stamp must be read from the clock of the device. Depending on the kind of instrument, or area of application, setting the clock may be legally relevant and appropriate protection means must be taken according to the strictness level to be applied.

The internal clock of a stand-alone measuring instrument tends to have a large uncertainty because there is no means to synchronize it with the global clock. But if the information concerning the time of measurement is necessary for a specific field of application, the reliability of the internal clock of the measuring instrument must be enhanced by specific means.

Example: (II) The reliability of the internal quartz-controlled clock device of a measuring instrument is enhanced by redundancy: a timer is incremented by the clock of the microcontroller which is derived from another quartz crystal. When the timer value reaches a preset value, e.g., 1 s, a specific flag of the microcontroller is set and an interrupt routine of the program increments a second counter. At the end of, e.g., one day the software reads the quartz-controlled clock device and calculates the difference in the seconds counted by the software. If the difference is within predefined limits, the software counter is reset and the procedure repeats; but if the difference exceeds the limits, the software initiates an appropriate error reaction.

6.4.2.4 Compatibility of operating systems and hardware; portability

The manufacturer must identify the suitable hardware and software environment. Minimum resources and a suitable configuration (e.g., processor, RAM, HDD, specific communication, version of operating system, etc.) necessary for correct functioning must be declared by the manufacturer and stated in the type approval certificate.

Technical means must be provided in the legally relevant software to prevent operation if the minimal configuration requirements are not met. The system must be operated only in the environment specified by the manufacturer for its correct functioning.

For example, if an invariant environment is specified for the correct functioning of the system, means must be provided to keep the operating environment fixed. This especially applies to a universal computer performing legally relevant functions.

Fixing the hardware, operating system, or system configuration of a universal computer or even excluding the use of an off-the-shelf universal computer has to be considered in the following cases:

- if high conformity is required (see Section 6.4.2.5 (d));
- if fixed software is required (e.g., Section 6.4.2.6 for traced software update);
- if cryptographic algorithms or keys have to be implemented (see Section 6.4.2.3).

6.4.2.5 Conformity of manufactured devices with the approved type

The manufacturer must produce devices and the legally relevant software which conforms to the approved type and the submitted documentation. There are different levels of conformity demands:

(a) identity of the *legally relevant functions* described in the documentation (6.1) of each device with those of the type (the executable code may differ);
(b) identity of *parts of the legally relevant source code*, and the rest of the legally relevant software complying with (a);
(c) identity of the *whole legally relevant source code*;
(d) identity of the *whole executable code*.

The relevant recommendation must specify which degree of conformity is suitable. This recommendation can also define a subset of these degrees of conformity.

Except for (d) there may be a software part with no conformity requirements if it is separate from the legally relevant part.

The means for making conformity evident, as described in Sections 6.4.1.1 and 6.4.2.1, must be provided.

Note: (a) and (b) should be applied in the case of a normal strictness level, and (c) and (d) should be applied in the case of a raised strictness level.

6.4.2.6 Maintenance and reconfiguration

Updating the legally relevant software of a measuring instrument in the field should be considered as

- a modification of the measuring instrument, when exchanging the software with another approved version;
- a repair of the measuring instrument, when reinstalling the same version.

A measuring instrument which has been modified or repaired while in service may require initial or subsequent verification, dependant on national regulations.

Software which is not necessary for the correct functioning of the measuring instrument does not require verification after being updated.

Only versions of legally relevant software that conform to the approved type are allowed for use (see Section 6.4.2.5). Applicability of the following requirements depends on the kind of instrument and is to set up in the relevant OIML recommendation.

It may differ, also depending on the kind of instrument under consideration. The following options are equivalent alternatives. This issue concerns verification in the field.

Verified update
The software to be updated can be loaded locally, i.e., directly on the measuring device, or remotely via a network. Loading and installation may be two separate steps (as shown in Figure 6.5) or combined into one, depending on the needs of the technical solution. A person should be on the installation site of the measuring instrument to check the effectiveness of the update. After the update of the legally relevant software of a measuring instrument (exchange with another approved version or reinstallation) the measuring instrument cannot be employed for legal purposes before a verification of the instrument has been performed and the securing means have been renewed (if not otherwise stated in the relevant OIML recommendation or in the approval certificate).

Traced update
The software is implemented in the instrument according to the requirements for traced update, if it is in compliance with the relevant OIML recommendation. A traced update is the procedure of changing software in a verified instrument or device, after which the subsequent verification by a responsible person at place is not necessary. The software to be updated can be loaded locally, i.e., directly on the measuring device, or remotely via a network. The software update is recorded in an audit trail. The procedure of a traced update comprises several steps: loading, integrity checking, checking of the origin (authentication), installation, logging, and activation.

(a) A traced update of software must be automatic. On completion of the update procedure the software protection environment must be at the same level as required by the type approval.

(b) The target measuring instrument (electronic device, subassembly) must have fixed legally relevant software which cannot be updated and which contains all of the checking functions necessary for fulfilling traced update requirements.

(c) Technical means must be employed to guarantee the authenticity of the loaded software, i.e., that it originates from the owner of the type approval certificate. If the loaded software fails the authenticity check, the instrument must discard it and use the previous version of the software, or switch to an inoperable mode.

Example: (II) The authenticity check is accomplished by cryptographic means such as a public key system. The owner of the type approval certificate (in general the manufacturer of the measuring instrument) generates an electronic signature of the software to be updated using the *secret key* in the manufactory. The *public key* is stored in the fixed software part of the measuring instrument. The signature is checked using the *public key* when loading the software into the measuring instrument. If the signature of the loaded software is correct, it is installed and activated; if it fails the check, the

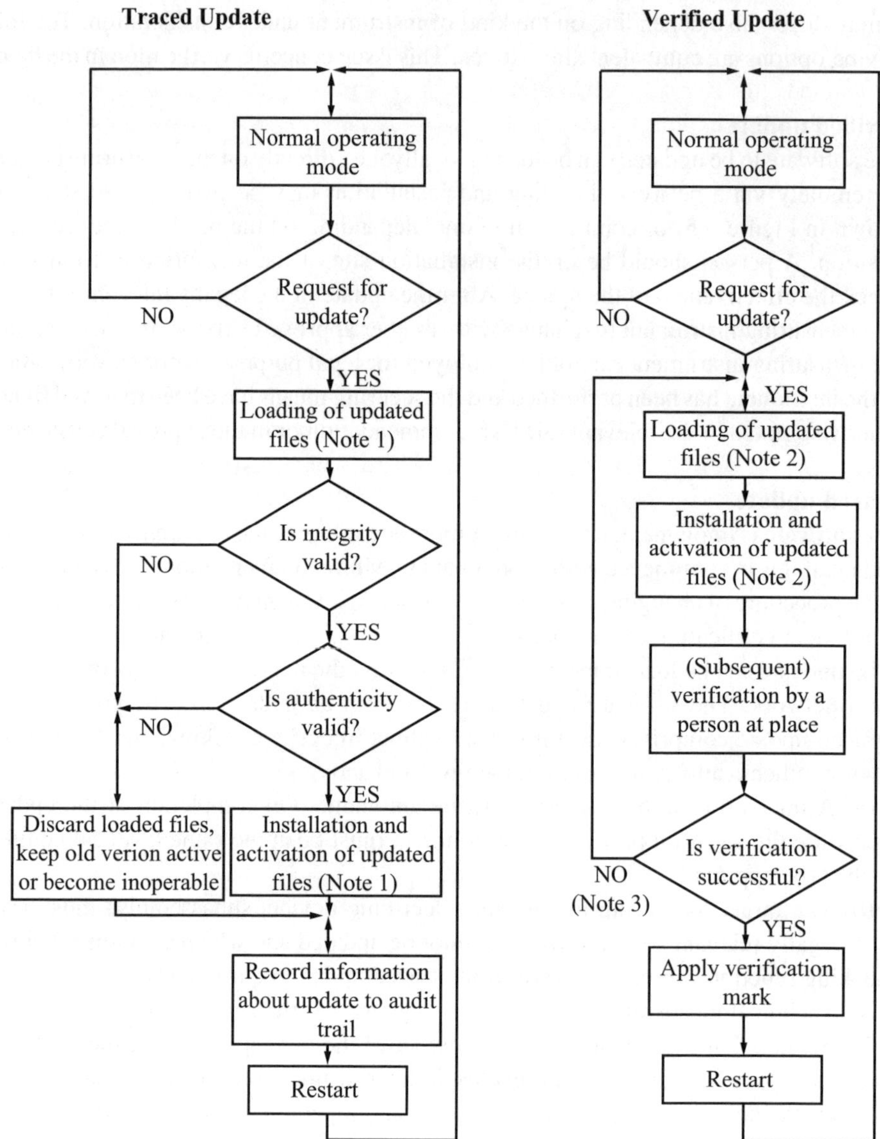

Figure 6.5. Software update procedure.

fixed software discards it and uses the previous version of the software or switches to an inoperable mode.

(d) Technical means must be employed to ensure the integrity of the loaded software, i.e., that it has not been inadmissibly changed before loading. This can be accomplished by adding a checksum or hash code of the loaded software and verifying it during the loading procedure. If the loaded software fails this test, the instrument must discard it and use the previous version of the software or switch to an inoperable mode. In this mode, the measuring functions must be prohibited. It must only be possible to resume the download procedure without omitting any step in the flow diagram for a traced update.

(e) Appropriate technical means, e.g., an audit trail, must be employed to ensure that traced updates of legally relevant software are adequately traceable within the instrument for subsequent verification and surveillance or inspection.

The audit trail must contain as a minimum the following information: success/failure of the update procedure, software identification of the installed version, software identification of the previous installed version, time stamp of the event, identification of the downloading party. An entry is generated for each update attempt regardless of its success. The storage device which supports the traced update must have sufficient capacity for ensuring the traceability of traced cpdates of legally relevant software between at least two successive verifications in the field/inspection. After reaching the storage limit for the audit trail, it shall be ensured by technical means that further downloads are impossible without breaking a seal.

Note: This requirement enables inspection authorities responsible for the metrological surveillance of legally controlled instruments to back-trace traced updates of legally relevant software over an adequate period of time (depending on national legislation).

(f) Depending on the needs and on the respective national legislation it may be required for the user or owner of the measuring instrument to give his consent for a download. The measuring instrument must have a subassembly/electronic device for the user or owner to express his consent, e.g., a button to be pressed before the download starts. It must be possible to enable and disable this subassembly/electronic device, e.g., by a switch which can be sealed, or by a parameter. If the subassembly/electronic device is enabled, each download has to be initiated by the user or owner. If it is disabled, no activity by the user or owner is necessary to perform a download.

(g) If the requirements in (a) through (f) cannot be fulfilled, it is still possible to update the legally nonrelevant software part. In this case the following requirements must be met:

- there is a distinct separation between the legally relevant and nonrelevant software, as according to Section 6.4.2.1;
- the whole legally relevant software part cannot be updated without breaking a seal;
- it is stated in the type approval certificate that updating of the legally nonrelevant part is acceptable.

Notes: (1) In the case of a traced update, updating is divided into two steps: "loading" and "installing/activating". This implies that the software is temporarily stored after loading without being activated, because it must be possible to discard the loaded software and revert to the old version, if the checks fail.

(2) In the case of a verified update, the software may also be loaded and temporarily stored before installation, but, depending on the technical solution, loading and installation may also be accomplished in one step.

(3) Here, only failure of the verification due to the software update is considered. Failure due to other reasons does not require reloading and reinstalling of the software, symbolized by the NO-branch.

The relevant OIML recommendation may require the setting of certain device-specific parameters available to the user. In such a case, the measuring instrument must be fitted with a facility to automatically and nonerasably record any adjustment of the device-specific parameter, e.g., an audit trail. The instrument must be able to present the recorded data.

Note: An event counter is not an acceptable solution.

The traceability means and records are part of the legally relevant software and should be protected as such. The software employed for displaying the audit trail belongs to the fixed legally relevant software.

6.4.3 Software validation methods

6.4.3.1 Overview of methods and their application

Selection and sequence of the methods given below (see Table 6.5) are not strongly assigned and may change in the procedure of validation from time to time.

6.4.3.2 Description of selected validation methods

Analysis of documentation, specification, and validation of the design (AD)

Application: This is the basic procedure which has to be applied in any case.

Preconditions: The procedure is based on the manufacturer's documentation of the measuring instrument. Depending on the demands this documentation must have an adequate scope.

(1) Specification of the externally accessible functions of the instrument in a general form (suitable for simple instruments with no interfaces except a display, all features verifiable by functional testing, low risk of fraud);

(2) Specification of software functions and interfaces (necessary for instruments with interfaces and for instrument functions that cannot be functionally tested and in case of increased risk of fraud). The description must make evident and explain all software functions which could have an impact on metrological features;

Table 6.5. Overview of the proposed selected validation methods.

Abbreviation	Description	Application	Preconditions, tools for application	Special skills for performing
AD	Analysis of the documentation and validation of the design	Always	Documentation	
VFTM	Validation by functional testing of metrological functions	Correctness of the algorithms, uncertainty, compensating and correcting algorithms, rules for price calculation	Documentation	—
VFTSw	Validation by functional testing of software functions	Correct functioning of communication, indication, fraud protection, protection against operating errors, protection of parameters, fault detection	Documentation, common software tool	—
DFA	Metrological data flow analysis	Software separation, evaluation of the impact of commands on the instrument's functions	Source code, common software tool (simple procedure), tools (sophisticated procedure)	Knowledge of programming languages. Instruction for the method necessary
CIWT	Code inspection and walk-through	All purposes	Source code, common software tool	Knowledge of programming languages, protocols, and other IT issues
SMT	Software module testing	All purposes when input and output can clearly be defined	Source code, testing environment, special software tools	Knowledge of programming languages, protocols, and other IT issues. Instruction for using the tools necessary

Note:: Text editors, hexadecimal editors, etc. are considered as "common software tools".

(3) Regarding interfaces, the documentation must include a complete list of commands or signals which the software is able to interpret. The effect of each command must be documented in detail. The way in which the instrument reacts to undocumented commands must be described;

(4) Additional documentation of the software for complex measuring algorithms, cryptographic functions, or crucial timing constraints must be provided, if this is necessary for understanding and evaluating the software functions;

(5) When it is not clear how to validate a function of a software program, the burden of developing a test method is placed on the manufacturer. In addition, the services of the programmer should be made available to the examiner for the purposes of answering questions.

A general precondition for examination is the completeness of the documentation and the clear identification of the EUT, i.e., of the software packages contributing to the metrological functions.

Description: The examiner evaluates the functions and features of the measuring instrument using the verbal description and graphical representations and decides whether or not they comply with the requirements of the relevant OIML recommendation. Metrological requirements as well as software-functional requirements (e.g., fraud protection, protection of adjustment parameters, nonallowed functions, communication with other devices, update of software, fault detection, etc.) must be considered and evaluated.

Result: The procedure gives a result for all the characteristics of the measuring instrument, provided that the appropriate documentation has been submitted by the manufacturer. The result should be documented in a section related to software in a software evaluation report included in the evaluation report format of the relevant OIML recommendation.

Complementary procedures: Additional procedures should be applied, if examining the documentation cannot provide substantiated validation results. In most cases validating the metrological functions by functional testing is a complementary procedure.

References: FDA, Guidance for FDA Reviewers and Industry Guidance for the Content of Premarket Submissions for Software Contained in Medical Devices, 29 May 1998; IEC 61508-7, 2000-3.

Validation by functional testing of the metrological functions (VFTM)

Application: Correctness of algorithms for calculating the measurement value from raw data, for linearization of a characteristic, compensation of environmental influences, rounding in price calculation, etc.

Preconditions: Operating manual, functioning pattern, metrological references and test equipment.

Description: Most of the approval and test methods described in the OIML recommendations are based on reference measurements under various conditions. Their application is not restricted to a certain technology of the instrument. Although it does not aim primarily at validating the software, the test result can be interpreted as a validation of some software parts, in general even the metrologically most important ones. If the tests described in the relevant OIML recommendation cover all the metrologically relevant features of the instrument, the corresponding software parts can be regarded as being validated. In general, no additional software analysis or test has to be applied to validate the metrological features of the measuring instrument.

Result: Correctness of algorithms is valid or invalid. Measurement values under all conditions are or are not within the MPE.

Complementary procedures: The method is normally an enhancement of the predecessor one. In certain cases it may be easier or more effective to combine the method with examinations based on the source code or by simulating input signals e.g., for dynamic measurements.

References: Various specific OIML recommendations.

Validation by functional testing of the software functions (VFTSw)

Application: Validation of e.g., the protection of parameters, indication of a software identification, software supported fault detection, configuration of the system (especially of the software environment), etc.

Preconditions: Operating manual, software documentation, functioning pattern, test equipment.

Description: Required features described in the operating manual, instrument documentation or software documentation are checked practically. If they are software controlled, they are to be regarded as validated if they function correctly without any further software analysis. Features addressed here are, e.g.,:

- normal operation of the instrument, if its operation is software controlled. All switches or keys and described combinations should be employed and the reaction of the instrument evaluated. In graphical user interfaces, all menus and other graphical elements should be activated and checked;

- effectiveness of parameter protection can be checked by activating the protection means and trying to change a parameter;
- effectiveness of the protection of stored data can be checked by changing some data in the file and then checking whether this is detected by the program;
- generation and indication of the software identification can be validated by practical checking;
- if fault detection is software supported, the relevant software parts can be validated by provoking, implementing, or simulating a fault and checking the correct reaction of the instrument;
- if the configuration or environment of the legally relevant software is claimed to be fixed, protection means can be checked by making unauthorized changes. The software should inhibit these changes or should cease to function.

Result: Software controlled feature under consideration is or is not OK.

Complementary procedures: Some features or functions of a software controlled instrument cannot be practically validated as described. If the instrument has interfaces, it is in general not possible to detect unauthorized commands only by trying commands at random. Besides this, a sender is needed to generate these commands. For normal validation the level method AD, including a declaration by the manufacturer, can cover this requirement. For an extended examination level, a software analysis such as DFA or CIWT is necessary.

References: FDA Guidance for Industry Part 11, August 2003; WELMEC Guide 2.3; WELMEC Guide 7.2.

Metrological dataflow analysis (DFA)

Application: Construction of the flow of measurement values through the data domains subject to legal control. Examination of software separation.

Preconditions: Software documentation, source code, editor, text search program, or special tools. Knowledge of programming languages.

Description: It is the aim of this method to find all parts of the software which are involved in the calculation of the measurement value or which may have an impact on it. Starting from the hardware port where measurement raw data from the sensor is available, the subroutine which reads them is searched. This subroutine will store them in a variable after possibly having done some calculations. From this variable the intermediate value is read by another subroutine and so forth until the completed measurement value is output to the display. All variables used as storage for intermediate measurement values and all subroutines transporting these values can be found in the source code simply by using a text editor and a text search program to find the variable or subroutine names in another source code file than the one currently open in the text editor. Other data flows can be found by this method, e.g., from interfaces to the interpreter of received commands. Furthermore circumvention of a software interface can be detected.

Result: This can be validated whether software separation is OK or not OK.

Complementary procedures: This method is recommended if software separation is realized and if high conformity or strong protection against manipulation is required. It is an enhancement to AD through VFTSw and to CIWT.

Reference: IEC 61131-3.

Code inspection and walk-through (CIWT)

Application: Any feature of the software may be validated with this method if enhanced examination intensity is necessary.

Preconditions: Source code, text editor, tools. Knowledge of programming languages.

Description: The examiner walks through the source code assignment by assignment, evaluating the respective part of the code to determine whether or not the requirements are fulfilled and whether or not the program functions and features are in compliance with the documentation. The examiner may also concentrate on algorithms or functions which he has identified as complex, error-prone, insufficiently documented, etc. and inspect the respective part of the source code by analyzing and checking it. Prior to these examination steps the examiner will have identified the legally relevant software part, e.g., by applying metrological data flow analysis. In general, code inspection or walk-through is limited to this part. By combining both methods, the examination effort is minimal compared to the application of these methods in the normal software production with the objective of producing failure-free programs or optimizing performance.

Result: Implementation compatible with the software documentation and in compliance or not in compliance with the requirements.

Complementary procedures: This is an enhanced method, additional to AD and DFA. Normally it is only applied in spot checks.

Reference: IEC 61508-7:2000-3.

Software module testing (SMT)

Application: Only if high conformity and protection against fraud is required. This method is applied when functions of a program cannot be examined exclusively on the basis of written information. It is appropriate and economically advantageous for validation of dynamic measurement algorithms.

Preconditions: Source code, development tools (at least a compiler), functioning environment of the software module under test, input data set and corresponding correct reference output data set or tools for automation. Skills in IT, knowledge of programming languages. Cooperation with the programmer of the module being tested is advisable.

Description: The software module being tested is integrated in a test environment, i.e., a specific test program module which calls the module under test and provides it with all necessary input data. The test program receives output data from the module under test and compares it with the expected reference values.

Result: Measuring algorithm or other tested functions are correct or not correct.

Complementary procedures: This is an enhanced method, additional to VFTM or CIWT. It is only profitable in exceptional cases.

Reference: IEC 61508-7:2000–3.

6.4.3.3 Validation procedure

A validation procedure consists of a combination of analysis methods and tests. The relevant OIML recommendation may specify details concerning the validation procedure, including:

(a) which of the validation methods described in Section 6.4.3 must be carried out for the requirement under consideration;

(b) how the evaluation of test results must be performed;

(c) which result is to be included in the test report and which is to be integrated in the test certificate.

In Table 6.6 two alternative levels, A and B, for validation procedures are defined. Level B implies an extended examination, compared to A. The choice of the A- or B-type validation procedures may be made in the relevant OIML recommendation – different or equal for each requirement, in accordance with the expected

- risk of fraud;
- area of application;
- required conformity to the approved type;
- risk of wrong measurement result due to operating errors.

6.4.3.4 Equipment being tested

Normally, tests are carried out on the complete measuring instrument (functional testing). If the size or configuration of the measuring instrument does not lend itself to testing as a whole unit, or if only a separate device (module) of the measuring instru-

Table 6.6. Recommendations for combinations of analysis and test methods for the various software requirements (acronyms defined in Table 6.5).

Requirement	*Validation procedure A* (normal examination level)	*Validation procedure B* (extended examination level)	*Comment*
Software identification	AD + VFTSw	AD + VFTSw + CIWT	Select "B" if high conformity is required
Correctness of algorithms and functions	AD + VFTM	AD + VFTM + CIWT/SMT	
		Software protection	
Prevention of misuse	AD + VFTSw	AD + VFTSw	
Fraud protection	AD + VFTSw	AD+VFTSw + DFA/CIWT/SMT	Select "B" in case of high risk of fraud
		Support of hardware features	
Support of fault detection	AD + VFTSw	AD + VFTSw + CIWT + SMT	Select "B" if high reliability is required
Support of durability protection	AD + VFTSw	AD+VFTSw+CIWT+ SMT	Select "B" if high reliability is required
	Specifying and separating relevant parts and specifying interfaces of parts		
Separation of electronic devices and sub-assemblies	AD	AD	
Separation of software parts	AD	AD + DFA/CIWT	
Shared indications	AD + VFTM/VFTSw	AD + VFTM/VFTSw + DFA/CIWT	
Storage of data, transmission via communication systems	AD + VFTSw	AD + VFTSw + CIWT/SMT	Select "B" if transmission of measurement data in open system is foreseen
The measurement value stored or transmitted must be accompanied by all relevant information necessary for future legally relevant use	AD + VFTSw	AD + VFTSw + CIWT/SMT	Select "B" in case of high risk
The data must be protected by software means to guarantee authenticity, integrity and, if necessary correctness of the information of the time of measurement	AD + VFTSw	/	
For a high protection level it is necessary to apply cryptographic methods	/	AD + VFTSw + SMT	

Table 6.6. (cont.)

Requirement	*Validation procedure A* (normal examination level)	*Validation procedure B* (extended examination level)	*Comment*
Automatic storing	AD + VFTSw	AD + VFTSw + SMT	
Transmission delay	AD + VFTSw	AD + VFTSw + SMT	Select "B" in the case of high risk of fraud, e.g. transmission in open systems
Transmission interruption	AD + VFTSw	AD + VFTSw + SMT	Select "B" in the case of high risk of fraud, e.g. transmission in open systems
Time stamp	AD + VFTSw	AD + VFTSw + SMT	
Compatibility of operating systems and hardware, portability	AD + VFTSw	AD+VFTSw+SMT	
Maintenance and re-configuration			
Verified Update	AD	AD	
Traced Update	AD + VFTSw	AD + VFTSw + CIWT/SMT	Select "B" in the case of high risk of fraud

ment is concerned, the relevant OIML recommendation may indicate that the tests, or certain tests, must be carried out on the electronic devices or software modules separately, provided that, in the case of tests with the devices in operation, these devices are included in a simulated setup sufficiently representative of its normal operation. The approval applicant is responsible for the provision of all the required equipment and components.

6.5 Type approval

6.5.1 Documentation for type approval

For type approval the manufacturer of the measuring instrument must declare and document all program functions, relevant data structures, and software interfaces of the legally relevant software part implemented in the instrument. No hidden undocumented functions are permitted to exist.

The commands and their effects must be completely described in the software documentation submitted for type approval. The manufacturer must state the completeness of the documentation of the commands. If the commands can be entered via a user interface, they must be completely described in the software documentation submitted for the type approval.

Furthermore, the application for type approval must be accompanied by a document or other evidence supporting the assumption that the design and characteristics of the software of the measuring instrument comply with the requirements of the relevant OIML recommendation, in which the general requirements of this document have been incorporated.

6.5.1.1 Typical documentation

Typical documentation (for each measuring instrument, electronic device, or sub-assembly) basically includes:

- a description of the legally relevant software and how the requirements are met:
  - list of software modules belonging to the legally relevant part, including a declaration that all legally relevant functions are included in the description;
  - description of the software interfaces of the legally relevant software part and of the commands and data flows via this interface, including a statement of completeness;
  - description of the generation of the software identification;
  - depending on the validation method chosen in the relevant OIML recommendation, the source code must be made available to the testing authority if high conformity or strong protection is required by the relevant OIML recommendation;
  - list of parameters to be protected and description of the means for protection;
- a description of the suitable system configuration and minimum required resources;
- a description of security means of the operating system (password, etc. if applicable);
- a description of the (software) sealing method(s);
- an overview of the system hardware, e.g., topology block diagram, type of computer(s), type of network, etc. Where a hardware component is deemed legally relevant or where it performs legally relevant functions, this should also be identified;
- a description of the accuracy of the algorithms (e.g., filtering of A/D conversion results, price calculation, rounding algorithms, etc.);
- a description of the user interface, menus, and dialogues;
- the software identification and instructions for obtaining it from an instrument in use;
- list of commands of each hardware interface of the measuring instrument/electronic device/sub-assembly including a statement of completeness;
- list of durability errors detected by the software, and if necessary for understanding, a description of the detecting algorithms;
- a description of data sets stored or transmitted;

- if fault detection is realized in the software, a list of faults which are detected and a description of the detecting algorithm;
- the operating manual.

6.5.2 Requirements for the approval procedure

Test procedures in the framework of the type approval, e.g., those described in OIML D 11:2004, are based on well-defined test setups and test conditions and can rely on precise comparative measurements. “Testing” and “validating” software are different activities. The accuracy or correctness of software in general cannot be measured in a metrological sense, although there are standards which prescribe how to “measure” software quality (e.g., ISO/IEC 14598). The procedures described here take into consideration both the legal metrology needs and also well-known validation and test methods in software engineering, but which do not have the same goals (e.g., a software developer who searches for errors but who also optimizes performance). As shown in Section 6.5.4 each software requirement needs individual adaptation of suitable validation procedures. The effort for the procedure should reflect the importance of the requirement in terms of accuracy, reliability and protection against corruption.

The aim is to validate the fact that the instrument to be approved complies with the requirements of the relevant OIML recommendation. For software-controlled instruments the validation procedure comprises examinations, analysis, and tests and the relevant OIML recommendation are to include an appropriate selection of the methods described below.

The methods described below focus on the type examination. Verifications of every single instrument in use in a particular field are not covered by those validation methods: see Section 6.5.3 (Verification) for more information.

The methods specified for software validation are described in Section 6.4.3. Combinations of these methods forming a complete validation procedure adapted to all requirements defined in Section 6.4.1 are specified in Section 6.4.5.

6.5.3 Verification

If metrological control of measuring instruments is required in a country, there must be a means to check in the field during operation the identity of the software, the validity of the adjustment, and the conformity to the approved type.

The relevant OIML recommendation may require carrying out the verification of the software in one or more stages, according to the nature of the considered measuring instrument.

Verification of a software must include:

- an examination of the conformity of the software with the approved version (e.g., verification of the version number and checksum);

- an test whether the configuration is compatible with the declared minimum configuration, if given in the approval certificate;
- a test whether the inputs/outputs of the measuring instrument are well configured in the software when their assignment is a device-specific parameter;
- a test whether the device-specific parameters (especially the adjustment parameters) are correct.

6.6 Assessment of severity (risk) levels

6.6.1 Brief review

By now a number of various approaches to the assessment of severity (risk) levels, extent of test rigidity, as well as to the selection of risk classes have been formed. They do not principally contradict each other.

The first approach set forth in a OIML document [181], uses only two such levels:

- (I) normal level of protection against falsification, as well as a level of confirming the compliance, reliability, and kind of measurement;
- (II) increased level of the seriousness of errors with enhanced countermeasures.

The second approach, indicated in the WELMEC Guide [541], assigns six risk classes for measuring instruments: A, B, C, D, E, and F. However, at present only three classes are normally applied: B, C, and D ("low", "middle" and "high"), and only in rare cases is the class E applied. Two risk classes, A and F, are left as "just in case", "in reserve", for an unexplored future time.

The third approach, adopted in Russia, operates with three degrees of severity of software tests: low, middle, and high.

Despite the difference in the terminology these approaches appear to be close to each other to a rather sufficient extent. Difference in understanding arises first of all, when answering the question: "Who has the right and duty to establish the mentioned levels of error seriousness, risk classes, and degrees of the severity of tests?"

However, it should be noted that, according to information mentioned in Section 6.2.2, the PTB and NPL representatives [1] propose the procedure of assessing risks based on ISO/IEC 16085 [233] and ISO/IEC 27005 [235], using the widely recognized approach for determining categories of risk (factors and level of risks) and facilities for minimizing them. The essence of these proposals is explained below.

The *algorithm of risk control* related to [235], is illustrated in Figure 6.6.

The *requirements for measurement software* must have information for enable

- determination of the category of risk on the basis of factors of risk specific for such a software with acceptable risk levels;

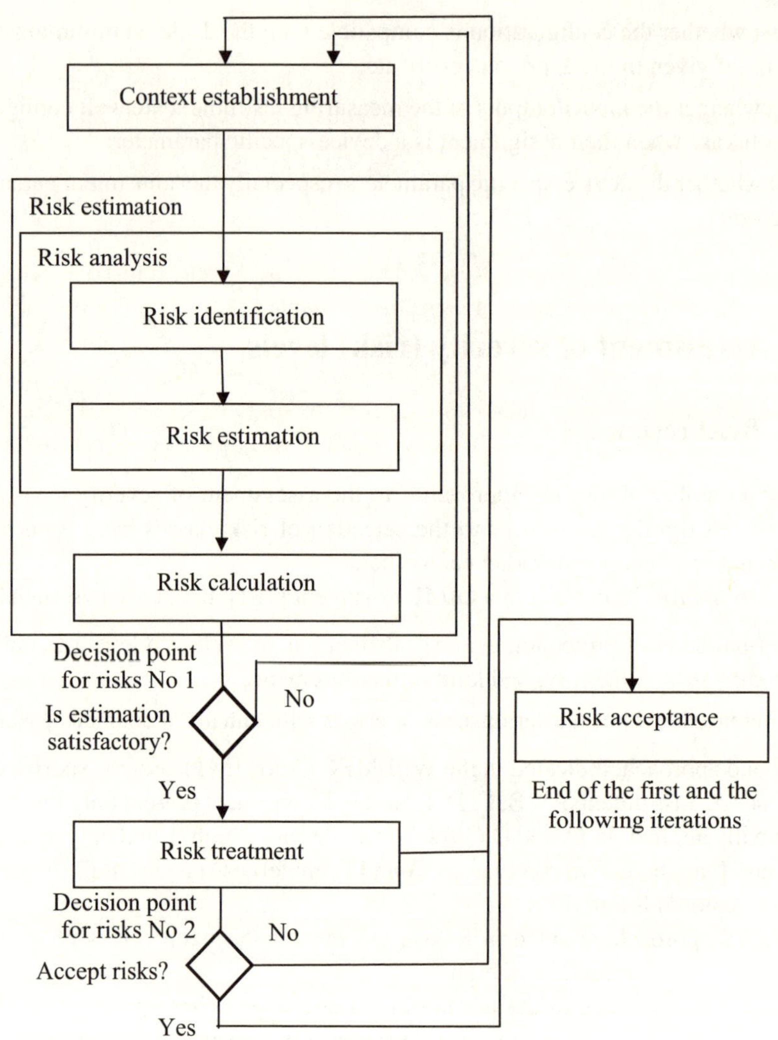

Figure 6.6. Control of risks related to standard ISO/IEC 27005.

- provision of each category of risk with characteristics which would point to measurement software;
- factors of risk must be traced to unified indices of risk, namely the Indices of measurement software (IMS);
- indication of each category of risk (IMS), what technical facilities must be used, as well as what level of activity there has to be for each process of a life cycle of software.

For *categories of risk* we distinguish:

- factors of risk, limited by three: the complexity level of control, level of calculation complexity, level of system integrity (undamaged state, safety and influence of environment). If necessary, they can be expanded to aspects specific for other fields;
- *levels of risk* supposed to limit the number of risk levels for each base factor of risk to four).

Characteristics for categories of risk. To determine a corresponding risk category of software, it is necessary to mark measurement-oriented characteristics for each factor of risk and for each level of risk.

Examples of the *risk factor "complexity of control"*: influence of software control functions on measurement processes; software influence on measurement results; amount and complexity of the software interaction with other software/hardware subsystems.

To create the procedure of risk control, the risk factors are traced to risk indices, i.e., to IMS. In accordance with a technical supposition the number of risk levels for basic IMS is limited to five (from 0 to 4).

Recommended facilities. The development guide which the authors recommend in [1] must provide technical devices and requirements acceptable for the process which are intended to be used for the corresponding IMS levels of each of selected processes of the life cycle.

A list of acceptable hardware for practical development and evaluation as well as for estimates of samples selected needs to be elaborated.

Indices of measurement software are illustrated in Table 6.7.

Technique recommended for designing is illustrated in Table 6.8.

Essential processes of programming includes: analysis of requirements, software design, software application, software test, maintenance of software in a healthy state.

When *developing risk estimates*, the following must be done: factors of risk and appropriate levels of risk, selection of a set of measurement software-oriented characteristics for each level of risk class/level, development of suggestions relative to IMS levels for all categories of risk, development of rules for designating technique that should be used for the appropriate IMS levels.

An approach of this type is being presently being developed. Therefore, in what follows the approaches considered in documents [181, 541] will be examined in more detail.

6.6.2 Assessment of severity (risk) levels according to OIML Document D 31

This section is intended as a guide for determining a set of severity levels to be generally applied for tests carried out on electronic measuring instruments. It is not intended

Table 6.7. Indices of measurement software.

<table>
<tr><th rowspan="2">System integrity</th><th rowspan="2">Calculation complexity</th><th colspan="4">Complexity of control</th></tr>
<tr><th>VL</th><th>L</th><th>H</th><th>VH</th></tr>
<tr><td rowspan="4">Very low (VL)</td><td>VL</td><td rowspan="4">?</td><td></td><td></td><td></td></tr>
<tr><td>L</td><td></td><td></td><td></td></tr>
<tr><td>H</td><td></td><td></td><td></td></tr>
<tr><td>VH</td><td></td><td></td><td></td></tr>
<tr><td rowspan="4">Low (L)</td><td>VL</td><td></td><td rowspan="4">?</td><td></td><td></td></tr>
<tr><td>L</td><td></td><td></td><td></td></tr>
<tr><td>H</td><td></td><td></td><td></td></tr>
<tr><td>VH</td><td></td><td></td><td></td></tr>
<tr><td rowspan="4">High (H)</td><td>VL</td><td></td><td></td><td rowspan="4">?</td><td></td></tr>
<tr><td>L</td><td></td><td></td><td></td></tr>
<tr><td>H</td><td></td><td></td><td></td></tr>
<tr><td>VH</td><td></td><td></td><td></td></tr>
<tr><td rowspan="4">Very high (VH)</td><td>VL</td><td></td><td></td><td></td><td rowspan="4">?</td></tr>
<tr><td>L</td><td></td><td></td><td></td></tr>
<tr><td>H</td><td></td><td></td><td></td></tr>
<tr><td>VH</td><td></td><td></td><td></td></tr>
</table>

Table 6.8. Technique recommended for designing.

Technique	*Levels of IMS*				
	0	1	2	3	4
Modeling of data flow		×			
Modeling control of data flow			×		
Relation of essences			×		
Unified language of modeling				×	
Z-communications					×
Transfers of states					×
Petri net					×
…	…				

as a classification with strict limits leading to special requirements as in the case of an accuracy classification.

Moreover, this guide does not restrict technical committees and subcommittees from providing severity levels which differ from those from the guidelines set forth in this document. Different severity levels may be used in accordance with special limits prescribed in the relevant OIML recommendations.

The severity level of a requirement has to be selected independently for each requirement.

When selecting severity levels for a particular category of instruments and area of application (trade, direct selling to the public, health, law enforcement, etc.), the following aspects can be taken into account.

(a) Risk of fraud:

- the consequence and the social and impact of malfunction;
- the value of the goods to be measured;
- platform used (built for a particular purpose or universal computer);
- exposure to sources of potential fraud (unattended self-service device).

(b) Required conformity:

- the practical possibility for the industry to comply with the prescribed level.

(c) Required reliability:

- environmental conditions;
- the consequence and the social and impact of errors.

(d) Interest of the defrauder:

- simply being able to commit fraud can be a sufficient motivational factor.

(e) The possibility of repeating or interrupting a measurement.

Throughout the requirements section various examples for acceptable technical solutions are given which illustrate the basic level of protection against fraud, of conformity, reliability, and of the type of measurement (marked with (I)). Where suitable, examples with enhanced countermeasures are also presented for considering a raised severity level of the aspects described above (marked with (II)).

The validation procedure and severity (risk) level are inextricably linked. A deep analysis of the software must be performed when a raised severity level is required, in order to detect software deficiencies or security weaknesses. On the other hand, mechanical sealing (e.g., sealing of the communication port or the module) should be considered when choosing the validation procedure.

6.6.3 Definition of risk classes according to the WELMEC Guide 7.2

6.6.3.1 General principles

The requirements of this guide [541] are differentiated according to (software) risk classes. Risks are related to software of the measuring instrument and not to any other risks. For convenience, the shorter term "risk class" is used. Each measuring instrument must be assigned to a risk class, because the particular software requirements to be applied are governed by the risk class to which the instrument belongs.

A risk class is defined by the combination of the appropriate levels required for software protection, software examination, and software conformity. Three levels, low, middle and high, are introduced for each of these categories.

6.6.3.2 Description of levels for protection, check, and compliance

The following definitions are used for the corresponding levels.

Software protection levels

Low: No particular protection measures against intentional changes are required.

Middle: The software is protected against intentional changes made by using easily-available and simple common software tools (e.g., text editors).

High: The software is protected against intentional changes made with sophisticated software tools (debuggers, hard disc editors, software development tools, etc).

Software examination levels

Low: Standard type examination functional testing of the instrument is performed. No extra software testing is required.

Middle: In addition to the low level examination, the software is examined on the basis of its documentation. The documentation includes the description of the software functions, parameter description, etc. Practical tests of the software-supported functions (spot checks) may be carried out to check the plausibility of documentation and the effectiveness of protection measures.

High: In addition to the middle level examination, an in-depth test of the software is carried out, usually based on the source code.

Software conformity levels

Low: The functionality of the software implemented for each individual instrument is in conformity with the approved documentation.

Middle: In addition to the conformity level "low", depending on the technical features, parts of the software shall be defined as fixed at type examination, i.e., alterable only with NB approval. The fixed part must be identical in every individual instrument.

High: The software implemented in the individual instruments is completely identical to the approved one.

6.6.3.3 Derivation of risk classes

Out of the 27 theoretically possible level permutations, only four, or at the most five, are of practical interest (risk classes B, C, D, and E, possibly F). They cover all of the instrument classes falling under the regulation of MID. Moreover, they provide a sufficient window of opportunity for the case of changing risk evaluations. The classes are defined in Table 6.9.

Table 6.9. Definition of risk classes.

Risk class	*Software protection*	*Software examination*	*Degree of software conformity*
A	Low	Low	Low
B	Middle	Middle	Low
C	Middle	Middle	Middle
D	High	Middle	Middle
E	High	High	Middle
F	High	High	High

6.6.3.4 Interpretation of risk classes

Risk class A: This is the lowest risk class. No particular measures are required against intentional software changes. Examination of software is part of the functional testing of the device. Conformity is required on the level of documentation. It is not expected that any instrument be classified as a risk class A instrument. However, by introducing this class, this possibility is kept open.

Risk class B: In comparison to risk class A, the protection of software is required on the middle level. Correspondingly, the examination level is moved up to the middle level. The conformity remains unchanged in comparison to risk class A.

Risk class C: In comparison to risk class B, the conformity level is raised to "middle". This means that parts of the software may be declared as fixed during type examination. The rest of the software is required to be conform on the functional level. The levels of protection and examination remain unchanged in comparison to risk class B.

Risk class D: The significant difference in comparison to risk class C is the moving of the protection level up to "high". Since the examination level remains unaffected at "middle", sufficiently informative documentation must be provided to show that the th protection measures taken are appropriate. The conformity level remains unchanged in comparison to risk class C.

Risk class E: In comparison to risk class D, the examination level is upgraded to "high". The levels of protection and conformity remain unchanged.

Risk class F: The levels with respect to all aspects (protection, examination, and conformity) are set to "high". Like risk class A, it is not expected that any instrument be classified as a risk F instrument. However, by introducing this class, the corresponding possibility is held open.

6.6.4 Determination of severity degrees of software tests in Russia

The COOMET Recommendation [123], in particular, establishes levels of requirements for software according to three basic criteria: test severity, degree of compliance, and protection. This Recommendation is intended for use in organizations developing software for measuring instruments, carrying out tests of measuring instruments with the purpose to approve the type, software validation, as well as applying this software according its direct designation.

The validation of software can be carried out by government centers for testing measuring instruments, authorized by the national agency for technical regulation and metrology to perform testing measuring instruments for the purpose of type approval, as well as bodies of the system of voluntary validation of measuring instrument software.

The requirements for the software of measuring instruments, which are within the sphere of government metrological control and supervision activities , as well as for software used in other measuring instruments, do not depend on their execution or functional designation. For each kind of requirement the following levels of requirements for software of measuring instruments have been established: low, middle and high.

Selection of levels is performed by the organization which carries out the software validation on the basis of coordination with the customer.

When establishing the levels of requirements, the specific technical features of measuring instruments and their destination are taken into account. Therefore, the requirements for software can be varied. The levels of requirements are shown in the tables below and determine:

- the degree of test severity (Table 6.10);
- the degree of compliance with validated software (Table 6.11);
- the level of protection (Table 6.12).

In view of the extremely large variety of software-controlled measuring instruments, it is possible to speak of an accepted methodology accepted and a standard approach

Table 6.10. Degree of software test severity.

Low	Software functions are validated in accordance with the validation program similar to that of usual tests of measuring instruments with the purpose of type approval.
Middle	Software is validated on the basis of the description of software functions presented by a developer or user. Software influence on measurement results is evaluated. Software facilities of identification and protection are also evaluated.
High	In addition to the usual testing carried out to determine the correctness of functions carried out, the software source code is checked. The object of the source code tests can be, for example, the realization of an algorithm for measurement data processing.

Table 6.11. Degree of software compliance of a measuring instrument with validated software.

Low	Parts of software having passed through the validation procedure have to be identical with respect to those which have been installed and fixed at its validation. The parts of a software not subjected to validation can be changed
Middle	In some cases, in addition to the "low" compliance level, caused by technical particularities, some additional parts of the software support, when used, can be designated as "not subject to changes".
High	In each measuring instrument there is software completely identical to the software installed and fixed at its validation.

Table 6.12. Levels of software protection of measuring instruments.

Low	No special protection of software under validation against inadmissible changes is required.
Middle	Software and data subjected to validation are protected against inadmissible changes done with simple programming facilities, for example, text editors.
High	Software and data subjected to validation are protected against inadmissible changes using special programming facilities (debugging monitors, hard disk editors, software support for developing software)

to metrological validation of software used in metrology. It is also necessary to keep in view that software validation of measuring instruments is a creative process which requires sufficiently high qualifications of all participants in this kind of metrological activity.

6.7 Summary

The subject of the book is so large in scope that it is practically impossible to cover it in one monograph. Therefore, the authors have selected only a number of separate fragments which are of current importance and relate to the field of their scientific interests.

Chapter 1 was devoted to some new aspects regarding the development of the international system of measurements, which began in 1999, when National Metrology Institutes (NMIs) signed the international document Mutual Recognition Arrangement (MRA) establishing the mutual recognition of national measurement standards, calibration, and measurement certificates issued by these NMIs [173].

The primary attention is given to the treatment of measurement data in the process of key comparisons of national measurement standards. Wide use of such comparisons for the purpose of establishing the degree of equivalence of national measurement standards and of experimentally confirming the calibration and measurement capabilities of NMIs in accordance with the MRA is evidence of the current importance of this problem.

In carrying out key comparisons a number of new metrology problems have arisen. This can be explained by the following. The establishment of the equivalence of national measurement standards is the first step in assuring the metrological traceability of measurement results to a reference (quantity value, measurement standard, primary procedure), which today is of special importance, taking into account the expansion of such concepts as "measurement" and "measurand".

The compatibility of measurement results is the most desirable property of measurement results regarding their practical use. Two experimental measurement procedures serve as an objective foundation for the compatibility of measurement results. These procedures are the comparison of primary measurement standards and calibration of measuring instruments.

The goal of any calibration is to provide the metrological traceability of measurement results to a primary measurement standard. Metrological traceability is based on a chain of subsequent calibrations, where each link in the chain introduces an additional uncertainty into the final measurement uncertainty. It should be emphasized that assurance of the traceability to a primary standard, in other words, the use of a calibrated measuring instrument in measurements, is the necessary but insufficient condition of the compatibility of measurement results. The application of validated measurement procedures is also required. The validation assumes the calculation of a measurement uncertainty under rated conditions in accordance with a recognized approach, which for today is the GUM [243]. The rated conditions correspond to the conditions of the intended use of measuring instruments. The comparisons of primary measurement standards with the goal of establishing their equivalence can be regarded as a final step in providing the metrological traceability of measurement results to a reference.

Some problems of evaluating measurement results when calibrating measuring instruments or comparing measurement standards have been considered. Comparisons of measurement standards and calibration of measuring instruments are the basic experimental procedures confirming the measurement uncertainties claimed by measurement and calibration laboratories. These procedures, accompanied by operation of quality management systems, supported by measurement and calibration laboratories, give an objective ground for the mutual recognition of measurement results obtained at these laboratories.

The models used in evaluation of key comparison data were analyzed, various approaches to expression of the degree of equivalence of measurement standards were discussed, including those which apply mixtures of distributions; the problem of linking regional key comparisons with CIPM key comparisons were treated in more detail.

In comparisons of measurement standards, routine calibration procedures are used. Working out a calibration procedure of measuring instruments implies formulating a measurement model and a design of a plan for a measurement experiment. Some general aspects of these issues were treated in Chapter 1.

Thus, in the Chapter 1 the problems of evaluating measurement results both in comparisons of measurement standards and in the calibration of measuring instruments and measurement standards, using the approach of measurements uncertainty to express the accuracy of measurements, were considered. Special emphasis was placed on the stage of modeling measurements in evaluating a KCRV, establishing the degree of equivalence of measurement standards, determining metrological characteristics of measuring instruments in their calibration.

Chapter 2 is devoted to the *systems of reproducing physical quantities units and transferring their sizes* (RUTS systems). The investigations described in this chapter are based on the system approach and on the use of the theoretical-set mathematical apparatus.

the classification of RUTS systems shows the necessity for expanding and improving the essence of a whole series of fundamental concepts of metrology connected with reproducing the measurement units and transferring their sizes.

The concepts "*full* RUTS system" and "*particular* RUTS system" were introduced, and the structure of a particular RUTS system was analyzed. Proceeding from the hierarchical structure of the RUTS system, in metrological measuring instruments three classes were singled out: *reference*, *subordinate,* and *auxiliary* metrological measuring instruments. Each of these classes carries a specific load, i.e., a metrological function.

The classification of RUTS system *according to the extent of centralization of the unit reproduction* has made it possible to distinguish four types of RUTS systems: systems with complete, multiple, and local centralizations in the national system of measurements, as well as with total decentralization.

The classification *according to the method of reproducing in time* has revealed two types of reference metrological measuring instruments (with continuous and discrete methods of reproduction).

Consideration was given to some aspects concerning the specific classification of RUTS systems (relative to concrete physical quantities), including the general nomenclature of measurable physical quantities, the problem of physical constants for RUTS systems, and the issue related to the RUTS systems for dimensionless quantities (coefficients). In particular, it has been shown that are about 235 physical quantities. The problem concerning the "*technical and economic efficiency*" of RUTS systems was also considered.

The investigations carried out on the *development of physical-metrological fundamentals of constructing RUTS systems* (in Section 2.2) have allowed the authors to arrive mainly at the following:

- to consider the *description of measurement* as both the simplest metrological system and process of solving a measurement problem. As a *system*, any measurement contains a definite set of controllable and uncontrollable elements. As a *process*, any measurement consists of three stages of successive transformation of the formal and informal type. Here, the role of a priori information in measurements is clearly seen;
- to show the general dependence of a measurement result on all its components, and to analyze the most considerable links among them, as well as to derive the general *equation of measurement correctness*;
- to investigate the general structure of a RUTS systems and determine its subsystems and elements, then on the basis of presenting the generalized element of the RUTS system, i.e., *metrological measuring instruments* in the form of measures, measurement instrument and transducers, to determine types of links possible between the elements (methods of transferring unit dimensions), and finally to show that the method of direct comparison (without any comparator) and method of indirect measurements cannot serve as the methods of transferring unit dimensions;
- to select the quality of measurements carried out in a national system of measurements (NSM) as the main criterion of the operation efficiency of all proper metrological systems (including RUTS systems), as well as to consider the main *quality indices*: accuracy, reliability, trueness, precision of measurements, and comparability of measurement results;
- the concept "*problematic situation*" when ensuring the measurement uniformity connected with the system "a manufacturer $\leftrightarrow$ a customer" was introduced; it is also shown that for a system of measurement uniformity assurance to be constructed the number of problematic situations hereby is important rather than the number of measurements carried out in an NSM;
- general estimates of NSM parameters and dimensions which describe this system (the 10-dimensions system with a great number ($\sim 10^{13}$) of realizations) and which are necessary for constructing a RUTS system are given. Since it is not possible to construct such a system in the general form, the main *principles* and *methods* of

constructing were described, according to which a general approach to the system can be used. Considerations of the *algorithms* which can be used for *constructing* the system at various stages were also given;

- metrological requirements for the means and methods of reproducing base and derived units of physical quantities were analyzed on the basis of the *reproduction postulates* also formulated by the authors;
- it was suggested to choose length, time, angle, mass, electric current force, temperature, and radionuclide activity as the basic quantities of the system of physical quantities, reflecting the modern physical world view and covering all fields of physics.

As an example for the theory of constructing a system of the unit reproduction in separate fields of measurements, we chose the field of measuring parameters of ionizing radiation (IRPM). The advantages and disadvantages as well as the perspectives of realizing a system of interrelated reference metrological measuring instruments in the field of the IRPM, based on a choice of the radionuclide activity and radiation energy in the capacity of base units of the system, which, in turn, rest upon fundamental constants and relative measurements of corresponding quantities, were analyzed in detail.

Thus, the following conclusion can be made. The theory developed and its fundamentals take into account the whole spectrum of problems connected with the problems of physical quantity unit reproduction, transfer of unit dimensions and construction of a corresponding RUTS system, as outlined above.

Chapter 3 was devoted to the error sources which in principle cannot be eliminated at any level of measurement technique. The potential accuracy of measurements is a problem which is constantly run across by both physicists-experimentalists carrying out high-precision physical experiments and metrologists developing measurement standards of physical quantities units. However, until now, unlike other problems of measurements, this problem has not received proper study and analysis in the metrology literature.

It is known that there is a certain type of physical knowledge about objects of the world surrounding us which forces us to speak of the principle limits of accuracy, with which we can obtain quantitative information about a majority of measurable physical quantities.

The first area of such knowledge is connected with the molecular-kinetic views concerning substance structure. On the basis of the systematic approach with use of the set-type mathematical apparatus introduced in Chapter 2, in considering the potential measurement accuracy it is necessary to analyze the influence of the following components: φ (a measurand as quality), o (an object under study as a measurand carrier), ψ (measurement conditions or totality of external influencing factors), $[\varphi]$ (a unit of a physical measurand), s (measuring instruments used to find a solution for a given measurement problem).

The estimates of physical measurand fluctuations were given using the examples of gas volume under a piston, the length of a metal rod, and electric voltage current caused by the thermal movement of charge carriers in a conductor.

Besides the natural limitations arising due to the fluctuation phenomena connected with a discrete structure of observed macroscopic objects, there are limitations to measurement accuracy which are of a more fundamental character. They are connected with the quantum-mechanical (discrete) properties of the behavior of microworld objects.

Using the example of evaluating the potential accuracy of joint measurements of the dependence between the electric current and time, a new, earlier unknown form of the Heisenberg's uncertainty relation was introduced which allows the restrictions of the accuracy of electric voltage measurements at a given accuracy of time measurements to be evaluated.

From the point of view of studying the potential accuracy of measurements, it is important to emphasize that all physical systems which have been studied have the finite extent both in space and time and all measurands or parameters are averaged with respect to this space volume and time interval and obey the Heisenberg relations.

However the same relations take place for the case of deterministic processes in macroscopic systems. The relation obtained in the example for an entirely deterministic process of continuous oscillations means that the degree of monochromatic ability of such a process is higher when the time interval of its realization and observation is more extended. Hence, it follows that the accuracy of determining the frequency of a quasi-harmonic process is inversely proportional to the time of its existence or observation.

Chapter 4 concerns the algorithms of data processing for the cases when the apparatus of the theory of mathematical statistics does not work.

It is known that for many practical problems this situation is typical when it is required to get an estimate of an unknown quantity on the basis of a rather small number of measurements. The evaluation algorithms related to the classical problem of three or two measurements were collected, systematized, and studied. Together with this, some cases of a greater number of measurements were considered.

In accordance with the classification of measurement errors, which was given in the chapter, errors were subdivided into six categories.

Special attention was given to crude errors of various origins, which are called mistakes, fails, noise, or signal misses.

The classification of the methods for evaluating results of repeated measurements was presented, whereby all evaluation methods were subdivided into groups: probabilistic, deterministic, heuristic, and diagnostic. Consideration was also given to the algorithms of optimal evaluation for the groups of the methods listed. In particular, the *probablistic* approach was characterized, which includes the maximum likelihood method, as well as the Markovian and Bayesian estimates.

Algorithms realizing the *deterministic* approach use various optimization criteria, in particular, the quadratic, module, minimax, power, as well as compound, combined, and other kinds of criteria.

It was shown that the principles of *heuristic* evaluation on the basis of definitions of means according to Cauchy and Kolmogorov lead to classical mean estimates for the case of two, three, and more measurements, such as an arithmetic, geometric, harmonic means, and some others. Moreover, linear, quasi-linear, and nonlinear estimates were also characterized. In particular, weighted arithmetic, and quadratic mean estimates, sample median, and some others also refer to these estimates.

When treating the *diagnostic* methods for getting estimates, the use of the *method of redundant variables* for increasing the evaluation accuracy as well as application of *algebraic invariants* are described. It is shown that the use of the redundant variables method leads to getting Markovian estimates, and using algebraic invariants it is possible to reject unreliable measurement values.

The estimates with rejection of one or two measurement values by either a maximum or minimum discrepancy were treated in detail. A comparison of various estimates, which differ in *evaluation algorithm character*, *method used for setting an algorithm* and *easiness of technical realization*, was made.

The principle of applying mean estimates for noise-immunity filtration of signals, the use of which leads to filters with finite memory, was described.

Great emphasis was placed on median filters and their characteristics, including optimization criteria, statistical properties, and root signals of filters. Consideration was also given to threshold and thresholdless versions of a diagnostic filter.

Algorithms of scalar quantity evaluation by three or two measurements were presented in the form of two tables, in which more than 100 different estimates are reflected. It should be noted that the list of these estimates can be continued, since the number of possible estimates is not limited, and the choice of one of them has to be made taking into account specific features of a particular practical measurement situation.

Chapter 5 illustrates the results obtained in this volume using as an example some magnetic recording instruments (MRI). The choice of such an unusual example is not accidental.

The specific features of MRI are:

- large and frequently uncertain (depending on the operator's actions) time gap between the moments of recording and reproducing a signal in both the mode of recording and mode of reproducing;
- time scale distortions of a reproduced signal due to oscillations, drift, and the non-nominal character of a magnetic carrier speed in both the mode of recording and mode of reproducing;
- the possibility of using the transposition of a tape-driving mechanism speed, thanks to which a spectrum of a signal recorded can be transformed.

As a consequence of theoretical and experimental investigations, a theoretical generalization of works aimed at determining the metrological characteristics of precise magnetic recording equipment for recording analogue electric signals and used as a part of information measurement systems (IMS) and measurement calculation complexes was realized.

On the basis of this generalization a number of new approaches for creation of the methods and means for measuring metrological characteristics of MRI were developed.

The results obtained provided a basis for the solution of the problem of metrological assurance of precise magnetic recording equipment, which is of great importance in connection with the wide use of MRI in IMS of various destinations.

Within the framework of the problem solution and concrete tasks which follow from it, the following results were obtained.

An algorithm of evaluating the quality of systems of metrological assurance under the conditions of incomplete and inaccurate data concerning their elements, connections, and properties, was developed, making use of the fuzzy-set theory. This algorithm was illustrated in an example of the project of an MRI verification scheme.

Taking into account the metrological status of MRI as an intermediate measuring transducer on the basis of an analysis of the equation for connection between the input and output signals of its measurement channel, a complex of its metrological characteristics which need to be normalized was shown. These characteristics include an error of transferring the signal over a channel, dynamic characteristics in the form of the transfer or pulse weight function, level of nonlinear distortions and time oscillations of the signal delay. Moreover, the metrological model of forming the resultant error was also grounded.

A new method for normalizing the dynamic characteristics of a measurement channel was proposed which is based on the application of Markov's parameters. The proposed method is alternative and equivalent with respect to the method based on applying moments of the pulse weight function. In addition to this method, one more method was developed which can be used for constructing functional dependencies for jointly measured quantities under the conditions of incomplete initial data. This method is based on solving the problem of multicriteria optimization.

For the first time anywhere (date of invention priority is 1973) a method for the experimental evaluation of the registration error of a signal in the MRI channel was developed, which is based on the use of the peculiar features of the test signal in the form of a quasi-random binary sequence of the maximum length and limited with respect to its spectrum according to the pass band of the channel under study.

Correlation methods and instruments for direct and differential measurements of values of the pulse weight functions of the MRI channel, taking into account special features of the equipment verified (number of channels, width of their pass band, different time sequence of recording and reproducing modes, availability of transposing the speed of the magnetic carrier movement and the possibility of separating a signal proportional to oscillations of the time delay) were presented. An analysis of the new method of coherent frequencies for the experimental determination of the nonlinearity of phase-frequency characteristics of MRI channels was performed.

A correlation method was developed for measuring nonlinear distortions of the signal of the "white noise" type in the MRI channel. This method is based on determining the pulse weight functions of the channel on high and low levels of the test signal. The development of methods and instruments for measuring oscillations of the time delay of signals in the channel allows both the oscillations and time delay drift to be estimated.

Various versions for the technical realization of instruments for measuring the MRI metrological characteristics were developed. Their novelty is confirmed by 18 author's invention certificates.

Eight types of measuring instruments, as developed by the authors, can be successfully used for experimental determination of the metrological characteristics of any equipment of precise magnetic recording and provide its verification and certification.

Chapter 6 is devoted to various aspects of solving the practical task of evaluating the accuracy of measurement results obtained using software for measurement data processing. Factors influencing the accuracy of a final result of measurement are discussed. A great attention is paid to issues of the metrological validation of modern software-controllable measuring instruments.

A great number of theoretical and practical publications as well as international and national normative documents are devoted to the solution of these problems. Among them are firstly Document OIML D 31 "General Requirements for Software Controlled Measuring Instruments" [181] and Document WELMEC 7.2 "Software Guide (Measuring Instruments Directive 2004/22/EC) [541], the basic theses of which underlie the content of this chapter.

At the same time, particular attention was given not only to the software of measuring instruments, but also to software for the simulation modeling of measurement experiments and infrastructural assurance of control, measurement, and metrology laboratories and centers.

Different approaches for the evaluation of precision parameters of software used in metrology, sources of uncertainty and methods of their evaluation in applying to the data processing software for obtaining a measurement result were the subject of speculation; an explanation was given to the scheme of validating the experimental data processing algorithms in measurements on the basis of the recommendation "National System of Ensuring Measurement Uniformity. Attestation of Algorithms and Software for Data Processing" [332], developed at the VNIIM.

A short review of the history, the current state of the matter, and the unsolved problems in this field in leading countries of the world was given.

Bibliography

[1] Advanced Mathematical and Computational Tools in Metrology. AMCTM VIII Conference, Paris, France (2008), AMCTM IX (2012). www.imeko-tc21.org.

[2] A. A. Afifi and S. P. Azen, *Statistical Analysis. A Computer Oriented Approach*, ed. II, Academic Press, New York San Francisco London, 1979.

[3] P. N. Agaletskii, Problems of measuring relations between quantities, in: *Measurement Techniques* **11**(4) (1968) 714–721.

[4] T. A. Agekian, *Osnovi teorii oshibok dlia astronomov i fizikov*, Nauka, Moskva, 1972 (in Russian).

[5] R. A. Agnew, Optimal Congressional Apportionment, *The Mathematical Association of America [Monthly* **115**, *April 2008]*, 297–303. http://home.comcast.net/~raagnew/Downloads/Optimal_Congressional_Apportionment.pdf.

[6] V. A. Aksenov, A. I. Viches, and M. V. Gitlic, *Tochnaya magnitnaya zapis'*, Energiya, Moskva 1973 (in Russian).

[7] V. Z. Aladjev and V. N. Haritonov, *General Theory of Statistics*, Fultus Corp., Palo Alto, CA, 2006.

[8] M. S. Aldenderfer and R. K. Blashfield, *Cluster Analysis* (2nd printing), Sage Publications, Inc., 1985.

[9] V. S. Aleksandrov, A. P. Sebekin, and V. A. Slaev, Osnovniye napravleniya razvitiya nauchnih issledovaniy v oblasti metrologii, in: *Trudi III sessii Mejdunarodnoy nauchnoy shkoli "Sovremenniye fundamental'niye problemi i prikladniye zadachi teorii tochnosti i kachestva mashin, priborov, system" TAPAQMDS'98*, pp. 103–104, St. Petersburg, 1998 (in Russian).

[10] Yu. I. Aleksandrov, E. N. Korchagina, and A. G. Chunovkina, New Generalization of Data on the Heat of Combustion of High-Purity Methane, *Measurement Techniques* **45**(3) (2002), pp. 268–273.

[11] A. I. Alekseev, A. G. Sheremet'yev, G. I. Tuzov et al., *Teoriya i primeneniye psevdosluchaynih signalov*, Nauka, Moskva, 1969 (in Russian).

[12] H. Alzer and S. L. Qiu, Inequalities for means in two variables. *Arch. Math.* **80**(2), (2003), 201–215.

[13] B. D. O. Anderson and J. B. Moore, *Optimal filtering*, Englewood Cliffs, NJ, 1979.

[14] T. W. Anderson, *An Introduction to Multivariate Statistical Analysis*, 3rd ed., John Wiley & Sons, Hoboken, New York, 2003.

[15] C. A. Andrews, J. M. Davies, and G. R. Schwarz, Adaptive data compression, *Proceedings of the IEEE* **55**(3) (1967), 267–277.

[16] M. Anwar and J. Pečarić, On log-convexity for differences of mixed symmetric means, *Mathematical Notes* **88**(5–6) (2010), 776–784.

[17] S. Aqaian, J. Astola, and K. Egiazarian, *Binary Polynomial Transforms and Nonlinear Digital Filters*, 1995.

[18] G. R. Arce and N. C. Gallagher, State Description for the Root Signal Set of Median Filters, in: *IEEE Trans Acoust., Speech and Signal Process.* **ASSP-30**(6) (1982), 894–902.

[19] O. I. Arkhangelskiy and L. A. Mironovskiy, Diagnosis of Dynamic Systems Using the Operator Norms, *Engineering Simulation* **13** (1996), 789–804.

[20] V. Ya. Arsenin and V. V. Ivanov, Freeing of the signal shape from distortions produced by equipment and transmission channel, *Measurement Techniques* **12** (1969), 101–108.

[21] V. O. Arutyunov and V. A. Granovskii and S. G. Rabinovich, Norm setting [standardization] and determination of the dynamic properties of measuring devices, *Measurement Techniques* **12** (1975), 25–37.

[22] J. Astola and G. Campbell, On Computation of the Running Median, in: *IEEE Trans. Acoust., Speech and Signal Process.* April, 1989.

[23] J. Astola and Y. Neuvo, An Efficient Tool for Analyzing Weighted Median Filters. *IEEE Trans. CAS II: Analog and Digital Processing* **41**(7) (1974), 487–489.

[24] J. Astola and Y. Neuvo, Optimal Median Type Filters for Exponential Noise Distributions, *Signal Process* **17**(2) (1989), 95–104.

[25] J. T. Astola and L. Yaroslavsky, Advances in Signal Transforms: Theory and Applications, *Hindawi*, 2007.

[26] R. M. Atahanov, D. S. Lebedev, and L. P. Yaroslavskiy, Podavleniye impul'snih pomeh v televizionnom priyemnom ustroystve, *Tehnika kino i televideniya* **7** (1971), 55–57 (in Russian).

[27] S. Atey, *Ustroystva zapisi na magnitnuyu lentu*, Energiya, Moskva, 1969 (in Russian).

[28] A. A. Avakyan, Improving methods of forecasting the need for measuring instruments in the national economy, *Measurement Techniques* **24**(6) (1981), 519–521.

[29] T. C. Aysal and K. E. Barner, Quadratic Weighted Median Filters for Edge Enhancement of Noisy Images, *IP* **15**(11) (2006), 3294–3310.

[30] S. A. Ayvazian, *Klassifikaciya mnogomernih nabliudeniy*, Statistika, Moskva, 1974 (in Russian).

[31] A. M. Azizov and A. N. Gordov, *Tochnost' izmeritel'nih preobrazovateley*, Energiya, Leningrad, 1975 (in Russian).

[32] V. A. Balalaev, A general description of measurement processes, *Measurement Techniques* **8**(8) (1975), 677–680.

[33] V. A. Balalaev, Problemi opisaniya i postroeniya metrologicheskih system, in: *Trudi VNIIM im. D.I. Mendeleyeva "Systemniye issledovaniya v metrologii"*, pp. 31–40, Leningrad, 1985 (in Russian).

[34] V. A. Balalaev, E. D. Koltik, V. V. Skotnikov, V. A. Slaev, and V. I. Fomenko, Analiz vremennih sootnosheniy parametrov izmeritel'nogo processa, in: *Tez. dokl. nauchno-tehnicheskogo soveschaniya "Voprosi dinamiki elektrometricheskoy apparaturi"*, pp. 3–6, Tartu, 1982 (in Russian).

[35] V. A. Balalaev, V. N. Romanov, and V. A. Slaev, Postroeniye funkcional'nih zavisimostey izmeriayemih velichin pri nepolnih ishodnih dannih, *Metrologiya* **(2)** (1985), 8–13 (in Russian).

[36] V. A. Balalaev, V. A. Slaev, and A. I. Siniakov, *Potencial'naya tochnost' izmereniy*, Nauch. Izdaniye, pod red. V. A. Slaeva, Professional, Sankt-Peterburg, 2005 (in Russian).

[37] V. A. Balalaev, V. A. Slaev, and A. I. Siniakov, *Teoriya system vosproizvedeniya edinic i peredachi ih razmerov,* Nauch. Izdaniye, pod red. V. A. Slaeva, Professional, Sankt-Peterburg, 2004 (in Russian).

[38] N. A. Balonin and L. A. Mironovskii, Flip Method for Obtaining Singular Functions of Hankel Operators and Convolution Operators, *Automation and Remote Control* **60**(12/2) (1999), 1523–1535.

[39] N. A. Balonin and L. A. Mironovskii, Linear Operators of the Dynamic System, *Automation and Remote Control* **61**(11/1) (2000), 1808–1818.

[40] N. A. Balonin and L. A. Mironovskii, Spectral Characteristics of the Linear Systems over a Bounded Time Interval, *Automation and Remote Control* **63**(6) (2002), 867–885.

[41] N. A. Balonin and L. A. Mironovsky, Solving Optimization Problems by System adjoint Operator Simulation, in: *Proc. of the Ist Intern. Conf. "Control of Oscillations and Chaos"*, pp. 553–556, St. Petersburg, Russia, 1997.

[42] A. N. Baranov and V. A. Slaev, Avt. svid. N°501366, MKI G01R 29/02. Sposob izmereniya perehodnoy harakteristiki apparaturi magnitnoy zapisi. – N°2069342/18-10, zayavl. 18.10.74; opubl. 30.01.76. – UDK 621.3.087, *Otkritiya. Izobreteniya* **4** (1976), 130 (in Russian).

[43] A. N. Baranov and V. A. Slaev, Avt. svid. N°508799, MKI G11D 27/36. Ustroystvo dlia izmereniya perehodnoy harakteristiki apparaturi magnitnoy zapisi. – N°2054591/18-10, zayavl. 26.08.74; opubl. 30.03.76. – UDK 681.84.083.8, *Otkritiya. Izobreteniya* **12** (1976), 127–128 (in Russian).

[44] A. N. Baranov, V. A. Slaev, and A. I. Sobolev, Avt. svid. N°538310, MKI G01R 29/00. Ustroystvo dlia izmereniya iskajeniy magnitnoy zapisi-vosproizvedeniya – N°2147706/10, zayavl. 23.05.75; opubl. 05.12.76. – UDK 621.317(088.8), *Otkritiya. Izobreteniya* **45** (1976), p. 167 (in Russian).

[45] A. N. Baranov, V. A. Slaev, and A. I. Sobolev, Metod opredeleniya pogreshnosti registracii signalov apparaturoy tochnoy magnitnoy zapisi, in: *Sovremenniye problemi metrologii*, vip. 6, pp. 28–29, Atomizdat, Moskva, 1977 (in Russian).

[46] Y. Bard, *Nonlinear Parameter Estimation*, Academic Press, San Diego, CA, 1974.

[47] S. I. Baskakov, *Radiotehnicheskiye cepi i signali*, Visshaya shkola, Moskva, 2000 (in Russian).

[48] B. Beizer, *Black-Box Testing*, John Wiley, New York, NY, 1995.

[49] B. Beizer, *Software Testing Techniques*, Van Nostrand, New York, NY, 1990.

[50] V. V. Belyakov and L. N. Zakashanskiy, Application of queuing-theoretic methods in determining service requirements for testing facilities, *Measurement Techniques* **19**(10) (1976), 939–942.

[51] M. Bencze, About Seiffert's mean, *RGMIA Research Report Collection* **3**(4) (2000), 681–707.

[52] J. S. Bendat and A. G. Piersol, *Engineering Applications of Correlation and Spectral Analysis*, John Wiley and Sons, Inc., New York, NY, 1980.

[53] J. S. Bendat and A. G. Piersol, *Random Data: Analysis and Measurement Procedures*, 4th ed., 2010.

[54] K. Berka, *Izmereniye: pohiatiya, teorii, problemi: Per. s chesch*, pod red. B. V. Biriukova, Progress, Moskva, 1987 (in Russian).

[55] J. Bernoulli, On *the law of large numbers: Part four of Ars Conjectandi*, trans. by O. Sheynin, NG Verlag, Berlin, Germany, 2005.

[56] B. Wichmann, G. I. Parkin and R. M. Barker, Software Support for Metrology Best Practice Guide No 1. Validation of Software in Measurement Systems, Version 2.1, March 2004, *NPL DEM-ES* **014**, 2007.

[57] P. J. Bickel and K. A. Doksum, *Mathematical Statistics: Basic Ideas and Selected Topics*, 2nd ed., vol. 1, updated printing, Pearson Prentice Hall, Upper Saddle River, NJ, 2007.

[58] P. J. Bickel and K. A. Doksum, *Mathematical Statistics: Basic Ideas and Selected Topics,* vol. II, Pearson Prentice Hall, Upper Saddle River, NJ, 2009.

[59] M. P. Bobnev, *Generirovaniye sluchaynih signalov*, Energiya, Moskva, 1971 (in Russian).

[60] T. Bocor, L. A. Mironovskiy, N. L. Mikhailov, and P. Rosza, A uniform algorithm for the transformation of MVS into canonical forms, *Linear algebra and its applications* **147** (1991), 441–467.

[61] Yu. L. Bogorodskiy, *Razreshayuschaya sposobnost' system magnitnoy zapisi*, pod red. A. F. Bogomolova, Energiya, Moskva, 1980 (in Russian).

[62] B. D. Bomshteyn, L. K. Kiselev, and E. T. Morgachev, *Metodi bor'bi s pomehami v kanalah provodnoy sviazi*, Sviaz', Moskva, 1975 (in Russian).

[63] N. A. Borodachev, *General problems of the theory of production accuracy*, AN USSR, Moskva, 1959 (in Russian).

[64] J. M. Borwein, Problem 10281. A Cubic Relative of the AGM, *Amer. Math. Monthly* **103** (1996), 181–183.

[65] J. M. Borwein and P. B. Borwein, Inequalities for compound mean iterations with logarithmic asymptotes, *J. Math. Anal. Appl.* **177** (1993), 572–582.

[66] J. M. Borwein and P. B. Borwein, *Pi and the AGM – a Study in Analytic Number Theory and Computational Complexity*, John Wiley & Sons, New York, 1987.

[67] J. M. Borwein and P. B. Borwein, The arithmetic-geometric mean and fast computation of elementary functions, *SIAM Rev.* **26**(3) (1984), 351–366.

[68] J. M. Borwein and P. B. Borwein, The way of all means, *Amer. Math. Monthly* **94**(6) (1987a), 519–522.

[69] G. E. P. Box, G. M. Jenkins, and G. C. Reinsel, *Time Series Analysis, Forecasting and Control*, 3rd ed., Prentice Hall, Englewood Clifs, NJ, 1994.

[70] V. B. Braginskiy, *Fizicheskiye experimenti s probnimi telami*, Fizmatgiz, Moskva, 1970 (in Russian).

[71] V. B. Braginskiy and A. B. Manukin, *Izmereniye malih sil v fizicheskih experimentah*, Nauka, GRFML, Moskva, 1974 (in Russian).

[72] D. A. Braslavskiy, Kvorum-elementi dlia ystroystv s funktcional'noy izbitochost'yu, in: *Systemi s peremennoy strukturoy i ih primeneniye v zadachah avtomatizatsii poleta*, pp. 217–225, Nauka, Moskva, 1968 (in Russion).

[73] D. A. Braslavskiy and A. M. Yakubovich, Optimal'noye preobrazovaniye neskol'kih priborov s uchetom pogreshnostey i otkazov, *Avtomatika i telemehanika* **10** (1968), 128–135 (in Russian).

[74] J. L. Brenner and R. C. Carlson, Homogeneous mean values: weights and asymptotics, *J. Math. Anal. Appl.* **123**(1) (1987), 265–280.

[75] L. N. Brianskiy, A. S. Doynikov and B. N. Krupin, *Metrologiya. Shkali, etaloni, praktika*, VNIIFTRI, Moskva, 2004 (in Russian).

[76] D. R. Brillinger, *Time Series: Data Analysis and Theory*, McGraw-Hill, 1981 (Reprint: Society for Industrial & Applied Mathematics, 2001).

[77] G. S. Britov and L. A. Mironovski, Testing Linear Finite Automata in the Case of Incomplete State Information, *Cybernetics* **12**(3) (1974), 4–9.

[78] K. Brown, http://www.mathpages.com/HOME/kmath130.htm.

[79] P. S. Bullen, *Handbook of Means and Their Inequalities,* Kluwer Acad. Publ., Dordrecht, 2003.

[80] P. S. Bullen, D. S. Mitrinovic and P. M. Vasic, *Means and Their Inequalities*, D. Reidel Publishing Co, Dordrecht, 1988.

[81] V. A. Burgov, *Fizika magnitnoy zvukozapisi*; Iskusstvo, Moskva, 1973 (in Russian).

[82] V. A. Burgov, *Teoriya fonogramm*, Iskusstvo, Moskva, 1984 (in Russian).

[83] F. Burk, The geometric, logarithmic and arithmetic mean inequalities, *Amer. Math. Monthly* **94**(6) (1987), 527–528.

[84] N. R. Campbell, *An account of the principles of measurement and calculation*, Longmans, Green, London, 1928.

[85] N. R. Campbell, *Physics – The elements (1920). Representations as Foundations of Science*, Dover, N.Y., 1957.

[86] P. J. Campion, J. E. Burns, and A. Williams, *A code of practice for the detailed statement of accuracy*, National Physical Laboratory, Her Majesty's Stationery Office, London, 1973.

[87] B. C. Carlson, Algorithms involving arithmetic and geometric means, *Amer. Math. Monthly* **78**(5) (1971), 496–505.

[88] B. C. Carlson, The logarithmic mean, *Amer. Math. Monthly* **79** (1972), 615–618.

[89] B. C. Carlson and J. L. Gustafson, Total positivity of mean values and hypergeometric functions, *SIAM J. Math. Anal.* **14**(2) (1983), 389–395.

[90] S. C. Chao, Flutter and Time Errors in Instrumentation Magnetic Recording, in: *IEEE Trans. on Aerospace and Electronic Systems*, vol. AES-2, no 2, 1966, pp. 214–222.

[91] S. C. Chao, The Effect of Flutter on a recorded Sinewave, in: *Proc. IEEE* vol. 53, no 7, 1965, pp. 726–727.

[92] V. M. Chernicer and B. G. Kaduk, *Preobrazovateli vremennogo masshtaba*, Sov. radio, Moskva, 1972 (in Russian).

[93] Y.-M. Chu, Y.-F. Qiu, M.-K. Wang, and G.-D. Wang, The optimal convex combination bounds of arithmetic and harmonic means for the Seiffert's mean. *J. Inequal. Appl.* V/2010 (2010), Article ID 436457, doi:10.1155/2010/436457.

[94] Y. M. Chu and W. F. Xia, Two sharp inequalities for power mean, geometric mean and harmonic mean. *J. Inequal. Appl.* (2009), Article No. 741923.

[95] A. G. Chunovkina, Determining Reference Values and Degrees of Equivalence in Key Comparisons, *Measurement Techniques* **46**(4) (2003), 409–414.

[96] A. G. Chunovkina, *Evaluation of Data in Key Comparisons of National Measurement Standards*, Professional, Sankt-Peterburg, 2009 (in Russian).

[97] A. G. Chunovkina, Introducing measurement uncertainty into methods of calibrating and checking measuring instruments, *Measurement Techniques* **51**(3) (2008), 341–343.

[98] A. G. Chunovkina, Measurement error, measurement uncertainty and measurand uncertainty, *Measurement Techniques* **43**(7) (2000), 581–586.

[99] A. G. Chunovkina, Methodological problems of realizing of accuracy algorithms according to the "Guide to the Expression of Uncertainty in Measurement", in: *Metrological Aspects of Data Processing and Information Systems in Metrology, PTB-IT–7*, pp. 146–155, Braunschweig and Berlin, Juni 1999.

[100] A. G. Chunovkina, Planning experiments that involve measurement and analyzing the experimental data with the construction of calibration curves of the analytical instruments, *Measurement Techniques* **41**(3) (1998), 294–300.

[101] A. G. Chunovkina, Sravneniye na etaloni i mejdunarodniye sravneniya, in: *Metrologiya i izmeritelna tehnika. Pod obschata redakciya na prof*, d.t.n. Hristo Radev, T. 1, 2008, pp. 546–568 (in Bulgarian).

[102] A. G. Chunovkina, Tasks in Estimating Measurement Accuracy, *Measurement Techniques* **44**(11) (2001), 1158–1160.

[103] A. G. Chunovkina, Usage of non-central probability distributions for data analysis in metrology, in: P. Ciarlini, M. G. Cox, F. Paese and G. B. Rossi (eds.), *Advanced mathematical and computational Tools in Metrology VI*, pp. 180–188, Wold Scientific Publishing Company 2004.

[104] A G. Chunovkina and A. V. Chursin, Certification of algorithms for determination of signal extreme values during measurement, in: *Advanced Mathematical Tools in Metrology III*, pp. 165–170, World Scientific Publishing Company 1997.

[105] A. G. Chunovkina and A. V. Chursin, "Guide to the expression of uncertainty in measurement" (GUM) and "Mutual recognition of national measurement standards and of calibration and measurement certificates issued by national metrology institutes" (MRA): some problems of data processing and measurement uncertainty evaluation, in: P. Ciarlini, M. G. Cox, E. Filipe, F. Pavese and D. Richter (eds.), *Advanced Mathematical and Computational Tools in Metrology V*; pp. 55–66, World Scientific Publishing Company 2001.

[106] A. G. Chunovkina and A. V. Chursin, Design of the measurement for the evaluation of calibration curve parameters, in: *Metrological Aspects of Data Processing and Information Systems in Metrology, PTB-IT-7*, pp. 111–118, Braunschweig and Berlin, 1999.

[107] A. G. Chunovkina and A. V. Chursin, Metrological certification of algorithms for turning-point position and value determination from measurements, *Measurement Techniques* **41**(8) (1998), 774–777.

[108] A. G. Chunovkina and A. V. Chursin, Selecting the parameters of a measurement experiment and a processing algorithm for identification of the dynamic characteristics of measurement devices, *Measurement Techniques* **36**(8) (1993), 872–874.

[109] A. G. Chunovkina, C. Elster, I. Lira and W. Woeger, Analysis of key comparison data and laboratory biases, *Metrologia* **45** (2008), 211–216.

[110] A. G. Chunovkina, C. Elster, I. Lira and W. Woeger, Evaluating systematic differences between laboratories in interlaboratory comparisons, *Measurement Techniques* **52**(7) (2009), 788–793.

[111] A. G. Chunovkina and I. A. Kharitonov, International comparisons as a tool for confirming the measurement and calibration capabilities of national metrological institutes, *Measurement Techniques* **50**(7) (2007), 711–714.

[112] A. G. Chunovkina, A. Link, and C. Elster, Comparison of Different Approaches for the Linking of RMO and CIPM Key Comparisons, http://mscsmq.vniim.ru/seminar-06-eng.html.

[113] A. G. Chunovkina and V. A. Slaev, Control of the parameter values on the basis of measurement results and associated uncertainties, in: *Proceedings of the 9-th International Symposium on Measurement Technology and Intelligent Instruments ISMTII-2009*, vol. 1, pp. 1-036–1-040, Saint Petersburg, 2009.

[114] A. G. Chunovkina, V. A. Slaev, and N. A. Burmistrova, Garmonizaciya mejdunarodnoy i otechestvennoy metrologicheskoy terminologii, *Mir izmereniy* **10** (2011), 36–39 (in Russian).

[115] A. G. Chunovkina, V. A. Slaev, A. V. Stepanov, and N. D. Zviagin, Estimating measurement uncertainty on using software, *Measurement Techniques* **51**(7) (2008), 699–705.

[116] A. G. Chunovkina and A. V. Stepanov, An algorithm for detecting drift in the value of the comparison standard when analyzing the data of key comparisons of national standards, *Measurement Techniques* **51**(2) (2008), 219–225.

[117] P. Ciarlini, M. Cox, F. Pavese, and G. Regoliosi, The use of a mixture of probability distributions in temperature interlaboratory comparisons, *Metrologia* **41** (2004), 116–121.

[118] P. Ciarlini and F. Pavese, Some Basic Questions in the Treatment of Intercomparison Data, *Measurement Techniques* **46**(4) (2003), 421–426.

[119] CIPM (MRA), *Mutual recognition of national measurement standards and calibration and measurement certificates issued by national metrology institutes*, Bureau International des Poids et Mesures, Sèvres, 1999.

[120] H. R. Cook, M. G. Cox, M. P. Dainton, and P. M. Harris, A Methodology for Testing Spreadsheets and Other Packages Used in Metrology, in: *NPL Report CISE* 25/99, September 1999.

[121] COOMET R/GM/14: 2006 Guidelines for data evaluation of COOMET key comparison, www.coomet.org.

[122] COOMET R/GM/19: 2008 Guidelines on COOMET supplementary comparison evaluation, www.coomet.org.

[123] COOMET R/LM/10: 2004 Recommendation, Software in Measuring Instruments: General Technical Specifications, www.coomet.org.

[124] I. Costin and G. Toader, Generalized inverses of power means, *General Mathematics* **13**(3) (2005), 61–70.

[125] D. A. Cox, Gauss and the arithmetic-geometric mean, *Notices of the Amer. Math. Soc.* **32**(2) (1985), 147–151.

[126] M. G. Cox, The evaluation of key comparison data: An introduction, *Metrologia* **39** (2002), 587–588.

[127] M. G. Cox, The evaluation of key comparison data, *Metrologia* **39** (2002), 589–595.

[128] M. G. Cox, The evaluation of key comparison data: determining the largest consistent subset, *Metrologia* **44** (2007), 187–200.

[129] M. G. Cox and P. M. Harris, Technical Aspects of Guidelines for the Evaluation of Key Comparison Data, *Measurement Techniques* **47**(1) (2003), 102–111.

[130] H. Cramer, *Mathematical Methods of Statistics*, 19th ed., Princeton University Press, 1999.

[131] D. Creisinger, Reducing Distortion in Analog Tape Recorders, *J. Audio Eng. Soc.* **23**(2) (1975), 107–112.

[132] E. L. Crow, Optimum Allocation of Calibration Errors, *Industr. Qual. Control* **23**(5) (1966).

[133] R. O. Curtis and D. D. Marshall, Why Quadratic Mean Diameter?, *West. J. Appl. For.* **15**(3) (2000), 137–139.

[134] Data Modeling for Metrology and Testing in Measurement Science, in: F. Pavese and A. B. Forbes (eds.), *Modeling and Simulation in Science, Engineering and Technology*, 2009.

[135] J. E. Decker, A. G. Steele, and R. J. Douglas, Measurement science and the linking of CIPM and regional key comparisons, *Metrologia* **45** (2008), 223–232.

[136] G. L. Devis, *Primeneniye tochnoy magnitnoy zapisi*, Energiya, Moskva, 1967 (in Russian).

[137] P. Dietz and L. R. Carley, Simple Networks for Pixel Plane Median Filtering, *IEEE Trans. CAS* **40**(12) (1993), 799.

[138] S. A. Doganovskiy and V. A. Ivanov, *Bloki reguliruyemogo zapazdivaniya*, GEI, Moskva – Leningrad 1960 (in Russian).

[139] V. A. Dolgov, V. A. Krivov, A. N. Ol'hovskiy and A. A. Grishanov, Perspective planning of measurement methods, *Metrologiya* **3** (1981), 27 (in Russian).

[140] E. F. Dolinskiy, Analysis of the results of checking gauges and instruments, *Measurement Techniques* **1**(3) (1958), 273–282.

[141] E. F. Dolinskiy, *Obrabotka rezul'tatov izmereniy*, Izd-vo standartov, Moskva, 1973 (in Russian).

[142] N. G. Domostroeva and A. G. Chunovkina [D.I. Mendeleev Inst.], COOMET.M.V-K1 key intercomparison of liquid viscosity measurement, *Metrologia* **45** (2008), Tech. Suppl. 07009.

[143] Z. Domotor and V. Batitskiy, The Analitic versus Representatinal Theory of Measurement: A Philosophy of Science Perspective, *Measurement Science Review* **8/1**(6) (2008), 129–146.

[144] L. I. Dovbeta, V. V. Liachnev and T. N. Siraya, *Osnovi teoreticheskoy metrologii*, SPbGETU "LETI", St. Petersburg, 1999 (in Russian).

[145] *Draft Measurement Canada Specification* (Metrological Software), 2002.

[146] A. T. Dyuzhin and Yu. A. Makhov, Application of magnetic recording systems in measuring and transmitting reference signals, *Measurement Techniques* **12**(11) (1969), 1505–1507.

[147] W. E. Eder, A Viewpoint on the Quantity "plane Angle", *Metrologia* **18** (1982), 1.

[148] N. Yu. Efremova and A. G. Chunovkina, Experience in evaluating the data of interlaboratory comparisons for calibration and verification laboratories, *Measurement Techniques* **50**(6) (2007), 584–592.

[149] B. Efron, Six questions raised by the bootstrap, in: R. LePage and L. Billard (eds.), *Exploring the Limits of Bootstrap*, Wiley, New York, 1992, pp. 90–126.

[150] B. Efron and R. J. Tibshirani, An Introduction to the Bootstrap, *Monografs of Statistics and Applied Probability* vol. 57, Chapman and Hall, Boca Raton, FL, 1993.

[151] A. N. Egorov and L. A. Mironovskiy, Use of Dynamic System Zeros in Engineering Diagnostics Problems, *Engineering Simulation* **14** (1997), 893–906.

[152] M. Eid and E. I. Tcvetkov, *Potencial'naya tochnost' izmeritel'nih avtomatov*, SZO MA, St. Petersburg, 1999 (in Russian).

[153] Ch. Eisenhart, Realistic Evaluation of the Presision and Accuracy of Instrument Calibration Systems, *J. Res. NBS, Eng. Instr.* **2** (1963).

[154] R. R. Harchenko (pod red.), *Electricheskiye izmeritel'niye preobrazovateli*, Energiya, Moskva – Leningrad, 1967 (in Russian).

[155] C. Elster, Calculation of uncertainty in the presence of prior knowledge, *Metrologia* **44** (2007), 111–116.

[156] C. Elster, A. G. Chunovkina, and W. Woeger, Linking of RMO key comparison to a related CIPM key comparison using the degrees of equivalence of the linking laboratories, *Metrologia* **47** (2010), 96–102.

[157] C. Elster, A. Link, and W. Woger, Proposal for linking the results of CIPM and RMO key comparisons, *Metrologia* **40** (2003), 189–194.

[158] C. Elster and B. Toman, Analysis of key comparisons: estimating laboratories' biases by a fixed effects model using Bayesian model averaging, *Metrologia* **47** (2010), 113–119.

[159] C. Elster, W. Woeger, and M. Cox, Analysis of Key Comparison Data: Unstable Travelling Standards, *Measurement Techniques* **48**(9) (2005), 883–893.

[160] D. J. Evans, N. Yaacob, A fourth order Runge-Kutta method based on the Heronian mean formula, *Int. J. Comput. Math.* **58** (1995), 103–115.

[161] S. E. Fal'kovich, *Ocenka parametrov signala*, Sov. radio, Moskva, 1970 (in Russian).

[162] A. N. Fedorenko, *Magnitnaya seysmicheskaya zapis'*, Nedra, Moskva, 1964 (in Russian).

[163] V. V. Fedorov, *Teoriya optimal'nogo experimenta*, Nauka, Moskva, 1971 (in Russian).

[164] L. Finkelstein, Fundamental concepts of measurement, in: *Measurement and Instrumentation, Acta IMEKO*, vol. 1 (1973), pp. 11–27, Budapest, 1974.

[165] P. Fishborn, *Teoriya poleznosti dlia priniatiya resheniy*, Nauka, Moskva, 1978 (in Russian).

[166] R. A. Fisher, *Statistical methods and scientific inference*, 3rd ed., Macmillan, New York, 1973 (reissued 1990).

[167] J. P. Fitch, E. J. Coyle and N. C. Gallagher, Root Properties and Convergence Rates of Median Filters, *IEEE Trans. Acoust., Speech Signal Process*, vol. ASSP-33(1) (1985), pp. 230–240.

[168] A. F. Fomin, O. N. Novoselov, and A. V. Pluschev, *Otbrakovka anomal'nih rezul'tatov izmereniy*, Energoatomizdat, Moskva, 1985 (in Russian).

[169] D. M. E. Foster and G. M. Phillips, Double mean processes, *Bull. Inst. Math. Appl.* **22**(11–12) (1986), 170–173.

[170] D. M. E. Foster and G. M. Phillips, General Compound Means, in: *Approximation Theory and Applications (St. John's, Nfld., 1984)*, Res. Notes in Math. 133, pp. 56–65, Pitman, Boston, London, 1985.

[171] D. M. E. Foster and G. M. Phillips, The arithmetic-harmonic mean, *Math. of Computation* **42**(165) (1984), 183–191.

[172] L. E. Franks, *Signal Theory*, Englewood Cliffs, New Jersey, Prentice-Hall, 1969.

[173] A. E. Fridman, *Osnovi metrologii. Sovremenniy kurs*, Professional, St. Petersburg, 2008 (in Russian).

[174] K. Fu, *Syntactical methods in pattern recognition*, Academic Press, New York, 1974.

[175] F. Gantmacher, *Matrix Theory*, Chelsea Publishing, New York, 1959.

[176] D. V. Gaskarov and V. I. Shapovalov, *Malaya viborka*, Moskva, 1978 (in Russian).

[177] C. F. Gauss, *Theoria combinationis observationum erroribus minimis obnoxiae. Werke* vol. 4, Göttingen, 1823 (English translation by C. W. Stewart, SLAM, Philadelphia, 1963).

[178] C. F. Gauss, *Theoria motus corporum coelestium in sectionibus conicis solem ambientium*, Hamburg, 1809 (English translation by C. H. Davis, Dover, 1963, reprinted 2004).

[179] I. M. Gelfand, On normed rings, *Doklady Akademii Nauk USSR* **23** (1939), 430–432.

[180] *General Principle of Software Validation, FDA Guidance for Industry*, U.S. FDA, 2002.

[181] *General Requirements for Software Controlled Measuring Instruments*, OIML D **31**, 2008.

[182] E. P. Gil'bo and I. B. Chelpanov, *Obrabotka signalov na osnove uporiadochennogo vibora (majoritarnoye i blizkoye k nemu preobrazovaniya)*, Sov. radio, Moskva, 1976 (in Russian).

[183] C. Gini, *Le Medie*, Unione Tipografico Torinese, Milano, 1958.

[184] C. Gini, *Variabilità e mutabilità. Lezioni di Politica*, Economica, 1926.

[185] M. V. Gitlic, *Magnitnaya zapis' signalov*, Radio i sviaz', Moskva, 1981 (in Russian).

[186] M. V. Gitlic, *Magnitnaya zapis' v systemah peredachi informacii*, Sviaz', Moskva 1978 (in Russian).

[187] S. W. Golomb (ed.), *Digital Communications with Space Applications*, Prentice-Hall, Englewood Cliffs, NJ, 1964.

[188] A. A. Golovan and L. A. Mironovskiy, Algorithmic control of a Kalman filter, *Automation and Remote Control* **54**(7/2) (1993), 1183–1294.

[189] A. V. Goncharov, V. I. Lazarev, V. I. Parhomenko and A. B. Shteyn, *Tehnikà magnitnoy videozapisi*, Energiya, Moskva, 1970 (in Russian).

[190] S. V. Gorbatsevich, E. F. Dolinskii and M. F. Yudin, The system of units and counting units, *Measurement Techniques* **21**(8) (1978), 40.

[191] *GOST 8.009-84, State System of Measurement Assurance, Normiruyemiye metrologicheskiye harakteristiki sredstv izmereniy*, Izd-vo standartov, Moskva, 1985 (in Russian).

[192] *GOST 8.027-01, State System of Measurement Assurance, Gosudarstvennaya poverochnaya skhema dlia sredstv izmereniy postoyannogo elektricheskogo napriajeniya i elektrodvijuschey sili*, Izd-vo standartov, Moskva, 2001 (in Russian).

[193] *GOST 8.057, State System of Measurement Assurance, Standards of physical units*, Basic provision (in Russian).

[194] *GOST 8.061-80, State System of Measurement Assurance, Poverochniye skhemi, Soderjaniye i postroeniye*, Izd-vo standartov, Moskva, 1980 (in Russian).

[195] *GOST 16263, State System of Measurement Assurance, Metrologiya, Termini i opredeleniya*, Izd-vo standartov, Moskva, 1973 (in Russian).

[196] *GOST 28195-89, Software Quality Evaluation, General Concepts* (in Russian).

[197] *GOST ISO/IEC 17025-2009, General requirements for the competence of testing and calibration laboratories.*

[198] *GOST R 8.129-99, State System of Measurement Assurance, Gosudarstvennaya poverochnaya skhema dlia sredstv izmereniy vremeni i chastoti*, Izd-vo standartov, Moskva, 1999 (in Russian).

[199] *GOST R 8.596-2002, State System of Measurement Assurance, Metrological Support to Measurement Systems: Basic Concepts* (in Russian).

[200] *GOST R 8.654-2009, State System of Measurement Assurance, Requirements to Software for Measurement Instruments* (in Russian).

[201] V. A. Granovskiy, *Dinamicheskiye izmereniya: Osnovi metrologicheskogo obespecheniya*, Energoatomizdat, Leningrad, 1984 (in Russian).

[202] V. A. Granovskiy, *Systemnaya metrologiya – metrologicheskiye systemi i metrologiya system*. CNIIEP, St. Petersburg, 1999 (in Russian).

[203] N. Greif and D. Richter, Software Validation and Preventive Software Quality Assurance for Metrology, in: *Data Modeling for Metrology and Testing in Measurement Science*, pp. 371–412, 2009.

[204] U. Grenander, *Lectures in Pattern Theory*, (vol. 1 *Pattern Synthesis*, 1976; vol. 2 *Pattern Analysis*, 1978; vol. 3 *Regular Structures*, 1981) Springer-Verlag, New York, Heidelberg, Berlin, 1976–1981.

[205] G. Gretcer, *Obschaya teoriya reshietok*, Mir, Moskva, 1982 (in Russian).

[206] K. Guan and H. Zhoun, The generalized Heronian mean and its inequalities, *Univ. Beograd, Publ. Elektrotehn. Fak., Ser. Mat.* **17** (2006), 60–75.

[207] Guidance for the Management of Computers and Software in Laboratories with Reference to ISO/IEC 17025/2005, *EUROLAB Technical Report* **2** (2006).

[208] B. N. Guo and F. Qi, A simple proof of logarithmic convexity of extended mean values, *Numer. Algorithms* **52** (2009), 89–92.

[209] P. H. Sydeham (ed.), *Handbook of Measurement Science*, J. Wiley and Sons Ltd., New York, 1982.

[210] R. Hamming, *Coding and Information Theory* 2nd ed., Prentice Hall, 1986.

[211] F. R. Hampel, *Robust Statistics: The Approach Based on Influence Functions*, Wiley Series in Probability and Statistics, Wiley-Interscience, 2005.

[212] J. Hajek, Z. Sidak, and P. K. Sen, *Theory of rank tests* 2nd ed., Academic Press, 1999.

[213] A. A. Harkevich, *Spektri i analiz*, GIFML, Moskva, 1962 (in Russian).

[214] P. A. Hästö, A Monotonicity Property of Ratios of Symmetric Homogeneous Means, *J. Ineqal. Pure and Appl. Math.* **3**(5) (2002), article 71.

[215] P. A. Hästö, Optimal inequalities between Seiffert's mean and power means, *Math. Inequal. Appl.* **7**(1) (2004), 47–53.

[216] H. Helmholtz, Schet i izmereniye, *Izvestiya Kazanskogo fiz.-mat. ob-va* **2** (1892) (in Russian).

[217] D. Hoffmann, *Handbuch Messtechnik und Qualitätssicherung,* 2. Stark bearb. Auflage, VEB Verlag Technik, Berlin, 1981.

[218] D. Hoffmann, Theoretical, physical and metrological problems of further development of measurement techniques and instrumentation in science and technology, *Acta IMEKO* (1979).

[219] W. E. Howard, Means Appearing in Geometrical Figures, *Mathematics Magazine* **76**(4) (2003), 292–294.

[220] T. S. Huang (ed.), *Two-Dimensional Digital Signal Processing II (Transforms and Median Filters)*, Springer-Verlag, 1981.

[221] T. S. Huang, G. T. Yang and G. Y. Tang, *A Fast Two-Dimensional Median Filtering Algorithm*, 1979.

[222] P. J. Huber, *Robust Statistics*, Wiley, 1981 (republished as paperback 2004).

[223] Ya. I. Hurgin and V. P. Yakovlev, *Finitniye funkcii v fizike i tehnike*, Nauka, Moskva, 1971 (in Russian).

[224] *IEC 61508, Functional of Electrical/Electronic/Programmable Electronic Safety-related Systems.*

[225] E. C. Ifeachor, B. W. Jervis, *Digital Signal Processing: A Practical Approach* 2nd ed., Pearson Education, 2002.

[226] C. O. Imoru, The power mean and the logarithmic mean, *Internat. J. Math. Sci.* **5**(2) (1982), 337–343.

[227] A. F. Ioffe, *Primeneniye magnitnoy zapisi*, GEI, Moskva, Leningrad, 1959 (in Russian).

[228] L. K. Isaev, The place of metrology in science system: On postulates, *Measurement Techniques* **36**(8) (1993), 853–854.

[229] G. V. Isakov, Problems in controlling system for transmitting physical quantity unit dimensions, *Measurement Techniques* **21**(2) (1978), 24–26.

[230] *ISO 5725 Accuracy (trueness and precision) of measurement methods and results*, parts 1–6, 1994–1998.

[231] *ISO/IEC 14598, Information Technology, Software Product Evaluation.*

[232] *ISO/IEC 15504, Information Technology, Process Assessment.*

[233] *ISO/IEC 16085, Systems and Software Engineering, Life Cycle Processes, Risk Management.*

[234] *ISO/IEC 25000, Software Engineering, Software Product Quality Requirements and Evaluation (SQuaRE), Guide to SQuaRE.*

[235] *ISO/IEC 27005, Information Technology, Security Techniques, Information Security Risk Management.*

[236] Yu. M. Ishutkin, *Izmereniye parazitnih moduliaciy i obuslovlennih imi iskajeniy v systemah zapisi-vosproizvedeniya zvuka*, LIKI, Leningrad, 1969 (in Russian).

[237] Yu. M. Ishutkin and V. V. Rakovskiy, *Izmereniya v apparature zapisi i vosproizvedeniya zvuka kinofil'mov*, Iskusstvo, Moskva, 1985 (in Russian).

[238] *IT-Related Challenges in Legal Metrology: International Workshop/237th PTB-Seminar*, Berlin, Germany, 2007 (www.ptb.de/de/suche/suche.html).

[239] V. A. Ivanov, Solution of measurement problems with application of the group theory, in: *Proceedings of the seminar "Fundamental problems of metrology"*, VNIIM, Leningrad, 1981 (in Russian).

[240] L. Jánoshy, *Theory and Practice of the Evaluation of Measurements*, Clarendon Press, Oxford, 1965.

[241] W. Janous, A note on generalized Heronian means, *Math. Inequal. Appl.* **4**(3) (2001), 369–375.

[242] J. Jaworski, *Matematyczne podstawy metrologii*, Wydawictwa Wakowo-Techniczne, Warszawa, 1979.

[243] *JCGM 100 (ISO/IEC Guide 98-3:2008), Guide to the Expression of Uncertainty in Measurement*, per. s angl. pod red. prof. V. A. Slaeva, VNIIM im. D.I. Mendeleyeva, St. Petersburg, 1999 (in Russian).

[244] *JCGM 101 (ISO/IEC Guide 98-3:2008/Supplement 1), Propagation of distributions using a Monte Carlo method*, per. s angl. pod red. V. A. Slaeva and A. G. Chunovkinoy, Professional, St. Petersburg, 2010 (in Russian).

[245] *JCGM 104 (ISO/IEC Guide 98-1:2009), An Introduction to the Expression of Uncertainty in Measurement and related documents*, per. s angl. pod red. d.t.n., prof. V. A. Slaeva and A. G. Chunovkinoy, Professional, St. Petersburg, 2011 (in Russian).

[246] *JCGM 200 (ISO/IEC Guide 99:2007), International vocabulary of metrology – Basic and general concepts and associated terms (VIM), per. s angl. i fr. / Vseros. nauch.-issled. in-t metrologii im. D.I. Mendeleyeva; Belorus. in-t metrologii. Izd. 2-e, ispr,* Professional, St. Petersburg, 2010 (in Russian).

[247] N. A. Jeleznov, *Nekotoriye voprosi teorii informacionnih elektricheskih system*, LKVVIA im. À.F. Ìîjayskogo, Leningrad, 1960 (in Russian).

[248] N. L. Johnson and F. C. Leone, Statistics and experimental design, in: *Engineering and physical sciences* vol. 1, New York, 1964; vol. 2, New York, 1966.

[249] V. M. Juravlev, *Metod izmereniy nelineynih iskajeniy s pomosch'yu polos shuma*, CKB kinoapparaturi, Leningrad, 1967 (in Russian).

[250] R. Kacker, Combining information from interlaboratory evaluations using a random effects model, *Metrologia* **41** (2004), 132–136.

[251] H. E. Kalman, Transversal filters, in: *PIRE* vol. 38, pp. 302–311, 1940.

[252] A. I. Kanunov, T. V. Kondakova, E. P. Ruzaev, and E. I. Tsimbalist, *Algoritm postroeniya optimal'nih ierarhicheskih poverochnih schemes. Standartizatsiya i izmeritel'naya tehnika*, vip. 4, p. 3, Krasnoyarsk, 1978 (in Russian).

[253] B. V. Karpyuk and M. P. Tsapenko, Ob izmeritel'nih informatsionnih systemah, *Avtometriya* **2** (1965), 18–25 (in Russian).

[254] A. N. Kartasheva, *Dostovernost' izmereniy i kriterii kachestva ispitaniy priborov*, Izd-vo standartov, Moskva, 1967 (in Russian).

[255] V. M. Kashlakov, *Sovremenniye metodi obespecheniya kachestva poverki sredstv izmereniy (obzor)*, Izd-vo VNIIKI, Moskva, 1985 (in Russian).

[256] G. I. Kavalerov and S. M. Mandel'shtam, *Vvedenie v informacionnuyu teoriyu izmereniy*, Energiya, Moskva, 1974 (in Russian).

[257] B. M. Kedrov, O klassifikacii nayk, in: *Filosofskiye voprosi sovremennoy fiziki*, Misl', Moskva, 1968 (in Russian).

[258] L. V. Khalatova, Accuracy estimation for automated computer-measurement processes, *Measurement Techniques* **28**(3) (1985), 209–211.

[259] I. A. Kharitonov and A. G. Chunovkina, Calculation of the reference value of key comparisons from the results of measurements made by different absolute methods, *Measurement Techniques* **50**(7) (2007), 715–719.

[260] I. A. Kharitonov and A. G. Chunovkina, Evaluating Data from Regional Key Comparisons, *Measurement Techniques* **48**(4) (2005), 327–335.

[261] I. A. Kharitonov and A. G. Chunovkina, Evaluation of regional key comparison data: two approaches for data processing, *Metrologia* **43** (2006), 470–476.

[262] E. A. Khatskevich, L. A. Konopel'ko, and A. G. Chunovkina, Metrological support to chromatographs for monitoring natural gas, *Measurement Techniques* **41**(7) (1998), 675–678.

[263] J.-O. Kim and C. W. Mueller, *Factor Analysis: Statistical Methods and Practical Issues* (11th printing), Sage Publications, Inc., 1986.

[264] W. R. Klecka, *Discriminant Analysis* (7th printing), Sage Publications, Inc., 1986.

[265] E. S. Klibanov and V. A. Slaev, Avt. svid. N°351166, MKI G01r. Ustroystvo dlia opredeleniya amplitudnih harakteristik chetirehpolusnikov, N°1489589/18-10, zayavl. 02.11.70; opubl. 13.09.72. – UDK 621.317.7.083.5(088.8), *Otkritiya. Izobreteniya* **27** (1972), 152 (in Russian).

[266] D. D. Klovskiy, *Teoriya peredachi signalov*, Sviaz', Moskva, 1973 (in Russian).

[267] N. I. Kl'uyev, *Informacionniye osnovi peredachi soobscheniy*, Sov. radio, Moskva, 1966 (in Russian).

[268] V. G. Knorring, Razvitiye reprezentacionnoy teorii izmereniy. *Izmereniye, kontrol', avtomatika* **11–12**(33–34) (1980), 3–9 (in Russian).

[269] A. Kofman, *Vvedeniye v teoriyu nechetkih mnojestv*, Radio i sviaz', Moskva, 1982 (in Russian).

[270] A. N. Kolmogorov, Sur la notion de la moyenns, *Atti Accad. Naz. Lincei. Rend.* **12**(9) (1930), 388–391.

[271] A. N. Kolmogorov and S. V. Fomin, *Elementi teorii funkciy i funkcional'nogo analiza*, Nauka, Moskva, 1972 (in Russian).

[272] E. D. Koltik and V. A. Slaev, Avt. svid. N°296054, MKI G01r 25/04, Ustroystvo dlia izmereniya izmeneniya sdviga faz, N°1346517/18-10, zayavl. 08.07.69; opubl. 12.02.71, – UDK 621.317.772(088.8), *Otkritiya. Izobreteniya* **8** (1971), 140 (in Russian).

[273] E. D. Koltik and V. A. Slaev, Avt. svid. N°288141, MKI G01r 25/04, Sposob izmereniya izmeneniya sdviga faz, N°1346518/18-10, zayavl. 08.07.69; opubl. 03.12.70. – UDK 621.317.373(088.8), *Otkritiya. Izobreteniya* **36** (1970), 81 (in Russian).

[274] L. A. Konopel'ko, Yu. A. Kustikov, T. A. Popova, E. A. Khatskevich, and A. G. Chunovkina, Methods of comparing standard gas mixtures, *Measurement Techniques* **43**(3) (2000), 296–297.

[275] G. A. Korn, *Modelirovaniye sluchaynih processov na analogovih i cifrovih mashinah*, Mir, Moskva, 1968 (in Russian).

[276] V. G. Korol'kov, *Ispitaniya magnitofonov*, Energiya, Moskva, 1965 (in Russian).

[277] V. A. Kotel'nikov, *Teoriya potencial'noy pomehoustoychivosti*, Gosenergoizdat, Moskva, 1956 (in Russian).

[278] E. P. Kotov and M. I. Rudenko, *Lenti i diski v ustroystvah magnitnoy zapisi*, Radio i sviaz', Moskva, 1986 (in Russian).

[279] A. V. Kramov and A. L. Semeniuk, *Vibor obraztsovih sredstv izmereniy (obzor)*, Izd-vo VNIIKI, Moskva, 1975 (in Russian).

[280] D. H. Krantz, R. D. Luce, P. Suppes, and A. Tversky, *Foundations of Measurement*, vol. 1–3, Academic Press, New York (1971–1990).

[281] M. Kraus and E. Voshni, *Informacionniye izmeritel'niye systemi*, Mir, Moskva, 1975 (in Russian).

[282] S. A. Kravchenko, The required reserve of accuracy between steps of a checking scheme in the field of phase measurements, *Measurement Techniques* **16**(1) (1973), 128–133.

[283] Ya. A. Krimshtein's, Optimization of the metrological networks structure, *Measurement Techniques* **20**(3) (1977), 340–343.

[284] Yu. A. Kudeyarov, *Certifying Instrument Software: A Textbook*, FGUP VNIIMS, Moskva, 2006 (in Russian).

[285] Zh. F. Kudriashova and A. G. Chunovkina, Expression for the Accuracy of Measuring Instruments in Accordance with the Concept of Uncertainty of Measurements, *Measurement Techniques* **46**(6) (2003), 559–561.

[286] O. A. Kudriavtsev, L. A. Semenov, and A. E. Fridman, Mathematical modeling system of assurance the measurement uniformity, in: *Proceedings of all-union meeting "Physical problems of accurate measurements,"* p. 8, Leningrad 1984 (in Russian).

[287] G. Kulldorff, *Estimation from Grouped and Partially Grouped Samples*, Wiley, New York, 1961.

[288] H.-J. Kunze, *Physikalische Messmethoden*, Teubner, Stuttgart, 1986.

[289] V. P. Kuznetcov, *Metrologicheskiye harakteristiki izmeritel'nih system*, Mashinostroyeniye, Moskva, 1979 (in Russian).

[290] V. A. Kuznetcov and G. V. Yalunina, *Osnovi metrologii*, Izd-vo standartov, Moskva, 1995 (in Russian).

[291] V. A. Kuznetcov, L. K. Isaev, and I. A. Shayko, *Metrologiya*, Standardinform, Moskva, 2005 (in Russian).

[292] R. Larsen, *Banach Algebras. An Introduction*, Dekker, New York, 1973.

[293] M. V. Laufer, *Izmereniye nestabil'nosti skorosti nositelia zapisi*, Sviaz', Moskva, 1980 (in Russian).

[294] M. V. Laufer and I. A. Krijanovskiy, *Teoreticheskiye osnovi magnitnoy zapisi signalov na dvijuschiysia nositel'*, Vischa shkola, Kiyev, 1982 (in Russian).

[295] R. L. Launer and G. N. Wilkinson (eds.), *Robustness in Statistics: Proceedings of a Workshop*, Academic Press, New York, 1979.

[296] E. B. Leach and M. C. Sholander, Extended mean values II, *J. Math. Anal. Appl.* **92**(1) (1983), 207–233.

[297] A. Lebeg, *Ob izmerenii velichin*, Uchpedgiz, Moskva, 1960 (in Russian).

[298] E. Baghdady (ed.), *Lectures on Communication System Theory*, McGraw-Hill, New York, 1961.

[299] R. Lee, *Optimal Estimation, Identification and Control*, The MIT Press, 1964.

[300] D. H. Lehmer, On the compounding of certain means. *J. Math. Anal. Appl.* **36** (1971), 183–200.

[301] A. M. Leroy and P. J. Roussecuw, *Robust Regression and Outlier Detection*, John Wiley & Sons, New York, 1987.

[302] M. J. Levin, Optimum estimation of impulse response in the presence of noise, in: *IRE Trans. on Circuit Theory*, vol. CT-7, pp. 50–56, 1960.

[303] E. S. Levshina and P. V. Novickiy, *Elektricheskiye izmereniya fizicheskih velichin (Izmeritel'niye preobrazovateli)*, Energoatomizdat, Leningrad, 1983 (in Russian).

[304] V. V. Liachnev, T. N. Siraya, and L. I. Dovbeta, *Osnovi fundamental'noy metrologii*, Elmor, St. Petersburg, 2007 (in Russian).

[305] T. P. Lin, The power mean and the logarithmic mean, *Amer. Math. Monthly* **81** (1974), 879–883.

[306] Yu. V. Linnik, *Metod naimen'shih kvadratov i osnovi teorii obrabotki nabl'udeniy*, Fizmatgiz, Moskva, 1958 (in Russian).

[307] V. V. Lipaev, *Software Quality*, Yanus-K, Moskva, 2002 (in Russian).

[308] I. Lira, Bayesian evaluation of comparison data. *Metrologia* **43** (2006), 231–234.

[309] I. Lira, Combining inconsistent data from interlaboratory comparisons. *Metrologia* **44** (2007), 415–421.

[310] I. Lira, A. Chunovkina, C. Elster, and W. Wöger, Analysis of key comparisons incorporating knowledge about bias, *IEEE T. Instrum. Meas.* **61** (2012), 2079–2084.

[311] I. Lira, C. Elster, and W. Wöger, Probabilistic and least-squares inference of the parameters of a straight-line model, *Metrologia* **44** (2007), 379–384.

[312] H. Liu and X.-J. Meng, The Optimal Convex Combination Bounds for Seiffert's Mean, *J. Inequal. Appl.* **1**(44) (2011), doi:10.1186/1029-242X-2011-44.

[313] V. Ya. Lojnikov, Axiomaticheskiye osnovi metrizacii, in: *Tez. dokl. II Vsesoyuznoy Konferencii po Teoreticheskoy Metrologii*, pp. 13–14, Leningrad, 1983 (in Russian).

[314] V. Lokesha, S. Padmanabhan, K. M. Nagaraja, and Y. Simsek, Relation between Greek means and various means, *General Mathematics* **17**(3) (2009), 3–13.

[315] V. Lokesha, M. Saraj, S. Padmanabhan, and K. M. Nagaraja, Oscillatory mean for several positive arguments, *J. Intelligent System Research* 2(2) (2008), pp. 137–139.

[316] B. Y. Long and Y. M. Chu, Optimal power mean bounds for the weighted geometric mean of classical means, *J. Inequal. Appl.* (2010), Article No. 905697.

[317] A. Lorenz, *Decay Data for Radionuclides used as Ñalibration Standards*, IAEA, Vienna, 1982.

[318] R. Lukac, Adaptive vector median filtering, *Pattern Recognition Letters* **24**(12) (2003), 1889–1899.

[319] L. Lucat, P. Siohan, and D. Barba, Adaptive and global optimization methods for weighted vector median filters, *Signal Processing: Image Communication* **17**(7) (2002), 509–524.

[320] R. D. Luce and P. Suppes, Representational theory, in: *Stevens' Handbook of Experimental Psychophysics*, vol. 4, Wiley, 2002.

[321] D. K. C. MacDonald, *Noise and Fluctuations: An Introduction*, Dover, 2006.

[322] *Magnitnaya zapis' zvuka. Sb. perevodnih materialov*, sostavitel' i redactor V. A. Burgov, Iskusstvo, Moskva, 1956 (in Russian).

[323] M. F. Malikov, *Osnovi metrologii. Ch. 1. Ucheniye ob izmerenii*, Kîmit°t pî delam mer i izmeritel'nih priborov pri SM SSSR, Moskva, 1949 (in Russian).

[324] J. C. Mallinson, The next Decade in Magnetic Recording, *IEEE Trans.* **M21**(16) (1985), 1217–1220.

[325] M. Marcus and H. Minc, *Survey of matrix theory and matrix inequalities*, Prindle, Weber & Schmidt, Dover, 1992.

[326] *Mathematics, statistics and computation to support measurement quality. International Seminar*, VNIIM, St. Petersburg, 2009, 2012.

[327] Measuring Instruments Directive (MID). Directive 2004/22/EC of the European Parliament and the Council on Measuring Instruments. *Official Journal of the European Union* **L135**(1) (2004).

[328] J. S. Meditch, *Stochastic Optimal Linear Estimation and Control*, McGraw-Hill, New York, 1969.

[329] J. K. Mericoski, Extending means of two variables to several variables, *J. Ineq. Pure and Appl. Math.* **5**(3) (2004), Article No. 65.

[330] Ch. Mi, *Fizika magnitnoy zapisi*, pod. red. V. G. Korol'kova, Energiya, Moskva, 1967 (in Russian).

[331] MI 83–76, *State system of measurement. Procedure of measuring parameters of verification schemes*, Izd-vo standartov, Moskva, 1976 (in Russian).

[332] MI 2174-91, *State System of Measurement Assurance. Certifying Algorithms and Data-Processing Software in Measurements: Basic Concepts* (in Russian).

[333] MI 2439-97, *State System of Measurement Assurance. Metrological Characteristics of Measurement Systems: Nomenclature. Principles for Regulation, Determination and Monitoring* (in Russian).

[334] MI 2440-97, *State System of Measurement Assurance. Methods of Experimental Determination and Monitoring for the Error Characteristics of Measurement Channels in Data-Acquisition Systems and Measurement Suites* (in Russian).

[335] MI 2955-2010, *State System of MeasurementMeasurement Assurance. A Typical Method of Certifying Software in Measuring Instruments and Sequence for Conducting It.* (in Russian).

[336] MI 3286-2010, *Proverka zaschiti programmnogo obespecheniya i opredeleniye ee urovnia pri ispitaniyah sredstv izmereniy v tseliah utverjdeniya tipa* (in Russian).

[337] MI 3290-2010, *Gosudarstvennaya systema obespecheniya edinstva izmereniy. Rekomendatciya po podgotovke, oformleniyu i rassmotreniyu materialov ispitaniy sredstv izmereniy v tseliah utverjdeniya tipa* (in Russian).

[338] N. P. Mif, *Modeli i otsenka pogreshnosti tehnicheskih izmereniy*, Izd-vo standartov, Moskva, 1976 (in Russian).

[339] L. A. Mironovskii, *Funktsional'noye diagnostirovaniye dinamicheskih system*, Izd-vo MGU-GRIF, Moskva – St. Petersburg, 1998 (in Russian).

[340] L. A. Mironovskii, Linear Systems with Multiple Singular Values, *Automation and Remote Control* **70**(1) (2009), 43–63.

[341] L. A. Mironovskii, *Modelirovaniye konechnomernih system. Momenti i markovskiye parametri*, LIAP, Leningrad, 1988 (in Russian).

[342] L. A. Mironovskii, The use of analytical redundancy in navigational measuring systems, *Measurement Techniques* 50(2) (2007), pp. 142–148 (Electronic Only).

[343] L. A. Mironovskii, Functional Diagnosis of Dynamical Systems – A survey, *Automation and Remote Control* **41** (1980), 1122–1143.

[344] L. A. Mironovskii, Functional Diagnosis of Linear Dynamic Systems, *Automation and Remote Control* **40** (1979), 1198–1205.

[345] L. A. Mironovskii, Functional Diagnosis of Nonlinear Discrete-Systems, *Automation and Remote Control* **50**(6/2) (1989), 838–843.

[346] L. A. Mironovskii, Hankel operator and Hankel functions of linear systems, *Automation and Remote Control* **53**(9/1) (1993), 1368–1380.

[347] L. A. Mironovskii and V. A. Slaev, Invariants in Metrology and Technical Diagnostics, *Measurement Techniques* **39**(6) (1996), 577–593.

[348] L. A. Mironovskii and V. A. Slaev, Optimal test signals as a solution of the generalized Bulgakov problem, *Automation and Remote Control* **63**(4) (2002), 568–577.

[349] L. A. Mironovskii and V. A. Slaev, Technical Diagnostics of Dynamic Systems on the Basis of Algebraic Invariants, in: *Proceedings of the III. Symposium of the IMEKO, TC on Technical Diagnostics*, pp. 243–251, IMEKO Secretariat, Budapest, 1983.

[350] L. A. Mironovskii and V. A. Slaev, *Algoritmi ocenivaniya rezul'tata tryeh izmereniy*, Professional, St. Petersburg, 2010 (in Russian).

[351] L. A. Mironovskii and V. A. Slaev, Ocenivaniye rezul'tatov izmereniy po malim viborkam, *Informacionno-upravliayuschiye systemi* 1 (2011). pp. 69–78 (in Russian).

[352] L. A. Mironovskii and V. A. Slaev, Root Images of Two-Sided Noise Immune Strip-Transformation, in: *Proceedings of the International Workshop on Physics and Mathematics IWPM 2011*, Hangzhou, PRC, 2011.

[353] L. A. Mironovskii and V. A. Slaev, *Strip-Method for Image and Signal Transformation*, De Gruyter, Berlin, 2011.

[354] L. A. Mironovskii and V. A. Slaev, Double-sided noise-immune strip transformation and its root images, *Measurement Techniques* **55**(9) (2012), 6–10.

[355] *Modern Instrumentation Tape Recording EMJ Technology Inc.*, England, 1978.

[356] *Monte Carlo Calculation of 3 Dimensional Geometry*, version 7.07, CNIIRTK, SPbGPU, St. Petersburg, 2007.

[357] P. Moulin, *L'enregistrement Magnetique d'Instrumentation*, Radio, Paris, 1978.

[358] V. I. Mudrov and V. L. Kushko, *Metodi obrabotki izmereniy (kvazipravdopodobniye otsenki)*, Sov. radio, Moskva, 1976 (in Russian).

[359] P. Muliere, On Quasi-Means, *J. Ineq. Pure and Appl. Math.* **3**(2) (1991), Article No. 21.

[360] K. Murugesan, D. P. Dhayabaran, E. C. H. Amirtharaj and D. J. Evans, A fourth order embedded Runge-Kutta RKCeM (4.4) method based on arithmetic and centroidal means with error control, *Int. J. Comp. Math.* **79**(2) (2002), 247–269.

[361] K. Murugesan, D. P. Dhayabaran, E. C. H. Amirtharaj, and D. J. Evans, Numerical strategy for the system of second order IVPs using RK method based on centroidal mean, *Int. J. Comp. Math.* **80**(2) (2003), 233–241.

[362] L. Narens, *Abstract measurement theory*, MIT Press, Cambridge, MA, 1985.

[363] E. Neuman, On Gauss lemniscate functions and lemniscatic mean, *Mathematica Pannonica* **18**(1) (2007), 77–94.

[364] P. E. Ng and K. K. Ma, A Switching Median Filter With Boundary Discriminative Noise Detection for Extremely Corrupted Images, *IP* **15**(6) (2006), 1506–1516.

[365] I. Newton, *Arithmetica universales*, 1707.

[366] L. Nielsen, Evaluation of measurement intercomparisons by the method of least squares, in: *Technical report DFM-99-R39*, Danish Institute of Fundamental Metrology, 2000.

[367] L. Nielsen, Identification and Handling of Discrepant Measurements in Key Comparisons, *Measurement Techniques* **46**(5) (2003), 513–522.

[368] P. V. Novickiy, *Osnovi informacionnoy teorii izmeritelnih ustroystv*, Energiya, Moskva, 1968 (in Russian).

[369] P. V. Novickiy and I. A. Zograf, *Ocenka pogreshnostey izmereniy*, Energoatomizdat, Leningrad, 1985 (in Russian).

[370] T. Nowicki, On the arithmetic and harmonic means, in: *Dynamical Systems*, pp. 287–290, World Sci. Publishing, Singapore, 1998.

[371] S. A. Orlovskiy, *Problemi priniatiya resheniy pri nechetkoy ishodnoy informacii*, Nauka, Moskva, 1981 (in Russian).

[372] P. P. Ornatskiy, *Teoreticheskiye osnovi informacionno-izmeritel'noy tehniki*, Vischa shkola, Kiyev, 1983 (in Russian).

[373] G. Ott, *Noise Reduction Techniques in Electronic Systems*, Wiley-Interscience, 1976.

[374] J. Pahikkala, On contraharmonic mean and Pythagorean triples, *Elemente der Mathematik* **65**(2) (2010), 62–67.

[375] V. G. Panteleev, V. A. Slaev, and A. G. Chunovkina, Metrological provision for image analyzers, *Measurement Techniques* **51**(1) (2008), 107–112.

[376] Ch. Pear, *Magnetic Recording in Science and Industry*, Reinhold Publishing Corp., New York, 1967.

[377] M. S. Pedan and M. N. Selivanov, An experience of applying GOST 8.057 in creation and development the standard base of the country, *Tr. metrolog. in-tov USSR "Obshchiye voprosi metrologii"* **200**(260) (1977), 112 (in Russian).

[378] S. Perreault and P. Herbert, Median Filtering in Constant Time, *IEEE Transactions on Image Processing* **16**(9) (2007), 2389–2394.

[379] V. B. Pestriakov, V. P. Afanasyev, V. L. Gurvic, et al., *Shumopodobniye signali v systemah peredachi informacii*, Sov. radio, Moskva, 1973 (in Russian).

[380] V. P. Petrov and Yu. V. Riasniy, Problems of constructing optimal verification schemes, *Proceedings of the VNIIM "Studies in the field of radio-engineering measurements"* **204**(264) (1976), 5 (in Russian).

[381] N. T. Petrovich and M. K. Razmahnin, *Systemi sviazi s shumopodobnimi signalami*, Sov. radio, Moskva, 1969 (in Russian).

[382] J. Pfanzagl, *Theory of measurement*, New York, 1968.

[383] V. P. Piastro, V. A. Slaev, G. P. Tcivirko, and B. A. Shkol'nik, Avt. svid. N°473125, MKI G01r 29/00, Sposob izmereniya iskajeniy pri magnitnoy zapici i vosproizvedenii, N°1898072/18-10, zayavl. 29.03.73; opubl. 05.06.75. – UDK 621.317.63(088.8), *Otkritiya. Izobreteniya* **21** (1975), 120 (in Russian).

[384] J. Piotrowsky, *Teoria Pomiarow*, Panstwowe Widawnictwo Naukowe, Warzawa, 1986.

[385] Ya. Piotrovskiy, *Teoriya izmereniy dlia injenerov*, Mir, Moskva, 1989 (in Russian).

[386] A. O. Pittenger, Inequalities between arithmetic and logarithmic means, *Univ. Beogard Publ. Elecktrotehn. Fak. Ser. Mat. Fiz.* (1981), 678-715, pp. 15–18.

[387] A. O. Pittenger, The symmetric, logarithmic and power means, *Univ. Beogard Publ. Elecktrotehn. Fak. Ser. Mat. Fiz.* (1981), 678-715, pp. 19–23.

[388] R. H. Prager, Time Errors in Magnetic Recording, *J. Audio Eng. Soc.* **7**(2) (1959), 81–88.

[389] W. K. Pratt, *Digital Image Processing* (4th ed.), Wiley, 2007.

[390] V. I. Pronenko and R. V. Yakirin, *Metrologiya v promishlennosti*, Tehnika, Kiyev, 1979 (in Russian).

[391] F. Qi and S. X. Chen, Complete monotonicity of the logarithmic mean, *Math. Inequal. Appl.* **10**(4) (2007), 799–804.

[392] F. Qi and B. N. Guo, An inequality between ratio of the extended logarithmic means and ratio of the exponential means, *Taiwanese J. Math.* **2**(7) (2003), 229–237.

[393] Y.-F. Qiu, M.-K. Wang, Yu-M. Chu, and G.-D. Wang, Two sharp inequalities for Lehmer mean, identric mean and logarithmic mean, *J. Math. Inequal.* **5**(3) (2011), 301–306.

[394] S. G. Rabinovich, *Evaluating Measurement Accuracy: A practical Approach*, Springer-Verlag, New York, 2010.

[395] S. G. Rabinovich, *Measurement Errors and Uncertainty. Theory and Practice*, Springer-Verlag, Berlin, 2005.

[396] S. G. Rabinovich, *Photogal'vanometricheskiye kompensacionniye pribori*, Energiya, Moskva – Leningrad, 1972 (in Russian).

[397] D. Rabrenovic and M. Lutovac, Minimum Stopband Attenuation of Cauer Filters without Elliptic Functions and Integrals, *IEEE Trans. CAS.* **40**(9) (1993), 618.

[398] C. R. Rao, *Linear Statistical Inference and its Applications* (2nd ed.), Wiley, 1972.

[399] M. Raissouli, Stabilizability of the Stolarsky Mean and its Approximation in Terms of the Power Binomial Mean, *Int. J. Math. Anal.* **6**(18) (2012), 871–881, http://www.m-hikari.com/ijma/ijma-2012/ijma-17-20-2012/raissouliIJMA17-20-2012-2.pdf.

[400] K. A. Reznik, Determination of a number of verification scheme steps, in: *Proceedings of metrology institutes of the USSR "General problems of metrology" ("Obschie voprosi metrologii")* 200(260), p. 106, Leningrad, 1977.

[401] K. A. Reznik, *Matematiko-statisticheskiy analiz system peredachi razmerov edinic fizicheskih velichin ot gosudarstvennih etalonov rabochim sredstvam izmereniy*. Kand. dis., NPO VNIIM im. D.I. Mendeleyeva, Leningrad, 1974 (in Russian).

[402] K. A. Reznik, Sootnosheniye mejdu pogreshnostiami obrazcovogo i poveriayemogo priborov, *Metrologiya* **4** (1971) (in Russian).

[403] RMG 29-99, *State System of Measurement Assurance. Metrologiya. Osnovniye termini i opredeleniya*, Izd-vo standartov, Minsk, 2000 (in Russian).

[404] F. S. Roberts, *Measurement theory*, Addison-Wesley, Reading, MA, 1979.

[405] V. N. Romanov and V. A. Slaev, Formalizaciya poniatiya kachestva metrologicheskih system na osnove nechetkih mnojestv, *Metrologiya* **1** (1985), 11–17 (in Russian).

[406] V. N. Romanov and V. A. Slaev, Obobscheniye zadachi mnogocelevoy optimizacii system na osnove nechetkih mnojestv, *Metrologiya* **12** (1985), 3–13 (in Russian).

[407] V. N. Romanov and V. A. Slaev, Principi obrazovaniya systemi osnovnih poniatiy metrologii, in: *Tez. dokl. Vsesoyuznogo nauchno-tehnicheskogo seminara "Teoreticheskiye problemi elektrometrii"*, pp. 10–14, Tartu, 1985 (in Russian).

[408] L. Rosenberg, The iteration of means, *Math. Mag.* **39** (1966), 58–62.

[409] L. G. Rosengren, Optimal Design of Hierarchy of Calibrations, *IMIKO VI*, Dresden, 1973, Sec. 1, Preprint, Berlin, 1973, p. 122.

[410] L. L. Rosine, Instrumentation Tape Recorders, *Electro-Technology* **83**(4) (1969), 67–72.

[411] G. D. Rossi, Probability in Metrology, in: *Data Modeling for Metrology and Testing in Measurement Science*, pp. 31–72, 2009.

[412] V. Ya. Rozenberg, *Vvedeniye v teoriyu tochnosti izmeritel'nih system*, Sov. radio, Moskva, 1975 (in Russian).

[413] N. A. Rubichev, *Ocenka i izmereniye iskajeniy radiosignalov*, Sov. radio, Moskva, 1978 (in Russian).

[414] N. A. Rubichev and V. D. Frumkin, Optimum structure of a calibration system, *Measurement Techniques* **13**(3) (1970), 315–320.

[415] B. Russell, *The Principles of Mathematics*, Norton, New York, 1937.

[416] T. L. Saaty, *Fundamentals of Decision Making and Priority Theory with the Analytic Hierarchy Process*, RWS Publications, Pittsburgh, PA, 1994.

[417] T. L. Saaty and P. K. Kevin, *Analytical Planning: The Organization of Systems*, Pergamon, 1985.

[418] L. K. Safronov and V. A. Slaev, Ispol'zovaniye fizicheskih yavleniy pri postroyenii preobrazovateley magnitnih poley v systemah kontrolia parametrov ob'ektov, in: *Tez. dokl. II Vsesoyuznogo soveschaniya "Tochniye izmereniya elektricheskih velichin: peremennogo toka, napriajeniya, moschnosti, energii i ugla fazovogo sdviga"*, pp. 23–24, Leningrad, 1985 (in Russian).

[419] E. Sage and J. Mels, *Theory of Estimation and its Applications in Communications and Control*, McGraw Hill, New York, 1971.

[420] J. Sándor, On certain inequalities for means, III, *Arch. Math. (Basel)* **76** (2001), 34–40.

[421] J. Sándor and E. Neuman, On certain means of two arguments and their extensions, *Int. J. Math. Sci.* **16** (2003), 981–993, doi:10.1155/S0161171203208103.

[422] I. J. Schoenberg, On the arithmetic-geometric mean, *Delta* **7** (1978), 49.

[423] G. H. Schultze, Time Displacement Errors in Instrumentation Tape Recording, in: *International Telemetring Conf.*, pp. 263–272, London, 1963.

[424] H.-J. Seiffert, Problem 887, *Nieuw Arch. Wisk.* **4**(11) (1993), 176.

[425] H.-J. Seiffert, Ungleichungen für einen bestimmten Mittelwert, *Nieuw Arch. Wisk.* **4**(13) (1995), 195–198.

[426] H.-J. Seiffert, Werte zwischen dem geometrischen und dem arithmetischen Mittel zweier Zahlen, *Elem. Math.* **42** (1987), 105–107.

[427] M. N. Selivanov, A. E. Fridman, and J. F. Kudryashova, *Kachestvo izmereniy: Metrologicheskaya spravochnaya kniga*, Lenizdat, Leningrad 1987 (in Russian).

[428] L. A. Semenov, Development of logical-mathematical model of a unit size transfer system from measurement standards to working measuring instruments, in: *Proceedings of the D.I. Mendeleyev Institute for metrology*, 1979 (in Russian).

[429] L. A. Semenov, V. A. Granovskiy, and T. N. Siraya, Obzor osnovnih problem teoreticheskoy metrologii, in: *Fundamental'niye problemi metrologii*, pp. 13–23, NPO "VNIIM im. D.I. Mendeleyeva", Leningrad, 1981 (in Russian).

[430] L. A. Semenov and T. N. Siraya, *Metodi postroeniya graduirovochnih harakteristik sredstv izmereniy*, Izd-vo standartov, Moskva, 1986 (in Russian).

[431] L. A. Semenov and N. P. Ushakov, Optimal disposition of reference measuring instruments for hydrophysical measurements, in: *Tr. metrolog. in-tov of the USSR "Issledovaniya v oblasti gidrofizicheskih izmereniy"*, p. 102, Energiya, Leningrad, 1976 (in Russian).

[432] L. A. Sena, *Edinitsi fizicheskih velichin i ih razmernosti*, Nauka, Moskva, 1977 (in Russian).

[433] H. G. Senel, R. A. Peters, and B. Dawant, Topological median filters, *IP* **11**(2) (2002), 89–104.

[434] R. G. Sharmila and E. C. H. Amirtharaj, Implementation of a new third order weighted Runge-Kutta formula based on Centroidal Mean for solving stiff initial value problems, *Recent Research in Science and Technology* **3**(10) (2011), 91–97.

[435] A. M. Shilov, Model of a system for transmitting unit dimensions from a standard to working measuring equipment, *Measurement Techniques* **17**(8) (1974), 1274–1278.

[436] K. P. Shirokov, Ob opredelenii zavisimostey mejdu fizicheskimi velichinami, in: *Trudi metrologocheskih institutov USSR*, vol. 130(190), pp. 23–26, Leningrad, 1971 (in Russian).

[437] K. P. Shirokov, Ob osnovnih poniatiyah metrologii, in: *Trudi metrologocheskih institutov USSR*, vol. 130(190), pp. 16–18, Lenigrad, 1971 (in Russian).

[438] I. F. Shishkin, *Teoreticheskaya metrologiya*, ch. 1. Obschaya teoriya izmereniy, 4th ed., Piter, St. Petersburg, 2010 (in Russian).

[439] I. F. Shishkin, *Teoreticheskaya metrologiya*, ch. 2. Obespecheniye edinstva izmereniy. 4th ed., Piter, St. Petersburg, 2011 (in Russian).

[440] V. P. Shkolin and A. N. Fogilov, Metodi postroeniya kosmicheskih BCVM (obzor), *Zarubejnaya radioelektronika* 3 (1978) (in Russian).

[441] W. Siebert, *Circuits, Signals and Systems*, MIT Press, Cambridge, MA – London, UK, 1986.

[442] T. N. Siraya, General metrological characteristics of group standards, in: *Proceedings of USSR metrological institutes "Theoretical problems of metrology"* 237(297), p. 17, Leningrad, 1979 (in Russian).

[443] R. Ya. Siropiatova, K voprosu o metodike experimental'nogo opredeleniya fazovoy harakteristiki trakta magnitnoy zapisi i vosproizvedeniya izmeritel'noy informacii, *Trudi MEI, seriya "Elektritehnika"* 57 (1964), pp. 201–213 (in Russian).

[444] Z. Skopets, Sravneniye razlichnih srednih dvuh polojitel'nih chisel, *Kvant* **2** (1971), 20–24. http://jwilson.coe.uga.edu/EMT668/EMAT6680.2000/Umberger/EMAT6690 smu/ (in Russian).

[445] V. V. Skotnikov, V. I. Fominykh and M. F. Yudin, Systematization of variables and designing sets of standards, *Measurement Techniques* **17**(9) (1974), 1446–1451.

[446] L. I. Slabkiy, *Metodi i pribori predel'nih izmereniy v experimental'noy fizike*, Nauka, Moskva, 1973 (in Russian).

[447] V. A. Slaev, Approaches to application of the Guide to the expression of uncertainty in measurement in Russia, *Measurement Techniques* **43**(5) (2000), 405–409.

[448] V. A. Slaev, Avt. svid. N°507828, MKI G01R 25/00, Ustroystvo dlia izmereniya iskajeniy nelineynosti fazochastotnih harakteristik, N°2069352/26-21, zayavl. 18.10.74; opubl. 25.03.76. – UDK 621.317.757(088.8), *Otkritiya. Izobreteniya* **11** (1976), 128 (in Russian).

[449] V. A. Slaev, Avt. svid. N°522513, MKI G11b 5/00, Ustroystvo dlia izmereniya iskajeniy zapisivayemih i vosproizvodimih v apparature magnitnoy zapisi signalov, N°2071791/10, zayavl. 04.11.74; opubl. 25.07.76. – UDK 534.852(088.8), *Otkritiya. Izobreteniya.* **27** (1976), 145 (in Russian).

[450] V. A. Slaev, Matematicheskiye aspekti teorii izmereniy, in: *Tez. dokl. Nauchno-tehnicheskoy konferencii "Datchik – 96"*, T. 1, Gurzuf, 1996, pp. 14–15 (in Russian).

[451] V. A. Slaev, Metrological problems of information technology, *Measurement Techniques* **37**(11) (1994), 1209–1216.

[452] V. A. Slaev, *Metrologicheskoe obespechenie apparaturi magnitnoy zapisi. Nauchnoe izdanie*, Mir i Sem'ya, St. Petersburg, 2004 (in Russian).

[453] V. A. Slaev, Normirovaniye dinamicheskih harakteristik sredstv izmereniy s ispol'zovaniyem markovskih parametrov, in: *Tez. dokl. nauchno-tehnicheskoy konferencii "Diagnostika, informatika, metrologiya, ekologiya, bezopasnost' – 96"*, pp. 132–133, St. Petersburg, 1996 (in Russian).

[454] V. A. Slaev, O meste teoreticheskoy metrologii v systeme nauk i ee predmetnoy oblasti, in: *Tez. dokl. konferencii "Diagnostika, informatika i metrologiya – 95"*, pp. 23–24, St. Petersburg, 1995 (in Russian).

[455] V. A. Slaev, O poverke analogovoy apparaturi magnitnoy zapisi, in: *Tez. dokl. Respublikanskoy nauchno-tehnicheskoy konferencii "Povisheniye effektivnosti traktov magnitnoy zapisi-vosproizveleniya"*, pp. 3–5, Vil'nus, 1976 (in Russian).

[456] V. A. Slaev, Principi razrabotki klassifikatora vidov izmereniy, in: *Trudi VNIIM im. D.I. Mendeleyeva "Systemniye issledovaniya v metrologii"*, pp. 41–63, Leningrad, 1985 (in Russian).

[457] V. A. Slaev, *Razrabotka sredstv izmereniy dlia ocenki metrologicheskih harakteristik apparaturi magnitnoy zapisi i issledovanie metodov povisheniya ee tochnosti* (Diss. na soisk. uchen. stepeni kand. tehn. nauk), VNIIM, Leningrad, 1973 (in Russian).

[458] V. A. Slaev, Teoreticheskaya metrologiya. Obespecheniye edinstva izmereniy: Kommentarii, *Mir izmereniy* **8** (2011), 20–21 (in Russian).

[459] V. A. Slaev, Zadachi metrologicheskogo obespecheniya apparaturi magnitnoy zapisi, in: *Tez. dokl. Respublikanskoy nauchno-tehnicheskoy konferencii "Perspectivi razvitiya tehniki magnitnoy zapisi i tehnologii proizvodstva magnitnih nositeley"*, pp. 62–64, Shostka, 1974 (in Russian).

[460] V. A. Slaev and A. G. Chunovkina, Improvement of Measurement Quality by Means of Planning the Measurement Procedure, in: *Abstracts of the III. International Conference on Problems of Physical Metrology "FIZMET'98"*, pp. 71–72, St. Petersburg, 1998.

[461] V. A. Slaev and A. G. Chunovkina, Sposobi virajeniya tochnostnih harakteristik etalonov, in: *Trudi XIX Nacional'nogo nauchnogo simpoziuma s mejdunarodnim uchastiyem "Metrologiya i metrologicheskoye obespecheniye 2009"*, pp. 26–30, Sozopol', Bolgariya, 10–14 sentiabria 2009 (in Russian).

[462] V. A. Slaev and A. G. Chunovkina, *Validation of Software used in Metrology: Reference Book* (monograph), pod red. V.A. Slaeva, "Professional", St. Petersburg, 2009 (in Russian).

[463] V. A. Slaev, A. G. Chunovkina, and A. V. Chursin, Increasing the quality of measurements by planning the measurement procedure, *Measurement Techniques* **42**(10) (1990), 919–925.

[464] V. A. Slaev, A. G. Chunovkina, and A. V. Chursin, Uncertainty in measurement: definition and calculation methods, *Measurement Techniques* **43**(5) (2000), 394–395.

[465] V. A. Slaev and V. A. Ovchinnikov, Avt. svid. N°541124, MKI G01R 25/04, Ustroystvo dlia izmereniya sdviga faz. N°2146614/21, zayavl. 20.06.75; opubl. 30.12.76. – UDK 621.317.772(088.8) , *Otkritiya. Izobreteniya* **48** (1976), 124 (in Russian).

[466] V. A. Slaev and L. K. Safronov, Aktivnoye ekranirovaniye precizionnih magnitnih preobrazovateley uglov i uglovih skorostey, in: *Poluprovodnikoviye ustroystva i termopreobrazovateli*, pp. 64–67, Leningrad, 1978 (in Russian).

[467] V. A. Slaev and L. K. Safronov, Primeneniye materialov s peremennimi magnitnimi svoystvami v ustroystvah magnitnoy zapisi, in: *Tez. dokl. XVI Vsesoyuznogo soveschaniya po magnitnim elementam avtomatiki i telemehaniki*, pp. 188–190, Nauka, Moskva, 1979 (in Russian).

[468] V. A. Slaev and L. K. Safronov, Problemi taktirovannoy potokochuvstvitel'noy magnitnoy zapisi i cifrovogo preobrazovaniya signalov pri sozdanii etalonov i obraztcovih sredstv izmereniy parametrov dvijeniya, in: *Trudi metrologocheskih institutov USSR "Issledovaniya v oblasti izmereniya parametrov dvijeniya"*, vol. 240(300), pp. 33–38, Energiya, Leningrad, 1979 (in Russian).

[469] V. A. Slaev and B. A. Shkol'nik, Avt. svid. N°566206, MKI G01R 27/28, Ustroystvo identifikacii dinamicheskih harakteristik chetirehpolusnika, N°2098439/21, zayavl. 22.01.75; opubl. 25.07.77. – UDK 621.317.757, *Otkritiya. Izobreteniya.* **27** (1977), 133 (in Russian).

[470] V. A. Slaev and A. I. Sobolev, Avt. svid. N°562863, ÌKI G11B 27/36, Ustroystvo dlia izmereniya perehodnih harakteristik magnitnoy zapisi, N°2146563/10; zayavl. 20.06.75; opubl. 25.06.77. – UDK 681.327.6, *Otkritiya. Izobreteniya.* **23** (1977) 154 (in Russian).

[471] V. A. Slaev and A. I. Sobolev, Avt. svid. N°513372, MKI G11B 5/00. Ustroystvo dlia opredeleniya perehodnih harakteristik mnogokanal'noy apparaturi magnitnoy zapisi. N°2079787/18-10, zayavl. 02.12.74; opubl. 05.05.76. - UDK 534.852 // Otkritiya. Izobreteniya. 1976, N°17. S. 162. (in Russian).

[472] V. A. Slaev and A. I. Sobolev, Avt. svid. N°556492, MKI G11B 37/36, Sposob izmereniya kolebaniy vremeni zapazdivaniya signalov, vosproizvedennih s podvijnogo nositelia, N°2172282/10, zayavl. 16.09.75; opubl. 30.04.77. – UDK 621.317.709:681.84(088.8), *Otkritiya. Izobreteniya.* **16** (1977), 150 (in Russian).

[473] V. A. Slaev, A. I. Sobolev, and A. N. Baranov, Avt. svid. N°585539, MKI G11B 31/00, Ustroystvo dlia izmereniya iskajeniy signalov, registriruyemih mnogokanal'noy apparaturoy magnitnoy zapisi, N°2379421/10-18, zayavl. 02.07.76; opubl. 25.12.77. – UDK 534.852(088.8), *Otkritiya. Izobreteniya.* **47** (1977), 163–164 (in Russian).

[474] V. A. Slaev, G. P. Tcivirko, and B. A. Shkol'nik, Avt. svid. N°442437, MKI G01r 23/02, Sposob opredeleniya perehodnih harakteristik, N°1831256/18-10, zayavl.

26.09.72; opubl. 05.09.74. – UDK 534.852(088.8), *Otkritiya. Izobreteniya.* **33** (1974), 125 (in Russian).

[475] V. A. Slaev, G. P. Tcivirko, and B. A. Shkol'nik, Avt. svid. N°474840, MKI G11b 11/00, Ustroystvo dlia izmereniy perehodnih harakteristik apparaturi magnitnoy zapisi, N°1893317/18-10, zayavl. 15.03.73; opubl. 25.06.75. – UDK 681.846.73, *Otkritiya. Izobreteniya.* **23** (1975), 122 (in Russian).

[476] V. A. Slaev, G. P. Tcivirko, and B. A. Shkol'nik, Avt. svid. N°547831, MKI G11B 27/36, Sposob izmereniya nelineynih iskajeniy v apparature magnitnoy zapisi, N°1876404/10, zayavl. 29.01.73; opubl. 25.02.77. – UDK 681.84.001(088.8), *Otkritiya. Izobreteniya.* **7** (1977), 158 (in Russian).

[477] V. A. Slaev, G. P. Tcivirko, and B. A. Shkol'nik, Metod opredeleniya dinamicheskih harakteristik, *Metrologiya* **1** (1975), 70–74 (in Russian).

[478] V. A. Slaev, N. D. Zviagin, and A. G. Chunovkina, Opit raboti VNIIM po attestacii programmnogo obespecheniya, ispol'zuyemogo v metrologii, in: *Tez. dokl. mejdunarodnogo nauchno-tehnicheskogo seminara "Matematicheskaya, statisticheskaya i komp'yuternaya podderjka kachestva izmereniy"*, pp. 124–126, St. Petersburg, 2009 (in Russian).

[479] Smaller Ph. Reproduce System Noise in Wide-Band Magnetic Recording Systems, in: *IEEE Trans. on Magnet.* vol. Mag. 1(4), pp. 357–363, 1965.

[480] A. I. Sobolev, V. A. Slaev, and A. N. Baranov, Avt. svid. N°669386, MKI G11B 5/00, Sposob izmereniya iskajeniy pri magnitnoy zapisi i vosproizvedenii, N°2505506/18-10, zayavl. 21.04.77; opubl. 25.06.79. – UDK 534.852(088.8), *Otkritiya. Izobreteniya.* **23** (1979), 168 (in Russian).

[481] V. I. Sobolev, *Informacionno-statisticheskaya teoriya izmereniy*, Mashinostroeniye, Moskva, 1983 (in Russian).

[482] G. N. Solopchenko, Principles of normalization, determination and control of error characteristics in data-and-measurement systems, *Measurement Techniques* **28**(3) (1985), 197–201.

[483] N. N. Solov'yev, *Izmeritel'naya tehnika v provodnoy sviazi*, ch. III. Izmereniya parametrov, harakterizuyuschih iskajeniya signalov sviazi, Sviaz', Moskva, 1971 (in Russian).

[484] E. D. Sontag, *Mathematical Control Theory*, Springer, New York, 1990.

[485] *Spravochnik po tehnike magnitnoy zapisi*, pod red. O. V. Porickogo and E. N. Travnikova, Tehnika, Kiyev, 1981 (in Russian).

[486] *Sredniye velichini*, Statistika, Moskva, 1970 (in Russian).

[487] A. P. Stahov, *Vvedenie v algoritmicheskujy teoriyu izmereniy*, Sov. radio, Moskva, 1977 (in Russian).

[488] A. P. Stahov, B. Ya. Lihtcinder, Yu. P. Orlovich and Yu. A. Storojuk, *Kodirovanie dannih v informacionno-izmeritel'nih sistemah*, Tehnika, Kiyev, 1985 (in Russian).

[489] Yu. S. Stark, *Teoriya izmereniy*, Moskva, 1973 (in Russian).

[490] A. Stepanov, Comparing algorithms for evaluating key comparisons with linear drift in the reference standard, *Measurement Techniques* **50**(10) (2007), 1019–1027.

[491] B. M. Stepanov and V. N. Filinov, Ob ocenke iskajeniy vremennogo masshtaba signalov systemami magnitnoy zapisi, *Izmerbntl'naya tehnika* **10** (1968), 51–54 (in Russian).

[492] O. A. Stepanov, *Osnovi teorii ocenivaniya s prilojeniyami k zadacham obrabotki navigacionnoy informacii*, ch. 1. Vvedeniye v teoriyu ocenivaniya, GNC RF CNII "Elektropribor", St. Petersburg, 2009 (in Russian).

[493] S. S. Stevens, On the theory of scales and measurement, *Science* **103** (1946), 667–680.

[494] K. B. Stolarsky, The power and generalized logarithmic mean, *Amer. Math. Monthly* **87**(7) (1980), 545–548.

[495] D. Stinson, *Combinatorial designs. Constructions and analyses*, Springer-Verlag, 2004.

[496] P. Suppes and J. Zinnes, Osnovi teorii izmereniy, in: *Psihologicheskiye izmereniya*, pp. 9–110, Mir, Moskva, 1967 (in Russian).

[497] L. M. Surhone, M. T. Timpledon and S. F. Marseken, *Winsorized Mean*, Books by request, Moscow, 2010.

[498] C. M. Sutton, Analysis and linking of international measurement comparisons, *Metrologia* **41** (2004), 272–277.

[499] V. S. Svintsov, Optimization of a system for transferring the unit size according to an economical criterion, *Metrologiya* **9**(13) (1980), 13 (in Russian).

[500] V. S. Svintsov, To the problem of modeling the process of measuring instruments operation, *Metrologiya* **4**(3) (1985), 3 (in Russian).

[501] Yu. V. Tarbeev and V. A. Balalaev, Sostoyaniye i perspektivi razvitiya teorii obespecheniya edinstva izmereniy, in: *Mat. II Vsesoyuz. soveshchaniya po teoreticheskoy metrologii "Fizicheskiye problemi tochnih izmereniy"*, p. 4, Energoatomizdat, Leningrad, 1984 (in Russian).

[502] Yu. V. Tarbeev, I. B. Chelpanov and T. N. Siraya, Metrological certification of algorithms for measurement-data processing, *Measurement Techniques* **28**(3) (1985), 205–208.

[503] Yu. V. Tarbeyev, Puti dal'neyshego razvitiya teoreticheskih osnov metrologii, in: *Trudi metrologicheskih institutov USSR*, vol. 237(297), pp. 3–8, Leningrad, 1979 (in Russian).

[504] Yu. V. Tarbeyev, Theoretical and practical limits of measurement accuracy, in: *2nd Symp. of the IMEKO Technical Committee on Metrology TC-8*, Budapest, 4 p.

[505] Yu. V. Tarbeyev, V. A. Balalayev, V. A. Slaev, V. S. Aleksandrov, and V. N. Romanov, Primeneniye systemnogo podhoda pri razrabotke dolgosrochnih prognozov i programm v oblasti metrologii, in: *Tez. dokl. III Vseakademicheskoy shkoli po problemam standartizacii i metrologii*, pp. 18–22, Tbilisi, 1985 (in Russian).

[506] Yu. V. Tarbeyev and L. I. Dovbeta, Soderjaniye metrologii i ee mesto v systeme nauk, in: *Fundamental'niye problemi metrologii*, pp. 4–12, NPO "VNIIM im. D.I. Mendeleyeva", Leningrad, 1981 (in Russian).

[507] Yu. V. Tarbeyev and V. A. Slaev, Status report on the metrological support and certification of sensors, *Measurement Techniques* **36**(11) (1993), 1197–1206.

[508] P. Taubert, *Abschätzung der Genauigkeit von Meßergebnissen*, VEB Verlag Technik, Berlin, 1985.

[509] E. I. Tcvetkov, *Osnovi teorii statisticheskih izmereniy*, Energoatomizdat, Leningrad, 1986 (in Russian).

[510] F. E. Temnikov, *Teoreticheskiye osnovi ihformacionnoy tehniki*, Energiya, Moskva, 1971 (in Russian).

[511] V. I. Tihonov, *Statisticheskaya radiotehnika*, Sov. radio, Moskva, 1966 (in Russian).

[512] G. Toader, Complementariness with respect to Greek means, *Automat. Comput. Appl. Math.* **13**(1) (2004), 197–202.

[513] G. Toader, Some remarks on means, *Anal. Numér. Théor. Approx.* **20** (1991), 97–109.

[514] G. Toader, Universal Means, *J. Math. Inequal.* **1**(1) (2007), 57–64.

[515] G. Toader, I. Costin and S. Toader, Invariance in Some Families of Means, in: *Functional Equations in Mathematical Analysis*, Springer Optimization and its Applications, vol. 52-2, pp. 709–717, Springer-Verlag, 2012.

[516] G. Toader and S. Toader, *Greek means and the Arithmetic-Geometric Mean*, RGMIA Monographs, Victoria University, 2005 (online: http://rgmia.vu.edu.au/monographs).

[517] G. Toader and S. Toader, Means and Generalized Means, *J. Inequal. Pure and Appl. Math.* **8**(2) (2007), Art. 45, 6 p.

[518] S. Toader, Derivatives of generalized means, *Math. Inequal. Appl.* **5**(3) (2002), 517–523.

[519] S. Toader and G. Toader, Greek means, *Automat. Comput. Appl. Math.* **11**(1) (2002), 159–165.

[520] J. Todd, The many limits of mixed means. II. The Carlson sequences, *Numer. Math.* **54**(1) (1988), 1–18.

[521] B. Toman and A. Possolo, Laboratory effects models for interlaboratory comparisons, *Accred Qual. Assur.* **14** (2009), 553–563.

[522] L. J. T. Tou, R. C. Gonzales, *Pattern Recognition Principles*, Addison-Wesley Publishing Company, London-Amsterdam-Dom Mills, Ontario Sydney Tokyo, 1974.

[523] E. N. Travnikov, *Mehanizmi apparaturi magnitnoy zapisi*, Tehnika, Kiyev, 1976 (in Russian).

[524] T. Trif, On certain inequalities involving the identric mean in n variables, *Studia Univ. Babes-Bolyai Math.* **46**(4) (2001), 105–114.

[525] A. A. Tsibina, O metodah ustanovleniya peremennih mejpoverochnih intervalov dlia obraztsovih i rabochih sredstv izmereniy, in: *Tr. SNIIM "Nadejnost' sredstv izmeritel'noy tehniki"*, vol. 4, Novosibirsk, 1970 (in Russian).

[526] A. A. Tsibina and A. M. Shilov, Metod opredeleniya strukturi poverochnoy schemi, *Measurement Techniques* 13(11) (1970).

[527] A. A. Tsibina, A. M. Shilov, and N. G. Nizovkina, Calculation of economic performance in the establishment of scientifically sound intervals between checks, *Measurement Techniques* **24**(8) (1981), 705–707.

[528] G. J. G. Upton, *The Analysis of Cross-Tabulated Data*, John Wiley & Sons Ltd, Bristol, 1978.

[529] US Patent 4.003.084, Int. Cl. G11Â 27/36, *Method of and means for testing a tape record/playback system*, J.C. Fletcher et al. – Appl. No. 561956, Filed 25.03.75, 11.01.77.

[530] H. L. Van Trees, *Detection, Estimation and Modulation Theory, Part I*, Wiley, New York, 1969; *Part II.*, Wiley, New York, 1971.

[531] G. I. Vasilenko, *Teoriya vosstanovleniya signalov: O redukcii k ideal'nomu priboru v fizike i tehnike*, Sov. radio, Moskva 1979 (in Russian).

[532] A. I. Vasil'ev, B. P. Zelentsov, and A. A. Tsibina, *O sushchnosti terminov "edinstvo izmereniy" i "edinoobraziye sredstv izmereniy"*, in: Nauchno-tehnicheskaya terminologiya (ref. inf.), No. 8, p. 6, Izd-vo VNIIKI, Moskva, 1974 (in Russian).

[533] A. I. Vasil'ev, B. P. Zelentsov, and A. A. Tsibina, System for controlling uniformity of measurements, *Measurement Techniques* **15**(4) (1972), 505–509.

[534] A. J. Viterbi, *Principles of coherent communication*, McGraw-Hill, 1966.

[535] L. I. Volgin, *Lineyniye elektricheskiye preobrazovateli dlia izmeritel'nih priborov i system*, Sov. radio, Leningrad, 1971 (in Russian).

[536] Yu. I. Vorontsov, *Teoriya i metodi makroskopicheskih izmereniy*, pod red. V. B. Braginskogo, Nauka, GRFML, Moskva, 1989 (in Russian).

[537] N. N. Vostroknutov, V. P. Kuznetsov, G. N. Solopchenko, and B. A. Frenkel', Object of metrological certification of algorithms and programs for processing of measurement data, *Measurement Techniques* **33**(7) (1990), 647–649.

[538] N. N. Vostroknutov, M. A. Zemel'man, and V. M. Kashlakov, Selection of standard facilities for periodic inspection using probabilistic criteria, *Measurement Techniques* (review) **20**(7) (1977), 954–959.

[539] M. K. Wang, Y. M. Chu, and Y. F. Qiu, Some comparison inequalities for generalized Muirhead and identric means, *J. Inequal. Appl.* (2010), Article No. 295620.

[540] S. Wang and Y. Chu, The Best Bounds of the Combination of Arithmetic and Harmonic Means for the Seiffert's Mean, Int. J. Math. Anal. **4**(22) (2010), 1079–1084.

[541] *WELMEC Guide 7.2. Software Guide* (Measuring Instruments Directive 2004/22/EC), Issue 5, 2011, http://www.welmec.org.

[542] D. R. White, On the analysis of measurement comparisons, *Metrologia* **41** (2004), 122–131.

[543] E. T. Whittaker and G. Robinson, *The Calculus of Observations: An Introduction to Numerical Analysis* (4th ed.), Dover Publications, 1967.

[544] V. E. Wiener and F. M. Cretu, Projektierung der Makrosystems für metrologische Versicherung, *Meas. und Instr. Acta IMEKO* **1** (1974).

[545] R. R. Wilcox and H. J. Keselman, Modern robust data analysis methods: Measures of central tendency, *Psychological Methods* **8**(3) (2003), 254–274.

[546] R. Willink, Forming a comparison reference value from different distributions of belief, *Metrologia* **43** (2006), 12–20.

[547] R. Willink, On the interpretation and analysis of a degree-of-equivalence, *Metrologia* **40** (2003), 9–17.

[548] R. Willink, Statistical determination of a comparison reference value using hidden errors, *Metrologia* **39** (2002), 343–354.

[549] A. Witkowski, Interpolations of Schwab–Borchardt Mean, *Research Group in Math. Inequal. and Appl.* **15** (2012), RGMIA/papers/v15/v15a2.pdf.

[550] W. Wöger, Information-Based Probability Density Function for a Quantity, *Measurement Techniques* **46**(9) (2003), 815–823.

[551] W. J. Youden, Experimentation and Measurement, in: *Vistas of Science: National Science Teachers' Association*, 1962 (reprinted by National Institute of Standards and Technology, 1997).

[552] L. A. Zadeh and J. R. Ragazzini, Optimum filters for detection of signal in noise, *PIRE* **40** (1952), 1223–1231.

[553] H. Zadeh and Ch. Dezoer, *Teoriya lineynich system*, Nauka, Moskva, 1986 (in Russian).

[554] A. I. Zaiko, *Tochnost' analogovih lineynih izmeritel'nih kanalov IIS*, Izd-vo standartov, Moskva, 1987 (in Russian).

[555] M. A. Zemel'man, *Metrologicheskiye osnovi tehnicheskih izmereniy*, Izd-vo standartov, Moskva, 1991 (in Russian).

[556] M. A. Zemel'man, A. P. Knyupfer, and V. P. Kuznetsov, Methods for normalizing the metrological characteristics of measuring devices, *Measurement Techniques* **12**(1), pp. 127–131; **12**(2), 250–255; **12**(3), 426–430.

[557] M. A. Zemel'man, V. P. Kuznetcov, and G. N. Solopchenko, *Normirovaniye i opredeleniye metrologicheskih harakteristik sredstv izmereniy*, Mashinostroeniye, Moskva, 1980 (in Russian).

[558] M. A. Zemel'man and N. P. Mif, *Planirovaniye tehnicheskih izmereniy*, Izd-vo standartov, Moskva, 1978 (in Russian).

[559] N. F. Zhang, H. Liu, N. Sedransk, and W. E. Strawderman, Statistical analysis of key comparisons with linear trends, *Metrologia* **41** (2004), 231–237.

[560] N. F. Zhang, W. Strawderman, H. Liu, and N. Sedransk, Statistical analysis for multiple artefact problem in key comparisons with linear trends, *Metrologia* **43** (2006), 21–26.

[561] T. M. Zolotova and M. A. Rozenblat, Bezinertsionnoye preobrazovaniye signalov neskol'kih priborov, optimal'noye po kriteriyu nadejnosti, *Avtomatika i telemehanika* **11** (1972) (in Russian).

[562] C. Zong and Y. Chu, An Inequality Among Identric, Geometric and Seiffert's Means, *Int. Math. Forum* **5**(26) (2010), 1297–1302.

Index